BASIC CHEMISTRY FOR THE LIFE SCIENCES

BASIC CHEMISTRY FOR THE LIFE SCIENCES

H. L. Helmprecht / L. T. Friedman

STATE UNIVERSITY OF NEW YORK
AGRICULTURAL AND TECHNICAL COLLEGE AT FARMINGDALE

McGraw-Hill Book Company

NEW YORK ST. LOUIS SAN FRANCISCO AUCKLAND BOGOTÁ DÜSSELDORF
JOHANNESBURG LONDON MADRID MEXICO MONTREAL
NEW DELHI PANAMA PARIS SÃO PAULO SINGAPORE SYDNEY TOKYO TORONTO

This book was set in Helvetica by Progressive Typographers.
The editors were Robert H. Summersgill and Michael LaBarbera;
the designer was J. E. O'Connor;
the production supervisor was Thomas J. LoPinto.
The drawings were done by Danmark & Michaels, Inc.
The cover illustration was done by Colos.
The printer was The Murray Printing Company;
the binder, The Book Press, Inc.

Library of Congress Cataloging in Publication Data

Helmprecht, H   L   date
  Basic chemistry for the life sciences.

    1. Chemistry.   I.   Friedman, L. T., date,
joint author. II. Title.
QD31.2.H44   540′.2′4574   76-3454
ISBN 0-07-027956-X

BASIC CHEMISTRY FOR THE LIFE SCIENCES

1234567890 MUBP 7832109876

CONTENTS

PREFACE

All of us who inhabit this earth, large creatures and small, are different from each other, and our differences help keep things interesting and lively. But there are similarities as well. Just as human beings have much more in common than the general arrangement of legs, arms, head, and all the rest, so, too, are there similarities between ourselves and other forms of life. In fact, the deeper we probe into living organisms the more an underlying similarity emerges. The common feature that finally comes into view is the general similarity of the chemical events through which living organisms function, whether a human being or an amoeba. Of course, this in no way makes us any the less human, nor does it make an amoeba any more than a one-celled bit of life. However, it does point up the remarkable fact that despite the myriad forms of life and the enormous differences between them, they all operate via substantially the same chemical mechanisms.

It also implies that some understanding of this chemistry is essential for those who will be working with living organisms. This is why this book was written.

The presentation of the subject matter is on a simple, descriptive level. No more than ordinary arithmetic or the very simplest algebra is used anywhere in the book. The first chapter introduces you to chemistry and invites you in. Because measurement is necessary in chemistry, as in all technical work, it is treated in the second chapter. In order to talk about the chemistry of life, we must first understand the underlying principles of chemistry itself, and this is what Chapters 3 to 9 deal with. Organic chemistry, which is especially relevant to living organisms, is treated in Chapters 10, 11, and 12. The last seven chapters present the basic elements of biochemistry, the chemistry of living matter.

The sketches and diagrams scattered throughout the book are there to visualize the textual material and to help the discussion along; we hope that they will serve that purpose for you. Since the mind needs exercise as well as the body, there are questions and problems at the end of each chapter. Sit down with pencil and paper and work them out.

You will note that some words or phrases are printed in **boldface.** These are technical terms that you should get to know and they are all included in the glossary at the end of the book. *Italicized* words are used for emphasis.

We hope that your travels through this text will be enjoyable and rewarding. The more we get to know and understand the chemical workings of living matter, the greater is our appreciation of chemistry and the greater is our respect for life.

H. L. Helmprecht / L. T. Friedman

THE SEARCHLIGHT OF CHEMISTRY
Open the Door and Come In

Once upon a time there was no "chemistry," nor were there any human beings. But at some time or times and at some place or places, human beings did appear on this green living planet called Earth and began to do what people are still doing today—act upon the world about them. It was then that what we call chemistry entered into their lives and into our history. For to act upon the world is to change it in some way, and where there is change, there is generally chemical change.

Put a seed into the ground, and if sun, soil, and water are favorable, there is growth. Agriculture became a way of life for prehistoric man. Press wet clay into a rectangular shape, let it bake in the sun, and the oldest man-made building material has been produced, with all its possibilities for human habitations and structures. Build a fire, that most celebrated achievement of early man, surrounded by myth and mystery, and energy itself has been conjured forth, to be trapped for use in endless ways. Millennia ago, two holes were dug into the windy side of a hill, one horizontal and the other vertical, meeting to form an L. A "charge" of reddish earth and charcoal was placed at the corner of the L, and a fire was started; as the wind swept through the

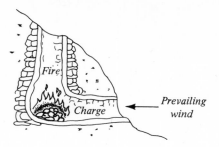

horizontal shaft, through the charge, and up the vertical shaft, the fire became hotter and hotter, and finally a "reaction" occurred between the red earth and the charcoal, and *iron* was released, strong but malleable, for tools, weapons, and structures.

THE PATH OF CHEMICAL CHANGE

Each of these events, and many more that marked the slow but cumulative progress of primitive man, is an example of chemical change. In each case, the starting materials are transformed into end products very different in appearance and behavior. These changes could be seen and used by the human beings who set them into motion; what they could not see were the chemical events within these changes. Soft, wet clay is not the same thing as a hard, dry brick. A seed is a long way from the plant that emerges from it. An inert log is quite different from the flames leaping from it, leaving only ashes at the end. And the crumbly mixture of red earth and black charcoal in the primitive furnace gives no hint of the iron it will release. As the chemist puts it, the composition of the **reactants,** the original materials, is not the same as the composition of the resulting **products.** But since the latter were undeniably derived from the former, some definite relation between them must exist. To seek out these relations is the heart of chemical thinking and effort.

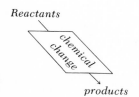

Before this change can be followed, if the path between reactants and products is to be traced (and this path can be one, two, or many steps), the **composition** of the original materials and of the end products must be known. Furthermore, if we look closely at these chemical changes, we can see in each case that **energy** was involved as a necessary part of the change: the energy of the sun's heat, or that of the fire, or even the energy contained *in* a material. In looking at a chemical change, therefore, the chemist essentially asks four questions:

1   What was the composition of the reactants before the change and of the products after the change?

2 What was the sequence of steps leading from reactants to products?

3 In which direction was there an energy flow as the reactants became products?

4 Why does one reaction occur and not another?

Put a match to a piece of paper and it burns; do the same with a piece of glass, and nothing happens. Why the difference?

TECHNOLOGY EVOLVES

Of course, these are the questions we ask today from the vantage point of about 10,000 years of human experience and effort. Prehistoric man could not possibly have raised such questions about the many chemical processes he initiated as he made himself increasingly at home on earth. And when we consider how few (as *we* see it) means were available to him, it is remarkable how well he succeeded. By about the time that recorded history evolved in the form of permanent inscriptions, paintings, drawings, and structures that indicate to later generations the thought and behavior of those who left those records, an impressive fund of knowledge and skill had been accumulated. This technology, with some extensions and additions, became the basis of day-to-day life for centuries to come, and included agriculture and animal husbandry; using and controlling fire, not only for heating and cooking, but also for producing pottery and extracting and working metals; weaving and dyeing textiles; the preparation of fermented beer and wines; making soap and cosmetics; glassmaking; tanning hides for leather; using bricks and masonry for building; to say nothing of the variety of tools and mechanical devices for carrying out these activities.

Much of this technology was chemical in nature, involving chemical reactions and processes, evolved and utilized by those early craftsmen and artisans whose creativity we can only admire. And yet, it was craftsmanship, not chemistry. Chemistry, like any science, requires speculation as well as practice, and for primitive man, the act of production was its own justification, its own explanation. It is at least 6000 years since copper was first extracted, probably from an ore called **malachite.** Could those who performed this eventful act wait for an explanation why the reddish metal flowed from the green, friable stone? At that moment it was the copper that served a purpose, not an explanation. When primitive humans did speculate, they directed their ideas to things that they could *not* control, to the unknown and unpredictable forces of what was for them still a hostile and dangerous world. From almost the very beginning of human life on earth, there emerges not only technology but also religion and magic, with all its consequences for human history.   The kind of ideas necessary to convert chemical practice into chemical science will come—but later.

THE NATURE OF MATTER: EARLY SPECULATIONS

When ancient Greek civilization appeared about the sixth century B.C. it was marked by a capacity for abstract and general thinking, as exemplified in the development of geometry.   (It should be recognized, however, that ancient Greece profited from the contributions of earlier civilizations, including those of Mesopotamia and Egypt.)   The speculations and the questionings of the ancient Greeks extended in many directions.   One of the questions raised was: What is the nature of matter, of what is it really composed?   In the fifth century B.C. the philosopher Democritus offered the idea that all matter is made up of atoms (*atomos*, Greek for not divisible).

This was interesting and provocative, but it represented an insight rather than a work of science.   It derived from no body of evidence, and there was no way of proving or disproving the idea.   However, the question and the answer *were* significant because it was an effort to express observable characteristics of matter in terms of underlying small particles.   When the atom reappeared again about 2000 years later as an answer to the same question, it was still speculation but under very different conditions.

There is a further point to remember regarding the atoms of Democritus.   Greece was a slave society, and the practitioners of chemistry, that is, those who did the work requiring chemical operations, were slaves and probably never even heard of an atom; their function was to produce, not to speculate.   There was a gap, therefore, between those who did the work and knew its details intimately and those who simply speculated and theorized.   Science can only suffer from such a separation or, at best, develop in a very uneven manner.   Chemistry, which depends so much on experiment, on physical proof, on relating fact to thought, and back again, showed very little development in ancient Greece.   The practice of chemistry in the pottery, the foundry, and wherever production called for chemical processes did, of course, continue, and whatever progress was made was due to the craftsmen and artisans themselves.

THE GLITTER AND ILLUSION OF GOLD

During the middle ages, chemical practices advanced slowly, although in the latter part of this period and even into the Renaissance there was an increased pace in mining and metalworking, probably stimulated by increased demand. A unique feature of the middle ages was the emergence of a group of chemical practitioners who were quite different in purpose and character from the chemical craftsmen of the time. These were the **alchemists,** perhaps the first professional chemists, equipped with an elaborate, if mysterious, theory. Unfortunately, their efforts were motivated by their overriding concern for converting "base" metal into gold. For the alchemist, anything that did not lead in that direction was not important, so that the useful results they obtained were few compared with the enormous efforts expended. When foundry workers mixed copper and zinc and produced what we call brass, the alchemists disdained it as "inferior gold." Nevertheless, the alchemists did leave their mark on chemistry; they investigated a variety of reactions, developed apparatus, and used symbols to represent chemical elements and compounds. Alchemy has been described as the "prelude to chemistry," but since its efforts could end only in frustration, these efforts became increasingly wrapped up in mystery and incomprehensibility, deteriorating rapidly. The coming of age of chemistry as a science was still some time away.

Some alchemical symbols

AN EARLY MEDICAL EFFORT

In the early sixteenth century, Paracelsus departed from the road taken by alchemy by applying chemistry to medicine. He tested chemical substances for their medical effects, advocated mineral baths, used opium and compounds of mercury and lead as pharmaceutical materials, and introduced alcoholic solutions, or **tinctures.** Although his contributions are controversial, he was moving in a fruitful direction, toward what we today call **biochemistry,** the chemistry of living matter.

Paracelsus was born in Switzerland about 1490; his real name was Theophrastus Bombastus von Hohenheim.

A NEW CLIMATE OF THOUGHT

In the latter part of the seventeenth century, and more rapidly and fully in the eighteenth century, chemical science assumed form and substance. A revolution in science, as well as major social changes (with more to come), had already occurred by the eighteenth century. There was a new climate of thought, one might even say a new necessity for thought, for an attitude that would not turn its back on practice but would accept it as an equal partner in the search for truth.

In fact, by now chemical practice was well in advance of chemical thinking and theory, with practice anticipating what theory had not yet expressed. A great reservoir of chemical experience and usage was at hand, waiting to be tapped. First, however, chemists had to close the gap between theory and practice. Speculation for its own sake, especially if what was being speculated could neither be proved nor disproved, was not enough. To be sure, it is by speculation that a theory is born, but its validity must be demonstrated by reality; it must be tested by its results.

In common with the other sciences, chemistry adopted what we call the **experimental method,** wherein experimental results are the basis for accepting or rejecting a theory. At the same time, chemistry also recognized that practice alone can be self-restricting; practical experience needs a deeper insight into what is happening than merely a description of what can be perceived by the senses. We know what we see, but is there more that the eye or the senses do not recognize? The observable may not be the whole story. Ignite a strip of wood, and it burns and disappears, leaving but a bit of ash. Ignite a strip of magnesium metal, and it also burns and disappears, but leaves a white powder that weighs more than the original metal itself. Different, and yet alike. What is occurring in both events that accounts for both the difference and the similarity? To "see" what cannot be seen by the eye (or, for that matter, by a microscope) requires the "eye of the mind," and chemists began to act out in their minds what might really be happening when a substance burns, when a heated mixture of iron ore and charcoal yields iron, when limestone dissolves in acid, when wine turns sour. Thinking has been described as action in rehearsal. Then again, *what is* matter composed of?

THE ATOMIC THEORY

Hydrogen,   *Oxygen,*
an element   *an element*

Water,
a compound

An important ingredient entered into the theory and practice of chemistry as it moved toward the modern era, the use of **quantitative methods.** Measuring, weighing, counting, taking the temperature—all must be done as accurately as possible. These procedures were important because they revealed quantitative and unvarying *regularities* in the composition of chemical substances, which in turn supported the concept of the atom as the unit of structure of matter.

By the end of the eighteenth century it was recognized that some chemical substances are elementary materials, or **elements,** that is, materials that cannot be reduced to simpler components. For example, hydrogen gas was combined with oxygen gas to produce water, leading to the conclusion that water is composed of hydrogen and oxygen. However, hydrogen and oxygen themselves could not be broken down further; i.e., they are elements. When water was analyzed, it was found to contain one part of hydrogen, by *weight*, to eight parts of oxygen. No matter where the water comes from or how often the analysis is made, the ratio of the weight of hydrogen in water to the weight of oxygen is always 1:8. When other materials were analyzed quantitatively for the proportions of their individual components, it was similarly found that weight ratio of these components is fixed and specific for a particular substance. If the substance is ammonia, composed of hydrogen and nitrogen, these elements are always present in a weight ratio of one part of hydrogen to almost five parts of nitrogen.

The general conclusion that the composition of any chemical material is specific to that material and is unvarying suggested that the component parts of the substance were themselves elementary, unvarying particles. When combined, these particles become that substance, and their individual weights add up to its total weight. This conclusion, called the **law of constant proportions,** was due largely to the work of A. L. Lavoisier (1743–1794), sometimes called "the father of modern chemistry." These elementary, unvarying particles are the atoms, so that there are hydrogen atoms, and oxygen atoms, and nitrogen atoms, and as many different kinds of atoms as there are elements. Since any sample of an element is made up of atoms of the same kind, any portion of hydrogen gas is made up of hydrogen atoms, and any portion of iron metal is made up of iron atoms, and so on. The concept of the atom as the unit of structure of matter served to clarify and unify a great deal of chemistry and opened the way for its extraordinary development during the nineteenth and twentieth centuries.

As originally formulated by the English chemist John Dalton in 1805, the atom was considered as a hard, indivisible, and indestructible particle. Much has happened in chemistry since that time, and we know today that the atoms themselves have an internal structure. Furthermore, atoms can be transformed from one kind to another, and, even more remarkable, entirely new ones can be and *have been* made. Ninety-two different elements have been found in nature, and therefore ninety-two different atoms; in addition, there are thirteen more man-made elements. However, the fact that atoms can *combine* with each other is their outstanding feature. It is the different atom *combinations* that account for the enormous varieties of matter, from the single atoms that make up helium gas, to the highly organized chemical structures of living organisms composed of thousands of atoms. When these combinations and arrangements of atoms undergo a change, the result is chemical change.

Like all the sciences, chemistry did not develop alone; it drew upon the findings of physics, mathematics, biology, and geology and supported them in turn. As the various sciences expanded in the nineteenth and twentieth centuries, they began to overlap, fusing purposes and methods for the common good. One such area that has yielded almost undreamed of results is biochemistry. Although chemists and physicists have provided a powerful insight into what *matter* is, there remained the even more intriguing question: What is *life?* The chemical basis of life is the subject of biochemistry, and although the early roots of biochemistry are in the nineteenth century, it is largely a creation of the twentieth century and its most notable achievements are less than 50 years old. In this short time biochemistry has penetrated deeply into such questions as how living organisms get their energy, how one generation transmits its characteristics to the next, and even how humans can manipulate the chemical factors that control living processes. Probably no other branch of chemistry is moving forward as rapidly or holds as much consequence for the future as biochemistry.

Today, chemistry can be likened to a huge and powerful searchlight. Every day, all over our world, it is being used to illuminate a myriad of problems—in producing goods and in the pollution resulting from that production; in growing food and its processing and distribution; in overcoming disease and maintaining health; in the deepest problems of the body and the mind.

Powerful and impressive as chemistry now stands, it was, after all, created by human beings, and human beings can make it serve their human potential. Let us hope that today's students will help see to it that this is done.

REVIEW QUESTIONS

1 In baking a cake, what are the reactants? (If you don't know, ask someone who does.) Is energy needed to convert the reactants to the product?

2 In baking a cake, what tools and equipment are needed, and what measurements are made?

3 When water freezes to ice on a cold day, is it a physical or a chemical change? Explain.

4 Is making an omelet a physical or a chemical change? Explain.

5 It was mentioned that the alchemists introduced symbols into chemistry and that since that time chemical symbols have become a kind of universal alphabet. Symbols are in use in many areas of daily life; can you think of some?

6 Would you say that it is a good thing or a bad thing that not *every* chemical material reacts with *every other* chemical material?

7 In the following list of distinct chemical materials or water solutions of a chemical material, check off those which are in your own home and write in the purpose each serves:

Boric acid _____   Table salt _____
Vinegar _____   Aspirin _____
Baking soda _____   Sugar _____
Alcohol _____   Milk of magnesia _____
Kerosene _____   Copper metal _____
Ammonia water _____   Oxygen gas _____
Sodium bicarbonate _____

2

MEASURING, THE METRIC SYSTEM, AND SCIENTIFIC NOTATION
Sizing It Up

Have you ever watched a small child at the beach busy with his pail and shovel? He fills the pail with sand, then dumps it out. He fills it again, and dumps it out again, only to fill it again, and so on and on—all with a serious and preoccupied air. This *is* serious business, filling what was empty and emptying what was full, arranging and rearranging things into new shapes and sizes, and slowly becoming aware of quantity, volume, and dimensions. The child is beginning to learn the language of measurement.

THE LANGUAGE OF MEASUREMENT

Of course, people are not alone in having a sense of distance or direction or time. Many animals are much better in using these abilities than we are. The annual migration of birds to and from distant feeding grounds is still a mystery; fish travel many miles to reach specific spawning areas at specific times. Watch a circus tiger jump through a burning hoop: it takes a most delicate sense of distance and timing. Nevertheless, so far as is known, only man has developed a **language of measurement** with the advantage that measurements can be communicated, compared, re-corded, and passed on. Just as languages differ from each other, so do the languages of measurement but even *more* so. Look at a book; a Frenchman would say *livre*, a Spaniard *libro*, a German *Buch*, and so on. Since all of these words represent the same thing, they are equivalent, and we can translate directly from one to the other. But now look at the line drawn in the margin. What is its length? We would say 2.5 inches, which is shorthand for saying that it is 2.5 times the **unit** of length we call an inch. On the other hand, the Frenchman, the Spaniard, and the German would say that the line is 6.35 centimeters, which again is a short way of saying that it is 6.35 times a unit of length they call a centi-meter. Not only are inch and centimeter different *words*, but they also represent different *units* of length, so that in order to trans-late from inches to centimeters (or vice versa), we must know how inches and centimeters compare with each other. Since 2.5 inches (in) and 6.35 centimeters (cm) represent the same length, 2.5 in = 6.35 cm, and

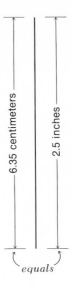

$$1 \text{ in} = \frac{6.35}{2.5} \text{ cm} = 2.54 \text{ cm}$$

CONVERSION FACTORS

The statement that 1 in contains 2.54 cm is the **conversion factor** between inches and centimeters. Since 1 in = 2.54 cm is an equality, the *ratio* of the two equal sides must be 1. However, this ratio can be written as two different fractions, equal to each other and to 1:

$$1 = \frac{1 \text{ in}}{2.54 \text{ cm}} = \frac{2.54 \text{ cm}}{1 \text{ in}}$$

Since each fraction equals unity, each can be used as a multiplier without changing the value of what is being multiplied. If some length is expressed in centimeters, multiplication by the *left-hand* fraction will convert it to inches, since the centimeters will cancel out. If the length is in inches, multiplying by the *right-hand* frac-tion will convert it to centimeters, since the inches term will cancel out. A few examples will illustrate this.

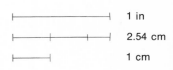

Problem

$$10 \text{ in} = ? \text{ cm}$$

The multiplier to use must cancel out inches and leave centime-ters, so the right-hand fraction above is used.

Solution

$$10 \text{ in} \times \frac{2.54 \text{ cm}}{1 \text{ in}} = 10 \times 2.54 \text{ cm} = 25.4 \text{ cm} \quad \textit{Ans.}$$

Problem

$$12.7 \text{ cm} = ? \text{ in}$$

The multiplier must cancel centimeters and leave inches; the left-hand fraction is used.
Solution

$$12.7 \text{ cm} \times \frac{1 \text{ in}}{2.54 \text{ cm}} = \frac{12.7}{2.54} \text{ in} = 5.0 \text{ in} \quad \textit{Ans.}$$

Problem

$$100 \text{ cm} = ? \text{ in}$$

Here again, the multiplier must eliminate centimeters and leave inches, so that the left-hand multiplier is used.
Solution

$$100 \text{ cm} \times \frac{1 \text{ in}}{2.54 \text{ cm}} = \frac{100 \text{ in}}{2.54} = 39.37 \text{ in} \quad \textit{Ans.}$$

The prefix *centi* means $^1/_{100}$, so that

$$1 \text{ centimeter} = {}^1/_{100} \text{ meter}$$

Multiply by 100:

$$100 \text{ cm} = 1 \text{ meter (m)}$$

Using the answer obtained for the above problem, we can write

$$100 \text{ cm} = 1 \text{ m} = 39.37 \text{ in}$$

This method of conversion can be used in changing over between any two units, so long as both are measuring the same *kind* of thing—weight, length, volume, or whatever. Changing from one kind of money to another is really a change of units. Suppose you want to convert $50 into Italian lira. In currency exchange the conversion factor is called the **rate of exchange,** and although this varies, we shall assume it to be $1 = 600 lira.

Again, there are two possible multipliers,

$$\frac{\$1}{600 \text{ lira}} = \frac{600 \text{ lira}}{\$1} = 1$$

Since dollars are being exchanged for lira, the suitable multiplier is 600 lira/$1, and the calculation is

$$\$50 \times \frac{600 \text{ lira}}{\$1} = 30,000 \text{ lira}$$

MEASURING IN THE METRIC SYSTEM

It was mentioned in the previous section that a Frenchman, a Spaniard, and a German would all specify 6.35 cm as the length of a line that we would call 2.5 in long. Since they all speak different languages, why do they use the same measurement language? Not long ago every country in Europe had its own system of measurement, and often there were differences between regions within the same country. This was true not only of the European nations but of *all* countries. So long as the various parts of the world had little to do with each other, this didn't matter much. A *kwan* was comfortably used in Japan, whereas a *pood* was convenient for Russia, and Egypt liked a *kantar*. All these are units of weight used in these countries, and the conversion factors to our pounds are shown in the margin. Similarly, oil in Sicily was measured out in *cuffiscos*, whereas Tangier used *kulas* for the same purpose, and we all know what is meant by a *hogshead* — or do we? These are all units of volume, and the conversion factors are shown in the margin.

1 kwan = 8.27 lb
1 pood = 36.11 lb
1 kantar = 99.03 lb

1 cuffisco = 5.6 gal
1 kula = 4.0 gal
1 hogshead = 63 gal

Many dozens of different units have been used throughout the world to designate length, weight, volume, and area, and special units were often used for particular materials such as wood, or precious stones or metals, or pharmaceuticals. If all these units were still in common use, it would put the Tower of Babel to shame and (more to the point) make the exchange of goods and technology enormously difficult and costly. Remember that the "translation" of measurement languages, meaning the conversion from one unit to another, is in one respect more difficult than ordinary translation since a numerical factor is always involved. As the world became smaller in the sense that it took less time to go from place to place, people became more interdependent and goods and ideas passed more and more frequently across national boundaries. It is interesting that the first common language to be devised was the measurement language called the **metric system.**

The metric system was developed and first accepted on the continent of Europe, where many nations have lived close to each other for centuries and where much of modern science and industry had its origins. Specifically, the metric system was formulated in France in 1790 and became the standard there for all measurements of length, weight, and volume in 1799. Since that time its use has spread to most of the nations of the world, and all nations accept it. Although the metric system was legally accepted in the United States in 1866 by an act of Congress, its use here is still mainly limited to scientific and technical work. However, there is good reason to expect that we, too, will follow the example of Great Britain, which is now shifting from its English system to the metric system.

THE BASIC UNITS OF THE METRIC SYSTEM

As science expanded, so did the metric system, and in time it came to include measurement units required by all branches of science. The comprehensive system of units that so evolved is now called the **Système International d'Unités** (SI). Essentially, it is the metric system grown up and sophisticated. The SI is subject to decisions made at meetings of the General Conference of Weights and Measures. The last meeting was the 13th, held in October 1967.

The measurements most frequently made in chemical work are determinations of length, volume, mass, time, and temperature. The **basic units** for these measurements in the metric system are listed in Table 2.1.

TABLE 2.1   Basic units of metric system

Measurement	Basic Unit	Abbreviation	Equivalent in English System
Length	Meter	m	39.37 in, or a "long yard"
Volume	Liter	l	1.057 qt, or a "fat quart"
Mass	Kilogram	kg	2.2 lb
Time	Second	s	1
Temperature	Degree Celsius	°C	A change of 1°C = a change of 1.8°F

LENGTH

The basic unit of length in the metric system, the **meter,** was originally intended to represent 1/10,000,000 of the distance from the equator to the North Pole. This would have left the actual length of the meter somewhat in doubt, since the equator-to-pole distance is difficult to establish, even with modern instruments. The standard representing a basic unit must itself be accurately determined, as well as stable and unvarying. For this reason the distance between two lines engraved on a platinum-iridium bar was accepted as the standard meter length and became the prototype for all meter lengths. The original bar was kept securely and at constant temperature in Sèvres, a town near Paris, and copies were given to those nations that used the meter as a standard of length. However, scientists prefer to relate basic units, where possible, to some **natural property** that is itself unchanging. The **wavelength** of light is such a property, and at an international conference in Paris in 1960 the meter was redefined in terms of this property. The experimental setup required is not too complicated for a well-equipped laboratory, so that the accuracy of a meter length can now be determined without reference to scratches on a bar kept in France.

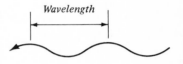

Wavelength

The meter is useful for measuring things such as a length of cloth, or the size of a room, or the distance of a race, or carpeting, and so on. But, it would be awkward, if not impossible, to use a meterstick to determine the thickness of a fine wire, the distance between two cities, or the dimensions of a dust particle. The size of the unit of measure should be compatible with what is being measured. A toothbrush is not used to sweep the floor or a broom to clean teeth. The metric system has a wide range of units of length, with the meter as the reference point, and you will see from Table 2.2 that the name of most units is made up of an identifying prefix attached to *meter;* note the 10 times ratio between successive units, except for the last three, which are used for very small distances.

In shifting between English and metric units of length suitable conversion factors are needed. Three common ones are shown in Table 2.3. The first factor, used for inch-to-centimeter conversion, was discussed before, and the others are found and used in the same way. A distance of 100 km is converted to miles by using the factor 1 km = ⅝ mi. Of the two multiplying ratios available from that factor, the right-hand one is used because it will cancel kilometers and leave miles. Hence,

$$100 \ \cancel{km} \times \frac{\frac{5}{8} \ mi}{1 \ \cancel{km}} = 100 \times \frac{5}{8} \ mi = 62\frac{1}{2} \ mi$$

TABLE 2.2   Metric units of length

Prefix	Value	Metric Unit of Length	Abbreviation	Conversion Factor
kilo	1000	Kilometer	km	1 km = 1000 m = 10^3 m
hecto	100	Hectometer†	hm	1 hm = 100 m = 10^2 m
deka	10	Dekameter †	dam	1 dam = 10 m = 10^1 m
		Meter	m	1 m = 10^0 m
deci	0.1	Decimeter	dm	1 dm = 0.1 m = 10^{-1} m
centi	0.01	Centimeter	cm	1 cm = 0.01 m = 10^{-2} m
milli	0.001	Millimeter	mm	1 mm = 0.001 m = 10^{-3} m
micro	0.000001	Micrometer ‡	μm	1μ = 0.000001 m = 10^{-6} m
nano	0.000000001	Nanometer§	nm	1 nm = 0.000000001 m = 10^{-9} m
		Angstrom	Å	1 Å = 0.0000000001 m = 10^{-10} m

† Not in common use.
‡ Formerly called a micron.
§ Formerly called a millimicron.

If a very accurate conversion is required, 0.621 would be used instead of ⅝ and the answer would be 62.1 mi.

In many instances, conversion between units of the *same* system are needed, and we are all familiar with changeovers from feet to inches, and to yards, and so on. In the metric system the conversion factors (Table 2.2) are all powers of ten, and this brings us to a method of calculation that is based on powers of ten, called **scientific notation.**

Scientific notation   This notation is based on the fact that any number can be written as the product of two numbers, the first of which is the **coefficient** and the second is a power of ten. For example, the number 300 can be written as 3×10^2, where 3 is the coefficient, or multiplier, and 10^2 is the power of ten. In general, 10^a is a power of ten whose exponent a can be any value, positive or negative. If a is *greater* than zero, 10^a is *greater* than 1; if a is

50 mi = ? km

$50 \text{ mi} \times \dfrac{1 \text{ km}}{\text{⅝ mi}} = 50 \times \text{⅝ km}$

$= 80 \text{ km}$ *Ans.*

TABLE 2.3   Common conversion factors

Conversion Factor	Multiplying Ratio		
2.54 cm = 1 in	$\dfrac{2.54 \text{ cm}}{1 \text{ in}}$	or	$\dfrac{1 \text{ in}}{2.54 \text{ cm}}$
1 m = 39.4 in	$\dfrac{1 \text{ m}}{39.4 \text{ in}}$	or	$\dfrac{39.4 \text{ in}}{1 \text{ m}}$
1 km = 0.621 mi $\approx$ ⅝ mi	$\dfrac{1 \text{ km}}{\text{⅝ mi}}$	or	$\dfrac{\text{⅝ mil}}{1 \text{ km}}$

Power of ten

$$10^4 = 10 \times 10 \times 10 \times 10 =$$

$$10^3 = 10 \times 10 \times 10 \quad =$$

$$10^2 = 10 \times 10 \quad\quad =$$

$$10^1 = 10 \quad\quad\quad\quad =$$

$$10^0 = 1 \quad\quad\quad\quad =$$

$$10^{-1} = \tfrac{1}{10} \quad\quad\quad =$$

$$10^{-2} = \tfrac{1}{100} \quad\quad =$$

$$10^{-3} = \tfrac{1}{1000} \quad\quad =$$

$$10^{-4} = \tfrac{1}{10000} \quad =$$

Decimal expression

10,000

1,000

100

10

0

0.1

0.01

0.001

0.0001

Every step up is a multiplication by 10

Every step down is a division by 10

FIGURE 2.1 Powers of ten.

less than zero, meaning negative, 10^a is *less* than 1; if a *equals* zero, then 10^a *equals* 1. Figure 2.1 illustrates this. Obviously, this list can be extended indefinitely upward, and downward, but enough is shown for us to outline some general rules for powers of ten.

1 If the exponent of 10 is a positive whole number such as 1, or 2, or 3, etc., *that* will be the number of zeros after the 1 in the decimal expression. In $10^4 = 10,000$ the exponent is 4 and the number of zeros after the 1 is also 4; in $10^2 = 100$ the exponent is 2 and the number of zeros after the 1 is also 2.

2 If the exponent of 10 is a negative whole number, $-a$, the position of the decimal is located by starting just to the right of 1, and moving to the *left* a times. For example, compare the exponent and the number of left-moving steps from 1 to the decimal position for each of the following:

$$10^{-1} = \tfrac{1}{10} = .1$$

$$10^{-2} = \tfrac{1}{100} = .01$$

$$10^{-3} = \tfrac{1}{1000} = .001$$

3 In *multiplying* 10^a by 10^b, where the exponents a and b are any two numbers, the exponents are *added*:

$10^2 \times 10^4 = 10^6$	since $2 + 4 = 6$
$10^2 \times 10^{-4} = 10^{-2}$	since $2 + (-4) = -2$
$10^{-2} \times 10^4 = 10^2$	since $-2 + 4 = 2$
$10^{-2} \times 10^{-4} = 10^{-6}$	since $-2 + (-4) = -6$

4 In *dividing* 10^a by 10^b, the exponents are *subtracted*, $a - b$:

$$\frac{10^2}{10^4} = 10^{-2} \qquad \text{since } 2 - 4 = -2$$

$$\frac{10^2}{10^{-4}} = 10^6 \qquad \text{since } 2 - (-4) = 6$$

$$\frac{10^{-2}}{10^4} = 10^{-6} \qquad \text{since } -2 - 4 = -6$$

$$\frac{10^{-2}}{10^{-4}} = 10^2 \qquad \text{since } -2 - (-4) = 2$$

Now let us see how powers of ten can be used in calculations. A **constant** is the unvarying numerical value of some physical or chemical characteristic, or property. One of the most important constants in science is the velocity of light, which has been established as 300,000 kilometers per second (km/s). Using powers of ten, we first write this as $3 \times 100{,}000$, where 3 is the coefficient and 100,000 is 10^5 (count the zeros), so that 300,000 km/s is now expressed as 3×10^5 km/s. Not only is this easier to write but also much easier to handle in calculations. For instance, to find the distance light will travel in 1 hour (h) you multiply 300,000 km/s by 60 s/min by 60 min/h. With all the zeros involved, this could become a bit sticky. With powers of ten the calculation becomes

$$(3 \times 10^5 \text{ km/s}) \times (6 \times 10 \text{ s/min}) \times (6 \times 10 \text{ min/h})$$

First the powers of ten are collected and multiplied, giving $10^5 \times 10 \times 10$; since the powers of ten are being *multiplied*, their exponents are *added*, and the result is 10^7. The coefficients are then collected and multiplied, $3 \times 6 \times 6 = 108$, so that the answer could be written 108×10^7 km/h. However, it is the practice in scientific notation that a coefficient should have a value somewhere between (but not including) 1 and 10, so that 108 is rewritten as 1.08×10^2. The final answer now is obtained from $1.08 \times 10^2 \times 10^7$; the exponents are added again, and we get 1.08×10^9 km/h.

Many chemical calculations involve very large numbers as well as very small ones. One such very large number when written in ordinary form, is

$$602{,}000{,}000{,}000{,}000{,}000{,}000{,}000$$

Using powers of ten, we write it as 6.02×10^{23}, a simpler and neater form.

From kilometers to angstrom units   Table 2.2, listing the metric units of length, begins with a kilometer, equal to 10^3 m and ends with an angstrom unit, equal to 10^{-10} m. The ratio between them

is therefore,

$$\frac{\text{Kilometer}}{\text{Angstrom unit}} = \frac{10^3 \ \cancel{\text{m}}}{10^{-10} \ \cancel{\text{m}}} = 10^{13}$$

Cross multiplying shows that a kilometer is 10^{13} times as big as an angstrom unit, so that obviously the two serve different measuring needs. Between them are all the other metric units of length, and it is helpful to consider where these units are being used, going from largest to smallest. As mentioned before, the **kilometer** is about ⅝ mi, and is used where we would use a mile, i.e., geographical distances, air and surface velocities, highway distances, and so on.

$$\frac{\text{km}}{\text{Å}} = \frac{10^{13}}{1}$$

Cross multiplying gives

$$\frac{1 \ \text{km}}{\text{Å}} = \frac{10^{13}}{1}$$

$$1 \ \text{km} = 10^{13} \ \text{Å}$$

The hectometer and dekameter that follow are not widely used, and we pass on to the basic unit of length, the **meter.** Its conversion factor was given before as 1 m = 39.37 in, making it about 10 percent more than our yard, that is, about 40 in. The meter is generally used where we use yards or feet. Our heights are given in feet and inches; meters are used in the metric system, generally to two decimal places. A person 5 ft 10 in tall would be 1.78 m, metrically speaking. Also, heights of mountains are in meters, so that Mt. Everest is 8882 m above sea level.

The **decimeter** is 0.1 m, making it close to 4 in, which is sort of betwixt and between, too small for large objects, and too big for really small ones. The decimeter has applications in technical work, but in general is used far less than its neighbor above it, the meter, or its neighbor below it, the centimeter.

The **centimeter** is 0.01 m, which makes it 0.3937 in, or close to 0.4 in. It is used where we would generally use an inch. Whereas our ordinary rulers and tape measures are in inches, with smaller markings for ¼, ½ in, and so on, their metric counterparts are in centimeters, with the smaller markings in millimeters, 10 to a centimeter. A metric tape measure is easier to read since the tenths are counted off directly, and there is no need to look for such units as ¼, or ⅜, or 5/16. Also, a length can be expressed either in centimeters or in millimeters; 3.4 cm can be read as 34 mm without any ambiguity.

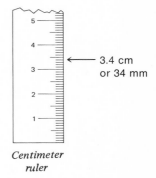

← 3.4 cm
or 34 mm

Centimeter ruler

The **millimeter** is 0.001 m and is the smallest unit used for objects measurable with the naked eye. Together the millimeter, centimeter, meter, and kilometer cover the range of measurements normally involved in day-to-day living, as well as in a great deal of technical work. Machine tools in metric countries are calibrated in millimeters, and engineering drawings are generally dimensioned in millimeters. Being 0.1 cm, 1 mm is about 0.04 in, but you might visualize it as just about the thickness of a dime. Squeeze 10 dimes together, and you have a centimeter; of course

A dime – about 1 mm *thick*

you also have a dollar, which is where we keep *our* decimal system.

From here on down, the unaided eye alone is not enough. The **micrometer** (μm) (formerly called micron) is used to measure such things as fine wires, hair, the size of cells, powders, and similar objects: here some degree of magnification is needed. Since 1 μm is 0.001 mm, imagine a dime sliced edgewise into a thousand parts. Since a millimeter is itself 10^{-3} m, a micrometer is $10^{-3} \times 10^{-3}$ m, or 10^{-6} m.

$$1 \text{ m} = 10^2 \text{ cm}$$
$$= 10^3 \text{ mm}$$
$$= 10^6 \text{ } \mu\text{m}$$
$$= 10^9 \text{ nm}$$
$$= 10^{10} \text{ Å}$$

The **nanometer** (nm) (formerly called a millimicron) is 10^{-3} μm, or $10^{-3} \times 10^{-6}$ m $= 10^{-9}$ m. This unit is small enough to be used for the wavelengths of visible light. From the violet to the red end of the visible spectrum the wavelength increases from about 350 to 700 nm. The smaller the wavelength, the higher the energy, so that the short ultraviolet radiation from the sun, with wavelengths below 350 nm, will "burn" the skin. Fortunately, our atmosphere shields us from most of this ultraviolet radiation. On the other hand, a substantial part of the sun's warmth comes to us as infrared radiation, with wavelengths over 700 nm. Infrared energy is also emitted by a warm, but not incandescent, body such as a hot plate. Also in the nanometer range are such things as viruses, microbes, and the macromolecules of living matter. These are not visible through ordinary microscopes, although electron microscopes have been enormously effective in these areas of biological research.

A burning candle, the glowing filament of an electric bulb, and the sun are all sources of radiant energy (heat or light), which is transferred across space.

The **angstrom unit** (Å), named for a Swedish physicist, is 0.1 nm, or 10^{-10} m, so that 10 billion of these units, laid end to end, will make up 1 m. Expressed in angstrom units, the visible spectrum from violet to red has wavelengths from about 3500 to 7000 Å, whereas x-rays have a wavelength of about 1 Å. The dimensions of atoms, molecules, and crystal structures are so small that they lend themselves to measurement in angstrom units.

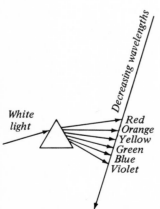

The smaller the wavelength of radiant energy, the greater is its energy.

Conversions between metric units of length Switching from one metric unit to another is a matter of multiplying or dividing by the suitable power of ten that is the ratio of the two units. The prefixes listed in Table 2.2 are the simplest guide, and a few examples will illustrate their use.

Problem

$$2.5 \text{ km} = ? \text{ m}$$

kilo- equals 1000×

The prefix *kilo* tells us that a kilometer is 1000 m, so that in this case it is only necessary to replace km by 1000 m.

Solution

$$2.5 \overset{\frown}{\text{km}} = 2.5 \times 1000 \text{ m} = 2500 \text{ m}$$

With powers of ten, this can be expressed as

$$2.5 \text{ km} = 2.5 \times 10^3 \text{ m} \quad Ans.$$

Problem

$$450 \text{ mm} = ? \text{ cm}$$

Keeping in mind that 1 cm = 10 mm, as shown in the sketch of the metric ruler on page 19, the multiplier to use is either 1 cm/10 mm or 10 mm/1 cm, both equal to 1. The first ratio is used since it cancels out mm and leaves cm.

Solution

$$450 \text{ mm} \times \frac{1 \text{ cm}}{10 \text{ mm}} = \frac{450}{10} \text{ cm} = 45 \text{ cm} \quad Ans.$$

Problem

$$5600 \text{ Å} = ? \text{ nm}$$

Here again it may be recalled that the angstrom unit is one-tenth as large as the nanometer just above it in Table 2.2. The conversion factor would then be 1 Å = 10^{-1} nm, or alternately, 10 Å = 1 nm, so that the multiplying ratio is either 10 Å/1 nm or 1 nm/10 Å. The latter ratio is used as the multiplier since it cancels Å and leaves nm.

nanometer = nm = 10^{-9} m

Solution

$$5600 \text{ Å} \times \frac{1 \text{ nm}}{10 \text{ Å}} = \frac{5600}{10} \text{ nm} = 560 \text{ nm} \quad Ans.$$

Problem

$$850 \text{ nm} = ? \text{ Å}$$

The same factor is used, but now the other form of the ratio must be applied since nanometers must cancel.

$$\left. \begin{array}{c} \dfrac{1 \text{ nm}}{10 \text{ Å}} \\[2mm] \dfrac{10 \text{ Å}}{1 \text{ nm}} \end{array} \right\} = 1$$

Solution

$$850 \text{ nm} \times \frac{10 \text{ Å}}{1 \text{ nm}} = 850 \times 10 \text{ Å} = 8500 \text{ Å} \quad Ans.$$

Each edge is 1 cm

1 *cubic centimeter* = 1 cm³
            = 1 ml

liter = l
milliliter = ml
1 ml = 10^{-3} l
1000 ml = 1 l

1 l = 1.057 qt

1 dl = $\frac{1}{10}$ l = 0.1 l
1 cl = $\frac{1}{100}$ l = 0.01 l
1 ml = $\frac{1}{1000}$ l = 0.001 l

VOLUME

The space occupied by a gas, a liquid, or a solid is its volume, and whereas length is one-dimensional and area two-dimensional, volume is three-dimensional. Volume must therefore be expressed in three-dimensional units, so that a cube whose three edges are all equal to 1 cm will have a volume of 1 cubic centimeter (cm³). One cubic centimeter is often shortened to 1 cc.

The cc is often used by the medical profession, but chemists throughout the world have long used the liter as the basic unit of volume, and chemical equipment is generally calibrated either in liters or milliliters. Since *milli-* means one-thousandth, a milliliter is one-thousandth (10^{-3}) of a liter. The use of the liter or milliliter is reasonable considering that chemists deal so often with liquids and gases, which may have a fixed volume without having fixed dimensions. In any case, there really is no confusion involved because the volume of 1 ml is exactly equal to 1 cm³, which would make 1 l = 1000 cm³.

A liter can be imagined as a large quart, and apart from its use by chemists, it is the unit of volume used in daily life in the metric countries, where it is used to measure milk, gasoline, wine, oil, and so on. Recipes refer to liters or to such subunits as milliliters, centiliters, or deciliters, instead of our quarts, pints, cups, tablespoons, teaspoons, and fluid ounces. The value of these subunits can be inferred from their prefixes.

Although the milliliter is the smallest unit of volume used for ordinary purposes, technical work may require extremely small volumes to be dispensed, in which case the unit used is the *microliter* (μl). Again, the prefix defines this unit as equal to 10^{-6} l, or 0.000001 l, so that 1 μl is one-millionth of a liter.

Since we have spoken of the smallest unit of volume in the metric system, we should also mention the large unit called the **hectoliter,** equal to 100 l. It is used where large volumes are involved, as in wholesale transactions of liquid materials such as wine, milk, or oil.

Conversions between units of volume are similar to those between units of length, except that the liter, rather than the meter, is the basic unit. A few examples follow.

Problem

$$2.4 \text{ l} = ? \text{ ml}$$

Since a *milli*liter is 1/1000 l, it follows that 1000 ml = 1 l, so that substituting 1000 ml in place of liters above gives the answer directly.

Solution

$$2.4\ l = 2.4 \times 1000\ ml = 2400\ ml \quad Ans.$$

Using exponents, we have 2.4 l $= 2.4 \times 10^3$ ml. The two answers are of course equivalent.

Problem

$$350\ ml = ?\ l$$

The relation 1000 ml $= 1$ l applies here as well, although it is easier to use when it is written as 1 ml $= 0.001$ l. Replacing milliliters by 0.001 l gives the answer.
Solution

$$350\ ml = 350 \times 0.001\ l = 0.350\ l \quad Ans.$$

Problem

$$350\ ml = ?\mu l$$

The relation between milliliters and microliters can be obtained from their prefixes: *milli-* $= 10^{-3}$ and *micro-* $= 10^{-6}$:

$$\frac{\text{Milliliter}}{\text{Microliter}} = \frac{10^{-3}}{10^{-6}} = 10^3$$

By cross-multiplying, 1 ml $= 10^3\ \mu l$, and this equality can be expressed as two multiplying ratios, both equalling 1.

$$\frac{1\ ml}{10^3\ \mu l} = \frac{10^3\ \mu l}{1\ ml} = 1$$

Since milliliters are to be canceled in the calculation, the right-hand ratio is the appropriate one.
Solution

$$350\ ml \times \frac{10^3\ \mu l}{1\ ml} = 350 \times 10^3\ \mu l$$

$$= 3.50 \times 10^2 \times 10^3\ \mu l$$
$$= 3.50 \times 10^5\ \mu l \quad Ans.$$

MASS

Although the terms **mass** and **weight** can be used interchangeably in many cases without harm, it is really incorrect, and the difference between them should be understood. The mass of an object is the quantity of matter in that object. The weight of that object is the gravitational **force** exerted by the earth on that mass. This is also the force exerted by the mass on the earth, since gravitational attraction is mutual. Being so much larger, the earth does not "feel it" as much. As mass increases, the gravitational force increases, so that for objects that are on the earth's surface, mass and weight are proportional, increasing and decreasing at the same rate. Obviously, if an object were to find itself on another planet, its "weight" would no longer be its "earth weight," and would be greater on big Jupiter and less on little Mercury; but in all cases its mass would be the same.

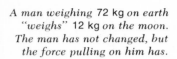

A man weighing 72 kg *on earth "weighs"* 12 kg *on the moon. The man has not changed, but the force pulling on him has.*

It is the **mass** of an object that is involved in chemical reactions, and it is the mass that is measured by a chemical balance, even if the operation itself is referred to as a "weighing." The basic unit of mass in the metric system is the **kilogram,** which is the quantity of matter in a platinum cylinder kept in Sèvres as the international standard. The National Bureau of Standards of the United States has a duplicate of its own as a reference standard.

Like length and volume units, there are multiple and fractional units based on the kilogram, and Table 2.4 lists the more important ones.

In the range of measurements used in daily living, the gram, the kilogram, and the metric ton are the more important. The metric ton is used for bulk cargo, freight, heavy equipment, and so on, and is 1.1 times our short ton of 2000 lb. The kilogram is the unit of the market place; beans, bacon, and bananas are sold in "kilos," or half kilos. The kilogram is also the unit of the weighing scale, and if the pointer reads 80 what is that in our pounds? The gram is obviously for small objects, and is the common unit of the chemical laboratory. There are 453.6 g in a

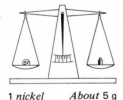

1 nickel   About 5 g

TABLE 2.4   Units of weight

---

1 metric ton = 1000 kg = 2204.6 lb
1 quintal = 100 kg = 220.46 lb
1 kilogram = 2.2046 lb
1 hectogram (hg) = 0.1 kg = 0.2205 lb
1 gram (g) = 0.001 kg = 0.035 oz avoirdupois
1 milligram (mg) = 0.001 g = 0.000001 kg
1 microgram (μg) = 0.000001 g = 10^{-6} g = 10^{-9} kg

---

pound, and, as shown above, 1 g is 0.035 oz. It might be helpful to keep in mind that a nickel weighs just about 5 g.

The **milligram** is used in technical work, and a standard analytical balance can measure mass to 0.1 mg. The microgram (μg), which is 0.001 mg, is for substances that even in trace amounts can make their presence felt, such as poisons, vitamins, or hormones.

Conversion between units of mass is carried on analogously to conversions between the other units discussed, and an article weighing 1.4 kg can be successively represented as follows.

Problem

$$1.4 \text{ kg} = ? \text{ g}$$

Solution The conversion factor is 1 kg $=$ 1000 g. Therefore,

$$1.4 \text{ kg} \times \frac{1000 \text{ g}}{\text{kg}} = 1400 \text{ g} \quad Ans.$$

Problem

$$1.4 \text{ kg} = ? \text{ mg}$$

Solution In the previous problem we found that 1.4 kg $=$ 1400 g. Now grams are to be converted to milligrams. As 1 mg $=$ 0.001 g, or 1000 mg $=$ 1 g, the multiplying ratio

$$\frac{1000 \text{ mg}}{1 \text{ g}} = 1$$

is obtained. Applying this multiplier

$$1.4 \text{ kg} = 1400 \text{ g} \times \frac{1000 \text{ mg}}{1 \text{ g}}$$

$$= 1.4 \times 10^3 \times 10^3 \text{ mg} = 1.4 \times 10^6 \text{ mg} \quad Ans.$$

Problem

$$1.4 \text{ kg} = ? \ \mu\text{g}$$

Solution Since *micro-* means one-millionth $= 10^{-6}$, one microgram $= 10^{-6}$ g, or $10^6 \ \mu$g $=$ 1 g. Applying $10^6 \ \mu$g/1 g as the multiplying ratio,

$$1.4 \text{ kg} = 1400 \text{ g} = 1.4 \times 10^3 \text{ g} \times \frac{10^6 \mu\text{g}}{1 \text{ g}}$$

$$= 1.4 \times 10^9 \mu\text{g} \quad Ans.$$

Before leaving the weight units, we should mention apothecary weights, which are used to formulate medical prescriptions. An apothecary weight often used is the *grain*. For example, the weight of aspirin in an aspirin tablet is 5 grains. The following conversion factors should be noted:

$$437.5 \text{ grains} = 1 \text{ oz avoirdupois}$$
$$15.4 \text{ grains} = 1 \text{ g}$$
$$0.0154 \text{ grain} = 1 \text{ mg}$$

Bowman holding bow and arrow. The bow is not deformed and has no potential energy.

HEAT AND TEMPERATURE

Up until the time of firearms, the bow and arrow was the principal weapon used for hunting. The bow and arrow is a simple but effective way of converting muscular energy into the kinetic energy of the arrow, meaning its energy of motion. Whenever some external resistance is overcome, as by lifting, pushing, bending, or any other way, work is being done and energy is being used up. Energy is the capacity for doing work. When the bowman draws the arrow back, he is doing work because the bow resists being bent. This energy input is stored in the bow as potential energy. It is called potential because it is usable but not *yet* used. When the string is released and the bow returns to its original shape, this potential energy is transmitted to the arrow, and off it goes. When the arrow finally stops, where is its kinetic energy? Is it lost? Energy can never be lost; it can only change form. If we imagine the arrow being released into a large body of water, the arrow will slow down and finally stop due to friction. But friction always results in heat, so that the work performed by the bowman, which was stored in the bow as potential energy and transmitted to the arrow as kinetic energy, ends up as *heat*.

Bowman drawing back bow and arrow, which now have potential energy.

In this example, work, or mechanical energy, was converted to heat, or thermal energy. However, all other forms of energy will also convert to heat, which is therefore the most *general* form of energy. In chemistry we speak of heat leaving or entering a **system,** which is some substance or some aggregate of substances, separated for the purpose of study. A system can be simple or complex; water in a beaker is a system, as is a steam engine. Everything outside the system is the **surroundings.** Using the simple example, let us visualize the water-beaker

Bowman releases arrow and potential energy is converted to kinetic energy.

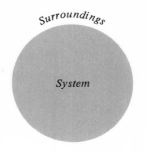

system being heated by a bunsen burner. The heat is coming from the surroundings, since the burner is outside the system. The longer the water receives heat from the flame, the hotter it gets and the higher its temperature. If the burner is removed, heat will now flow from the water to the cooler air. The flow of heat is therefore always *from hotter to cooler,* from a warmer object to a cooler object. All this is obvious enough, but it is obvious only because we have seen it happen over and over again and never the opposite. The fact that heat runs "downhill," from hotter to cooler, is a central principle of science that reaches a long way into chemistry.

As heat flows into some mass of matter, we can imagine the level of heat rising, just as though water were running into a vertical pipe closed at its lower end (Figure 2.2). As the water level in the pipe goes up, the water pressure goes up; as the heat level in an object rises, the temperature rises and we can visualize temperature as a measure of **heat level,** or heat pressure. Pressure and temperature are examples of what are called **intensity factors,** of which the voltage of an electric current is another example. They serve as the push behind the flow, whether a flow of water, or a flow of heat, or a flow of electricity. The temperature change that results when heat flows in, or out, of some object depends upon:

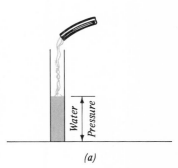

1 How much *heat* is entering or leaving
2 How much *mass* is being heated
3 The *kind* of mass being heated

If the mass of material is 1 g and the kind of material is pure water, the amount of heat required to change the temperature 1 Celsius degree is called a **calorie.** The Celsius temperature scale is used in the metric system, and the calorie is the unit of heat of the metric system. Since heat is the most general form of energy, the calorie is also the unit of energy in whatever form it may appear.

If the mass of water being heated were increased, there would be a proportional increase in the calories needed to cause a 1°C change; similarly, if the temperature change were other than 1°C, there would be a proportional change in the number of calories required. The general relation between the quantity of water being heated, the temperature change, and the calories involved is therefore

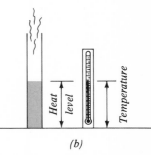

FIGURE 2.2 (*a*) As water keeps flowing in, water pressure keeps increasing. (*b*) As heat keeps flowing into a body, its heat pressure, or temperature, keeps increasing.

Although the calorie is widely used throughout the world, the SI unit of energy is the joule (J); 4.18 J = 1.00 cal.

The symbol Δ is used to mean "the change in. . . ."

$$\text{Calories} = \text{grams of water} \times \Delta T_c$$

where ΔT_c means the temperature change in Celsius degrees.

For most purposes the mass of 1 ml of water can be taken as equal to 1 g. Since liquids are more often measured by their volume than by their mass, the expression given above can be rephrased as follows,

$$\text{Number of calories in or out} = \text{ml } H_2O \times \Delta T_c$$

When 50 ml *of water is heated from* 20 *to* 80°C, *the calorie intake is* 50 × (80 − 20) = 50 × 60 = 3000 cal.

If ΔT_c is negative, the water has lost calories. If ΔT_c is positive, the water has gained calories.

Heat capacity So far, only water has been considered as losing or gaining heat and changing in temperature accordingly. Suppose some other material were used. Would the results be the same? Definitely not, because the rate at which a material heats up or cools down is characteristic of *that* material. The rate at which water changes its temperature with heat input or outgo is based upon the meaning of the calorie: one calorie per one gram per one Celsius degree change. This heating rate is called the **heat capacity,** and for water its value is 1. Water has a very high heat capacity compared with most materials, which means it heats up and cools down more slowly. Water is taken as the point of reference for the heat capacity of other materials, and the experimental value, compared with the value of 1 for water, is called the **specific heat capacity** c of the material, or simply the specific heat.

The relative heat capacities for several materials are listed in Table 2.5, and in each case the calories needed to cause a *change* of 1°C in 1 g of the substance is given. The specific heat of each substance is the ratio of that value to the value of 1 for water.

The specific heat can now be included in the expression relating calories with mass and the temperature change of a given

TABLE 2.5 Specific heat of selected materials

Material	Calories to Change 1g of Material by 1°C	Specific Heat c
Water	1	1
Magnesium	0.25	0.25
Air	0.24	0.24
Aluminum	0.21	0.21
Soil	0.20	0.20
Iron	0.11	0.11
Copper	0.09	0.09
Gold	0.03	0.03

substance:

$$\text{Calories (in or out)} = \text{mass (in grams)} \times \Delta T_C \times c$$

For the same mass of material and the same caloric input, the temperature change is *inverse* to the value of the specific heat. An example will illustrate.

Problem A mass of 5000 g of water (5 l), and an equal mass of 5000 g of magnesium metal both absorb 20,000 cal of heat. What is the temperature change of the water and of the magnesium? *Solution*

H_2O: $20,000 \text{ cal} = 5000 \text{ g} \times \Delta T_c \times 1$

$$\Delta T_c = \frac{20,000}{5000} \times 1 = 4°C \quad Ans.$$

Mg: $20,000 \text{ cal} = 5000 \text{ g} \times \Delta T_c \times 0.25$

$$\Delta T_c = \frac{20,000}{5000} \times \frac{1}{0.25} = \frac{4}{0.25} = 16°C \quad Ans.$$

The calorie is used in a great deal of scientific work. When it is considered that a gram is a small mass and that a 1°C change is also small, it can be understood that a calorie is a very small quantity of heat. Where larger amounts of heat are involved, as in nutrition, the kilocalorie, equaling 1000 cal, is used; the kilocalorie is also called a "large calorie."

In the English system the British thermal unit (Btu) equals 252 cal.

Fahrenheit and Celsius temperatures Since temperature can be thought of as the level of heat, a scale of reference is needed to measure changes in heat level. This is called a temperature scale, and the intervals, or steps, on the scale are called **degrees.** We are all familiar with the Fahrenheit scale of temperature measurement, and with degrees Fahrenheit. However, whereas the Fahrenheit scale is still in use in the English-speaking nations, the metric system countries use the Celsius scale. Both scales are based on the same reference points, the freezing point and the boiling point of pure water, each of which represents a fixed and reproducible heat level, or temperature. Since both the Fahrenheit scale (°F) and the Celsius scale (°C) have the same reference points, it should be possible to change over from one to the other if proper consideration is given to the numerical differences between them. There are two such differences: (1) the numbers assigned to the freezing and boiling points are different on the two scales, and (2) the size of the steps, or degrees, on the

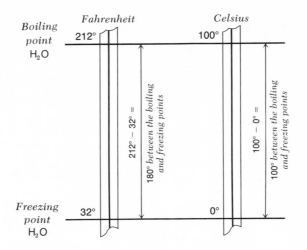

FIGURE 2.3 Fahrenheit and Celsius scales. Because there are 100 degrees between the two reference points, the Celsius scale was formerly called the centigrade scale.

two scales are not the same. These are illustrated in Figure 2.3.

What adjustments are required in converting between the two scales? There are 180 steps, or degrees, in going from the freezing to the boiling point of water on the Fahrenheit scale but only 100 on the Celsius scale, so that each Celsius degree represents a larger step up or down than a Fahrenheit degree. That is, since

$$180 \text{ Fahrenheit degrees} = 100 \text{ Celsius degrees}$$

$$18°F = 10°C \quad \text{and} \quad 1.8°F = 1°C$$

This means that for every movement of *one* degree up or down the Celsius scale, an equal movement up or down the Fahrenheit scale will be represented by 1.8°F (Table 2.6). In converting from the two scales, this adjustment must therefore be made.

The freezing point of water is 32° on the Fahrenheit scale and 0° on the Celsius scale; the boiling point is 212°F and 100°C. Any temperature on either scale can therefore be located some number of degrees above or below a reference point of that scale. The value of zero for the freezing point of water on the Celsius scale is convenient since all positive values are above it and all negative values are below it.

To get to some value $X°$ on the Celsius scale you move up X steps from 0°C. On the Fahrenheit scale the same starting point

TABLE 2.6

Change, °F	Change, °C
1.8	1.0
9	5
18	10
180	100

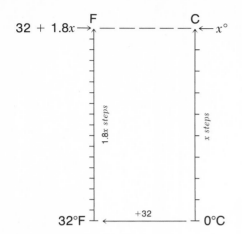

$32 + 1.8x \rightarrow$

F

C

$\leftarrow x°$

1.8x steps

x steps

$32°F$ $\xleftarrow{\quad +32 \quad}$ $0°C$

FIGURE 2.4   From Celsius to Fahrenheit.

is 32°F, and to get to the same level as $X°C$ you move up $1.8X$ steps above 32°F. In switching from a temperature whose value is given in degrees Celsius to the corresponding value in degrees Fahrenheit, you first multiply the numerical value of degrees Celsius by 1.8, then add the number 32; the result is the temperature in degrees Fahrenheit. This changeover can be expressed as follows:

$$°F = 32 + 1.8°C$$

Although this equation was derived by going from Celsius degrees to Fahrenheit degrees, it can be used in changing from either side, and a few examples are shown.

Problem

$$20°C = ?°F$$

Solution   The Celsius value is substituted in the equation:

$$°F = 32 + 1.8 \times 20 = 32 + 36 = 68$$
$$20°C = 68°F \quad Ans.$$

Problem

$$86°F = ?°C$$

Solution   The Fahrenheit value is substituted in the equation.

$$86 = 32 + 1.8°C$$

from which, by transposing,

$$86 - 32 = 1.8°C$$

and therefore $1.8°C = 54$.   Dividing gives

$$°C = \frac{54}{1.8} = \frac{540}{18} = 30$$

$$86°F = 30°C \quad Ans.$$

Problem

$$-40°F = ?°C$$

Solution Again substituting the Fahrenheit value,

$$-40 = 32 + 1.8°C$$
$$-40 - 32 = 1.8°C \quad or \quad 1.8°C = -72$$

Dividing gives

$$°C = \frac{-72}{1.8} = \frac{-720}{18} = -40$$

$$-40°F = -40°C \quad Ans.$$

Notice that the numerical values are the same on both scales; this is the only value for which this happens.

Problem

$$14°F = ?°C$$

Solution Substituting the Fahrenheit value,

$$14 = 32 + 1.8°C$$
$$14 - 32 = 1.8°C$$
$$-18 = 1.8°C$$
$$°C = \frac{-18}{1.8} = \frac{-180}{18} = -10$$
$$14°F = -10°C \quad Ans.$$

TABLE 2.7

If °C Is	Then °F Is
0	32
5	41
10	50
15	59
20	68
25	77
?	?
?	?

Keeping in mind that a change of 5°C equals a change of 9°F, it is easy to work up a simple conversion table like Table 2.7.   We start at 0°C, which equals 32°F; from then on, every 5° increase in the Celsius column is accompanied by a corresponding increase of 9° in the Fahrenheit column. It would be a good idea to continue the °C and °F columns going down (−5°C = ?°F) as well as up. Try it!

Absolute temperature In some areas of technical work an absolute temperature scale is required. The Fahrenheit and Celsius scales are not absolute in the sense that they have no visible lower limit that serves as *the* starting point and below which you cannot go. Such an absolute temperature scale has been developed on the basis of theoretical considerations, and is generally called the Kelvin scale. Instead of degrees the steps of this scale are called kelvins (K). Figure 2.5 compares the Fahrenheit, Celsius, and Kelvin scales, and you will note that absolute zero, or 0 K, equals −273°C; also, as the Kelvin scale goes up 273 steps, so does the Celsius scale; that is, the Kelvin and Celsius steps are equal in size. We can visualize the absolute zero of temperature as that minimum where the particles of even the simplest gas stop moving, the condition of complete immobility.

DENSITY

Whereas specific heat relates mass, heat, and temperature, **density** relates *mass* and *volume*. We sometimes speak of one thing being "heavier" than another, e.g., "lead is heavier than iron," or "oil is lighter than water," but what is actually being compared are the different densities of these materials. Density is the rela-

Which is heavier—the feathers or the iron?

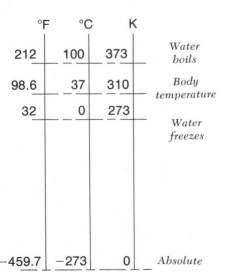

°F	°C	K	
212	100	373	*Water boils*
98.6	37	310	*Body temperature*
32	0	273	*Water freezes*
−459.7	−273	0	*Absolute zero*
°F	°C	K	

FIGURE 2.5 The temperature scales. In the absolute Kelvin scale the degree symbol is omitted.

TABLE 2.8   The ratio of mass to volume

Mass, g	Volume, ml	Ratio M/V, $\frac{g}{ml}$
22.6	2.0	11.3
113.0	10.0	11.3
4.52	0.40	11.3
21.64	2.8	11.3

tion between the *mass* of an object and its *volume*, the ratio of what it weighs to the volume it occupies.   That is,

$$\text{Density} = \frac{\text{weight (in grams)}}{\text{volume (in milliliters)}}$$

and the greater the density, the greater this ratio becomes.   For example, if an object made of lead is weighed (in grams) and its volume determined (in milliliters), the *ratio* of that weight to that volume will be 11.3 g/ml.   Note that the actual masses and volumes could be any number of values depending on the size of the piece used.   It *could* be as shown in Table 2.8.   The ratio, which is density, is constant.

Remember that
1 cm³ = 1 ml

In the English system density is expressed in pounds per cubic foot (lb/ft³).

Since density relates the mass and the volume of a material, it can be used to calculate the mass of an object from its volume or to determine its volume if the mass is known.   That is, since density = mass/volume, it follows that

$$\text{Mass} = \text{density} \times \text{volume} \qquad (2.1)$$

$$\text{Volume} = \frac{\text{mass}}{\text{density}} \qquad (2.2)$$

Some problems will illustrate the use of these relations.

Problem   An irregularly shaped object made of pure aluminum weighs 92.4 g.   The density of aluminum is 2.70 g/ml.   What is the volume of this object?

Solution   Since the mass and density are known, Eq. (2.2) can be used to calculate the volume:

$$\text{Volume} = \frac{92.4 \; \cancel{g}}{2.70 \; \cancel{g}/ml} = \frac{92.4 \; ml}{2.70} = 34.3 \; ml \quad Ans.$$

Problem   The density of gasoline is about 0.70 g/ml.   What is the weight of 40 l of gasoline?

Solution Since we have the density and volume, Eq. (2.1) will give the mass:

$$\text{Mass} = 0.70 \frac{g}{\cancel{ml}} \times 40,000 \; \cancel{ml} = 28,000 \; g = 28 \; kg \quad \textit{Ans}.$$

SPECIFIC GRAVITY

It was mentioned in the last problem that the density of gasoline is 0.70. This makes gasoline 0.70 times as dense as water, which weighs 1 g/ml and so has a density of 1. The value 0.70 is called the **specific gravity** of gasoline, and is numerically the same as its density but without any units. The units are absent because specific gravity is a ratio, the ratio of the density of a material to the density of water. The density of corn oil, as an example, is 0.92 g/ml, from which its specific gravity is simply

$$\frac{0.92 \; \cancel{g/ml}}{1 \; \cancel{g/ml}} = 0.92$$

A quart of corn oil, which is 1.06 l, or 1060 ml, will therefore weigh 0.92 as much as the same volume of water, or 1060 g $\times$ 0.92 = 975.2 g. On the other hand, the specific gravity of gold is 19.3, so that 1 ml of gold weighs 19.3 g compared with the 1 g that 1 ml of water weighs.

The specific gravity of liquids is often determined by using a hydrometer, a glass float with a weighted bottom and calibrations on its stem (Figure 2.6). There are many kinds of hydrometers, depending on the application. They can be calibrated to measure the percent sugar in a solution, or the density of urine, or the concentration of a salt solution, or the change in the specific gravity of the acid in your battery as it gets used up.

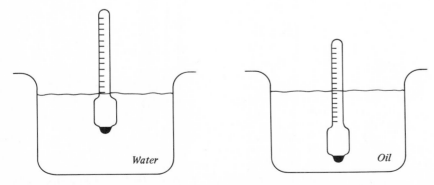

FIGURE 2.6 Hydrometer. It goes up in denser liquids and down in lighter liquids.

Although units, measurements, and calculations are not in themselves chemistry, they are an essential part of it; they help put the pieces together and make them fit. And now to the business of chemistry itself.

REVIEW QUESTIONS

$$1 \text{ in} = 2.54 \text{ cm}$$
$$= ? \text{ m}$$
$$1 \text{ ft} = ? \text{ m}$$

Atmospheric dust particles act as "nuclei" for raindrops. Dust particles in mines can cause lung disease.

$$1 \text{ ml} = 10^{-3} \text{ l}$$
$$1 \,\mu\text{l} = 10^{-6} \text{ l}$$

1 **a** What do a millimeter, milliliter, milligram, and millisecond all have in common? **b** What does each unit measure?

2 **a** What do a meter, micrometer, millimeter, kilometer, and centimeter all have in common? **b** Arrange these units in an increasing sequence, starting with the smallest and ending with the largest. **c** If someone's height is 1.78 m, what is it in centimeters? What is it in millimeters?

3 It was mentioned before that Mt. Everest is 8882 m above sea level. How much is that in feet?

4 Dust is no more than finely divided solid particles, but it is surprisingly important in cloud formation, rainfall, and as an industrial health hazard. Dust particles vary considerably in size, but often they are less than 0.1 μm (or 0.1 micron) across. How much is that in angstrom units? In millimeters?

5 **a** The top of the broadcasting antenna at the top of the Eiffel Tower rises 320 m above Paris. How much is that in yards? **b** The Eiffel Tower rests on four masonry piers arranged in a square, 330 $\times$ 330 ft. What is the size of the square in meters? **c** The open lattice structure of the Tower offers little wind resistance, and it has been estimated that even a hurricane would cause the Tower to sway no more than 22 cm. How much is that in inches?

6 The speedometer of an automobile leaving Rome by way of the Autostrada reads 96. How fast is the car going in miles per hour?

7 **a** A liter of wine contains how many milliliters? **b** How many quarts are there in 1 l of wine? **c** If there are 4 cups to a quart, how many cups of wine in 1 l?

8 The unit of volume used in the United States petroleum industry is the barrel (bbl), which equals 42 United States gallons. How many liters are there in 1 bbl?

9 **a** How many liters are there in 850 milliliters? **b** How many liters in 1600 ml? **c** How many milliliters in 55 μl?

10 A man steps on a weight scale in Munich and the pointer stops at 75. How much does he weigh in pounds?

11 How much would a 10-lb bag of potatoes weigh in Mexico City?

12 A man weighing 150 lb on earth would weigh over 2 tons on the sun. Explain.

13 **a** How many grams are there in 0.85 kg? **b** How many milligrams in 0.85 kg? **c** How many kilograms in 2300 g?

14 **a** $1.75 \times 10^4 \,\mu$g equals how many milligrams? **b** $1.75 \times 10^4 \,\mu$g equals how many grams? **c** 0.04 mg equals how many micrograms?

15 It was mentioned before that an aspirin tablet contains 5 grains, apothecary weight, of aspirin. If you took two tablets, how many grams of aspirin did you take?

16 **a** Chemical glassware for quantitative use often bears the inscription 20°C as the temperature of optimum accuracy. What is this in degrees Fahrenheit? **b** The standard temperature for many chemical calculations is often taken as 25°C. How much is this in °F? In kelvins (absolute scale)?

17 Birds are warm-blooded animals whose body temperatures are higher than those of mammals. For example, the tiny chickadee can maintain a temperature of 41°C even in cold weather. How does this compare to our own normal temperature of 98.6°F?

18 **a** The temperature reading on a warm day in Washington is 95°F. What is that in °C? **b** One of the colder cities in the United States is Bemidji, Minnesota, where a winter temperature of −22°F is not so unusual. What is this in °C?

19 As spring progresses to summer, would you expect a lake or the soil around the lake to warm up more rapidly? Refer to Table 2.2 to support your answer.

20 A beaker of water contains 200 ml of water at a temperature of 22°C. The water is heated to *just* the boiling point. How many calories of heat did the water absorb? How many kilocalories?

21 A hot metal bar is cooled by dropping it in a pail of water. The volume of water is 11 l, and its temperature is 68°F. **a** In which direction will heat flow? **b** When thermal equilibrium is reached the temperature of the water is 86°F. What is the temperature of the bar? **c** How many calories did the water absorb in coming to 86°F? How many calories did the bar lose in coming to this same temperature?

22 10 g of water, 10 g of aluminum, and 10 g of copper, all at the same initial temperature of 25°C, each absorb 200 cal of heat. Using the data of Table 2.2, calculate the temperature rise of each material.

23 The following data were obtained in determining the densities of a number of materials. Calculate the density of each material.

Material	Mass, g	Volume, ml	Density
A	17.6	4.2	?
B	9.3	3.8	?
C	128.7	10.5	?
D	65.0	8.8	?

24 The specific gravities of several materials are given here. For each material, calculate that volume, in milliliters, which will weigh 100 g.

Materials	Specific Gravity	Volume in ml Weighing 100 g
Corn oil	0.9	?
Water	1.0	?
Brick	2.3	?
Aluminum	2.7	?
Steel	7.9	?
Gold	19.3	?

3

THE ATOM AND ITS STRUCTURE
The Plus and Minus at the Heart of the Matter

What is **matter?** It can be many things in many ways: hard, soft, sharp, smooth, strong, weak, living, and nonliving. It would seem that no single answer is possible and that there are many kinds of matter. This is both true and untrue. On the basis of many years of thought and practice, science presents us with the powerful idea that although one piece of matter may differ in many ways from some other piece of matter, they are nonetheless both made up of the same fundamental units and that the differences between them derive from differences in the number and arrangement of these units.

ELEMENTS AND ATOMS

Imagine that a piece of aluminum is being divided into smaller and smaller pieces. Ultimately, a particle will be obtained, which, if divided *further*, would no longer be aluminum. This smallest particle that can still be identified as aluminum is the aluminum **atom.** As mentioned in Chapter 1, the idea of an atom is very old, but today we have come so close to the atom that it is treated more as a fact than as an idea. To begin with, experimental evidence indicates that atoms are incredibly small. For example, 1 cm³ of aluminum contains 6×10^{23} atoms, all of which are alike. By probing aluminum metal with x-rays, we have even learned how these atoms are packed together to make a solid piece of matter. Aluminum is an **element,** composed of characteristic aluminum **atoms.** You are already familiar with a good many elements: iron, which is made up of iron atoms; copper, made up of copper atoms; oxygen, made up of oxygen atoms; you can probably think of others.

Being composed of different atoms, no two elements behave in the same way; they have different **physical properties,** such as color, melting point, density, or electrical conductivity. They also have different **chemical properties;** that is, the way atoms interact with each other, and with matter generally, varies from atom to atom. Some atoms are good joiners, others are loners, and many are somewhere in between. At the temperatures and pressure that prevail on earth, most of the elements are solid materials, with their atoms closely packed in specific arrangements. Only five elements are liquids at or near room temperature. The most familiar of these is liquid mercury, generally used in thermometers and barometers.

STATES OF MATTER

Atoms in a solid

In a liquid the atoms are still close to each other, but their arrangement is less orderly than in solids, and since they are not held in fixed positions, they can flow past each other. A dozen elements exist as gases under ordinary conditions: not very many, but very important ones, like the oxygen of the atmosphere, which is essential to our existence.

Atoms in a liquid

The gases in our atmosphere are held there by the gravitational pull of the earth. In a gas, such as helium, the atoms are far apart and very mobile, so that despite their small size, they occupy whatever volume is available to them.

Of the 105 different elements known, 92 are found in nature, the others were made in the laboratory. Since each element is different, there are 105 different atoms and because scientists can now make atoms, there may be more than 105 by the time you read this.

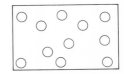

Gaseous helium atoms

An understanding of these differences and of the reasons for them was made possible by basic scientific discoveries in the late nineteenth and early twentieth centuries, which showed that all atoms have an internal structure. Far from being simple, uniform spheres, atoms are made up of smaller units, and it is the specific number and arrangement of these **subatomic particles** that distinguish one atom from another. The detailed study of the atom has revealed many subatomic particles, and the closer science gets to the atom, the more such particles are found. Fortunately, only three subatomic particles seem to be directly involved in determining the structure and chemical behavior of the atom—the **proton,** the **neutron,** and the **electron.** It is time to make their acquaintance.

PROTONS, NEUTRONS, AND ELECTRONS

Since the atom is the unit of structure of matter, its subatomic particles must themselves represent matter, and, as such, they must possess **mass.** All matter possesses mass, and all matter occupies space. If this were not so, there would be no way of knowing whether matter were there or not. With the atom as small as it is, its component parts must be correspondingly smaller and lighter. The mass of the proton has been determined as being about 10^{-24} g, which is so far removed from human experience that it is difficult to conceive. It is more convenient, and just as useful for the moment, to assign to the proton an arbitrary mass of 1 **atomic mass unit** (amu). By comparison, the mass of the neutron is also 1 amu, while that of the electron is only $1/1837$ amu. In an atom, therefore, the proton and the neutrons are the heavy particles that make up the bulk of the atom's mass, whereas the electrons are so light that they can often be neglected when atomic mass is being considered. Although the electrons are small in mass they are very *active*, and the chemical behavior of an atom is largely determined by the number and arrangement of its electrons. An atom, therefore, is not simply a random collection of protons, neutrons, and electrons but an *organized* arrangement whose major features are shown schematically in Figure 3.1.

At the center of an atom is its **nucleus,** which contains all its protons and neutrons. Although the volume of the nucleus is trivial compared with the total volume of the atom, it contains practically its entire mass. Around this very small, very dense nucleus are the electrons. They occupy practically all the atom's volume while contributing very little to its mass. The region of space occupied by the electrons that surround the central nucleus is therefore the atomic volume, and in Figure 3.1 the boundary of that spherical volume is shown as the outer circle. However, this bounding surface should not be considered as a rigid spherical shell; from the standpoint of modern theory it is a geometrical rather than a physical surface, within which there is a

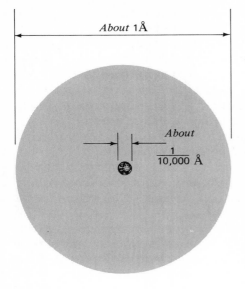

About 1Å

About

$$\frac{1}{10,000} \text{ Å}$$

FIGURE 3.1 The structure of the atom. This schematic representation of the atom would be difficult to draw to scale. Even if the central volume, which represents the nucleus, were drawn no bigger than ·, which is about 1 mm across, the outer circle would be 10 m in diameter, or almost 11 yd. Within this large volume, and even beyond it, are the electrons. Although the nucleus is relatively insignificant in size, it contains all the protons and neutrons.

high *expectancy* of finding the electrons. Before pursuing this point further, there is an important question to answer: Why do the electrons surround the nucleus and stay with it? Why don't the electrons go their way and the nucleus its way? If this happened, there would be no atoms. But the fact is that on this planet atoms are generally stable structures; otherwise matter might be one thing today and another tomorrow. This brings us to the most important single feature of the atom: Its protons and electrons carry equal but opposite electric charges. A proton has a unit of **positive charge,** and an electron has a unit of **negative charge.**

Although the modern picture of charged electrons and protons emerged toward the end of the nineteenth century, the observation that objects can become "charged" goes back to the ancient Greeks. As long ago as 600 B.C. they noted that when amber (a fossilized resin from a type of pine now extinct) is rubbed with wool, it attracts light objects and particles in its vicinity. The Greek word for amber was *elektron,* hence our term. Actually, this occurs quite often; "charged" behavior results when a glass rod is rubbed with silk, when paper is passed through the rolls of a printing press, when you walk across a carpet, when a nylon dress rubs up against the wearer. You can probably think of other examples. Charged behavior can be annoying and even dangerous. If the charge gets big enough, it may generate a spark; in the clouds the charge can get so big that it will generate

As an indication of stability, helium gas atoms at ordinary temperatures collide with each other at a speed about that of a bullet from a gun; all that happens is that they bounce apart as quickly as they collide.

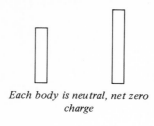

Each body is neutral, net zero charge

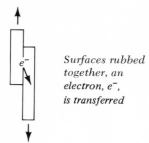

Surfaces rubbed together, an electron, e⁻, is transferred

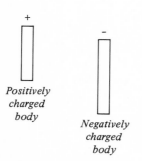

Positively charged body

Negatively charged body

FIGURE 3.2 Transfer of electrons.

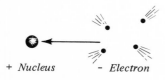

+ *Nucleus*   - *Electron*

Pulling together versus pulling away

an enormous spark—lightning. This is what Benjamin Franklin, with his key, string, and kite, was looking for.

When charged objects were studied more closely, it was soon recognized that some rubbing or sliding contact between two objects is always involved. Further, it was also found that *both* the objects rubbed together become charged, never one alone. This suggested that something is being transferred from one object to the other, resulting in a charge on both. By the end of the nineteenth century, it was recognized that what is indeed being transferred is the smallest unit of negative electricity, an electron. Before being rubbed together (Figure 3.2), each object has an equal number of negative and positive units of electricity, adding up to *zero* charge. As a whole, each object is therefore **neutral,** neither negative nor positive in charge.

Remember that electrons are very light and very active and are in the outer volume of the atom. As the surface of one object rubs against the surface of the other, there is a transfer of some electrons, with the result that one object has an excess of electrons and the other a corresponding deficiency. What one loses the other gains. Since electrons carry a negative charge, the gainer becomes negatively charged and the loser becomes positively charged. Which object becomes negative and which positive depends on how strongly each holds onto its electrons. For example, when a glass rod is rubbed with silk, electrons leave the glass, which thereby becomes positive, and move to the silk, which becomes negative. Apparently, glass holds its electrons less tenaciously than silk. Evidence from many directions led to the far-reaching conclusion that all matter contains electrically opposite particles and that the unity of these opposites makes for the stability of the atom, the togetherness of the positive nucleus and the negative electrons around it. At the same time, *opposing* this togetherness is the energy of motion of the electrons, their tendency to move off and away. If the electrons lacked this energy of motion, they would collapse onto the nucleus and the size of the atom would be the size of the nucleus. In that case each of us would be reduced to about the size of a dust particle. Happily, that is not the case, so that while the attraction between positive protons and negative electrons continues to pull them together, the energy of motion of the electrons also continues to move them apart. The upshot is a balance between two opposing tendencies, resulting in a stable atom whose tiny positive nucleus is enveloped in a relatively enormous volume in which its electrons buzz around. That opposites attract is not only a common figure of speech but a guiding tenet of science. If we add to this the observation that particles carrying the *same* charge *repel* each other, we have the basis for a great deal of chemistry. The forces acting between charged particles, whether

TABLE 3.1

	Mass, amu	Electric Charge	Symbol	Place in Atom
Proton	1	+1	p^+	In nucleus
Neutron	1	0	n	In nucleus
Electron	$1/1837$	−1	e^-	Around nucleus

attraction or repulsion, are called **electrostatic forces.** The closer the particles, the greater the electrostatic force. The farther apart the particles, the smaller the force acting between them.

The third fundamental unit is the neutron, which, as its name indicates, has no electric charge and therefore contributes only mass to the atom. However, the neutron is not as dull a fellow as may appear, and under certain circumstances it is very lively indeed. Neutrons can be **radioactive,** decomposing spontaneously into a positive proton while emitting a negative electron. This occurs in the nucleus of certain radioactive atoms and also whenever neutrons are outside the nucleus, on their own. This changeover, which is believed to involve more than the simple separation of a proton-plus-electron unit, indicates that the neutron is more complex than its neutrality of charge would suggest. The neutron is also the particle involved in setting off the chain reaction required for the explosion of an atomic bomb, as well as for the generation of energy in an atomic reactor.

Interesting and important as these matters are in themselves, in the chemistry of living organisms that is our concern, the neutron is for the most part a weighty bystander rather than an active participant. The characteristics of the three fundamental units of the atom are shown in Table 3.1.

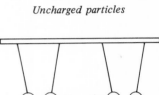

Uncharged particles

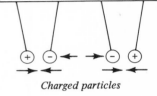

Charged particles

ATOMIC NUMBER AND ATOMIC WEIGHT

We are now in a position to begin filling in some of the major features of the atom.

The nucleus contains practically the entire mass of the atom and is the positively charged center of the atom. The charge on the nucleus is simply the number of protons in that nucleus. Each proton carries one positive charge. If an atom contains only one proton in its nucleus, which is true for hydrogen, the simplest of all the atoms, the charge on the nucleus is +1. Aluminum has 13 protons in its nucleus, and the charge on that nucleus is +13. Uranium has 92 protons in its atomic nucleus, so that the charge on the nucleus is +92. The number of protons, that is, the number of positive charges in the atom nucleus,

The exact amu values are proton = 1.00728, neutron = 1.00867.

serves to identify and characterize that atom and is called its **atomic number** Z. Therefore Z is 1 for hydrogen, 13 for aluminum, and 92 for uanium. In general, atomic number Z = number of protons (+). Using their atomic numbers, we can arrange all the different atoms in sequence, $Z = 1$, $Z = 2$, $Z = 3$, and so on. The **atomic weight** (more properly the atomic mass) of an atom is simply the sum total of the mass of its component protons, neutrons, and electrons. For practical purposes, the mass of both the proton and the neutron equals 1 amu, and the electrons are so light that they can be ignored. Then

Atomic mass = number of protons + number of neutrons

In arranging the different atoms in a logical sequence, we now have two characteristic numbers. The first, the atomic number, is the essential guide, increasing regularly by 1 from the hydrogen atom onward. The atomic weight also increases with Z, but not as regularly, since the number of neutrons in the respective nuclei does not increase regularly. Finally, each of the different atoms has been assigned an identifying symbol so that, for example, H stands for hydrogen, Al stands for aluminum, and U stands for uranium. A full listing of atomic numbers and atomic weights and the symbol for each atom are given on the inside back cover.

All the atoms of a given element have the same atomic number, and therefore the same number of positive protons and negative electrons. However, not all the atoms of an element will have the same atomic mass, which, as shown above, is the sum of the number of protons plus the number of neutrons. For any element the number of protons in its atoms is constant, whereas the number of neutrons may differ by one or two, so that those atoms having more neutrons will be correspondingly heavier, and those having less neutrons will be lighter. Atoms of an element having the same atomic number but different atomic weights are called **isotopes** and, in fact, most elements are mixtures of two or more isotopes. This means that the atomic weight of an element is the arithmetic average of the weights of its isotopes. If you look at the table of atomic weights you will see that this results in numerical values that contain decimal fractions rather than simple whole numbers.

ELECTRON SHELLS AND ORBITALS

An atom is a neutral, electrically balanced assembly of protons, neutrons, and electrons. The neutrons carry no charge; therefore, for every positive proton in the nucleus there must be a negative electron in the space around the nucleus. That is, although every atom contains positive and negative charges in its structure, its *net* charge is *zero*. For a neutral atom, therefore,

Number of protons (+) in nucleus =
number of electrons (−) around nucleus

In addition to their opposite charges, protons and electrons differ in other important ways. The proton is heavy and stays put in the nucleus unless it becomes violently disturbed from the outside. In a stable atom the protons and neutrons live together quietly in the nucleus; however, the nucleus is by no means simple, and this behavior should not be taken as meaning a lack of energy. For our purposes we can consider the proton as a homebody. Being so much smaller and lighter, an electron is more easily moved around; it is more mobile and more responsive to external stimuli in the form of heat, light radiation, or energy in general. Further, the electrons occupy much more space than protons and are much freer to move about; under certain conditions they will even leave home. In this atomic community, it is the active electrons that are the units involved in "doing business" with other atoms. This is equivalent to saying that electrons are particles characterized by high energy.

It is not difficult to see that a particle can be small and light and still occupy a considerable volume. Suppose we attach a small weight, say a button, to a string. When it simply hangs motionless, it occupies a negligible volume, but if it is spun around in all directions, it carves out a spherical volume whose radius is the length of the string. Further, if the spinning speed increases, thereby increasing the energy of the system, the button will try to move out still further, and if the string can stretch, a larger volume will be occupied at higher energy levels.

In an atom, the "string" that holds the negative electrons to the positive protons in the nucleus is their electrostatic attraction since, as has been pointed out, opposites attract. If this mutual attraction were the *only* factor present, then, as pointed out earlier, the electrons and protons would approach each other and the atom size would shrink. But even as the rotating button has an energy that tends to move it outward, the electrons also have energy that tends to increase the size of the space they occupy. However, every atom in its normal, undisturbed state has a given size. This is called the **ground state,** since no smaller size of the atom is known. At this ground state, the electrons of that atom must possess some specific energy that opposes the pull of the protons; that is, a stable balance has been established between two opposing tendencies, which is what is meant by **equilibrium.** A neutral atom is therefore a stable equilibrium system.

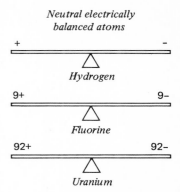

Neutral electrically balanced atoms

In each case the atom has a net zero charge

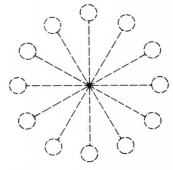

The analogy between a rotating weight and the energy of electrons cannot be carried too far. Modern atomic theory developed by applying the principles of **quantum mechanics,** which incorporates new ways of looking at the physical world. However, the formulation of quantum mechanics is largely in mathematical terms, which makes it inaccessible to most of us. Nonetheless, its results are very much present in our science and technology and in chemistry in particular.

Up to now, we have spoken of electrons as being energetic and as being around the nucleus of the atom. But where is "around," and does the space occupied by electrons relate to their degree of energy? What happens when the number of protons (and therefore electrons) increases from 1, to 10, to 20, and so on? Can more and more electrons crowd into the same space? These are important questions for chemistry, and in very simplified form the answers are as follows.

One of the consequences of quantum mechanics is the conclusion that it is not possible to specify completely and accurately the particular volume of space that the electrons of an atom occupy. However, it is possible to describe the size and shape of that region in space around the nucleus where there is about a 90 percent *likelihood* of finding the electrons. This region of space is called an **orbital.** There is therefore some probability of finding an electron *outside* an orbital, but this probability decreases sharply with increasing distance from the nucleus.

In order to occupy some given orbital, an electron must have the amount of energy required for that orbital. Beginning with the ground level of energy, which is the minimum, the energy of an atom and its electrons is associated with a number n, called the **principal quantum** number; the value of n may be equal to 1, 2, 3, 4, ...

At the ground level, where $n = 1$, the following rules apply:

1 The orbital, or the region of space occupied by the electrons, is spherical with the nucleus at the center.
2 The electrons occupying this orbital have the lowest energy.
3 No more than *two* electrons can be accommodated in this orbital because there is no room for more.

Since electrons repel each other, they need sufficient room in which to keep their distance. When $n = 1$, the orbital is the smallest, and its occupancy limit is two electrons.

Atoms having more than two protons, and hence more than two electrons, can place two electrons in this ground-level orbital but must find additional housing for their third, fourth, fifth, etc., electrons. To reach this housing, or orbitals, there must be a jump in

All electrons carry a negative charge, and like charges repel each other.

90 *percent probability of finding electrons in this volume*

Nucleus

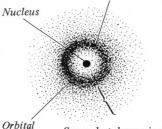

Orbital

Some, but decreasing, likelihood of finding electrons

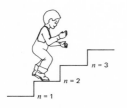

$n = 3$
$n = 2$
$n = 1$

energy to the next level, where $n = 2$. It is a feature of quantum theory that going from one energy level to another is analogous to a succession of jumps, up or down, like a child jumping up and down steps. There is no in-between position; the electron is either at $n = 1$ and has least energy or at $n = 2$, or at $n = 3$, and so on, each with successively higher energies. Also, as the energy level goes up, the corresponding orbitals become larger. Whereas the ground-level orbital can take no more than two electrons, when $n = 2$ the orbitals can hold up to eight electrons, and at the $n = 3$ level the orbitals can accommodate up to eighteen electrons. The larger the house, the more people can live there, even if they do find each other repulsive.

What are referred to here as energy levels, related to the principal quantum number n, are also called "shells," a name which harks back to an earlier formulation of atomic theory, in which the electrons were visualized as traveling in fixed spherical orbits around the nucleus. Although this conception, which was due to the great Danish scientist Niels Bohr, was a historic step forward, it failed to satisfy other requirements of chemical theory and observation and was replaced by the less rigid but more useful picture of electrons in orbitals, regions of space around the nucleus where these electrons are most likely to be found. However, the term shell is still in use, and Table 3.2 relates the quantum number n to the identifying **shell letter** and to the number of electrons that each shell can accommodate. In current theory, this is the number of electrons that the orbitals of a given energy level can hold.

At energy levels above the ground level, electrons may occupy both spherical and nonspherical orbitals. The number of possible orbitals at some given value of n is equal to n^2. Consequently, at the ground level, when $n = 1$, the number of orbitals is 1; if $n = 2$, a total of four orbitals is available; if $n = 3$, there are nine orbitals, and so on. For our purposes, we shall consider only the orbitals when n equals 1 and when it equals 2. At the energy level $n = 1$ there is only one orbital within which an electron can be found, and, as mentioned earlier, it is spherical. The single electron of the hydrogen atom will be found in this spheri-

TABLE 3.2

Energy Level	Shell Letter	Electron Capacity	Number of Orbitals
1	K	2	1
2	L	8	4
3	M	18	9
4	N	32	16

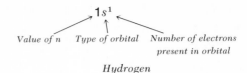

$1s^1$

Value of n   Type of orbital   Number of electrons
present in orbital

Hydrogen

cal orbital, which by convention is called the s orbital. All s orbitals are spherical. A simple symbolism is used to relate the value of n, the type of orbital occupied, and the number of electrons in that orbital. If the atom receives a suitable input of energy, its electron will be excited and may jump into the next higher energy level, $n = 2$, so that its symbolic designation would then be $2s^1$. The $1s$ and $2s$ orbitals are both spherical, but the $2s$ is larger. Having more energy, the electron occupies more space. Since all physical systems tend to a state of *least* energy, which is their *stable* state, the excited hydrogen atom will in time (if not excited further) drop *back* down to $1s^1$, at the same time releasing the energy it had received.

When $n = 2$, there are four orbitals available. Of these, the first is again the spherical s orbital, and the other three, of slightly higher energy, are called p orbitals, and their shape is shown in Figure 3.3. You will note that there are three sets of p orbitals in

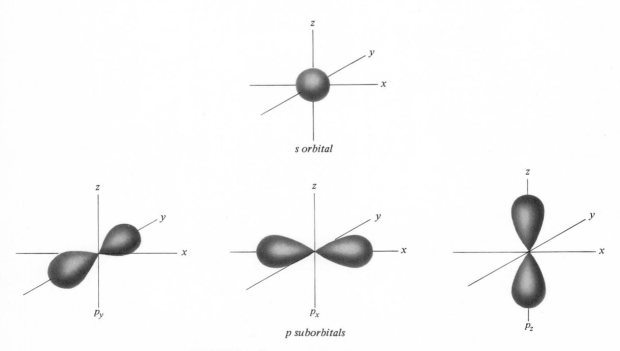

FIGURE 3.3 Representation of s and p orbitals.

accordance with their orientation in space, that is, in the direction of the x, y, and z axes.

In order to distinguish between them, the three p orbitals, or suborbitals, are identified as p_x, p_y, and p_z. It was mentioned before that the p orbitals for a given energy level n have a somewhat higher energy than the s orbital for the same n level. Although this conclusion derives from the mathematics of the situation, it can be intuitively recognized that this is to be expected. The s orbital is spherical, which is the simplest geometry a region of space can assume. A sphere imposes no directional preference; if you start at the center, all directions are equivalent. Furthermore, a sphere is the most economical way of enclosing some given volume of space; any other shape will require more surface to enclose the same volume. The p orbitals, whose two lobes roughly resemble a balloon that has been pinched to a point at the center, impose directional restrictions on the region of space the electrons in them may occupy. In general, it takes some energy to impose or to change direction. In a very descriptive way, the more complex shape of the p orbitals is the reason for their somewhat higher energy compared with the simpler s orbital.

We can now begin to work out the electron arrangements of the different atoms, the point of departure being the atomic number Z. It was noted before that hydrogen, the simplest and lightest atom, consists of one proton and one electron in a spherical orbital. This arrangement is specific to hydrogen, and we can imagine building up a hydrogen atom by putting one electron in a spherical region of space around one proton. Taking the other atoms in order of increasing value of Z, we now want to place their electrons in suitable orbitals around the nucleus. This is important not only because it would provide a floor plan of each atom but also because the specific electron arrangement is the clue to the chemical behavior of the atom. However, in arranging the electrons around the nucleus there are three guiding rules to be followed.

1 No more than *two* electrons may occupy a given orbital. An orbital with two electrons is full up, no more allowed; with one electron an orbital is half full, and with no electrons it is empty. An electron in an atom creates a magnetic effect as if it were spinning about its own axis, so that we speak of **electron spin.** If an orbital is full, its two electrons must have opposite spins. If the spin of one electron is considered clockwise, the spin of the second electron in the orbital will be counterclockwise. The two directions of spin are symbolically represented as ⇅.

The buildup of electron arrangements of successive atoms is often referred to by the German word for buildup, Aufbau.

Filled p suborbital

Half-filled p suborbital

Empty p suborbital

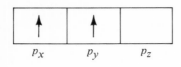

p_x p_y p_z

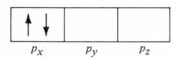

p_x p_y p_z

Not allowed! With p_y and p_z empty, p_x may not be full.

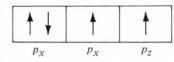

p_x p_x p_z

2 Oribitals are filled in sequence, beginning with the one of least energy, and going up. Remember, no more than two electrons to an orbital.

3 When filling orbitals of equal energy, such as the three suborbitals that constitute the p orbital, there is no doubling up until it is unavoidable. If, for example, two electrons are to be housed in the p orbital, they will be distributed among the p_x, p_y, and p_z suborbitals as shown in the margin. This is an example of Hund's rule, which states that two electrons may not occupy a suborbital until the other suborbitals contain one electron.

The two electrons *could* have been put in the same suborbital, but the energy of the system is lower if the electrons are in separate suborbitals. Furthermore, you will note that they have parallel spins, which is also the preferred condition for minimum energy. However, if four electrons are to be accommodated, the situation will be as shown. There is no alternative here except to put the fourth electron in with another, in which case their spins are opposite. Summarized in Table 3.3 is the relation between energy level n (and the shell letter associated with it), the orbitals (and suborbitals) available at that n, and the electrons that can be accommodated.

ELECTRON CONFIGURATION

The location of the electrons of an atom in its shells and orbitals is called its **electron configuration.** We have already discussed the hydrogen atom and the location of its single electron in the $1s$ orbital. Let us now consider how the configuration of carbon can be established. Carbon has six protons and therefore six electrons. Where will these six electrons be found? Remember that the orbital of lowest energy is filled first, then successively higher ones. From Table 3.3 the first orbital to be filled will be

TABLE 3.3

n	Shell	Orbital	Suborbitals		Electron Capacity
1	K	s			2 = total in K shell, or n = 1 level
2	L	s		2	= 8 total in L shell, or n = 2 level
		p	p_x 2	6	
			p_y 2		
			p_z 2		

TABLE 3.4

Atomic Number	Element	K Shell 1s	L Shell 2s	2p_x	2p_y	2p_z
1	H	↑				
6	C	↑↓	↑↓	↑	↑	

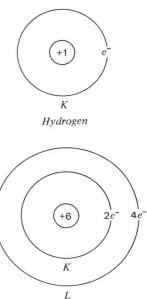

K
Hydrogen

K

L

Carbon

the s orbital associated with the ground level of energy, $n = 1$ or the K shell. Two of the carbon's six electrons therefore go into this $1s$ orbital, which becomes filled, and the two electrons in it must therefore have opposite spins. The next energy level, $n = 2$, permits two orbitals, the $2s$ and the $2p$. The $2s$ orbital, being somewhat lower in energy than the $2p$, is filled first, and the two electrons it holds must again have opposite spin. The two electrons still remaining to be housed have their choice of $2p_x$, $2p_y$, and $2p_z$. Arbitrarily (since the three are equivalent) we assign the last two electrons to $2p_x$ and $2p_y$. For the stable condition of minimum energy, their spins must be parallel. In Table 3.4 the electron configurations of carbon and hydrogen are represented.

TABLE 3.5   Electron configurations of the first 20 atoms

Atomic Number	Element Symbol	K Shell 1s	L Shell 2s	2p_x	2p_y	2p_z	M Shell 3s	3p_x	3p_y	3p_z	N Shell 4s
1	H	↑									
2	He	↑↓									
3	Li	↑↓	↑								
4	Be	↑↓	↑↓								
5	B	↑↓	↑↓	↑							
6	C	↑↓	↑↓	↑	↑						
7	N	↑↓	↑↓	↑	↑	↑					
8	O	↑↓	↑↓	↑↓	↑	↑					
9	F	↑↓	↑↓	↑↓	↑↓	↑					
10	Ne	↑↓	↑↓	↑↓	↑↓	↑↓					
11	Na	↑↓	↑↓	↑↓	↑↓	↑↓	↑				
12	Mg	↑↓	↑↓	↑↓	↑↓	↑↓	↑↓				
13	Al	↑↓	↑↓	↑↓	↑↓	↑↓	↑↓	↑			
14	Si	↑↓	↑↓	↑↓	↑↓	↑↓	↑↓	↑	↑		
15	P	↑↓	↑↓	↑↓	↑↓	↑↓	↑↓	↑	↑	↑	
16	S	↑↓	↑↓	↑↓	↑↓	↑↓	↑↓	↑↓	↑	↑	
17	Cl	↑↓	↑↓	↑↓	↑↓	↑↓	↑↓	↑↓	↑↓	↑	
18	Ar	↑↓	↑↓	↑↓	↑↓	↑↓	↑↓	↑↓	↑↓	↑↓	
19	K	↑↓	↑↓	↑↓	↑↓	↑↓	↑↓	↑↓	↑↓	↑↓	↑
20	Ca	↑↓	↑↓	↑↓	↑↓	↑↓	↑↓	↑↓	↑↓	↑↓	↑↓

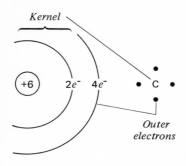

Kernel

+6   2e⁻   4e⁻   C

Outer electrons

Shorter representations of the electron configuration for hydrogen and carbon are shown in the margin. In these diagrams the electrons in each shell are lumped together. Following the same rules, the electron configurations for the first twenty atoms are shown in Table 3.5. These 20 atoms are the most common. An understanding of their chemical behavior begins with an understanding of their electron configuration.

As the next chapter will show, the outer electrons of an atom are those which are involved in chemical reactions. It is therefore common practice to represent an atom as composed of a **kernel,** surrounded by its outer electrons; this is illustrated for the carbon atom. The kernel is the carbon atom less its four outer electrons.

REVIEW QUESTIONS

1 How can you arrange four golf balls so that each one would be touching the other three? (This is an example of the close packing of atoms referred to on page 39.)

2 **a** Which two subatomic particles discussed in this chapter are found in the nucleus of the atom? **b** How do these two particles compare in mass and charge?

3 **a** Where in the atom is the third subatomic particle found? **b** How does it compare with the other two in regard to mass and charge?

4 Several examples of the charged behavior obtained when objects rub up against each other are given on page 42. Can you think of some other examples?

5 The marginal sketch on page 42 shows an electron leaving one object and going to another, so that both objects become oppositely charged. Is it likely that a proton, rather than an electron, would be transferred? Explain.

6 Oxygen has an atomic number (Z) of 8 and its atomic weight is 16. **a** How many protons are there in the nucleus of the oxygen atom? **b** How many neutrons would you expect to find in the oxygen nucleus? **c** How many electrons surround the oxygen nucleus? **d** What is the net charge on the oxygen atom?

7 Complete all columns for each element shown in the table below. (As an example, the values for the beryllium atom are shown in full.)

Element	Atomic Number	Number of Neutrons	Atomic Weight	Number of Electrons
Beryllium	4	5	9	4
Helium	2		4	
Carbon	6	6		
Sodium	11		23	
Phosphorus	15	16		

8 It has been pointed out that electrons occupy orbitals around the atom nucleus, and that the energy of the electrons occupying a given orbital is associated with a quantum number n. **a** When $n = 1$, the energy level is least, and there is only one orbital available for electron occupancy. What is the shape of this orbital, and how is it identified? **b** When $n = 2$, how many orbitals are available for electron occupancy? What are the shapes of these orbitals, and how are they identified? **c** Up to how many electrons can there be in any one orbital?

9 The H atom has one proton, and therefore one electron. **a** When the atom is in its ground state of least energy, its electron is represented as $1s^1$. What does each of the three symbols designate? **b** What does the representation $2s^1$ mean for the hydrogen electron? Is the H atom in the ground state?

10 The electron configurations for the hydrogen and carbon atoms are shown on page 51. Fill in the electron configurations for the atoms between H and C, namely helium, lithium, beryllium, and boron, atomic numbers 2, 3, 4, and 5, respectively. Also write the electron arrangement of these atoms using the shell representation as given for hydrogen and carbon on page 51.

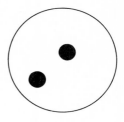

11 If two electrons were dropped into a ring, as shown, and couldn't get out, what positions do you think they would assume? What positions relative to each other would three electrons assume?

12 Using the information provided in Table 3.5, write the electron arrangement, by shells, of the following atoms. (For each atom use only as many shells as are needed.)

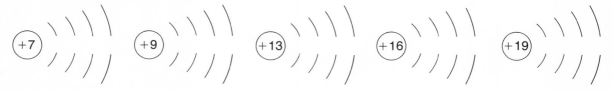

+7	+9	+13	+16	+19
Nitrogen	*Fluorine*	*Aluminum*	*Sulfur*	*Potassium*

ATOM BONDING
How Little Differences Become Big Differences

By the end of the nineteenth century, about two-thirds of the naturally occurring elements had been identified, and many of their chemical and physical properties were known. As more and more was learned of the various elements, it was recognized that there were similarities, as well as differences, between them. Although all atoms have their own specific identity, they also show family resemblances, similarities in chemical and physical behavior. If the elements are arranged in order of increasing atomic number, it can be seen that elements having similar properties are spaced periodically, that is, they appear at regular intervals in the sequence. This periodic recurrence of similar properties suggests that some underlying condition within the atom is repeating itself regularly. What is this underlying factor?

THE PERIODIC TABLE

Beginning at the turn of the century, and extending for about three decades into the present century, the deeper understanding of the atom was sparked by a burst of remarkable advances in physics. These new ideas led to the concept of the nuclear atom discussed in the previous chapter and to the recognition of atomic structure as the key to the behavior of the atom. Although the atomic number Z was established as the identifying feature of the different atoms, it was the actual arrangement of the electrons around the nucleus that served to explain the specific chemical characteristics of that atom. On this basis, atoms found to have similar properties would be expected to have some similarity of electron arrangement. If the listing of the elements in order of increasing atomic number shows a regular repetition of chemical behavior, it implies that there is also a repetition of some feature of electron configuration. The common feature of electron configuration shared by *atoms having similar properties* is that they all have the *same number of outermost electrons*. These outermost electrons, which are the most active, as well as the most weakly held, are called the **valence electrons** of the atom. In some cases, an active, weakly held electron will leave home; in all cases the valence electrons are those which enter into the atom-to-atom bonds through which stable combinations of atoms are formed.

The related factors of atomic number, electron configuration, and the similarities and differences of the chemical properties of the atoms are combined in the periodic table. The full table is shown on the inside front cover, but for a closer look at how chemical behavior relates to electron configuration, we shall consider the first 20 atoms arranged in increasing order of atomic number in Table 4.1. The three elements marked with a single asterisk, helium, neon, and argon, are members of the same chemical group, all of which are gases, and all conspicuously inert, with almost no capacity for joining or interacting with other atoms. Their atomic numbers are 2, 10, and 18 respectively, so that the repeating interval, or **period,** is 8. The three elements that follow the three inert gases, namely, lithium, sodium, and potassium, each marked with a double asterisk, are also members of a chemical family called the **alkali metals.** Since their atomic numbers are 3, 11, and 19 respectively, they also have a period of 8. However, although each member of this group is only one atomic number removed from one of the inert gases, they are very far apart in chemical behavior. Whereas the inert gases are the most unreactive atoms, the alkali metals are among the most reactive. To help understand the reason why one proton makes such a big difference, we rewrite the vertical list as horizontal sequences of eight atoms, or periods, as shown in Figure 4.1. Note that each period begins with an alkali metal and ends with an inert gas. Although hydrogen is shown with the alkali metals because, like them, it has a single outer electron, it is not a

The periodic table developed from the work of a Russian chemist, Dimitri Mendeleev and a German chemist, J. L. Meyer. They used atomic weights, since it was only later that the English scientist H. G. Moseley showed that the atomic number is more fundamental.

TABLE 4.1

Element	Z
Hydrogen	1
Helium*	2*
Lithium **	3**
Beryllium	4
Boron	5
Carbon	6
Nitrogen	7
Oxygen	8
Fluorine	9
Neon *	10*
Sodium **	11**
Magnesium	12
Aluminum	13
Silicon	14
Phosphorus	15
Sulfur	16
Chlorine	17
Argon *	18*
Potassium **	19**
Calcium	20

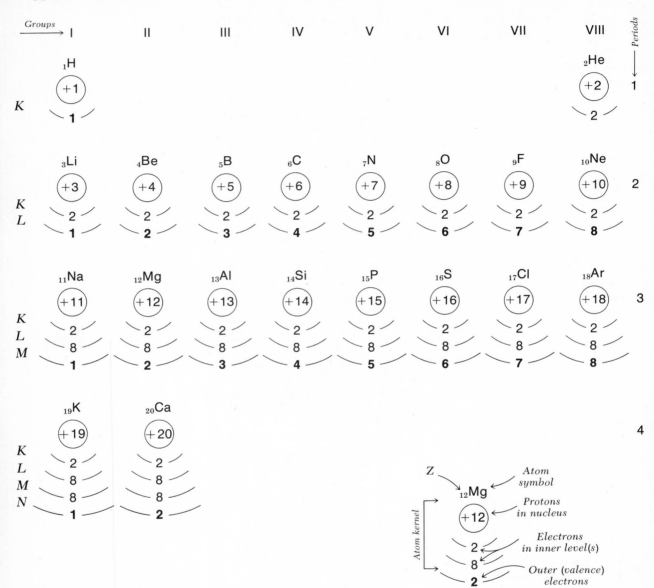

FIGURE 4.1 The first 20 elements of the periodic table.

member of that group. As the simplest atom, hydrogen is unique, and although it has many friends among the other atoms, it has no real relatives.

There is a great deal of information in Figure 4.1, both particular and general, but it must be read properly to be used to best advantage. Let us begin with the alkali metals, which are listed in group I of the periodic table because their atoms all have *one* outer electron; this is the common feature of their electron configurations and the reason for their chemical similarity. The elements in the next vertical column, $_4$Be, $_{12}$Mg, and $_{20}$Ca, all have two outermost electrons, and therefore belong to group II, called the **alkaline-earth metals.** As one moves farther to the right, the number of outer electrons in the atoms of each group increases to a maximum of eight. The one exception is the helium atom, which can have no more than two electrons in its orbital because it has only two protons in its nucleus. Each horizontal sequence of atoms is a period, so that each atom is at the intersection of a vertical group and a horizontal period. For example, carbon, $_6$C, is in group IV and period 2.

When one follows a period from left to right, the chemical scene changes progressively in several ways. To illustrate this we shall follow the changes in going left to right along period 3, from $_{11}$Na to $_{18}$Ar, as shown in Figure 4.2. Note that the value of the atom **electronegativity** is included. The electronegativity of an atom is an index of how strongly an atom will tend to acquire and keep electrons; it is a major consideration when two or more atoms combine to form a **compound.** From left to right along period 3 the electronegativity increases.

Alkali metals

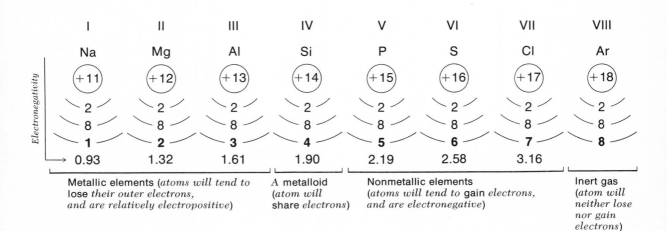

FIGURE 4.2 The atoms of period 3.

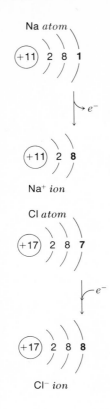

Na *atom*

(+11) 2 8 **1**

→ e⁻

(+11) 2 **8**

Na⁺ *ion*

Cl *atom*

(+17) 2 8 **7**

→ e⁻

(+17) 2 8 **8**

Cl⁻ *ion*

Na   Cl
atom   *atom*

e⁻

Na⁺ → ←Cl⁻

₂He (+2) 2

₁₀Ne (+10) 2 **8**

₁₈Ar (+18) 2 8 **8**

K L M

Chemically speaking, a metal is electropositive and a nonmetal electronegative. An atom like sodium, which holds onto its one outermost electron weakly, as indicated by its small electronegativity, will tend to surrender that electron to some other atom with a strong attraction for electrons, such as chlorine. When that happens, only 10 negative electrons are left around the 11 positive protons in the sodium nucleus, with the result that the atom now has a net positive charge of +1. Sodium is therefore strongly metallic, even if it is physically soft. As for the chlorine atom, its tendency to gain an additional electron makes it electronegative. After transfer it has 18 negative electrons around 17 positive protons in the nucleus. As the result of the large electronegativity difference between them, the transfer of an electron from sodium to chlorine leaves the former with a +1 charge and the latter with a −1 charge. The two neutral atoms have become respectively a **positive sodium ion,** and a **negative chloride ion,** and being oppositely charged, they attract each other.

Of the eight atoms in Figure 4.2, sodium is the most electropositive, or metallic, while chlorine is the most electronegative, or nonmetallic. From left to right the atoms are decreasingly metallic, so that silicon, standing in the middle, has both metallic and nonmetallic features. It is called a **metalloid.** Further along the period, phosphorus and sulfur are distinctly nonmetallic. The last element in the period, argon, is one of the inert gases which make up group VIII. The inert gases are also called the noble gases, presumably because they do not join with "ordinary" atoms.

The difference between intensely reactive chlorine and very inert argon is one proton and one electron, but the result is a very large difference in chemical behavior. A material that resists change is stable, so that it must be inferred that argon and all the inert gases, have high stability. The common feature of the electron configurations of the inert gases is that they all have *filled orbitals.* The two electrons of helium fill its 1s orbital at the *K* energy level; neon has eight *additional* electrons filling its 2s and 2p orbitals at the *L* level, and argon has eight *more* in the 3s and 3p orbitals of its *M* level.

Apparently, possession of completely filled atomic orbitals is related to chemical stability, which for inert gases is so great that they are the only chemical species that "live alone," not even joining up with themselves. Since filled orbitals make for stability and unfilled orbitals do not, the seven other atoms of this period, each with an incomplete outer shell, should enter into those chemical combinations which will result in filled outer shells containing eight electrons, and in general this is so. Although this **octet principle** is not *always* adhered to, it is nonetheless a very

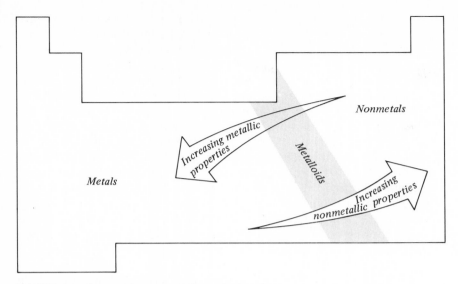

FIGURE 4.3 General features of the periodic table.

useful guide in understanding the formation of bonds between atoms. Except for the inert gases, all the atoms on earth are found in combination with each other or with other atoms. Whenever a chemical reaction takes place, some atom-to-atom bonds are broken and others are formed. In general, bonding is stable if it leads to filled outer shells for the atoms being joined.

The changes noted in going from left to right along period 3 are, in the main, typical of the other periods of the periodic table, as summarized in Figure 4.3.

STABILITY AND EQUILIBRIUM

Of all of the many kinds of matter on this planet, only the inert gases are found as single atoms; everything else is made up of combinations of atoms. Some atoms combine in pairs, forming diatomic molecules, two atoms joined together and acting as an individual unit. For example, hydrogen, oxygen, and nitrogen exist as diatomic gases: H_2, O_2, and N_2, respectively. These are stable molecules; hydrogen, oxygen, and nitrogen are not normally found as individual atoms under the conditions of this planet, and when individual atoms of these elements *are* formed, they spontaneously combine into diatomic molecules. The stability of these molecules is also indicated by the fact that they resist being dissociated, or separated, into two individual atoms.

H_2

O_2

N_2

Let us visualize two hydrogen atoms *before* they unite to become a hydrogen molecule, H_2. Each consists of a central nucleus, which is simply one proton, around which an electron is buzzing within and even beyond its spherical $1s$ orbital. When the two atoms join, their two $1s$ orbitals *overlap* and merge, forming one **molecular orbital** (see Figure 4.4).

Although the two electrons occupy the entire orbital, they spend most of their time between the two attractive protons on either side. It is the attraction of both protons for both electrons, and vice versa, that keeps the molecule together. But how close is together? The protons are shown as being 0.75 Å apart. This distance is called their **equilibrium spacing;** it is the position to which the protons will return if they are pulled some distance apart or pushed closer together. But to pull them apart requires work to overcome the electrostatic attraction acting between each proton and the electrons shared between them. Also, to *push* them together closer would also require work to overcome the increased *repulsion* between the two protons, both positive in charge, as they get closer to each other. The equilibrium spacing of 0.75 Å between the protons is therefore the distance at which two opposing tendencies *balance out.* One tendency is for both protons to move closer to the electrons between them, and therefore closer to each other, while the other tendency is to move far-

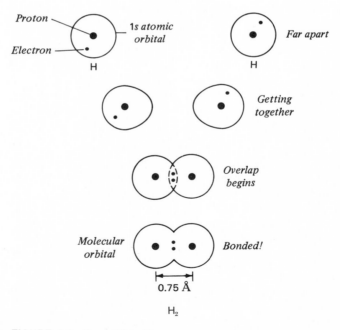

FIGURE 4.4 From 2H to H_2.

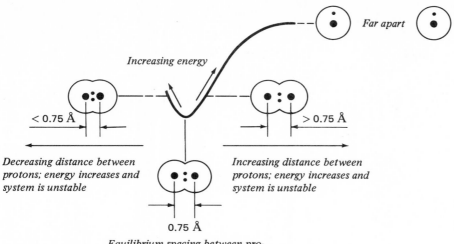

Increasing energy

Far apart

< 0.75 Å

> 0.75 Å

Decreasing distance between protons; energy increases and system is unstable

Increasing distance between protons; energy increases and system is unstable

0.75 Å

Equilibrium spacing between protons; energy of the molecule is at minimum

FIGURE 4.5 Energy change versus departure from equilibrium condition for H_2 molecule. Any change in the proton spacing from the equilibrium value of 0.75 Å results in an increase of energy of the molecule. If the molecule is squeezed together, the proton-proton repulsion causes a rapid rise in energy. The energy also increases if the protons are pulled farther apart, rising to a maximum, as seen at the right. When the atoms become completely separated their energy is at a maximum.

ther apart from each other because of the mutual repulsion of like charges. When the two opposing tendencies balance, the system is in **equilibrium.**

Any displacement of the protons that would make the molecule smaller or larger would require an energy *input*, and thus increase the energy of the system. The rate at which this energy increase occurs is shown in Figure 4.5. The essential point to note is that any displacement from the equilibrium spacing results in an increase in energy. If the constraint that keeps the two kernels too far apart or too close together is removed, they will return to their equilibrium position of least energy and greatest stability.

A system in equilibrium can therefore be described as follows:

1 The component parts of the system occupy specific positions and form a specific arrangement. For the simple H_2 molecule the equilibrium distance between protons is 0.75 Å, and the molecular arrangement is linear.
2 At equilibrium the system is stable and will resist change.
3 At equilibrium, the energy of a system is at a minimum.

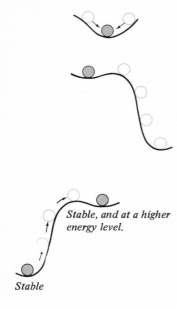

Stable, and at a higher energy level.

Stable

Having said that equilibrium, stability, and minimum energy are interdependent, we must add that stability is a relative matter. The ball in the little hollow is in equilibrium at the lowest point of the curve and is stable; if it is displaced to the left or to the right, the ball will roll back to its original low point. However, if the scene widens and the hollow is next to the crest of a hill, the ball is stable only so long as it remains in the hollow. If it is pushed just over the peak, the ball will roll down to a new and lower energy level, to a *new* equilibrium position.

When the two hydrogen atoms discussed before were far apart and "unaware" of each other, *each* was a system in equilibrium and each was stable. When the two atoms come together, they form a single system which spontaneously reduces its energy to the minimum level shown in Figure 4.5, and the two kernels arrive at the equilibrium spacing of 0.75 Å between them.

In an overall and descriptive way, it can be said that the search for equilibrium is behind the formation of stable bonds between atoms and their combination into new chemical entities. Atom-to-atom bonding and the resulting combinations are therefore not random events but arrangements of atoms that are stable. If it were not stable, the new arrangement would not last. Further-more, if this combination occurs of its own accord, that is, *spontaneously*, the system formed will also be at a new and *lower* level of energy. However, it is also possible for chemical changes to occur that result in a new and stable arrangement of atoms having a *higher* energy level if an *input* of energy lands them in a favorable energy hollow, as shown in the sketch. Once in this upper hollow, the system assumes a minimum energy position.

KINDS OF BONDING

Although there are many atoms, there are, broadly speaking, two ways in which they combine with each other. In the first, electrons are *shared* between atoms. This is called **covalent bonding:** the atoms, including their valence electrons, *co*operate. If the atoms being bonded are of the same kind, each has the same claim on the shared electrons; the sharing is *equal*. However, if dissimilar atoms are being joined, one of them will be more effective in pulling on these electrons. Since the pull on the negative electrons comes from the positive protons in the nucleus, the atom with more protons will pull harder. Also, the size of the atom is a factor, and different atoms have different sizes. The result is that electron sharing between *unlike* atoms is generally *unequal*, in which case it is called **polar covalent bonding.**

The second method of combination is when electrons are *transferred* from one atom to another. One atom *loses* one or more electrons, and the other *gains* one or more electrons. The loser will now have more protons (+) than electrons (−) and becomes a **positive ion;** the gainer will now have more electrons (−) than protons (+) and becomes a **negative ion.** The transfer is such that both atoms (now ions) achieve filled outer shells. Being oppositely charged, the two ions attract each other and form a stable union. This is called **ionic bonding** since it is the *ions* that are being held together.

The atoms of metallic materials, such as copper, iron, and aluminum, are held together by **metallic bonding.** Metallic bonding has some similarity to covalent bonding in that the outer valence electrons are shared. However, there is the important difference that in metals these shared electrons are much freer to move about, and it is their greater mobility that accounts for the well-known fact that metals are good *conductors of electricity and heat.*

COVALENT BONDING BETWEEN IDENTICAL ATOMS

The simplest case of covalent bonding joins two atoms of the same kind together, for example, the two hydrogen atoms discussed before. The single electron of a hydrogen atom occupies the lowest energy level, which is therefore half full, since it can accommodate two electrons. When the two hydrogen atoms join up, their two electrons (one from each atom) are *shared* between them, resulting in a filled outer shell for *each* atom. Each atom contributes an electron. Sharing this electron pair gives each partner a filled outer shell. The shared pair of electrons constitutes the covalent bond between the two atoms, and it is conventional to represent this covalent bond by a short line between the atoms being joined. The joining of two hydrogen atoms to form a hydrogen molecule can therefore be represented as

$$H\cdot + \cdot H \longrightarrow H\ddot{(\cdot)}H \qquad H-H \quad \text{or} \quad H_2$$

Two separate hydrogen atoms join — They are bonded by a pair of shared electrons — To form a molecule — *structural formula* — *molecular formula*

Other atoms besides hydrogen form diatomic molecules by covalent bonding, among them chlorine, oxygen, and nitrogen. Using the same type of representation as above, we show the formation of these bonds for the three atoms schematically in Table

TABLE 4.2   Electron-dot structures

Individual Atoms	Joined Atoms	Pairs of Electrons Shared	Structural Formula	Molecular Formula
:C̈l· ·C̈l:	:C̈l()C̈l:	One pair; single bond	Cl—Cl	Cl_2
.Ö: :Ö.	.Ö()Ö.	Two pairs; double bond	O=O	O_2
:N: :N:	:N(:::)N:	Three pairs; triple bond	N≡N	N_2

4.2.   The outer shells involved are the L shell of oxygen and nitrogen and the M shell of chlorine, whose s and p orbitals are full when they together hold *eight* electrons.

From Table 4.2 you can see that Cl_2 has a single covalent bond joining its two atoms, O_2 has two covalent bonds, and N_2 has three covalent bonds, corresponding respectively to the number of pairs of electrons shared.   By sharing electron pairs, each atom of the diatomic molecules has achieved a full set of eight electrons in its outer shell.

In the example of covalent bonding between similar atoms, the molecules formed were all diatomic and all gases.   However, covalent bonding between similar atoms can also result in crystalline solids having enormous numbers of atoms.   Examples of elements that form crystalline solids are carbon, silicon, and germanium.   These atoms all belong to group IV of the periodic table and, having four valence electrons, can form four covalent bonds with four other atoms of the same kind.   The spatial arrangement is **tetrahedral,** where the atom in the center sees a similar atom at each of the four corners.   Each of the four corner atoms in turn sees four atoms around it in the same way, and the structure continues outward indefinitely.   This crystalline pattern is called a **diamond structure** since diamond consists of carbon atoms bound together in this geometry.   Carbon atoms also form graphite.   The difference between a hard, glittering diamond and black, slippery graphite is in the difference in their bonding (see Figure 4.6).

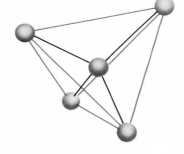

BONDING BETWEEN DISSIMILAR ATOMS

Although covalent bonding between similar atoms is important, the more general situation is covalent bonding between *dissimilar* atoms.   The atom which is most often involved in covalent bonding, both with itself and with other atoms, is carbon.

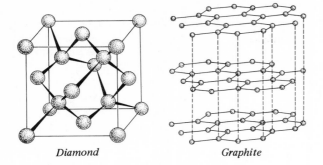

Diamond   *Graphite*

FIGURE 4.6 (*a*) Diamond and (*b*) graphite.

Although the bonding of carbon with other atoms will be discussed more fully in Chapters 5 and 10, the essential features will be considered at this point. Let us consider covalent bonding between carbon and hydrogen. For simplicity, the following conventions will be used:

1 As noted before, the kernel is the entire atom structure *exclusive* of its valence electrons.

2 For clarity, the carbon electrons will be shown as dots and those of hydrogen atoms as small crosses.

3 The electron configuration of carbon being $\left(+6\right)$ 2 4, its electron-dot structure will be $\cdot\dot{\underset{\cdot}{C}}\cdot$, where the C represents the kernel.

For the carbon atom to achieve a full valence shell of eight electrons, it must share four electrons with four hydrogen atoms, one with each, and the net result is shown in Table 4.3.

TABLE 4.3 Covalent bonding

Electron-Dot Structure 1 Carbon + 4 Hydrogens → Joined atoms	Pairs of Shared Electrons	Structural Formula	Molecular Formula
$\cdot\dot{\underset{\cdot}{C}}\cdot$ + H×H (H above/below) → H C H (H above/below)	Four pairs; four single bonds	H—C—H (H above/below)	CH_4

As can be seen from the electron-dot structure of the union between carbon and the four hydrogens, not only does C have a filled outer shell of eight electrons but each of the hydrogens has its full quota of two electrons in its electron shell. The product of this union of C and H is called a **compound,** meaning a stable combination of two or more different atoms; the molecular formula of this compound is CH_4, as shown, and its name is methane. At ordinary temperatures, methane is a gas, which becomes liquid only at $-162°C$. Since it is drawn on a flat surface, the structural formula of methane as shown misrepresents its spatial structure, which is tetrahedral. Chemists sometimes use models to show the spatial arrangement of atoms in a compound, and a stick-and-ball model of the methane molecule is shown here.

In an analogous manner, carbon can form covalent bonds with chlorine, leading to carbon tetrachloride, a compound whose molecular formula is CCl_4. The true structure is again tetrahedral, with chlorine atoms in place of hydrogens.

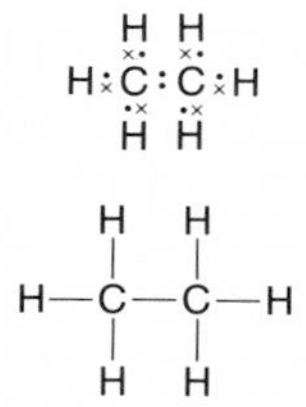

It has already been mentioned that carbon bonds covalently with itself to form a crystalline solid. What is more important is the fact that carbon atoms bond with each other in sequence while *also* bonding with other atoms such as hydrogen, oxygen, and nitrogen. This is the beginning of the chemistry of living matter, and will be treated more fully later. However, an illustration of what happens is shown in the electron-dot and the structural formulas for ethane, C_2H_6.

Methane and ethane are the first two members of a **homologous** (same structure) class of compounds called **alkanes.** They all consist of carbon and hydrogen only; all the bonds are single and covalent, and the carbon atoms link to each other. Thus the next higher alkane would have a three-carbon chain, and so on and on.

POLAR COVALENT BONDING

In the hydrogen molecule discussed before, the equilibrium distance between the two protons is 0.75 Å, and the most probable position of the two electrons they share is halfway between them. This type of covalent bonding is called nonpolar because the molecule that results is electrically the same in both directions. That is, if you could stand at the geometrical center of the molecule, the number of electrons and of protons to the left would be the same as the number to the right. The same is true for the diatomic molecules discussed earlier. However, this is *not* the case when *dissimilar* atoms are joined by sharing a pair of electrons. Since there is sharing, the bonding is still covalent but it is no longer an equal sharing. The even-handed distribution of electric charges has given way to a one-sidedness, or polarity, of electric charge. The reason is that unlike atoms will have a different number of protons in the nucleus and will be different in size, both of which influence their electronegativity, or attraction for electrons. The atom with the larger electronegativity is more effective in pulling the shared electrons toward itself and away from the other atom. The result is a polar structure, and the greater the electronegativity difference between the two atoms being bonded, the greater the polarity of the bond. If you had an "electron divining rod" (no patent pending) and used it to traverse the molecule formed, you would get more electron "signals" at one end and fewer at the other; that is, one end of a polar structure is relatively negative, and the other end is relatively positive. This type of bonding is called polar covalent, and the equilibrium position of the shared electrons is no longer equidistant between the two atoms but favors the direction of the more electronegative atom.

An example of polar covalent bonding is the gas molecule formed between hydrogen and chlorine called hydrogen chloride, HCl. Since the chlorine atom is more electronegative than hydrogen, the electron pair shared between them is displaced preferentially toward the Cl, as shown. For clarity this displacement is exaggerated, and the hydrogen and chlorine electrons are shown as a small cross and as dots, respectively. The shift of the paired electrons toward the Cl kernel makes the molecule relatively negative at its Cl end and relatively positive at its H end. It is the convention to indicate relatively positive and negative parts of a polar covalent molecule by the symbol δ, so that the polar covalent HCl molecule can be schematically represented as $\delta^+ \ \delta^-$. Now let us imagine that a number of HCl molecules are placed in an electric field (see sketch). Because they are polar, the HCl molecules line up in the electric field so that each relatively positive end faces the negative electrode and each relatively negative end is nearer to the positive electrode. This alignment is the condition of minimum energy, since to turn the HCl molecules around would require work. The amount of work, or energy input, required is related to the **dipole moment** of the molecule. The dipole moment of a molecule, which can be measured experimentally, is therefore a quantitative indication of the **degree of polarity** of a molecule. It will vary from zero for a nonpolar structure such as Cl_2 to increasing values with increasing polarity of the molecule.

In determining dipole moments, the unit of measurement is the debye. Table 4.4 lists the dipole moments of a number of molecules. Let us see what these values tell us. The zero dipole

H :C̈l:

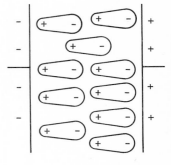

TABLE 4.4 Dipole moments of selected molecules

Molecule	Dipole Moment, debyes
Water, H_2O	1.8
Sulfur dioxide, SO_2	1.7
Ammonia, NH_3	1.5
Hydrogen sulfide, H_2S	1.5
Hydrogen chloride, HCl	1.0
Carbon dioxide, CO_2	0
Methane, CH_4	0
Carbon tetrachloride, CCl_4	0
Hydrogen, H_2	0
Oxygen, O_2	0

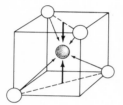

Each pair of electrons shared between C and the H's are equally displaced towards C. The resultant displacements ↓ and ↑ are equal and opposite, so that CH_4 has a zero dipole and is nonpolar.

moments for the hydrogen and oxygen molecules are to be expected, since they share electrons equally, but what of CH_4 and CCl_4? Since carbon is more electronegative than hydrogen and will pull electrons to itself more strongly, why does methane have a zero moment? On the other hand, CCl_4 also has a zero dipole, although here the chlorine atom is more electrovalent than carbon. The answer lies in the geometry of these molecules, which, as shown before, is tetrahedral, with the four bonds extending *symmetrically* from the carbon in the center to the hydrogens or the chlorines at the corners. The shared electrons are indeed shared unequally *but* to the same extent in all directions; the symmetry of the structure serves to cancel out the dipole moments that exist at each of the four bonds. Here, as in many other situations, the shape or geometry of a molecule affects its chemical and physical behavior.

A similar consideration applies to carbon dioxide. Its electron-dot structure is shown in the margin. Here the oxygens, due to their greater electronegativity, have shifted the position of the shared electrons toward them but to the *same* extent and in *opposite* directions, so that the dipoles cancel each other. The zero dipole here is due to the linear structure of CO_2. On the other hand, SO_2 has a substantial dipole moment, although its molecular formula is analogous to that of carbon dioxide. Obviously, it cannot have a linear structure, and its geometry is inferred to be as shown.

Similarly, the high dipole moment of water indicates that its bonding is polar covalent and nonlinear. Its electron-dot and structural formulas are shown. Water is a unique liquid, and its peculiarities, which are largely the result of its polar structure, are of great importance to chemistry and biology. This will be discussed further in Chapter 6.

$$\overset{\times\times}{_+\ddot{O}} : C : \overset{\times\times}{\ddot{O}}_+$$

$$O=C=O$$

Covalent and polar covalent bonding are the characteristic types of bonding present in the organic compounds of living matter. Polar covalent structures are of special importance in the long and complex molecules often found in living tissue and play an important role in their biological functioning. The fact that a polar structure contains both negative and positive ends makes it possible for the negative end of one structure to hook up with the positive end of another. Although these hooking forces are relatively weak, they are often critical in forming those specifically organized structures that are so important in the chemistry of life processes.

IONIC BONDING

Table 4.4 showed the range of values normally found in polar covalent compounds, with about 2 debyes the highest. However, if potassium fluoride, KF, which is a solid at ordinary temperatures, is heated until it becomes a gas (like the molecules listed in the table), its dipole moment is 8.6, which is more than fourfold the highest value for the polar covalent molecules. Since the difference in electronegativity between covalently bonded atoms is considered the reason for the shifting of the shared electron pair toward the more electronegative atom, it appears that in the case of KF (and other compounds like it) the K and the F are so far apart in their affinity for electrons that a *new* situation exists. The potassium atom *loses* the single electron in its outer shell to the highly electronegative fluorine atom, which then uses it to build up its outer valence shell to a full eight. Instead of an equal or unequal sharing of electrons there is an electron "takeover" from the K to the F. This is precisely the process by which sodium atoms and chlorine atoms react, as discussed on page 58.

But there is still a grand finale. The K$^+$ and F$^-$ ions, or Na$^+$ and Cl$^-$ ions, with their opposing charges, attract each other, as all oppositely charged particles do. This is called **electrostatic attraction.** In this case, since the particles are ions, it is referred to as **ionic attraction.** To separate the mutually attracting ions *from* each other would require work and thus raise the energy of the system. On the other hand, to join them together would minimize the energy, and join they do to form a stable compound KF (potassium fluoride) or NaCl (sodium chloride). Since the bonding is between ions, this is called **ionic bonding.** These are ionic compounds since they are made up of ions.

To get a fuller picture of ionic bonding and ionic compounds refer back to period 3 shown in Figure 4.2. You will note the following:

Electron transfer and ionic bonding

1 Each atom is neutral, with as many positive protons as negative electrons.

2 Each atom has its inner shells filled, with a full quota of two electrons in the first energy level and a full eight in the second level.

3 The number of electrons in the outermost shell increases regularly from 1 for Na to 8 for Ar. The inert gas of this period, argon, is the only atom with a filled outermost shell.

4 The electronegativity of the atoms, or their attraction for electrons, increases consistently in going from left to right, except for argon.

Keeping in mind that the electron transfer between atoms, which leads to ionic bonding, also results in filled outer shells for the atoms involved, we may ask which of the atoms shown will do business with each other. Argon is ruled out: It is full and contented. The first atom, Na, resembles the potassium atom, K, in the previous example. It also belongs to group I and has one electron in its outer shell. In fact, Na could combine with F just as well as K did, in which case NaF, sodium fluoride, would be formed. Fluorine is in group VII; below it in this period the group VII atom is chlorine, Cl, which also has seven electrons in its outer shell. It is easy to see that Na and Cl will form an ionic compound quite analogous to the formation of KF from K and F.

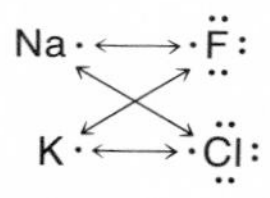

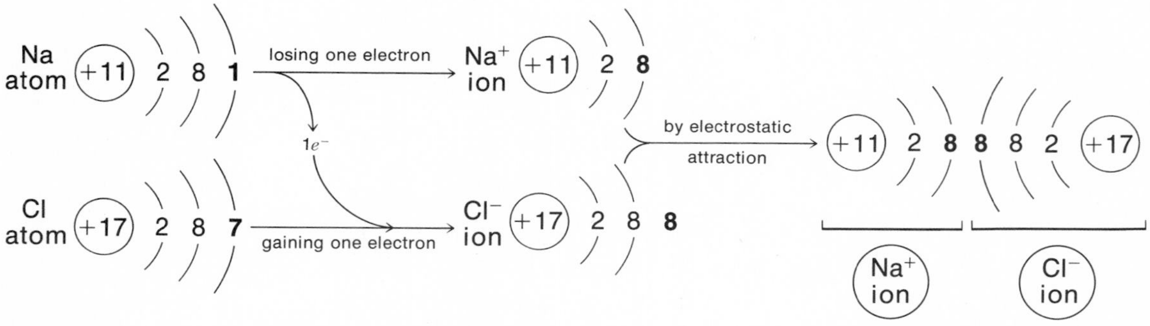

NaCl
(sodium chloride)

The product formed, sodium chloride or NaCl, is ordinary table salt, one of the most common chemical compounds in daily use. From looking at the formula NaCl, one might get the impression that table salt is made up of one atom of sodium and one atom of chlorine. But sodium is a *violently* reactive metal, and chlorine is a yellow-green, *poisonous* gas. Is *that* what we put on our eggs in the morning? Obviously not. The compound NaCl is not two atoms pasted together but two *ions* joined by the electrostatic attraction between them. In other words, when two atoms unite the result is not an averaging out of the properties of each. When the atoms of a solid element combine with the atoms of a gaseous element, the product is not something halfway between a solid and a gas. The product of the union of Na and Cl resembles neither atom; the methane formed from a carbon and four hydrogen atoms is quite different from both carbon and hydrogen. This is the unique feature of atomic bonding; the emergence of new features in the product not present in the original atoms.

At ordinary temperatures, ionic compounds are solid materials, and we all know table salt as a white, hard crystalline solid. It is a characteristic of crystalline solids that they are built up by the regular repetition in three dimensions of a specific unit of structure. You have probably noticed that a wallpaper design can be the repetition of some pattern in two dimensions, and the same is true of fabrics or of tiled flooring. In each case, some smallest unit of repetition can be isolated. For NaCl the unit of repetition

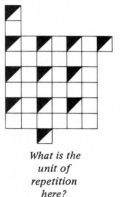

What is the unit of repetition here?

is $(Na^+ \; Cl^-)$, where the Na^+ ion is smaller than the Cl^- ion because it has only two electron shells compared with three for Cl^-. This unit simply adds to itself in three dimensions, that is, left and right, up and down, and front and behind, always keeping the sequence of positive to negative to positive to negative.

FIGURE 4.7 NaCl lattice.

As seen in Figure 4.7, the underlying geometry is the **lattice** structure shown, where the lattice points are the centers of the sodium and chloride ions. The ions around these centers touch each other, forming a close-packed crystal. If you look at the Na^+ ion in the center of the lattice, you will observe that the ion sees six Cl^- ions around it; similarly each Cl^- ion has six Na^+ ions as next-door neighbors. Since there are as many chloride ions as there are sodium ions, any salt particle is electrically neutral. If salt particles are observed under a low-power microscope, they are seen to be cubes, which is consistent with their lattice structure.

Having considered Na, let us go to the second atom in that period, namely Mg, magnesium. From Figure 4.2, you see that this atom has two electrons in its outer shell. Can this atom bond with fluorine or with chlorine in the same way as Na and K? Yes—and no. Yes, it can bond ionically like Na or K, and no because with *two* outer electrons to give away, instead of one, Mg must unite with *two* Cl (or F) atoms.

Schematically, this can be represented as shown in Figure 4.8. By losing its two outer electrons to the more electronegative chlorines, Mg becomes Mg^{2+} ion, and each of the two Cl atoms becomes Cl^-. The electrostatic attraction now brings together, in stable union, the Mg^{2+} ion and the two Cl^- ions to form $MgCl_2$. Again, the ions achieve a full quota of eight electrons in their outer shells.

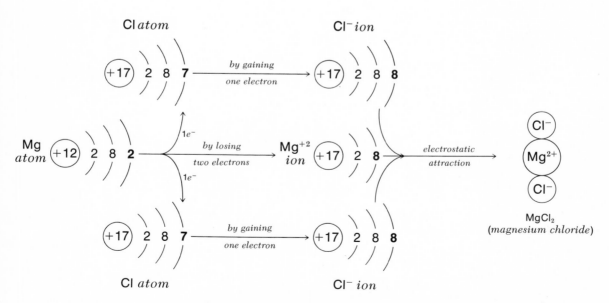

FIGURE 4.8 Schematic representation of magnesium bonding with chlorine.

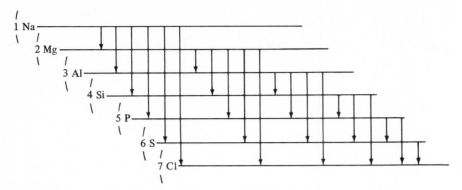

FIGURE 4.9 Diagrammatic representation of electronegativity differences.

Summarizing:

Using the atoms of period 3, let us draw some general conclusions regarding ionic bonding. The greater their difference in electronegativity, the more readily two different atoms will form an ionic bond and the more completely ionic the compound will be. When the atoms are arranged as in Figure 4.9, a rough (*not to scale*) indication of electronegativity *differences* between them can be seen. The longer the arrow between atoms, the greater their electronegativity difference and the more readily they will form ionic bonds.

When atoms join in atomic bonding, they form complete outer electron shells as a result of transfer of an electron or electrons. The atom that loses electrons uncovers a lower, fully filled shell; the atom that gains electrons uses them to fill up its incomplete electron shell.

Using Figure 4.9 as a guide, we can see that Na will bond most readily with Cl and quite readily with S. Generally, group I atoms bond ionically with group VI and group VII atoms of the same period, or other periods. Since Na has only one outer electron, it will form NaCl with Cl but Na_2S with S since the latter needs two electrons to complete its outer shell. Two Na atoms, each contributing one electron, will therefore bond with S. Mg will also bond with Cl, and, as shown before, the compound formula will be $MgCl_2$. It will bond less readily with S, and the compound formula will be MgS, since Mg will donate two electrons, which is just what S needs to fill up its outer shell. In general, group II atoms will bond with group VII and group VI atoms. Al will form an ionic compound with Cl, and its formula will be $AlCl_3$; this is because Al will donate its three outer electrons and therefore needs three Cl atoms to accept them all, each Cl taking only one. Again, group III atoms will tend to bond with group VII atoms, although with considerably less enthusiasm than atoms of groups

I or II. Atoms of groups IV and V prefer covalent to ionic bonding.

As can be seen, ionic bonding is rather limited in its combinations, restricted as it is by the requirement of a high electronegativity difference between the bonded atoms. Nevertheless, ionic bonds are of great importance in every area of chemistry, and though it occurs much less frequently in living matter than covalent and polar covalent bonds they are essential in many life processes.

The following conventions and definitions regarding ionic bonding and ionic compounds are important to our discussion throughout the rest of the book.

1 Ionic compounds are often referred to as **inorganic compounds** or materials. At the same time, inorganic compounds often contain structures that include both ionic and polar covalent bonding.

2 In naming an ionic compound the positive ion is named first, then the negative ion; generally, the negative ion derived from a single atom is given an -*ide* ending. Examples: $CaCl_2$ is calcium chloride; MgO is magnesium oxide; Na_2S is sodium sulfide.

3 A subscript indicates the number of ions present in that compound. If no subscript is present, it is assumed to be 1. That is, $CaCl_2$ consists of one calcium ion and two chloride ions. The subscripts are specific to the compound, and cannot be altered.

4 Those atoms which tend to lose electrons in forming ionic bonds, for example, Na and Mg, are called **electropositive** atoms since they become positively charged after losing one or more electrons. The atoms of groups I, II, and often III are electropositive. By the same token, atoms that gain electrons, for example, Cl, F, and O, are called **electronegative** atoms since they become negative ions by gaining one or more electrons. Atoms of groups VI and VII are especially electronegative and those of group V are less so.

5 The *loss* of one or more electrons is called **oxidation.** The *gain* of one or more electrons is called **reduction.** When Na loses an electron to become Na^+ ion, it has been oxidized; when Cl gains that electron to become Cl^- ion, it has been reduced. Oxidation and reduction go together; they are the "put" and "take" of electron transfer.

6 The term **valence** is often used to indicate the combining capacity of an atom. It can simply be taken as the charge on an ion after the atom has lost or gained electrons. Na becomes Na^+; therefore, its valence is +1. Mg becomes Mg^{2+},

therefore its valence is +2; Cl becomes Cl⁻, therefore its valence is −1; 0 becomes O^{2-}, therefore its valence is −2, etc.

MIXED BONDING—COMPLEX IONS

Many inorganic materials contain complex ions, or **radicals,** which differ from a simple ion in being a charged *group* of atoms rather than a charged single atom. The several atoms of a complex ion are themselves bonded together by covalent bonding and become ions because they need more electrons to complete their outer shells than they have among them. One important complex ion is the sulfate ion, SO_4^{2-}, whose ionic formula means that it contains two more electrons than the sulfur atom and four oxygen atoms can together provide. The electron-dot structure is shown here; keep in mind that both O and S are in group VI, so that as neutral atoms they have *six* outer electrons each. This structure represents a distinct chemical entity, and the −2 charge applies not to any part of the complex ion but to the entire structure as a whole.

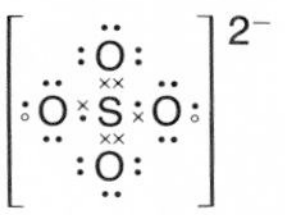

Without the two electrons donated by an electropositive atom or atoms there would not be enough electrons to fill the outer electron shells of the five atoms involved. *With* these two "foreign" electrons, filled structures are obtained, and stable covalent bonding results, but these two electrons are in excess, and the sulfate structure has a net −2 charge. Students sometimes wonder where the two extra electrons came from. When we encounter a complex ion, the extra electron or electrons are already there and there is no way of knowing what supplied them. Suffice it to say that some other atom or atoms contributed

TABLE 4.5 Formula and charge of some complex ions

Complex Ion	Ion Formula	Charge	Typical Compounds Containing This Complex Ion
Hydroxide	OH⁻	−1	$NaOH$, $Ca(OH)_2$
Sulfate	SO_4^{2-}	−2	Na_2SO_4, $MgSO_4$
Nitrate	NO_3^-	−1	KNO_3, $Ca(NO_3)_2$
Carbonate	CO_3^{2-}	−2	Na_2CO_3, $MgCO_3$
Bicarbonate	HCO_3^-	−1	$NaHCO_3$, $Ca(HCO_3)_2$
Phosphate	PO_4^{3-}	−3	Na_3PO_4, $Ca_3(PO_4)_2$
Monohydrogen phosphate	HPO_4^{2-}	−2	K_2HPO_4, $MgHPO_4$
Acetate	CH_3COO^-	−1	CH_3COONa, $(CH_3COO)_2Ca$
Cyanide	CN⁻	−1	$NaCN$, $Mg(CN)_2$
Ammonium	NH_4^+	+1	NH_4Cl, $(NH_4)_2SO_4$
Borate	BO_3^{3-}	−3	H_3BO_3

$MgSO_4$	*magnesium sulfate*
Na_2SO_4	*sodium sulfate*
$CaSO_4$	*calcium sulfate*
H_2SO_4	*hydrogen sulfate*
$Al_2(SO_4)_3$	*aluminum sulfate*

them. Since covalent bonds are stable, a complex ion behaves like a single unit and participates as such in chemical reactions.

Just as there are many chlorides, there are many sulfates. In each case the compound is electrically neutral, which requires that the total positive charges from the sulfate ion's partner (Ca^{2+}, $2Na^+$, etc.) equal the total negative charges from the SO_4^{2-} ion or ions. The mix and match must always result in *net* zero charge on the compound. In working this out for aluminum sulfate, note that three sulfate ions are present, and two aluminum ions.

All of these compounds have *mixed* bonding, that is, ionic and covalent. Far from being uncommon, mixed bonding is found widely, both in living and nonliving matter.

Table 4.5 lists some important complex ions and the charge they carry when they are bonded to an oppositely charged ion in inorganic compounds. Get to know them well.

FORMULA WEIGHT

When atoms join to form compounds, the mass of the compound is the sum of the masses of all of its atoms. The mass of NaCl is the sum of the mass of the Na atom (23) and the Cl atom (35.5), totaling 58.5. Similarly, the mass of CH_4 is the sum of the mass of the carbon atom (12) and the four H atoms (4×1), totaling 16. The mass of any chemical molecule is, therefore, the total of the masses of its component parts. This sum is called the *molecular weight*, but since it is actually calculated from the chemical formula of the compound, it is preferable to refer to it as the **formula weight.** On this basis, the formula weight of the OH^- ion is $16 + 1 = 17$, and the formula weight of the SO_4^{2-} ion is $32 + 4(16) = 96$.

REVIEW QUESTIONS

1 Because there are similarities in their chemical behavior, the elements lithium, sodium, and potassium are classed together with the alkali metals. What accounts for their similar chemical nature?
2 Referring to the electron configurations shown for the atoms shown in Figure 4.2, explain why Na is the most electropositive and Cl the most electronegative. (*Hint:* Which has more protons in the nucleus?)
3 Referring to Figure 4.1, which atom in period 2 would you consider to be a metalloid? In the same period, which atom is most metallic, and which is most nonmetallic? Explain.
4 On the electronegativity scale, fluorine has a value of 4.0, making it the most electronegative atom. By reference to Figure 4.1, explain

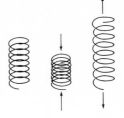

why F is somewhat more electronegative than Cl, even though the latter has eight more protons in its nucleus. (*Hint:* Which positive nucleus is closer to its outer electrons? The attraction between $+$ *and* $-$ charges goes up sharply the closer they are. Also, which atom has more electrons? These electrons will tend to repel an additional, incoming electron.)

5 Fluorine forms a diatomic molecule, F_2. Show the electron dot structure of F_2. Why is the bonding between the two atoms of all diatomic molecules nonpolar covalent?

6 The spring shown in the margin can be considered as being in equilibrium, since of itself it will not change in size nor shape. If the spring were forced, by being pushed or pulled, to change its size, would it be in equilibrium? How is this analogous to the energy increase shown for the H_2 molecule as shown in Figure 4.5?

7 The hydrogen sulfide molecule shown here is polar because on the electronegativity scale, H is 2.2 and S is 2.6. Draw the electron dot structure of H_2S and indicate the relatively positive and relatively negative ends of the molecule.

8 The atoms of the following molecules are all joined by covalent bonding: Cl_2, NH_3, CH_4, H_2O, and H_2. For each molecule indicate **a** whether the bonding is polar or nonpolar and **b** whether the molecule itself is polar or nonpolar. **c** If the molecule has polarity, indicate the relatively positive and relatively negative parts.

9 It was mentioned before that F is the most electronegative atom. That being the case, would you expect the HF molecule to be more or less polar than the HCl molecule, whose electron dot structure is on page 67?

10 On page 65, the electron dot structure of carbon was shown as represented here, that is, as the C kernel surrounded by its four valence electrons. In a similar manner show the electron dot structures of Li, B, N, Ar, Al, P, and Ca.

11 The sketch of page 70 shows the formation of the ionic compound NaCl. Which atom is undergoing oxidation and which is undergoing reduction?

12 When Mg combines with two Cl atoms as shown on page 72, which atom or atoms are being oxidized and which are being reduced?

13 Write the formula and the name of the ionic compound formed **a** between Ca and F, O, Cl, and S, respectively, and **b** between Na and F, O, Cl, and S, respectively.

14 What is the difference between a simple ion and a complex ion?

15 Complete each of the following sentences, supplying the formula of each of the two ions that make up the compounds. As shown in the examples, include the number of ions and the charge on the ion.

The ions of $CaCl_2$ are	Ca^{2+} and	$2Cl^-$.
The ions of $Mg(HCO_3)_2$ are	Mg^{2+} and	$2HCO_3^-$.
The ions of $MgSO_4$ are	and	
The ions of NaOH are	and	

The ions of H_2SO_4 are   and

The ions of Na_2SO_4 are   and

The ions of $NaNO_3$ are   and

The ions of $Ca(NO_3)_2$ are   and

The ions of NH_4Cl are   and

The ions of Na_3PO_4 are   and

The ions of Na_2CO_3 are   and

The ions of CH_3COONa are   and

COVALENT AND IONIC BONDING AND STRUCTURES
The Inside Story

LOOKING BACKWARD

The greatest mystery story took place about 3 billion years ago, give or take a few hundred million years, on an Earth that neither looked, tasted, nor smelled the way it does today—even if there was no one to see, taste, nor smell it, since there was absolutely no life. An eminent scientist has called it The Strange Case of the Self-duplicating Molecule, and although it may never be completely solved, scientific work over the past few decades has provided a rough outline of how our rather undistinguished planet, astronomically speaking, came to be the *land of the living*. Understandably, the story is tentative and incomplete, but it was probably something like the scenario that follows.

THE OPENING SCENE

The earth was much warmer, its seas were shallower, and its atmosphere quite different from what it is today. The atmosphere contained hydrogen gas. This is called a **reducing** atmosphere, the opposite of our oxidizing atmosphere. Because the

Today, our atmosphere is an oxidizing atmosphere: It contains oxygen.

earth was warmer, there was more water vapor in the atmosphere than there is now. Furthermore, where today an outer layer of ozone absorbs most of the ultraviolet radiation from the sun, then there was no ozone in the atmosphere, leaving it transparent to these high-energy rays. From the point of view of *now*, it was hot, humid, and deadly, but from the point of view of *then* it was just right, and things began to happen.

NEW AND DIFFERENT

That reducing atmosphere of long ago probably contained methane, CH_4; ammonia, NH_3; hydrogen sulfide, H_2S; water vapor, H_2O; and hydrogen gas, H_2. These are all simple molecules, a long way from the complex structures found in living matter. However, given a suitable input of *energy*, these small molecules can be combined into a variety of larger ones (Figure 5.1). The energy required for this **synthesis,** or building up, of larger mole-

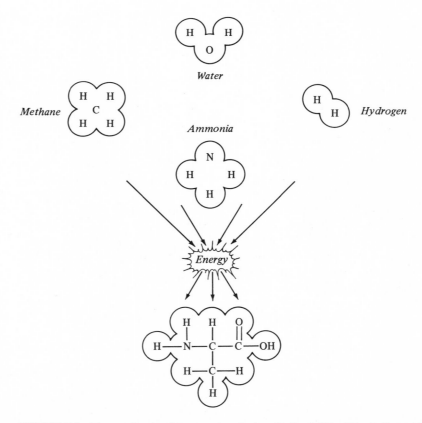

FIGURE 5.1 The synthesis of larger chemical units from CH_4, NH_3, H_2O, and H_2.

cules from smaller ones could have come from a variety of sources (Table 5.1).

However this synthesis may have taken place, the results were *new* chemical entities differing in important respects from the original raw materials, and we can consider these differences in terms of before and after (Table 5.2).

THE WATERY CRADLE

Though born in a gaseous atmosphere, the new carbon-containing molecules found their home in the water, collecting in protected coves, lakes, or lagoons. In these waters they found new chemical participants, the **inorganic minerals** from the solid crust of the earth. Some were relatively simple in composition, such as sodium chloride, NaCl, whose great solubility makes the seas salty; others were more complex. This mix of carbon-containing compounds, dissolved minerals, and warm water added up to what one pioneer worker, J. B. S. Haldane, described as a "hot thin soup." In this soup, with the cumulative changes that took place over millions of years, the chemical stuff of life began to take form.

Water served well as the staging area for the assembly of the life molecules. To begin with, as in any good soup, it provided a place for the ingredients to mix, blend, and interact. The organic, carbon-containing compounds provided the modules for further assembly into larger structures; the dissolved minerals provided **ions** for critical linkages and strategic services; water itself provided polar molecules to fill in or shore up as required; time provided opportunity, and changes in temperature and acidity provided the proper circumstances. When the primitive ready-to-live macromolecules were finally formed, they may have looked different from those of today, but they probably included early versions of our modern proteins and nucleic acids, the essential chemical structures of living matter.

TABLE 5.1 Energy sources

Heat of earth
Ultraviolet radiation
Lightning discharges
Cosmic rays

Body weight is 60–65% water

A 10 day old infant may be 77 % water by weight

TABLE 5.2

Molecules Before	Molecules After
Small	Larger
Gases	Solids
Contain no more than two different atoms	Contains many different atoms
One type of bond per molecule	Many different bonds per molecule
In these simple molecules no one atom was more important than any other	Carbon atom is principal atom of these more complex chemical structures

TABLE 5.3 Principal elements in the human body

Element	Percent	Element	Percent
Oxygen	65	Chlorine	0.15
Carbon	18	Magnesium	0.05
Hydrogen	10	Iron	0.004
Nitrogen	3	Zinc	0.0033
Calcium	1.5–2.0	Rubidium	0.0017
Phosphorus	1.0–1.1	Copper	0.00015
Potassium	0.35	Manganese	0.00013
Sulfur	0.25	Iodine	0.00004
Sodium	0.15	Cobalt	Trace

When living matter finally did emerge, its watery environment took on still another role: it protected the young protoplasm from the lethal effects of the abundant ultraviolet radiation, the same energy that may have helped its formation at an earlier stage.

CHEMICAL PAST BECOMES CHEMICAL PRESENT

The story is ended, but from Table 5.3, listing the major elements found in the human body, we can recognize past events. The composition of the carbon-containing molecules synthesized from the atmospheric gases is reflected in the values for carbon, hydrogen, and nitrogen, which together with oxygen account for 96 percent of the total.

As for most of the other atoms, they first entered into the chemistry of life with the dissolved minerals of that hot thin soup. However, some of these, such as calcium and iron, are most likely of more importance to us bony, warm-blooded mammals than they could have been to those one-celled bits of life that were the first living organisms. This tells us that the end of our story is the beginning of another: the evolution of living matter from its inception, and of the parallel changes in the air, the seas, and the surface of our now living earth.

Having had a backward look at the atomic ingredients of living matter, let us now turn to a consideration of how these atoms are organized into structures.

FITTING BONDS TO ATOMS

Atoms become organized into chemical structures by bonding, in the course of which some of their outer electrons take up new equilibrium positions. The specific bond formed depends on the specific atoms being joined. As discussed in Chapter 4, the

Water is 89% oxygen.

Atomic weight: 1   16   1

Total atomic weight of atoms in water = 18

Percentage of oxygen in molecule = 16/18 or 89%

All living organisms contain the same elements, but the proportions vary from species to species. Vertebrates have more calcium than bacteria, and green plants use magnesium where animals use iron.

primary factor determining the outcome is the difference in electronegativity between the two atoms. When two atoms of the
same element are being joined, the electronegativity difference
between them is zero and the bond formed is covalent. Since
equilibrium is reached when opposing tendencies balance, and
since the opposing tendencies are equal because the two kernels
exert the same pull on the electron pair between them, the equilibrium position of the electrons is exactly between the two
kernels. The electron scene is therefore the same in both directions, so that the bond and the molecule have zero polarity.

When two *dissimilar* atoms are joined, the electronegativity difference between them imposes an *unequal* pull on the shared
pair of electrons, whose equilibrium position is therefore closer to
the atom kernel exerting the larger positive pull. The bond
formed is called **polar covalent** because it has a greater concentration of negative electrons in one direction and a correspondingly smaller concentration of electrons in the other. The
bond is relatively *negative* at the electron-heavy end and relatively

:F̈·   *and*   ·F̈:
bond covalently

:F̈:F̈:   *or*   (∘ : ∘)
F₂ *molecule*

H·   ·C̈l:
*Hydrogen atom and chlorine
atom bond covalently*

H :C̈l: *or*   δ⁺(• : ∘)δ⁻
Polar HCl *molecule*

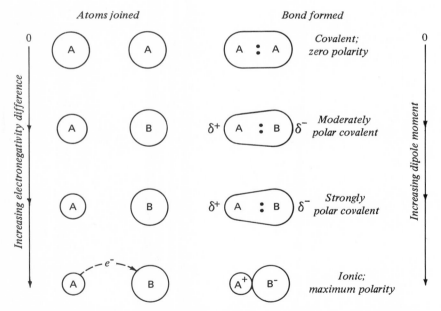

FIGURE 5.2 From covalent to polar covalent to ionic bonding. For the sake of
simplicity the difference in the electronegativity of the atoms joined is indicated by
the difference in their size. When the same atoms A and A are joined, the electronegativity difference is zero and the resulting covalent bond has zero polarity.
When dissimilar atoms A and B at the bottom join, their electronegativity difference is so great that an electron is transferred from A to B, resulting in the formation of ions and of an ionic bond.

δ⁺ or δ⁻ *indicates a partial positive or negative charge.*

positive at the electron-light end.   The greater the electronegativity difference between two atoms, the greater this imbalance will be and the greater the bond polarity.

Although a polar covalent molecule has more electrons at one end and fewer at the other, as a *whole* the structure is *neutral.* However, as electronegativity difference between the atoms bonded increases and polarity increases accordingly, a point is reached were the inequality finally becomes a complete shift of an electron to the more electronegative atom.   Instead of each atom contributing an electron to form a shared pair, there is an electron **transfer** from the less electronegative atom, which thus becomes a positive *ion,* to the more electronegative atom, which becomes a negative *ion* (see Figure 5.2).

Thus, the sequence of increasing polarity follows increasing electronegativity difference; it begins at one end with simple covalent bonding, proceeds through increasingly polar covalent bonding, and ends with ionic bonding.   A covalent bond has no polarity and strictly speaking occurs only between atoms of the same element.   An ionic bond has maximum polarity, and the atoms joined are far apart in their electronegativity.   Between the two extremes is the wide range of polar covalent bonding, which is the type of bonding most often found in living matter.   As the polarity of covalent bonding increases, the molecules become less covalent and increasingly ionic.   As pointed out in Chapter 4, the quantitative measure of polarity is the dipole moment.

It is instructive to follow the change in polarity, and hence in

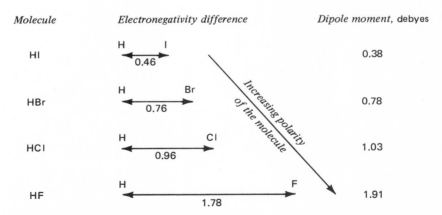

FIGURE 5.3 Dipole moment and electronegativity differences of hydrogen halide molecules. The degree of polarity, as indicated by the dipole moment of the molecule, follows the increasing values of the electronegativity difference between the atoms joined. HCl and HF are both strongly polar and therefore have both covalent and ionic characteristics.

dipole moment, of the molecules that hydrogen forms with iodine, bromine, chlorine, and fluorine. All these atoms belong to the halogen family (group VII), with seven electrons in their outer shell. Figure 5.3 summarizes the data.

BONDING BY ORBITAL OVERLAP

Figure 4.4 showed two hydrogen atoms forming a bond by the overlap of their respective $1s$ orbitals. Orbital overlapping is the general mechanism of covalent bonding between atoms. The overlap between an orbital of one atom and an orbital of the other is the bridge between them, and the more they overlap, the stronger the bridge and the stronger the bond. This fusion of the two atomic orbitals results in a single **molecular orbital.** This region in space around the two joined kernels is now occupied by the shared pair of electrons. In the hydrogen molecule the two shared electrons no longer "know" to which atom they belong since both protons are equally attractive and equally distant. Although the electrons can be found over the entire region of the molecular orbital, they are usually found around each kernel (which in the case of hydrogen is simply the proton) and along the line between them, as shown here schematically.

Since covalent bonding occurs by the overlap of two atomic orbitals, the kind of bond formed depends on the orbitals that participate. The two general types of bonds are referred to as σ and π bonds.

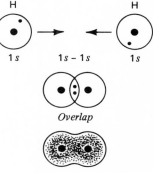

H_2 *molecular orbital; the electron "cloud" indicates that the two shared electrons are most often around and between the two kernels.*

$\sigma = sigma$
$\pi = pi$

THE SIGMA BOND

Of the two types of bonds, the σ is the more common and the more stable. The simplest way to form a σ bond is to cause overlap of the spherical s orbital of one atom with the s orbital of another atom, as in the formation of the H_2 molecule. Generally, each of the orbitals involved is half-full, so that each contributes one electron to the molecular orbital. With two electrons the resulting molecular orbital is full, since like any other orbital it can hold no more than two electrons. The molecular orbital is called the **bonding orbital,** and its two electrons are the **bonding electrons.**

Although the two bonding electrons occupy the entire molecular orbital, their most common whereabouts is on a line between the two positive atomic kernels. If we think of the electrons in the orbital as a cloud, the cloud is densest around and between the two atom kernels, as was shown for the H_2 molecule. This is the situation when the two atoms being bonded are *alike*, that is,

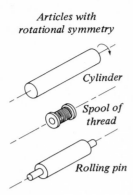

Articles with rotational symmetry

Cylinder

Spool of thread

Rolling pin

in pure covalent bonding. When two *dissimilar* atoms are being joined, the shared electrons will find one of the two more attractive and will therefore be displaced to some degree in *that* direction. The electron cloud will still be around and between the two atom kernels, but it will be thinner at one end and thicker at the other.

Although the spherical symmetry of the two original orbitals is lost when they overlap and join, the resulting σ orbital is symmetrical around the line between the two atom kernels. An example of an object having symmetry around an axis is an ordinary spool of thread. As you rotate the spool around a spindle through its center, the shape it sweeps out is unchanged. If you rotate a σ orbital around the axis of the molecule, which is the line passing through the two atom kernels, its shape will also remain the same. The σ orbital is therefore characterized by **rotational symmetry.**

In many cases a σ bond, with its characteristic rotational symmetry, is obtained by the overlap of an s orbital of one atom with a p orbital of another atom. This is the mechanism whereby the HF, HCl, HBr, and HI molecules of Figure 5.2 were formed. The overlap and merging of the atomic orbitals of H and F to form the HF molecule, which is generally typical of polar covalent molecules, are shown in simplified form in Figure 5.4.

The molecule formed has several features that characterize, to a greater or lesser degree, polar covalent molecules in general.

1 The shared pair of electrons is in the bonding orbital, which is therefore filled. As a result of the sharing, both joined atoms achieve full outer shells.
2 The bonding electrons are closer to the more electronegative atom.
3 Because of the preference of the shared electrons for one atom, the molecule is polar, as indicated by its dipole moment. Because F is so electronegative, HF is strongly polar.

SIGMA AND PI BONDS BY OVERLAP OF p SUBORBITALS

The distinguishing feature of σ bonds is their rotational symmetry. This feature is absent in π bonds, whose symmetry is more restricted. Generally, the more limited the symmetry of a structure, the higher the energy associated with it and the less stable it is relative to more symmetrical structures.

It was pointed out in Chapter 3 that the three p suborbitals of a given energy level, or shell, have an energy somewhat higher than the simpler s orbital of the same shell. A sphere is the simplest and most symmetrical three-dimensional shape. A fly crawling

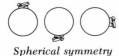

Spherical symmetry

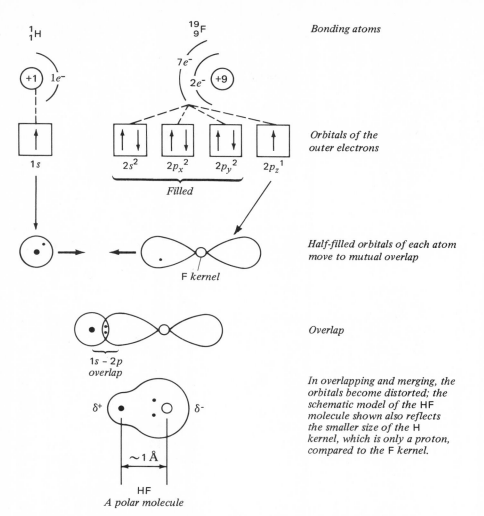

Bonding atoms

Orbitals of the
outer electrons

Half-filled orbitals of each atom
move to mutual overlap

Overlap

In overlapping and merging, the
orbitals become distorted; the
schematic model of the HF
molecule shown also reflects
the smaller size of the H
kernel, which is only a proton,
compared to the F kernel.

FIGURE 5.4 Formation of the HF molecule. The molecular orbital formed by the merging of the $1s$ orbital of hydrogen and the $2p_z$ of fluorine has rotational symmetry about the line between the kernels, so that the bond is a σ bond. The molecule is linear since the two kernels (● = atom kernels) are essentially two points, and two points determine a straight line.

on a ball always experiences the same curvature and always sees the same scene, no matter which way he moves. A sphere is also the most *economical* shape. If a sphere, say a balloon, is squeezed out of shape, it will stretch to a larger surface area, although the amount of air inside remains the same. In other words, when the balloon is spherical, it has the *least surface area* and is most relaxed, with equal forces pushing in all directions in-

To and from
the center all
directions
are the same

Not all
directions
are the
same

side and outside the sphere. Any change in that spherical shape will require effort, increasing its energy. The three p suborbitals are higher in energy than the corresponding s orbital because they have direction and different shape. The difference between the three orbitals lies only in their directions, which are mutually perpendicular, so that they all have the same energy. In Figure 5.5 the p_x, p_y, and p_z suborbitals are shown separately for clarity; when the three are superimposed on each other, they represent the region in space around the atom kernel that could accommodate six electrons (two per orbital) having that level of energy.

The figure shows the three p suborbitals of two identical atoms side by side. Let us first look at the facing lobes of the two p suborbitals at the top. As the atoms continue to approach each other, the two orbitals overlap to become one molecular orbital. Since the atoms are alike, the electrons are shared equally between them. The line joining the two atom kernels passes through the center of the molecular orbital, which therefore has rotational symmetry about the centerline. The molecular orbital resulting from the overlap of two in-line p_x orbitals is thus a σ type. Note the two smaller regions on either side of the atom kernels. This means that though the electrons *may* be found in these regions, the probability is much larger that they will be found in the section between the two kernels. This is consistent with the idea that the concentration of negative charge, that is, of electrons, pulls the positive kernels together. Bonding is still a matter of plus and minus; the orbitals are really a map showing where these pluses and minuses spend their time.

The two lower sketches in Figure 5.5 represent the p_y and p_z orbitals, which are equivalent. In both cases, the lobe axis of one atom is parallel to the lobe axis of the other atom. When the atoms come closer together, the lobes approach each other, not nose to nose, as in the case of the p_x orbitals, but *side* to *side*, like two parallel dumbbells rolling toward each other. The only difference between p_y and p_z orbitals is the fact that they are 90° apart; otherwise, the same results and conditions apply. As the atoms come close, each lobe reaches sideways for its parallel opposite, so that overlap and merging occur *above* and *below* the atom kernels. For the p_y and p_z suborbitals, the net results in each case are two banana-shaped regions on opposite sides of the line connecting the atom kernels. The two regions together constitute the molecular orbital, and together they contain the two bonding electrons. Having been formed from parallel p orbitals, each such bonding orbital is called a *pi*, or π, orbital, and the bond is a π bond; it differs from the σ type in lacking rotational symmetry.

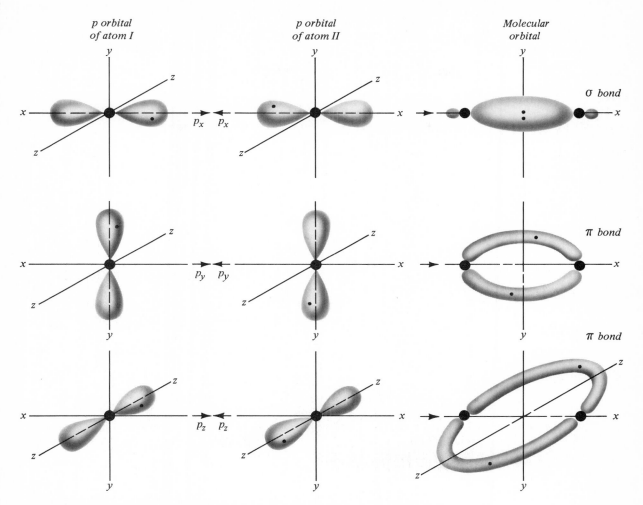

FIGURE 5.5 The molecular bonding orbitals that can form by the overlap of the atomic p orbitals of two atoms.

THE DOUBLE BOND

Generally, the σ bond is more stable than the π type, and if only one pair of electrons is available for sharing between two atoms, only the σ type will form. However, when *two* pairs of electrons are being shared between the atoms, the *second* bond will be of the π variety. This means that a double bond between two atoms is made up of one σ and one π bond. Although in structural formulas a double bond is shown as two parallel lines that look alike, the two bonds are *not* really alike. This is a matter of con-

A═B

Double bond equals
σ *bond* + π *bond.*

siderable importance. A double bond is more *chemically active* than a single bond between the same two atoms. Given the opportunity, the double bond (one σ, one π) breaks up, leaving only a σ bond. We shall come to this again; for the moment let us think of the double bond as a site of possible chemical action.

REARRANGING THE ORBITALS; HYBRID ORBITALS

The geometrical shape of a covalent molecule results from the directions of its molecular orbitals. These are formed by the overlap and merging of atomic orbitals, which are generally half-filled. This implies that the number of bonds that an atom can form is limited to the number of its half-filled orbitals. But this is not the case, and many atoms form more bonds than they have half-filled atomic orbitals. The most notable example is carbon, whose electron configuration shows only *two* half-filled orbitals, whereas the chemistry of carbon and carbon compounds rests on the fact that it forms *four* bonds. How does the carbon atom manage to do this? The carbon atom (and others as well) internally rearranges its orbitals, somewhat like rearranging the rooms of a house to make it more livable: knock down a wall or two, put up some new ones, and so on. For a house, this costs money; for an atom this costs energy. The expenditure is well worth it, because when the improvements are completed, the atom can form more bonds, each of which results in an energy decrease.

The carbon atom has four outer electrons, two in the filled $2s$ orbital and one each in the $2p_x$ and $2p_y$ suborbitals. By promoting *one of the two* electrons of the $2s$ orbital into the unoccupied $2p_z$ orbital, the carbon atom finds itself with *four* orbitals, each with one electron. These orbitals are neither $2s$ nor $2p$, but four new hybrid orbitals, and since one s and three p orbitals were involved in their formation, they are referred to as sp^3 hybrid orbitals. Each of the four hybrids is half-filled, and since there is no distinction between them, they are equivalent. This equivalence expresses itself also as a similarity of orbital shape and direction, as shown in Figure 5.6.

The four hybrid orbitals, each at an angle of 109.5° to each other, point to the four corners of a tetrahedron, with the carbon kernel at its center. The tetrahedral arrangement not only serves to make the four orbitals directionally equivalent but also *minimizes* the *electron repulsion* between them by keeping the electron pair in each orbital as far from other pairs as possible. This makes for minimum energy and therefore for stability.

With four sp^3 orbitals available, carbon can bond with four hydrogen atoms. As shown in Figure 5.6, the resulting methane

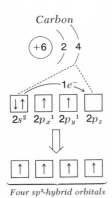

Carbon

Four sp³-hybrid orbitals

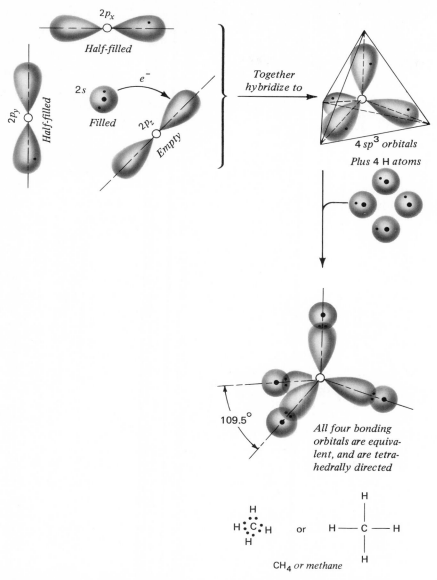

FIGURE 5.6 The formation of the four sp^3-hybrid orbitals of carbon and its bonding with four hydrogen atoms. Carbon forms bonds with other atoms and with itself via the four sp^3 orbitals obtained by hybridizing its one s orbital and three p orbitals. The four bonds it forms with hydrogen are slightly polar, but because of symmetry the molecule is nonpolar. The tetrahedral arrangement of the bonds minimizes the repulsion energy between the four pairs of shared electrons.

molecule is tetrahedral in shape, with the four hydrogen atoms arranged at the corners of a tetrahedron and the carbon atom at its center. In general, a tetrahedral architecture can exist around a central atom that provides four sp^3 hybrid orbitals.

TABLE 5.4 Major elements in the crust of the earth

Element	Percent
O	46.60
Si	27.72
Al	8.13
Fe	5.00
Ca	3.63
Mg	2.09
Na	2.83
K	2.59

THE MANY FACES OF THE sp^3 ORBITALS

Many chemical materials are built up via sp^3 orbitals. Consider the silicon atom, just below carbon in group IV. Just as carbon is the backbone atom of living matter, silicon is the primary atom responsible for the hard crust of the earth (Table 5.4). Its most common and most versatile chemical manifestation, the silicate structure, is tetrahedral, with the silicon atom in the center bonding to four oxygens via four sp^3 orbitals. In providing each atom with eight outer electrons, the covalent structure has more electrons than protons, so that as a whole it is a negative ion. Internally the SiO_4^{4-} ion is held together by sp^3 covalent bonds; externally it bonds with positive ions such as Al^{3+}, Mg^{2+}, and Fe^{2+}. The mixture of covalent and ionic bonding makes possible a tremendous diversity of structures, from fibrous asbestos, to slippery talc, to hard quartz. All these and many others are built up of repeating silicate units that form chains, rings, sheets, and three-dimensional structures. In general, the product is a **network** structure, with the tetrahedral silicate ions forming the sequences and the positive ions serving to bind and coordinate them. Whereas silicates constitute a large part of the solid ground beneath us, including the soil, the diamond is the most glamourous of the network structures. The diamond is an extended network of carbon atoms connected to each other by the overlap of four tetrahedrally directed sp^3 orbitals of each carbon. The covalent bond is itself strong, and in even the smallest diamond there are an enormous number of covalent bonds. Combined with the reinforcing effects of the tetrahedral arrangement, this fact makes diamond the hardest substance known. A hard material can scratch a softer one, but not vice versa, so that diamonds are very useful for cutting and grinding and wherever hardness and resistance are essential.

There are many other instances of tetrahedral, or nearly tetrahedral, bonding involving sp^3 orbitals. In some cases, the angles between the bonds depart somewhat from the 109.5° required by the geometry of the tetrahedron, indicating the presence of some modifying circumstance. Two interesting and important examples are the structures of ammonia, NH_3, and water, H_2O.

Group IV

Carbon

Silicon

A silicate chain

NITROGEN AND AMMONIA

Of the five outer electrons of the nitrogen atom, three are unpaired, each half filling a $2p$ suborbital. This suggests that the
three half-filled $2p$ orbitals would be used for bonding. *If* that
were so, the angles between the three bonds would be the angles

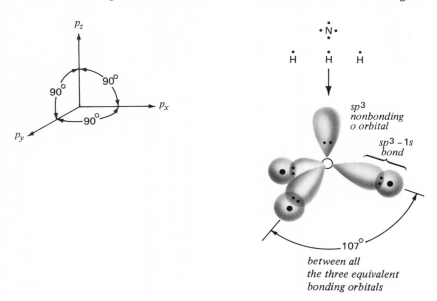

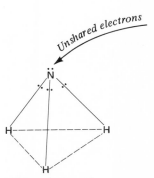

*Trigonal pyramid
structure of the
ammonia molecule*

between the three suborbitals, that is, 90°. The experimental evidence, however, does not agree; when nitrogen bonds with three
hydrogen atoms to form ammonia, NH_3, the three N—H bonds are
equivalent but the angle between them is about 107°, which is
quite close to the tetrahedral condition. The bonding of nitrogen
is interpreted as being via four sp^3 hybrid orbitals, analogous to
the way carbon bonds but with a difference. In carbon, all four
sp^3 orbitals form a bond, and are bonding orbitals; in nitrogen,
only the three half-filled orbitals participate in bonding, joining
with three hydrogens as shown, while the fourth is a nonbonding
orbital, housing the pair of unshared electrons.

The geometry of the ammonia molecule is that of a trigonal pyramid, or a three-sided pyramid, with the three hydrogens at the
triangular base and the nitrogen at the apex. The three N—H
bonds are equivalent, the result of three similar sp^3-plus-1s
overlaps, while the fourth sp^3 hybrid projects upward, holding its
lone pair of nonbonding electrons. It is this pair of electrons that
acts to modify the tetrahedral arrangement by taking up more
space than the bonding pairs. However, this lone pair is important for another reason: It can pick up another hydrogen (*not* a

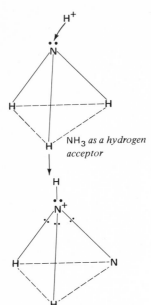

NH₃ *as a hydrogen acceptor*

*Tetrahedral bonds
of the resulting
NH₄⁺ ion*

hydrogen atom because the orbital cannot take a third electron, but a hydrogen *ion*), a bare positive proton, which buries itself in the two-electron cloud within the sp^3 orbital. That is, a nitrogen atom that has three bonds may take on a hydrogen ion at its fourth sp^3 orbital so that it acts as a **hydrogen-ion acceptor.** As we shall see later, a material that accepts hydrogen ions is considered an alkaline, basic, material. When ammonia, NH_3, takes on a hydrogen ion, the result is a charged structure, ammonium ion, NH_4^+, which, being a positive ion, can form ionic bonds with negatively charged ions. In living tissue, nitrogen often bonds with carbon as well as with hydrogen, and if in so doing it forms *four* bonds via four sp^3 orbitals, the nitrogen atom will carry a positive charge. Keep this in mind: It becomes important later on.

The shape of the water molecule is also explained in terms of sp^3 orbitals. The six outer electrons of oxygen plus two more

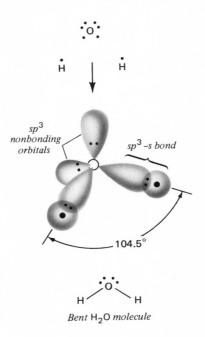

Bent H_2O *molecule*

from the two hydrogens make a total of eight. These eight electrons are considered as filling four sp^3 orbitals, two being bonding orbitals joining the oxygen to the two hydrogens and two being *nonbonding* orbitals containing the two lone pairs of the oxygen. As in ammonia, the nonbonding orbitals modify the tetrahedral geometry somewhat, so that the angle between the bonding orbitals is reduced to 104.5°.

OTHER HYBRID ORBITALS

In addition to sp^3 orbitals, which are associated with tetrahedral structures, other hybrid orbitals involved in bonding give rise to different spatial arrangements. This can be illustrated by the atomic orbitals of the beryllium and boron atoms and the hybridized orbitals they use for bonding. Figure 5.7 shows the changeover from atomic to hybrid orbitals and the resulting geometries of the structures formed with beryllium and boron as central atoms. In both cases, the bonding partner is chlorine, whose only half-filled suborbital overlaps the half-filled hybrid orbitals of Be or B to form the molecular orbitals.

Note that the straight-line shape of the $BeCl_2$ molecule and the triangular shape of the BCl_3 molecule serve to minimize the repulsion between the shared electrons in the bonding orbitals; this is a general requirement for putting a stable chemical structure together. The relation between bonding orbitals and the geometrical shape of a number of covalently bonded structures is shown in Figure 5.8.

When some other atom replaces a hydrogen in the methane molecule, the tetrahedral bonding arrangement is generally not changed significantly. For example, chlorine (like other members of the halogen family) will bond with carbon, and if we begin with CH_4 and imagine a Cl atom successively replacing a hydrogen, a series of compounds will result:

$$CH_4 \longrightarrow CH_3Cl \longrightarrow CH_2Cl_2 \longrightarrow CHCl_3 \longrightarrow CCl_4$$

CH_4 and CH_3Cl are gases at room temperature. The others are volatile liquids that are useful as solvents; CCl_4 is carbon tetrachloride. $CHCl_3$ is chloroform, an early anesthetic. The tetrahedral bonding habit of carbon is maintained when it bonds with itself (recall the diamond structure), but it also applies when carbon-to-carbon chains form. Let us imagine a methane-minus-one hydrogen structure, symbolically represented by CH_3— and named a **methyl group.** If two methyl groups were to join, the unemployed sp^3 hybrid orbitals of the two carbons would merge to form the bonding orbital between them. The molecule formed is called **ethane,** and its bonding can be represented as shown in Figure 5.9.

If you could stand at either carbon, you would see four tetrahedrally directed bonds 109° apart. All the bonds have rotational symmetry about the axis and are of the σ type. If a three-carbon chain were visualized as being formed by replacing one of the hydrogen atoms of ethane by another methyl group, the third carbon would bond to the second by another sp^3-to-sp^3 overlap

$_{17}Cl$ $(+17)$ 2 8 7

↑↓	↑↓	↑↓	↑
$3s^2$	$3p_x{}^2$	$3p_y{}^2$	$3p_z{}^1$

Half-filled

Methane

Methyl group

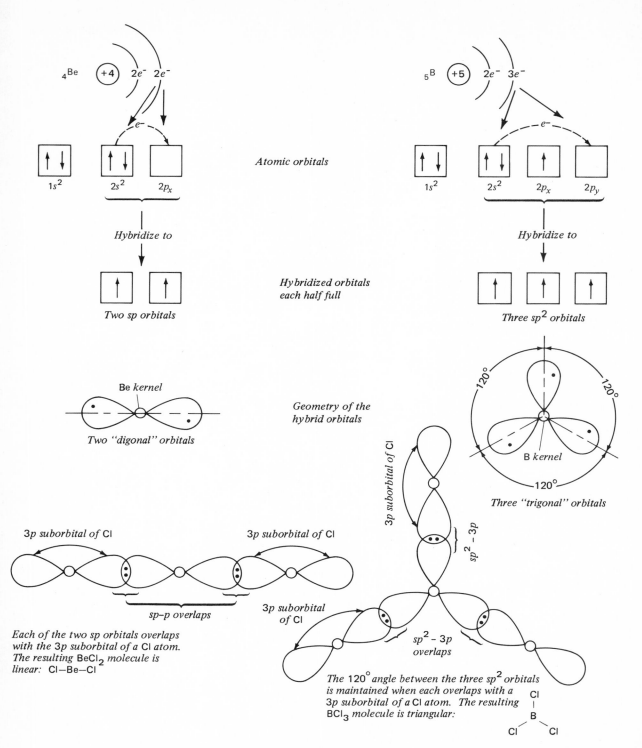

Atomic orbitals

$_4$Be ⊕ +4 $2e^-$ $2e^-$

$1s^2$ $2s^2$ $2p_x$

e^-

Hybridize to

Hybridized orbitals
each half full

Two sp orbitals

$_5$B ⊕ +5 $2e^-$ $3e^-$

$1s^2$ $2s^2$ $2p_x$ $2p_y$

e^-

Hybridize to

Three sp^2 orbitals

Geometry of the
hybrid orbitals

Be kernel

Two "digonal" orbitals

120°   120°

B kernel

120°

Three "trigonal" orbitals

3p suborbital of Cl   3p suborbital of Cl

3p suborbital of Cl

sp^2–$3p$

sp–p overlaps

3p suborbital
of Cl

sp^2–$3p$
overlaps

Each of the two sp orbitals overlaps
with the 3p suborbital of a Cl atom.
The resulting $BeCl_2$ molecule is
linear: Cl—Be—Cl

The 120° angle between the three sp^2 orbitals
is maintained when each overlaps with a
3p suborbital of a Cl atom. The resulting
BCl_3 molecule is triangular:

Cl
|
B
Cl   Cl

FIGURE 5.7 Linear and triangular molecules from sp and sp^2-hybrid orbitals.

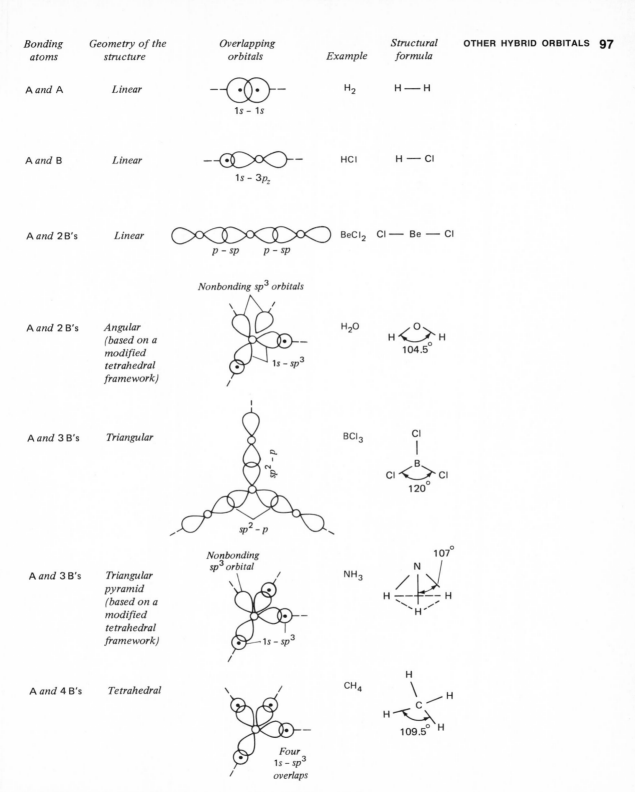

FIGURE 5.8 Some covalent structures and their bonding orbitals.

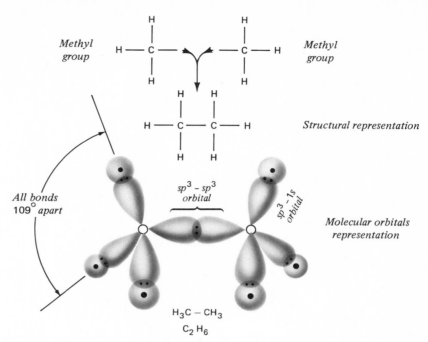

Methyl group

Methyl group

Structural representation

All bonds 109° apart

$sp^3 - sp^3$ orbital

$sp^3 - 1s$ orbital

Molecular orbitals representation

$H_3C - CH_3$

C_2H_6

Figure 5.9 The bonding in ethane, C_2H_6. Ethane is formed by the overlap of one sp^3 orbital from each of two methyl groups. Since the sp^3-sp^3 molecular orbital has rotational symmetry, the bond is sigma, and the two methyl groups can rotate around it.

Angle between bonds is 109°

Propane

Butane

Pentane

The carbon chains are usually written in a straight line for simplicity, but the angle between all bonds is the tetrahedral 109°.

C_2H_4

and the tetrahedral bonding relation would persist. Obviously, methyl groups can be added successively, forming a **homologous** (same structure) series. At each carbon the bonding directions are tetrahedral, and the total structure is considered *nonpolar*, as methane is. Long carbon chains of this sort are very common and very important.

CARBON-TO-CARBON DOUBLE-BOND FORMATION

All the molecular bonding orbitals discussed so far resulted in single bonds of the σ type. But what of double bonds? We pointed out before that the two bonds of a double bond are not equivalent, one being a σ and the other a π type. To illustrate this, let us consider the bonding of ethylene, C_2H_4, as a typical case. From the structural formula for ethylene shown in the margin, it can be seen that each carbon atom has three atoms around it. This recalls the trigonal arrangement beryllium used in forming three equivalent bonds with chlorine. In ethylene the trigonal arrangement has another condition superimposed. We can imagine that each carbon first forms three hybrid-sp^2 orbitals, using three of its four valence electrons. Each of the three sp^2 hybrids holds one valence electron and is arranged about 120° apart in a common plane, which is the normal trigonal geometry. The diagram illustrates the bonding condition up to this point: Each carbon uses two of its sp^2 orbitals to overlap with a $1s$ orbital from each of two hydrogens; this leaves a third sp^2 orbital from each carbon, which overlap nose to nose, to form a σ bond between the two carbons. However, each carbon still has *one more* valence electron, each in an *unhybridized* $2p$ orbital. For clarity, the diagram is rewritten in conventional structural form, where each bonding line represents a pair of shared electrons and on which the half-filled unused p orbitals are superimposed. As shown in Figure 5.10, the upper and lower lobes of these p orbitals overlap sideways, forming a π bond between them; the end result is a double bond between the two carbon atoms, one σ, one π.

Single and double bonds differ in several ways. A σ bond (all single bonds are the σ type) has rotational symmetry around the bond axis, allowing the parts joined by the bond to rotate relative to each other. The two CH$_3$— groups of ethane can rotate relative to each other around the axis between them. However, the π portion of a double bond lacks rotational symmetry, so that in ethylene the CH$_2$— groups are fixed relative to each other; any relative rotation would strain the π bond, thus increasing the energy of the molecule. Structures with double bonds are therefore more fixed in their spatial arrangement, and the specific location of a double bond, or bonds, in a molecule helps determine the chemical behavior of the molecule.

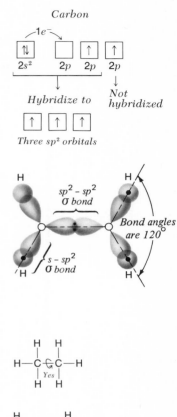

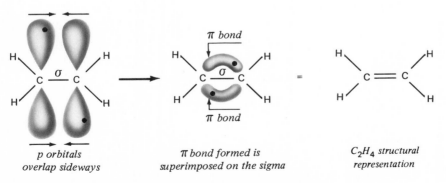

FIGURE 5.10 The σ and π bonds of the double bond of ethylene.

FROM DOUBLE TO SINGLE CARBON BONDS

A double bond is chemically more reactive than a single bond. The electron pair of the less stable π part of a double bond will unpair if the electrons have the opportunity to take part in single-bond formation. This may occur when such atoms as hydrogen or chlorine, with one unpaired valence electron, are available for bonding. What happens in the case of ethylene is shown in Figure 5.11; since the action is at the pair of electrons that constitute the π bond, these are made heavier in the figure. This reaction is called an **addition reaction** since atoms are added to the ethylene, and since the added atoms are hydrogen, it is called **hydrogenation.** The result is the conversion of double-bonded ethylene into single-bonded ethane. Not only hydrogen but chlorine, iodine, or even water can be added in such a reaction. In each case the electron pair of the π bond separates, and each electron finds a partner for σ bonding, leaving behind a single bond where there was a double bond before.

SATURATED AND UNSATURATED BONDING

Long hydrocarbon chains, found in fats and oils, vary in length and in the *number* of double carbon-to-carbon bonds present. If a structure contains *no* double bonds, it is referred to as **saturated,** since its single bonds cannot add any more atoms, *its bonding capacity is full.*

If the chain contains one or more double bonds, it is **unsaturated** to that extent, since at each double bond two additional atoms (hydrogen, iodine, or others) can be added. The degree of unsaturation of a fat or an oil can therefore be measured quantitatively by determining how many additional atoms it can take up. Iodine is commonly used in the laboratory for the purpose, so that the more unsaturated an oil is, that is, the more double bonds it has in its structure, the greater the amount of iodine it will react with.

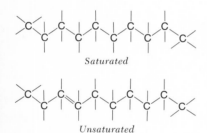

Saturated

Unsaturated

More unsaturated

Ethylene

Hydrogen

Ethane

FIGURE 5.11 The hydrogenation of ethylene to ethane. The addition of two hydrogen atoms across the double bond converts ethylene into ethane. In general, the hydrogenation of an alkene changes it to an alkane.

TABLE 5.5

Bond	Bond Length, Å	Relative Bond Strength, kcal/mole
C—C	1.54	83
C=C	1.34	146
C≡C	1.20	200

Triple bonds also occur, although not often in organic molecules. The three bonds of a triple bond are not equivalent, consisting of one σ and two π types. Examples of triple bonding are acetylene, C_2H_2, and the nitrogen gas molecule, whose two atoms are united by three bonds.

From Table 5.5 you can see that double and triple bonds between carbon are stronger than single bonds (but *not* twice or three times as great) and that they are shorter than single bonds.

Bonds characterized by:
1 *Strength (energy necessary to break bond)*
2 *Length (distance between centers)*
3 *Angle between bonds*

H—C≡C—H
Acetylene, C_2H_2

N≡N N_2
Nitrogen molecule

IONIC COMPOUNDS

Solid ionic compounds are held together by the electrostatic attraction between their positive and negative ions. The resulting arrangement maximizes the attraction between oppositely charged ions and minimizes the repulsion between similarly charged ions. For any given ionic compound a regular pattern is present, in which negative and positive ions are as close to each other as possible, since the closer they are together the greater the electrostatic attraction between them; at the same time, the similarly charged ions are spaced as far apart as possible. For each ionic compound, such as KF, $MgCl_2$, or CaO, there is one specific arrangement that achieves the objectives of maximum attraction and minimum repulsion, and this arrangement repeats itself throughout the structure. A repeating arrangement of atom positions is characteristic of crystalline materials and ionic compounds are generally crystalline.

The structure of NaCl, or table salt, which is typical of many ionic compounds was shown on page 71. There is really no molecule present, if by molecule is meant a specific entity having a given number and type of atoms. Instead, the formula NaCl represents the unit of repetition, which extends itself in three dimensions to form a network of ions. The network is often referred to as a space lattice, and the details of the lattice geometry vary with the particular size and kind of ions involved. Any

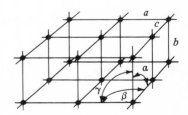

Portion of a space lattice. The distances a, b, and c, and the angles α, β, and γ, fix the lattice geometry.

network is inherently a strong structure because a force applied to it at any point will tend to become distributed over the interlocking network instead of continuing to be concentrated at one area.

Physically, ionic compounds are hard, rather brittle, and with high melting points. The melting point of sodium chloride, 801°C, is quite characteristic; some other melting points are shown in Table 5.6. Note that the melting point increases when the number of transferred electrons increases from 1 to 2. This is analogous to the greater strength of the covalent bond when the number of shared electrons goes from one to two to three pairs.

A major feature of ionic compounds is their ability to conduct an electric current when they are *in water solution* and when they are heated until they become *molten liquids*. In both cases, the ions become mobile and able to move in response to an electrical push, or voltage. A solid ionic compound cannot conduct a current because its ions are held in *fixed* positions in the crystalline lattice. The ions oscillate around the fixed positions, but they cannot leave them. When the solid is heated to the liquid state, the lattice arrangement is largely undone and the electrons are free to move; that is, they will carry the current. What is really being carried is electrons, because an electric current is a flow of electrons. If NaCl is the ionic compound, the Cl^- ion will give up an electron at the positive electrode, and the Na^+ ion will take an electron at the negative electrode.

When NaCl is mixed with water, it dissolves and the Na^+ and Cl^- ions separate, or dissociate. Remember that the water molecules are strongly polar, with a relatively negative end and a relatively positive end. The water molecule may be called two-faced, showing its positive side to the negative chloride ion and its negative side to the positive sodium ion. By dissociating, the Na^+ and Cl^- ions are expressing a preference for the company of water molecules over each other's company. However, if the water evaporates, leaving the Na^+ and Cl^- ions behind, the ions will *reassociate* and reform the solid NaCl crystal. Ionic compounds are called **electrolytes** because of their ability to conduct an electric current when dissolved in water. This is an important distinction between ionic compounds and nonionic compounds, one that can be experimentally determined. Figure 5.12 shows a schematic setup for distinguishing an electrolyte from a nonelectrolyte.

If the liquid used is *not* water but some nonpolar liquid, the NaCl will not dissolve and no current will be carried. To the extent that ionic compounds dissolve, they dissociate, or **ionize.** This verb is something of a misnomer, since they are already ionic; in water the ions simply *let go* of each other.

TABLE 5.6

Compound	Melting Point, °C
KCl	776
CaF_2	1360
CaO	2580

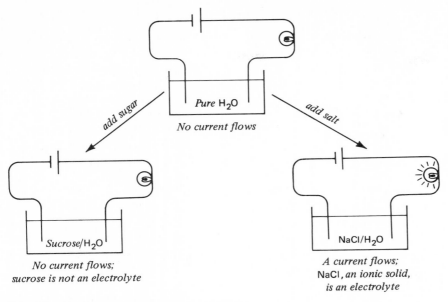

FIGURE 5.12 Electrolyte versus nonelectrolyte.

Reactions between ionic solids are really reactions between *their ions* and therefore must take place in aqueous solution. In a watery environment the ions are free and mobile and can interact and react. In living tissue simple ionic solids do not really exist *as such;* there are no solid particles of NaCl in the body, but there *are* Na$^+$ ions, and there *are* Cl$^-$ ions. It all began in that hot, thin soup.

SECONDARY BONDS

Covalent and ionic bonds are known as primary bonds and are the strong bonds that hold chemical substances together. However, there are weaker, secondary bonds that are also important. One type is hydrogen bonding, discussed in the next chapter. Still another is a very weak atom-to-atom attraction called **van der Waals forces,** which are always present. The detailed analysis of this type of bond is complex, but in a general way it can be described as an attractive force between atoms in *different* molecules. For example, consider the hydrogen molecules shown in the margin. *Within* each molecule the bond uniting the hydrogen atoms is covalent. At the same time, the atoms in each molecule have a small but persistent interest in the atoms of the

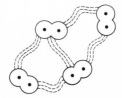

Symbolic representation of van der Waals forces between hydrogen molecules.

other molecules. It becomes significant when (1) the temperature is low, so that the molecules move very slowly, and (2) when the molecules are large and many atoms are involved. Under these circumstances, van der Waals forces become important. This molecular attraction can convert a gas into a liquid when the temperature is reduced or into a solid when the molecules are large enough or the temperature low enough to intensify the effect of the van der Waals forces. But these forces are inherently weak, and the structures they yield, even when solid, lack the hardness and stability of solids formed by covalent and ionic bonding.

THE OUTSIDE OF THE INSIDE

We now turn from a consideration of the inside to a view from the outside, to the behavior of physical materials. Do they exist as gases, liquids, or solids? How do they react to heat or an electric current?

COVALENT SUBSTANCES

Depending on its composition and structure, a covalent compound can be a gas, a liquid, or a solid at room temperature.

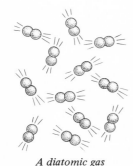

A diatomic gas

Covalent gases We have pointed out that the atoms or molecules like H_2, O_2, N_2, and CH_4 are held together by covalent bonds. At room temperature these substances exist as gases. This means that each molecule chases around by itself, paying practically no attention to its neighbors. Together the molecules occupy whatever volume is available. If a molecule collides with another or with the wall of its container, it simply bounces off. These molecules can exist in the gaseous state at room temperature for two reasons: (1) They are all small and light and therefore move very quickly; they have a high energy of motion, a high kinetic energy, which tends to keep them separate. (2) The only attractive forces between them are the van der Waals forces, which are not only very minute but also decrease rapidly with increasing distance. The net result is that whatever small attraction the molecules may have for each other is more than overcome by their high kinetic energy. Obviously, if the gas molecules are made to slow down, say by lowering the temperature, the attractive forces become more effective in bringing them together. But at room temperature the life-style of a gas is expansive, disorganized, and energetic. The gas we know best, the one most important for us, is actually a mixture of gases, the air (Table 5.7). The gravity of the earth keeps the atmosphere

TABLE 5.7 Composition of air, volume %

N_2	78
O_2	21
Ar	~1
CO_2	~0.03
H_2O	0–2†

† Variable.

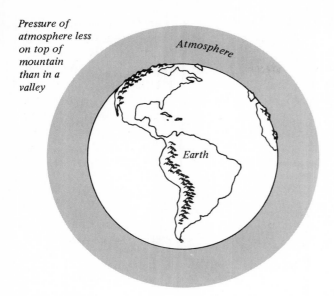

Pressure of atmosphere less on top of mountain than in a valley

FIGURE 5.13 The pressure of the earth's atmosphere varies as the topography varies.

wrapped around it like a blanket, and the weight of the blanket is the pressure of the atmosphere. At high altitudes the air is thinner, and the pressure is less (Figure 5.13).

When you put air in a tire, you are really squeezing it in, and it squeezes back accordingly. A gas can be squeezed because its molecules are far apart; put them in a smaller volume, and they push back that much harder; that is, the pressure increases. Push them still closer together, and the gas molecules get so close that they condense into a liquid. A great deal of pressure is needed, and liquefied gases (like liquid propane or liquid nitrogen) are stored in special high-strength tanks.

Because gas molecules are mobile, they respond readily to energy input or output, including temperature changes. As a gas gets hotter, its molecules move about more rapidly, and if the gas is not allowed to expand, the pressure builds up. This is what happens to the air in a tire that is driven fast and for a long distance on a hot day. If the pressure buildup is more than the tire walls can take, there is a blowout.

Because gas molecules are energetic and free to move, they *diffuse* readily. Open a gas jet at one end of a room and the odor is likely to be recognized soon at the other end—fortunately. The sense of smell is very sensitive, and enters into such matters as the taste of foods, the mating habits of moths, and the social life of ants, not to mention all the "smell-goods" of the cosmetics industry. Gas molecules are small enough and mobile enough to intrude into liquids and even solids. The importance of the solubility of gases in liquids can be appreciated when we consider that fish breathe the *dissolved* oxygen in water. If the dissolved oxygen in a lake is depleted by pollution, the fish die. The degree of solubility of a gas in water depends on pressure and temperature. The higher the gas pressure, the more gas dissolves in the liquid. On the other hand, if the temperature goes up and the gas molecules get livelier, they tend to resist being confined to the liquid and will prefer escaping; that is, the gas solubility decreases.

The fizz in champagne is caused by the dissolved carbon dioxide gas bubbling out of the liquid when the cork pops out and the pressure is reduced. One reason for chilling champagne is to slow the gas molecules down to keep them from leaving too fast.

A more serious effect of a gas bubbling out of solution when the pressure is reduced may occur when people are working under high pressure, as in underwater tunnels or deep-sea diving. At these higher pressures, the amount of nitrogen gas dissolved in blood and other body tissues increases. If the worker is brought back to atmospheric pressure too quickly, the excess nitrogen comes out of solution too fast and forms bubbles in the tissues and bloodstream. The result, which is both painful and dangerous, is called **the bends.** Decompression, therefore, must be slow.

Covalent liquids The only familiar substance that is liquid at room temperature and is *not* covalently bonded is mercury, which is a metal and an element. But at ordinary temperatures such common liquids as water, oils, gasoline, antifreeze, alcohol, carbon tetrachloride, and many others consist of molecules held together by covalent bonds, polar covalent bonds, and mixtures of the two. In turn, these molecules associate with each other as a loose but close aggregation, the liquid state. Let us look at some of the features of covalent liquids.

Compared with a gas, the molecules of a liquid are much closer together and occupy much less room, so that a liquid is a condensed system. When a gas is liquefied by cooling, as when steam turns to water, it condenses.

Because its molecules are close to each other, a liquid cannot be compressed as a gas can. In fact, for all practical purposes liquids are incompressible. This fact makes them useful for transmitting a force. When you push down on the brake pedal of a car with hydraulic brakes, you are pushing a column of liquid (through a tube), which then pushes the brake surface against the wheel.

The forces that bring the molecules together into the liquid state are either van der Waals forces or derive from the polarity of the molecules. The van der Waals forces are very weak but exist between all the molecules, whether covalent or polar covalent. Compared with a gas, the liquid molecules are larger and more complex. Because they are heavier, they move more slowly, and because they are longer and more irregularly shaped, they cannot bounce off each other as lightly as the smaller, more spherical gas molecules. When these more cumbersome molecules meet, there is greater opportunity for contact and interaction, which increases and accumulates the van der Waals forces between them. The net result is a coming together of the molecules to form the liquid condition. The importance of size can be seen by considering methane and hexane (see margin). Both are essentially nonpolar, and both are chemically in the same class of alkane compounds, but the smaller, rounder methane is a gas at room temperature, whereas the heavier hexane molecules are a liquid. Since the van der Waals forces are weak, in the absence of other intermolecular forces, the liquid will have a low boiling point and will be **volatile,** or easily vaporized. For example, the boiling points of chloroform and of carbon tetrachloride are 61 and 77°C, respectively, and the odor of each of these volatile compounds is easily detected at room temperature.

Superimposed upon the van der Waals forces there may be other attractive forces arising from the *polarity* of the molecules. Hydrogen bonding has already been mentioned, and we shall meet other kinds as well. In general, *polar* molecules will tend to *associate*, as illustrated by the two compounds in the margin. From their similar formula weights, you would expect both to be either liquids or gases, but nonpolar butane is a gas at room temperature, while acetone, whose polar carbonyl group adds "togetherness" to its molecules, is a liquid.

Whatever the bonding forces between the covalent molecules of a liquid may be, they are weak forces, so that the joining up in a liquid state is not rigid; there is a constant changing of position, a tumbling and sliding of molecules over molecules, with no getting away from each other.

Covalent solids Most of the hundreds of thousands of covalent substances are solids at room temperature. The molecules of solids are often large and complex in arrangement, and the interaction between them provides the cohesive forces that hold them together.

The most general cohesive forces between the molecules are the van der Waals forces, which become increasingly significant as the molecules become longer. In plastics van der Waals forces are the principal intermolecular bonding forces. In many

Methane, a gas

Hexane, a liquid

Butane, FW = 58, gas

a polar group

Acetone, FW = 58, liquid

An imaginary room temperature liquid; a jostling, restless crowd of covalent molecules stuck with each other.

plastics, such as in polyethylene and Teflon, the carbon-to-carbon bonding in the long molecular chains is consistently *single* bonding. This makes for stability and inertness, so that these plastics are very useful where such qualities are needed, as in containers, seals, or other conditions requiring chemical inactivity. The linearity and regularity of these chains also allow for some degree of crystallinity, that is, a repeating pattern of structure, but real crystallization is generally incomplete or absent. These very long chains of organic plastics are built up from small molecules by using a variety of temperatures, pressures, acidity conditions, catalysts, additives, and the entire apparatus of chemical technology.

On the other hand, the far more complex solid substances of living tissue are put together at practically constant temperature, and instead of being inert, they are uniquely active and reactive. Although van der Waals forces are present here as well, and although the covalent bond is the most common atom-to-atom union, the solid substances of living matter use the entire gamut of bonding, including polar covalent, hydrogen bonding, and even ionic bonding—often in the *same* structure.

HYDROPHILIC AND HYDROPHOBIC BONDING

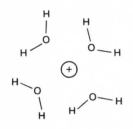

Water molecules around a positive ion. The relatively negative oxygen end points to the + ion.

For the macromolecules of living matter one type of bonding is inadequate; it would cramp their style. They need as many chemical talents and abilities as are available, whether coming from the strong, primary types of bonding, or from the weaker, secondary attractive forces between molecules, such as interaction of dipoles, van der Waals forces, and hydrogen bonding.

Another secondary bonding results from the **hydrophilic** (water-loving) or **hydrophobic** (water-hating) character of a molecule. The friendly polar molecules of water are friendly only toward other *polar* structures, including ions, the ultimate in polarity. Hydrophilic structures, whether polar molecules or ions, find that the polar water molecules cling to them; in a sense it can be said that the water is **bound** to them. Where an ion is concerned, the orientation of water molecules is as shown in the sketch; the ion is surrounded by water molecules oriented in accordance with its charge, so that an ion can be considered as very hydrophilic. As we shall see later, the proteins play a central role in the chemistry of life; their molecules are very large and contain many polar and ionic sites. The protein molecule is therefore hydrophilic, and water molecules will cling to protein structures and surround them. At the same time, protein molecules may also contain nonpolar groups, toward which water is *not* friendly. These groups are called hydrophobic because of the mutual incompatibility of their nonpolar structure and the strongly polar water. When two hydrophobic groups meet in water, their common dislike of water and their rejection by water bring them together in a hydrophobic bond. Hydrophilic molecules, therefore, contain polar or ionic structures (or both) around which the polar water molecules will orient themselves, plus to minus. On the other hand, water will squeeze out nonpolar structures which therefore join together in a hydrophobic bond.

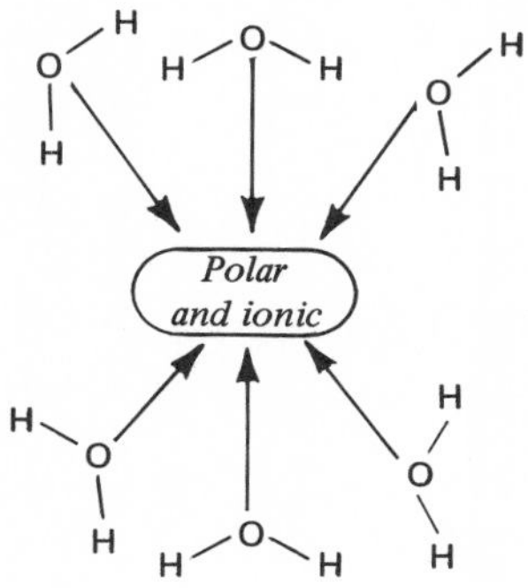

Hydrophilic:
protected by H₂O

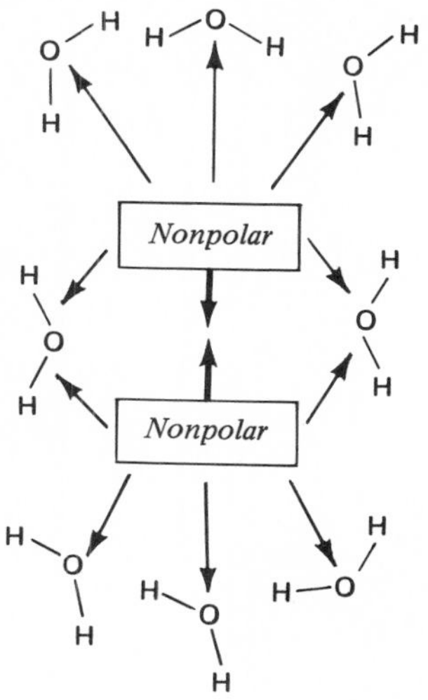

Hydrophobic:
rejected by H₂O

As a general class, molecular solids are inherently *weak*, because the molecule-to-molecule forces that bind them together are for the most part the weaker secondary forces. When ionic bonds are present along with a covalent structure, as in living tissue, they contribute structural characteristics rather than significant strength. Covalent solids are therefore mechanically weak, with low melting points, and in many cases they will disintegrate chemically before reaching a true melting point. For example, if you put two "look alikes," such as table salt (sodium chloride) and table sugar (sucrose) together on a griddle and heat them over a burner, the sucrose will soon melt, and then char while the sodium chloride will remain unchanged. And yet both these materials are crystalline solids. Sodium chloride is, of course, an ionic compound, and ionic compounds are all crystalline solids. Covalent solids may have varying degrees of crystallinity depending on the extent to which the molecules can be arranged in a regular and repeating pattern.

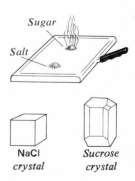

Sugar

Salt

NaCl
crystal

Sucrose
crystal

REVIEW QUESTIONS

1 There is reason to believe that the conditions under which life first emerged on this planet would be deadly to us today. What were these conditions, and why would they be intolerable now?

2 What essential factor was needed to synthesize the small molecules present in the earth's early atmosphere into larger molecules? Besides being larger, what new and important characteristic did these synthesized molecules have?

3 What were the general ingredients of that hot thin soup in which it is believed living matter developed on this planet?

4 Approximately what percent of the human body is water? What percent of water, H_2O, is oxygen?

5 When two atoms of the same kind form a covalent bond between them, the shared pair of atoms constituting the bond is generally found equidistant between the two atoms. Explain why.

6 When two *dissimilar* atoms are joined by a covalent bond, the shared electron pair favors one atom or the other and the bonding is referred to as *polar* covalent. Explain why the shared electrons prefer one atom over the other and illustrate by showing a sketch of the bond formed between **a** H and F and **b** H and O.

7 Why is ionic bonding considered the extreme limit of polar covalent bonding?

8 From the data in Table 5.8, which substance has its atoms held together by **a** ionic bonding, **b** polar covalent bonding, **c** covalent bonding?

9 Figure 5.3 shows the formation of the polar HF molecule from the overlap and merging of the $1s$ orbital of the H atom and the $2p_z$ orbital of the F atom, each orbital being half-filled. Show a sketch of two F atoms joining to form an F_2 molecule. What orbitals are being merged? Is the bond polar?

10 From a consideration of Figure 5.4 explain why the two atoms of the nitrogen gas molecule, N_2, are joined together by *three* covalent bonds. Of these three, how many are σ and how many are π bonds?

11 The electron configuration of carbon as a free atom is shown in the margin. When carbon enters into chemical combination, its electron configuration becomes modified as shown in the lower arrangement. **a** How does the second arrangement derive from the original arrangement in the free C atom? **b** What is the advantage of the modified configuration? **c** What is the name given to this changeover of electron configuration, and how are the new orbitals symbolized?

12 In the ammonia molecule, NH_3, three of the four available sp^3 orbitals of N have merged with the $1s$ orbitals from three H atoms to form three N—H covalent bonds. What does the *fourth sp^3* orbital do? What may occur at this fourth sp^3 orbital?

13 How many nonbonding sp^3 orbitals are there in the structure of the water molecule, and how does this affect the angle between the two covalent O—H bonds? Is the bonding between O and H polar or

TABLE 5.8 Dipole moments

Substance	Debyes
Oxygen gas, O_2	0
Hydrogen sulfide, H_2S	0.97
Sodium fluoride, NaF	8.16

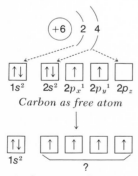

$1s^2$ $2s^2$ $2p_x{}^1$ $2p_y{}^1$ $2p_z$

Carbon as free atom

$1s^2$?

Carbon in combination

nonpolar? How do the two pairs of unshared electrons at the top of the oxygen atom contribute to the polarity of the water molecule?

14 *Without* reference to Figure 5.5, draw the structure of CH_4 showing the spatial arrangement of its four sp^3 orbitals merged with the $1s$ orbitals from four H atoms. Check your sketch against Figure 5.5 and make any necessary changes. Would the geometry of the structure change if the hydrogen atoms of CH_4 were successively replaced by a Cl atom, resulting in turn in CH_3Cl, CH_2Cl_2, $CHCl_3$, and CCl_4?

15 Referring to Figure 5.6, how do the linear structure of $BeCl_2$ and the triangular structure of BCl_3 serve in each case to minimize the repulsion between the shared electrons in the bonding orbitals?

16 Starting with methane, CH_4, show how the successive replacement of a hydrogen atom by a methyl group will result in the homologous series of compounds ethane, propane, butane, etc.

17 How do the two bonds of the carbon-carbon double bond of ethylene differ from each other? When ethylene is hydrogenated, what product is formed? Does this new compound contain two different kinds of bonds?

18 What is meant by saturated bonding? Unsaturated bonding?

19 What is meant by the strength of a bond?

20 Discuss the properties of ionic compounds in terms of mechanical strength, melting point, and crystalline structure.

21 How do most covalently bonded solids (including polar covalent solids) compare with ionic solids in mechanical strength, melting point, and crystalline structure?

22 **a** Why won't *solid* NaCl (or any similar ionic material) conduct an electric current, whereas a water solution of NaCl will conduct? **b** If the amount of dissolved NaCl in solution is increased, would you expect the electrical conductivity to increase, decrease, or remain the same? **c** What is an electrolyte?

23 The most common electrolyte we ingest is table salt, NaCl. Is it present in the body as solid NaCl? Explain.

24 Generally speaking, chemical structures are held together by two broad classes of bonds, namely, the strong *primary* bonds and the weaker *secondary* bonds. **a** Which are the strong primary bonds? **b** Which are the weak secondary bonds?

25 Under what conditions do the very weak van der Waals' forces become effective in bringing individual molecules together and condensing them into a liquid or solid state?

26 Whether a covalently bonded molecule will be a gas, a liquid, or a solid at room temperature depends largely on its size, weight, and polarity. How does increasing molecular weight influence the physical state of the substance? How does increasing polarity of the molecule influence its physical state?

27 **a** What is the dictionary definition of diffusion? **b** Why does a gas diffuse rapidly? **c** If a drop of ink is added gently to a beaker of water, will the ink diffuse rapidly through the water? Explain. What could be done to speed up the diffusion? **d** When you put sugar in coffee what must the sugar do before it can diffuse?

28 The extent to which a gas dissolves in a liquid is influenced by pres-

sure and temperature, and the fact that gases like oxygen, nitrogen, and carbon dioxide dissolve in water is biologically important. How will the solubility of these and other gases be affected by **a** an increase in pressure and **b** by an increase in temperature?

dipole-dipole attraction

29 Dipole-dipole attraction occurs between molecules that have a polar structure. **a** Which of the four structures shown in Figure 5.2 would you expect to exert the strongest dipole-dipole force? **b** Since dipole-dipole attraction acts to bring the molecules together, which of the four hydrogen halide molecules in Figure 5.3 would you expect to have the highest boiling point? (*Hint:* At the boiling point the molecules become separated.) **c** In addition to dipole-dipole interaction between molecules of the same kind, there is also dipole-dipole attraction between dissimilar molecules if they are polar. On this basis, would you expect ammonia, NH_3, to have good solubility in water? See Table 4.1 for the dipole moments of a number of molecules. **d** Would you expect carbon tetrachloride, CCl_4, to be water-soluble?

30 That oil and water don't mix is a common and correct observation. On the other hand, the fact that table salt dissolves readily in water is also a matter of common experience. Using the terms hydrophobic and hydrophilic, discussed in this chapter, explain why oil will not mix with water no matter how hard you stir or agitate the two liquids whereas salt starts to dissolve as soon as it is mixed with water.

WATER AND WATER MIXTURES
"Mix It with Water . . ."

Water is the indispensable liquid of life; it is in water that the chemical substances of living matter meet, mix, and interact. Water also serves as our internal transport system and our outer cooling system. Water is inside the cells and between the cells. On the face of this planet there is more water than anything else: *on* the earth as seas, rivers, and lakes; just *below* the earth as underground water: just *above* the earth as clouds and water vapor. We ourselves are almost two-thirds water. Yet this most common material is quite extraordinary.

About 75 percent of the earth's surface is covered by water to an average depth of 2 mi; water is also present as ground water.

The unique properties of water which set it apart from other liquids derive largely from two features of its composition and structure:

1   The water molecule is strongly polar.
2   Water molecules undergo slight but significant self-ionization into a hydrogen ion, H^+, and hydroxide ion, OH^-.

THE POLAR WATER MOLECULE

As described in the last chapter, the water molecule is built around the oxygen atom, whose four sp^3 orbitals hold the six outer electrons of the atom. The two half-filled orbitals bond

113

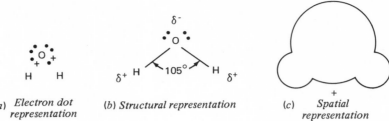

(a) *Electron dot representation* (b) *Structural representation* (c) *Spatial representation*

FIGURE 6.1 The polarity of the water molecule. (*a*) The molecule is polar because (1) the oxygen atom has two unshared pairs of electrons, shown at top, and (2) the electron pair shared between oxygen and each hydrogen atom is displaced toward the oxygen and away from the hydrogen atoms. (*b*) The molecule has a bent structure, with an angle of 105° between the two oxygen-to-hydrogen bonds. The electron unbalance is expressed by the symbolism δ^+ (relatively positive) and δ^- (relatively negative). (*c*) As a three-dimensional structure, the molecule is lopsided geometrically and electrically, the larger top being negative relative to the more positive and smaller lower part.

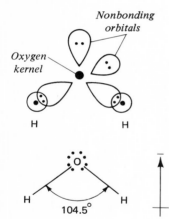

with two hydrogens by overlap with their respective $1s$ orbitals. The remaining four outer electrons are housed in the two remaining sp^3 orbitals, each of which takes two electrons. The negative charges in these two nonbonding orbitals disturb the normal tetrahedral geometry so that the bonding angle is 104.5°. There are two contributions to the polarity of the molecule: (1) the displacement of the shared bonding electrons toward the more electronegative oxygen and away from the hydrogens and (2) the two pairs of electrons in each of the oxygen's two nonbonding sp^3 orbitals. Together, the result is a pile-up of negative charge in the direction of oxygen, leaving the hydrogen ends relatively positive. Of all common liquids water is the *most polar*, with consequences that are far-reaching (Figure 6.1).

HYDROGEN BONDING

Because they are polar, water molecules are always *pulling toward each other*. The relatively negative oxygen end of one molecule and the relatively positive hydrogen end of another molecule attract each other, so that a cohesive force is always acting between water molecules. This electrostatic attraction between adjacent molecules, called **hydrogen bonding,** is illustrated in Figure 6.2.

Since water molecules are always in motion, it can be imagined that they are constantly changing partners. Although the hydrogen bond is weak compared with covalent and ionic bonds, it is the crucial factor that accounts for many of the unusual properties of water.

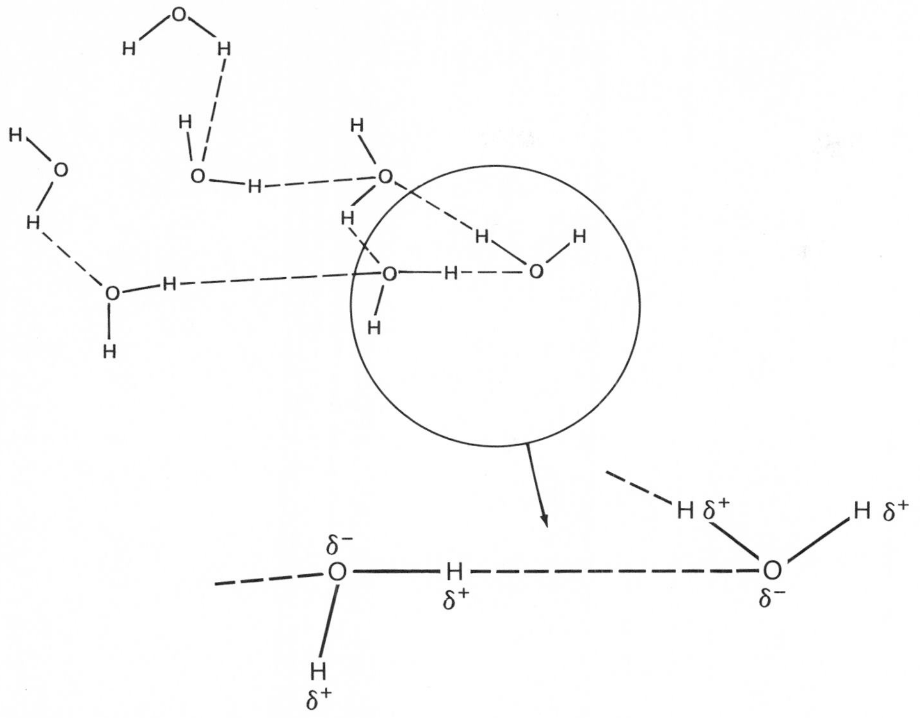

FIGURE 6.2 Hydrogen bonding between water molecules results from the bonding of the relatively positive hydrogen atom of one molecule and the relatively negative oxygen of another molecule. Since the water molecules are always in motion, it can be imagined that they are constantly changing partners.

THE HIGH BOILING POINT OF WATER

Covalent molecules of about the same weight generally share the same physical state, so that small molecules tend to be gases at room temperature, larger ones join together as liquids, and still heavier and bulkier ones form solids. If we look at Table 6.1, relating formula weight to boiling point, water is seen as conspicuously out of line, being the only molecule listed that is a *liquid* at room temperature. All the other molecules are gases, with boiling points far below room temperature. Although water molecules are smaller and lighter than most of the other molecules shown, the hydrogen bonds acting between them cause them to cling together and to cohere as a liquid. It is this built-in togetherness that makes water liquid at room temperature. Were it not for its hydrogen bonds, water would *also* be a gas, and what a difference that would make!

TABLE 6.1

Molecule	Formula Weight	Boiling Point, °C
CH_4	16	−161
H_2O	18	+100
C_2H_6	30	−89
SiH_4	32	−112
H_2S	34	−61

THE HIGH HEAT OF VAPORIZATION OF WATER

When liquid is brought to its boiling point, it is ready to begin boiling. However, actually separating molecule from molecule and converting the liquid into a gas requires the bonds between all the molecules to be completely ruptured. The heat energy in calories needed to accomplish this for 1 g of liquid is called the **heat of vaporization.** It varies for different materials. Of all common liquids water has the highest heat of vaporization because of the energy it takes to undo the hydrogen bonding between its molecules. Evaporation takes place at all temperatures. Water in a dish will evaporate at room temperature, and in so doing will cool the air around it. Although the human body never gets hot enough to boil water, water is constantly vaporizing from it in the form of water vapor from the breath and perspiration from the skin. This draws a great deal of heat from the body and cools it down. The body calls upon this useful mechanism to dispose of excessive heat generated by a fever, heavy physical exertion, or hot weather.

THE HIGH SURFACE TENSION OF WATER

Surface tension can be observed in a drop of water hanging from a faucet and falling as a sphere. Since a sphere is the most economical of all shapes for enclosing space, a spherical drop has less surface area than any other form the water could assume. In other words, water spontaneously reduces its surface area to a minimum because in so doing it also reduces its total *energy*. Any surface is a region of high energy, because inside a liquid each particle has neighbors on all sides to which it is attracted. At the surface the particles form fewer associations because they

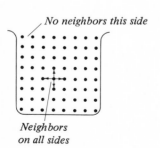

No neighbors this side

Neighbors on all sides

It's lonelier at the surface.

have fewer neighbors, since there are none on the outside. Remember that the formation of bonds, even weak ones due to van der Waals forces, results in a *decrease* in the energy of a system, which is why two hydrogen atoms join to form a stable H_2 molecule. With fewer bonds available to them, the molecules at the surface of a liquid have more energy than molecules in the interior. Since equilibrium and stability are associated with minimum energy, a liquid will reduce its outer surface as much as possible, which is why liquids fall as droplets and soap bubbles are spherical: least surface and least energy.

In its tendency to maintain minimum energy, a liquid will *resist* an increase in its surface area. For example, if a steel needle is laid very carefully on the surface of water, the needle will be supported even though it is almost 8 times denser than water. Being partly immersed in the water, the needle is forcing some increase in the surface area of the water. In turn, the water is resisting any further intrusion that would further increase its surface and its energy. Two glass plates with a water film between them will

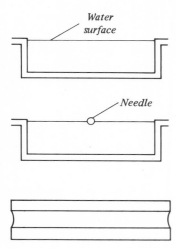

Water surface

Needle

resist separation due to the adhesion between water and glass and the cohesion between water molecules. Resistance to an increase in surface area is called surface tension, and since water molecules cohere strongly because of the hydrogen bonds between them, water has the highest surface tension of common liquids. For example, a drop of water hanging from an eyedropper is larger and heavier than a drop of alcohol hanging from the same eyedropper. The higher surface tension of water supports a heavier load than the alcohol because the surface water molecules cling together tighter, as if they constituted a stronger skin.

Water   *Alcohol*

High surface tension reduces the ability of a liquid to wet a surface; the liquid tends to ball up and shrink away and will not fill corners and crevices. Since water is often the medium in which medicinals, cosmetics, and food materials are dispersed, its high surface tension can be a problem. As the carrier of these substances, water must be able to reach in and *wet* the surfaces to be acted upon. This requires a reduction in its surface tension. Materials that will do that are called **surface-active agent** or **surfactants** because they displace the water molecules at the surface, in effect producing a new and weaker surface. There are many surface-active agents for many purposes, best-known being soaps and detergents.

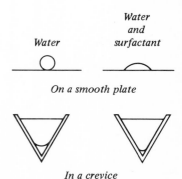

Water   *Water and surfactant*

On a smooth plate

In a crevice

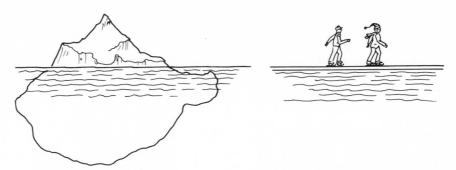

FIGURE 6.3 At 0°C, the density of ice is less than that of water. Therefore, we can see the top of an iceberg or go skating on a lake.

WATER EXPANDS WHEN IT FREEZES

With very few exceptions, liquids contract as they cool and freeze, so that the solid is denser than the liquid from which it crystallizes. Water is an exception. In cooling down from room temperature, water follows the normal pattern of decreasing volume, but at 4°C it reverses itself, expanding as it goes down to its freezing point and solidifies. At 0°C the density of ice is 0.92 g/ml, and ice will float in water, whose density at this temperature is just about 1 g/ml. For this reason, we can speak of the tip of the iceberg and go skating on the lake (Figure 6.3). If the solid ice were denser than the water around it, the iceberg would never be seen and the ice in the lake would sink to the bottom as soon as it was formed. What would happen to the fish in lakes and streams if water froze from the bottom *up*? What of the effect on climate as ice accumulated on the bottom of lakes and seas in colder regions?

The expansion of water as it freezes in the cracks and crevices of rocks exerts tremendous pressures, splitting and fissuring them, contributing to rock weathering and soil formation. The expansion of water as it turns to ice is also related to its hydrogen bonds. At ordinary temperatures, hydrogen bonding between H_2O molecules is effective in keeping them together as a liquid but does not keep them in fixed positions, so that they turn, roll, and push each other around. As the temperature falls to 4°C and below, the molecules slow down considerably and begin to take positions in line with their hydrogen bonds. Whereas previously these bonds were constantly being made and broken as the molecules moved about, so that at any moment the number of effective bonds was less than the number of possible bonds, at low enough temperatures, all hydrogen bonds become stationary and effective. At 0°C, ice crystallizes and the regimentation is prac-

In the winter, freezing water also cracks piping and unprotected automobile engines.

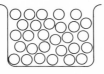

Liquid H_2O

Solid H_2O
more open, less dense, structure

tically complete, with each molecule in a fixed position relative to its neighbors. The oxygen atom of each molecule is now surrounded by four other oxygen atoms in a tetrahedral arrangement. The solid structure that results is more open internally than the liquid from which it crystallized and, being less dense than the liquid, will float.

WATER, THE POLAR SOLVENT

Water is by far our most important solvent, because it is omnipresent and because it dissolves more things better than any other liquid. Here again, the polarity of the water molecule accounts for its action as a solvent. When a material dissolves in water, its component molecules or ions are, in effect, expressing a preference for the company of water molecules over each other's company. Water will therefore dissolve substances which respond to the polarity of the H_2O molecules, that is, substances which are themselves polar, such as ionic compounds. Sodium chloride is a hard ionic solid, but in water the Na^+ and Cl^- ions let go of each other to become free ions around which the polar water molecules cluster. Figure 6.4 is a schematic representation of how the friendly water molecules come to surround each ion.

Note that in Figure 6.5 water is the **solvent** and the sodium chloride is the **solute;** in general, a solute-solvent system is a

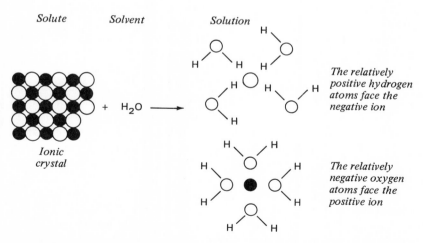

Solute *Solvent* *Solution*

Ionic crystal + H_2O →

The relatively positive hydrogen atoms face the negative ion

The relatively negative oxygen atoms face the positive ion

FIGURE 6.4 Dissolution of an ionic solid in water. The clustering of water molecules around dissolved particles is known as *solvation*, and serves to keep the particles in solution. If the particles are ions, then the water molecules arrange themselves around each ion in accordance with its charge. (● = + ion; ○ = − ion.)

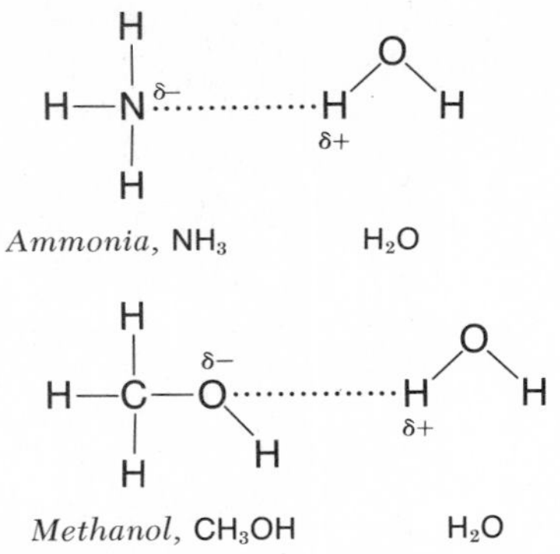

Ammonia, NH$_3$ H$_2$O

Methanol, CH$_3$OH   H$_2$O

FIGURE 6.5 NH$_3$ and CH$_3$OH in water solution.

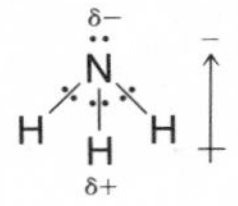

NH$_3$

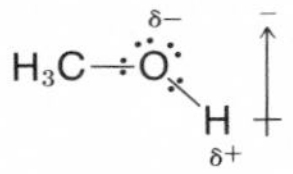

CH$_3$OH

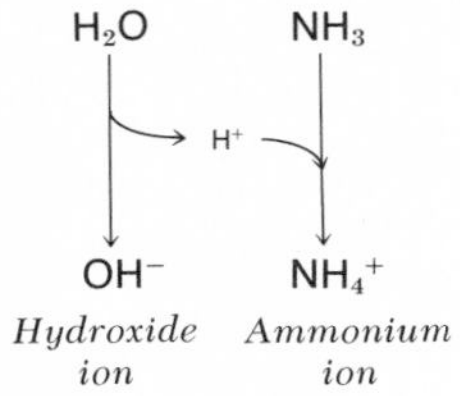

Hydroxide   Ammonium
ion   ion

solution, in this instance an NaCl–H$_2$O solution. Although many ionic compounds have high or moderate solubility in water, their degree of solubilities varies widely and some ionic materials are conspicuously insoluble.

Water also dissolves nonionic substances if they have polar characteristics. One such example is ammonia gas, NH$_3$; its structure was discussed in Chapter 5, and its polarity is shown here. Another example is the organic molecule methyl alcohol, CH$_3$OH, which has a polar —O—H group. Both these polar materials are soluble in water because they form hydrogen bonds with the water molecules (Figure 6.5).

In both cases, the particles in solution are molecules, although they behave differently. We learned in Chapter 5 that NH$_3$ will take protons from water molecules. In losing an H$^+$ ion, H$_2$O becomes an OH$^-$ ion, and in gaining it, NH$_3$ becomes an NH$_4^+$ ion. An NH$_3$–H$_2$O solution therefore contains charged particles which will conduct an electric current. On the other hand, dissolved methyl alcohol has practically no tendency to ionize, and the solution is neutral. Organic compounds are soluble in water to the extent that they have polar groups in their structure, and in solution they exist as undissociated molecules. Methyl alcohol can dissolve in water in all proportions, that is, in any ratio from 0 to 100 percent. Liquids that dissolve in each other in this way are called **miscible.**

If an organic material has no polar groups, or if only a small part of its total structure is polar, it will not be soluble in water. For example, such substances as gasoline, oils, and carbon tetrachloride are nonpolar and water insoluble. Even if a mixture of nonpolar oil and polar water is stirred or beaten vigorously, so that oil droplets form, the water will reject them and the oil will collect as a uniform layer over the heavier water. However, nonpolar materials will dissolve in each other. This is the basis for

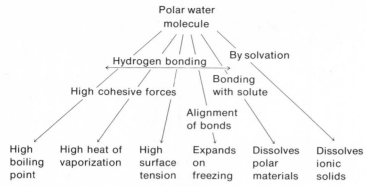

FIGURE 6.6 The polar water molecule.

dry cleaning, in which volatile organic solvents remove organic stains, such as grease and oil. It is also the basis for much industrial chemistry.

In a schematic way the relation between the polarity of the water molecule and some of the unusual properties of water can be recapitulated as shown in Figure 6.6.

Two other distinctive properties of water were discussed in Chapter 2: its high density and high specific heat. The high specific heat of water partly explains why the watched pot never boils. But this also works to our advantage. Each of us is about two-thirds water, and the high specific heat of water acts to minimize temperature fluctuations, and so helps keep body temperature more or less constant.

THE SELF-IONIZATION OF WATER

The polar covalent bonds that hold the two hydrogens and the oxygen of the water molecule together are strong bonds, and considerable energy is needed to separate these atoms from each other. However, there is a very small but significant tendency for a hydrogen *ion* to be jarred loose from one molecule and be picked up by another. Leaving its electron behind it, the positive hydrogen ion jumps aboard the attractive electron cloud of the neighboring molecule. This proton transfer is shown in the margin and also expressed in the equation

$$H_2O \ + \ H_2O \ \rightleftharpoons \ H_3O^+ \ + \ OH^- \qquad (6.1)$$

Water molecule I   Water molecule II   Hydronium ion   Hydroxide ion

An **equation** connects the two sides of a reaction, one side showing the reacting substances and the other side the substances produced. In general, equations use the language of chemical symbols. The two arrows connecting the two sides of the equation are signposts carrying two messages.

If only one arrow is shown in an equation, it means that all the reactants have become products.

Message 1 The reaction has proceeded until there is no further change; that is, the reaction has come to equilibrium. At equilibrium, a certain proportion of the reacting materials, or **reactants,** has been converted into a certain amount of **products.** Since equilibrium has been reached, the ratio of products to reactants is fixed and is characteristic of that reaction.

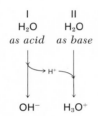

Message 2 The shorter arrow points to the component that is present in smaller amount in the equilibrium mixture (in this equation the products); the longer arrow points to the component present in larger amount in the equilibrium mixture (in this equation the reactants).

Equation (6.1) tells us that each proton transfer between two water molecules results in the formation of one hydronium ion, H_3O^+, and one hydroxide ion, OH^-, which is the mechanism for the self-ionization of water. The extent of this self-ionization is extremely small, with only one proton transfer in about 555 million water molecules. With such a small ratio of ions present, the electrical conductivity of pure water is negligible, so that water is itself not an electrolyte. Although small, the self-ionization of water is of great importance because it explains how water acts as a host to two general classes of materials that are chemically *opposite* in nature, **acids** and **bases.** An acid is any chemical entity that *donates* a proton, and a base is any chemical entity that *accepts* a proton. In the proton transfer between two water molecules as expressed in Eq. (6.1), the two reacting water molecules are alike, but they behave quite differently. Molecule I donates a proton and so becomes a hydroxide ion, OH^-, whereas molecule II accepts this proton and becomes a hydronium ion, H_3O^+. Since water molecules can function both as proton donors and proton acceptors, water is both an acid and a base. The special feature of pure water is that it is an acid and a base to the *same* extent, since each proton transfer involves one proton donor and one proton acceptor. The net result is that pure water is neither acid nor basic; it is *neutral.* Although the dual acid-base nature of water is balanced when it is pure, it can be tipped in either direction by adding acidic or basic materials.

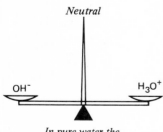

In pure water the number of OH^- ions and H_3O^+ ions are equal.

ACIDIC AND BASIC SOLUTIONS IN WATER

When an acidic material (a proton donor) is added to pure water, the water responds by accepting the protons, thus acting as a base. On the other hand, if the added material is a base, or proton acceptor, the water responds by giving up protons, acting as an acid. A solution of acid in water therefore contains an excess of H_3O^+ ions, which is the condition for an acidic solution; the solution of a base in water contains an excess of OH^- ions, which is the condition for a basic solution. If we represent a *general* acid as HA and a *general* base as B, their respective interactions with water are

$$\overset{\frown{H^+}}{HA} + H_2O = H_3O^+ + A^- \quad \text{or} \quad \overset{\frown{H^+}}{HCl} + H_2O \longrightarrow H_3O^+ + Cl^-$$

$$\overset{\frown{H^+}}{B} + H_2O = OH^- + BH^+ \quad \text{or} \quad \overset{\frown{H^+}}{NH_3} + H_2O \rightleftharpoons OH^- + NH_4^+$$

Next to the generalized reactions are specific reactions, where hydrochloric acid, HCl, is the acidic material and NH_3 is the basic material. Note that the $=$ sign is used for the general reactions where the extent to which reactants convert to product are not specified. Several conclusions can now be drawn regarding solutions of acidic and basic materials in water.

For a material to be an acid, or proton donor, an ionizable H^+ ion must be present in its structure for donation to water. (CH_4 is not an acid because the hydrogens are strongly held and not ionizable.) When water accepts these protons, the number of hydronium ions in the solution accordingly increases over that of the hydroxide ions.

For a material to be a base, or proton acceptor, its composition and structure must permit it to accept a proton from water. As discussed in Chapter 5, the NH_3 molecule is a base because it has a pair of unshared electrons in its fourth sp^3 orbital that is highly attractive to a proton. When the water molecules surrender protons to the receptive base, the number of hydroxide ions in the solution increases over that of the hydronium ion. It should be noted that a base is also called an **alkali;** the two terms are equivalent.

With water as the solvent, an acid solution is characterized by the presence of an excess of H_3O^+ ions over OH^- ions, and an alkaline solution is characterized by an excess of hydroxide ions over hydronium ions. Preparatory to the discussion of acids and bases in Chapter 7, two additional points must be made.

H_2O

$\downarrow H^+$ *Proton donor, or acid*

H_3O^+

H_2O

Proton acceptor, or base H^+

OH^-

Some proton donors

$HCl \longrightarrow H^+ + Cl^-$

$H_2CO_3 \rightleftharpoons H^+ + HCO_3^-$

$NH_4^+ \rightleftharpoons H^+ + NH_3$

Some proton acceptors

$CO_3^{2-} + H^+ \rightleftharpoons HCO_3^-$

$PO_4^{3-} + H^+ \rightleftharpoons HPO_4^{2-}$

H_3O^+ VERSUS H^+

The notation H^+ (aq) *is also used to remind us that the hydrogen ion is in water medium and is hydrated, where* (aq) *stands for aqueous.*

The difference between the hydronium ion, H_3O^+, and the hydrogen ion, H^+, is a water molecule, and since water is the common medium, the difference is often ignored and the shorter term is used. Although it must be understood that free H^+ ions do not exist as such in a water medium, except when in motion between different chemical entities, the shorter, simpler term and symbol will be used here as well. When the hydrogen ion, H^+, is referred to, keep in mind that it exists as H_3O^+.

THE HYDROXIDE ION

$$NaOH \longrightarrow Na^+ + \mathbf{OH^-}$$
$$Ca(OH)_2 \longrightarrow Ca^{2+} + 2\mathbf{OH}$$
$$KOH \longrightarrow K^+ + \mathbf{OH}$$

The addition of any basic material to water results in an excess of OH^- over H^+, and this characterizes every basic solution in which water is the solvent. When dissolved in water, the class of ionic compounds called hydroxides increases the number of OH^- ions, not by interacting with water as bases do but because they contain OH^- ions in their structure. As shown here, these hydroxides dissociate in water, releasing OH^- ions directly to the solution and increasing its basicity. Since these hydroxides do what all bases do in water, that is, increase the number of hydroxide ions, hydroxides are considered as basic materials. In fact, sodium hydroxide, NaOH, is the base most commonly used in chemical laboratories.

MIXING IT WITH WATER

If a handful of soil is added to a liter or so of water and mixed thoroughly, the resulting mixture will contain particles over a wide range of sizes, from the finest to the coarsest. On the basis of their behavior, these particles can be placed in three different categories.

Ions found in soil:
K^+, Ca^{2+}, Na^+, Cl^-, SO_4^{2-}, Mg^{2+}, NO_3^-, Fe^{3+}, HPO_4^{2-}

Particles in true solution These are the smallest and consist of ions or molecules roughly 0.5 to 10 Å in diameter. Examples are the ionic minerals in a soil which may dissolve to form a true solution.

Colloidal particles in soil are clay and humus.

Particles in colloidal dispersion These are intermediate in size and may be present either as aggregates of many molecules clumped together, or as individual but huge macromolecules. The size range is wide, from about 10 to 1000 Å across. The dispersion of colloidal solids in water is called a sol. The soil particles having a diameter between about 10 and 1000 Å will therefore be present as a colloidal sol.

-oid-

Suspended particles in soil: sand and silt.

Suspended particles This coarser material, generally over 1000 Å in diameter, can be separated out both visually and physically. Sand in water is an example of suspended particles.

The size of a particle has a considerable effect on its behavior. Some of the criteria applied to these three size groups are as follows:

1 Will gravity cause the particles to settle out?
2 Will the particles be collected on a filter or pass through it?
3 Are the particles visible?
4 What is the appearance of the water mixture?
5 Will the particles interfere with or reflect a beam of light passing through the water mixture?

Soil and water as a muddy mixture

Keeping these questions in mind, let us look at the soil-water mixture. It looks like muddy water and obviously contains insoluble particles because they can be seen. In time, these coarse particles will settle out, the larger ones first and the finer ones last, so that eventually two **phases** will be seen, two kinds of matter that can be distinguished from each other. A faster separation of the two phases can be made by filtration, as shown in Figure 6.7.

Soil and water as two separated phases

The coarse, insoluble particles which were in suspension in the mixture are retained on the filter paper. This portion is called the **residue.** The liquid portion, which should be clear, passes through the filter, and is called the **filtrate.** We know that the

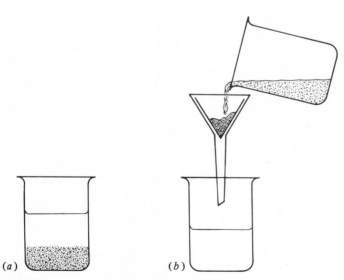

FIGURE 6.7 Filtration of a soil-water mixture. (*a*) With time the coarse particles settle out and the liquid above them can be separated from the undissolved solids by pouring off, or *decanting*, it. More often the separation is made by filtration. (*b*) In filtering the soil-water mixture, the coarse particles are retained on the filter paper; this is called the *residue*. Particles in colloidal dispersion and in solution pass through the filter and are in the filtrate in the beaker.

mixture contained suspended particles, but what of the filtrate? Does it contain particles in true solution? Particles in colloidal dispersion? Both? Appearance offers no clue, since in neither size groups are the particles visible to the naked eye. Therefore they disappear from sight in water. In both cases, the clear liquid may or may not be colored. A clear liquid might therefore be simply water or a true solution or a colloidal dispersion. However, if a beam of light is now passed through the soil-water filtrate, the light will show up as a continuous, cloudy path through the liquid. This **Tyndall effect** confirms the presence of colloidal particles in the liquid. Although particles in the colloidal size range are too small to be seen by the eye, or even by ordinary microscopes, they are large enough to scatter light that hits them, that is, individually to reflect the light. The Tyndall effect is not uncommon. It can be seen when narrow rays of sunlight enter a room, say through venetian blinds. Otherwise invisible dust particles signal their presence by tiny flashes, each flash a reflection of the bright sunlight from a particle surface. In a colloidal dispersion the many tiny reflections add up to the cloudy streak that

FIGURE 6.8 The Tyndall effect. The beam of light passes unseen through the solution but can be seen in the colloidal dispersion.

signals the presence of colloidal particles. The molecules and the ions of a pure liquid or a true solution are too small to reflect light, which passes through the liquid unimpeded and unseen (Figure 6.8).

Now that the Tyndall effect has confirmed that the soil-water mixture contains colloidal particles, the next question is: Does it also contain a true solution? Particles in solution may be ions or molecules. If dissolved ions are present, the electrical conductivity of the solution will be increased accordingly, and testing the soil filtrate for conductivity will clearly indicate the presence of dissolved ions. Further specific tests would also show the presence of dissolved organic molecules and inorganic ions. Soil is a natural and a complex material, with particles in each of the three size categories. In living organisms true solutions and colloidal dispersions are the environment in which chemical activity takes place.

TRUE SOLUTIONS

When we speak of a solute dissolving in a solvent, what is meant is a **true solution,** with the dissolved solute present (but invisible) as ions or relatively small molecules. A true solution is a single phase, a body of matter that looks the same throughout, although it may contain two or more components. In principle, gases, liquids, and solids can each dissolve in other gases, liquids, and solids, so that nine possible solution combinations, or systems, are possible. However, since water is *the* liquid of life, we shall concern ourselves mostly with water solutions, touching upon other systems and solvents as required.

THE BEHAVIOR OF WATER SOLUTIONS

When any solute dissolves in water, the mixture may look the same, but it will not behave the same in a number of respects.

Electrical conductivity If the solute is ionic, the component ions will dissociate and the solution will conduct a current. If the solute is nonionic, such as sugar, there will be no increase in electrical conductivity, but there will still be other changes.

Density In general, the density of a solution will differ from that of pure water. If the solute is denser than water, the specific gravity of solution will be greater than 1, the value for water. For this reason, seawater is denser than fresh water, and it is easier to float in the ocean than a lake. On the other hand, mixing ethyl

$$Density = \frac{mass, g}{volume, ml}$$

Density of water = 1 for most purposes.

Specific gravity =
$$\frac{mass\ of\ some\ volume\ of\ material}{mass\ of\ same\ volume\ of\ water}$$

alcohol (specific gravity 0.79) with water makes a solution less dense than water.

Melting and boiling points If a solute such as NaCl is added to water, the boiling point is increased. NaCl is a **nonvolatile** solute because it has a very high boiling point compared with that of water. Other nonvolatile solutes will have the same effect. Adding salt to water in cooking makes the water boil at a temperature above 100°C, and adding salt to an icy sidewalk in the winter melts the ice, because with the added salt water begins to freeze below 0°C.

Physiological effects Water may be essential to life, but watch out for what is added to it. Just a pinch of sodium cyanide added to a glass of water makes a lethal and quick-acting poison. We cannot drink seawater because it contains about twice as much NaCl as can be eliminated through the kidneys in the urine.

Osmotic pressure All liquids have **vapor pressure,** a tendency for the molecules to escape the liquid state and fly off as individual gas molecules. When pure water is just melting from ice, its vapor pressure is least; when it is at 100°C and ready to boil, its vapor pressure is at a maximum, just equal to atmospheric pressure. If a nonvolatile solid is dissolved in pure water at some temperature, the vapor pressure of the solution will be *less* than that of the water itself, as though the dissolved particles were holding the water molecules back in their effort to escape. Imagine a volume of pure water beside a volume of water solution, the two separated by a partition, or membrane, that allows the passage of water molecules but not of solute particles. Because water itself has a higher vapor pressure than a *solution* of water, its molecules will push against the membrane harder than those in the solution, with the result that there will be a net flow of H₂O molecules *from* the water *to* the solution. This phenomenon is called **osmosis,** and the intensity of the push that is moving water molecules across the semipermeable membrane to the water *solution* is measured by the *counterpush* that will just *stop* their movement. This counterpush is the **osmotic pressure** of that water solution (Figure 6.9), and its magnitude will vary with the solution concentration. The greater the solution concentration, the greater is its osmotic pressure and the more readily will water molecules move across a semipermeable membrane *to* that solution. If a semipermeable membrane separates two solutions of *unequal* concentration, the solution of higher concentration will have the higher osmotic pressure, and there will be a net flow of water *from* the less concentrated *to* the more concentrated

Relative vapor pressure

Water   *Water solution*

Net flow of water

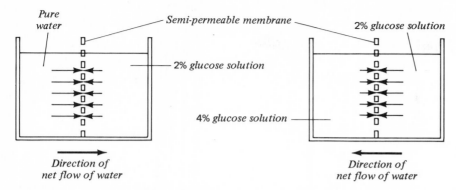

Pure water — Semi-permeable membrane — 2% glucose solution

2% glucose solution

4% glucose solution

Direction of net flow of water   *Direction of net flow of water*

FIGURE 6.9 Flow of water across a membrane by osmotic pressure. The sizes of the arrows represent the relative vapor pressures of the liquids shown. The lower the concentration the higher the vapor pressure, so that the net transfer of water across the membrane is from the less to the more concentrated solution.

solution. The flow of H_2O molecules is from the *more* watery *to* the *less* watery solution.

The transfer of water molecules from a less to a more concentrated solution is one more example of a spontaneous effort to move toward equilibrium. If the 2% glucose solution in Figure 6.9 continues to lose water to the 4% solution, both will become 3% solutions. The two solutions will then be **isotonic** with respect to each other, and there will be no net flow of water either way.

ISOTONIC, HYPERTONIC, AND HYPOTONIC

For the normal functioning of cells in living organisms and of the organism as a whole, a proper balance must be maintained between the **intracellular fluid** within the cell and the **extracellular fluid** outside it. This balance is achieved largely through osmosis, which is set in motion by the differences in the concentration of inorganic ions in these fluids, especially Na^+ and K^+ ions. Drinking seawater causes a buildup of Na^+ ions, which by accumulating in the extracellular fluid, makes it **hypertonic,** or too strongly concentrated relative to the fluid inside the cells. Osmotic pressure would then move water out of the cell, drying it up. The need to replenish this lost water means that the more seawater you drink, the thirstier you get.

An osmotic imbalance can also be dangerous in the bloodstream. Human red blood cells are isotonic with respect to a 0.9% solution of NaCl, so that in a solution of this concentration they neither lose nor gain water. If the solution becomes hypertonic, with more than 0.9% NaCl, osmotic pressure will move water through the membrane of the cell into the more concentrated

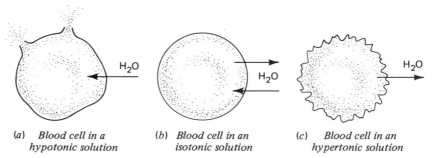

(a) *Blood cell in a hypotonic solution*   (b) *Blood cell in an isotonic solution*   (c) *Blood cell in an hypertonic solution*

FIGURE 6.10 Red blood cells under hypotonic, isotonic, and hypertonic conditions. (*a*) The concentration of the salt is less in the solution than in the cell, so water enters the cell, which swells until it bursts. This is called *hemolysis*. (*b*) Water passes in and out of the cell at equal rates and the cell remains normal. (*c*) The concentration of the salt is greater in the solution than in the cell, so water leaves the cell, which shrinks. This is called *crenation.*

solution, and the cell will shrivel up. Similarly, if the solution around the cell is **hypotonic,** containing less than 0.9% NaCl, osmosis will push water *into* the cell, which will burst (Figure 6.10).

When medication and solutions are injected directly into the bloodstream of a patient, it is important that they be isotonic solutions, or else the crenation or hemolysis shown in Figure 6.10 could occur.

PLANTS, PICKLES, AND PRESERVES

The push behind the osmotic transfer of water across a membrane can be quite high, and this serves plant life right at the root tips. A plant gets the energy it needs from the sun through its leaves, and it gets water and inorganic nutrients from the soil through its root system. The damp soil provides a dilute water solution, and the root membranes are permeable to water and small ions; osmotic pressure does the rest, pushing water and ions into the more concentrated fluids within. Osmotic pressure is persistent and stubborn, and roots are also persistent and stubborn.

Osmosis is also a potent mechanism in the treatment and preservation of foods. Pickling is the most obvious, salt again acting as the osmotic agent. Since the pickling brine is strongly hypertonic, water leaves whatever food is being pickled, which therefore shrinks. Microorganisms present are also in a pickle because they, too, give up water, and thus become inactivated. For the same reason, a very high sugar content, such as fruit preserves, jellies, and honey are prevented from spoiling. Although

Microorganism

H₂O

Pickling brine, a concentrated salt solution

a mold may form *on* them, generally microorganisms will not grow in them, giving up water to what is, in effect, a highly concentrated solution.

SOLUTIONS: UNSATURATED, SATURATED, AND SUPERSATURATED

Suppose you add so much salt to water that an undissolved excess settles to the bottom, even after thorough stirring. There are now two types of water mixtures present: a true solution, containing Na^+ and Cl^- ions, and a suspension of solid salt particles. The two are in equilibrium; that is, neither will increase or decrease with time. It is a dynamic equilibrium because at any moment a few solid particles may be dissolving, to which a few dissolved ions respond by uniting to form solid particles. The rate at which solid particles go into solution is equal to the rate at which dissolved ions precipitate from solution. This means that there is no net change and the amount of salt in solution remains constant. The true solution can be separated from the suspended solid salt either by decantation or by filtration. Decantation is faster, but filtration is more complete and gives better separation.

Saturated NaCl *solution*

Undissolved NaCl

Residue on filter paper is undissolved NaCl

Filtrate of saturated NaCl *solution*

The clear filtrate obtained is a saturated solution of NaCl, which contains 37 g NaCl and 100 g (or 100 ml) of water. This is the quantitative measure of the solubility of salt in water. The water cannot dissolve more, and even the smallest additional amount would appear as a solid precipitate. Any salt solution containing less than 37 g of salt per 100 ml of water is therefore **unsaturated,** because more solute can be added before the point of saturation is reached.

Usually, but not always, the solubility of a solute in water increases with increasing temperature, so that a solution saturated at a higher temperature will have more solute in it than a solution saturated at a lower temperature. If a hot, saturated solution is cooled down to room temperature, the lower solubility of the solute at the lower temperature usually manifests itself by the crystallization of the excess. However, with some solutes, if the cooling is slow and without agitation and the solution is free of particles, the excess solute sometimes does not crystallize out. Such a solution is called **supersaturated.** Since it is not at equilibrium, the solution is **metastable,** that is, given a little encouragement in the form of a jolt or a scratch on the inner surface of the container or a particle of the solute itself, it will change to the more stable equilibrium condition, i.e., crystals will appear. Both inorganic and organic substances can form supersaturated water solutions.

As a practical matter, sweet foods such as jellies and chocolates may undergo crystallization of the sucrose, or table sugar, they contain; this is apparently due to the formation of a supersaturated solution during processing, which slowly reverts to the more stable condition.

HARD WATER

When rainwater starts down from the clouds, it is as pure as it will ever be. In its path to our faucet it is exposed to gases, soil, rock, and other water. The more of certain minerals that water picks up and keeps in solution, the **harder** it becomes and the more problems it presents in use. The most common and serious offenders are the Ca^{2+}, Mg^{2+}, and Fe^{3+} ions, accompanied by bicarbonate, HCO_3^-, sulfate, SO_4^{2-}, and chloride, Cl^-, ions. The positive ions, such as Ca^{2+} and Mg^{2+}, react with dissolved soap in water to form the unpleasant slippery scum known as ring around the bathtub. Dissolved soap in water yields a long carbon-to-carbon chain with a negative charge at one end; its reaction with the Ca^{2+} ion can be represented as shown in Figure 6.11.

The result is that the negative ions of the soap are used up in combining with the positive ions of the hard water before they can go to work as a cleaning agent. Not only is this a waste of soap (and the harder the water the more the waste), but it is also a nuisance in washing clothes because the unpleasant precipitate sticks to them and acts like more dirt. To make hard water soft, the positive ions must be removed. There are three general ways of doing this.

Boiling This is effective only if the negative ions are bicarbonate, and the reactions for Ca^{2+} and Mg^{2+} are

$$Ca(HCO_3)_2 \xrightarrow{boil} CaCO_3 \downarrow + CO_2 \uparrow + H_2O$$

$$Mg(HCO_3)_2 \xrightarrow{boil} MgCO_3 \downarrow + CO_2 \uparrow + H_2O$$

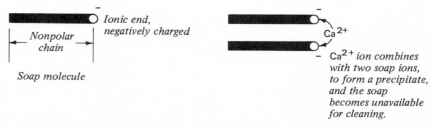

FIGURE 6.11 The precipitation of soap by Ca^{2+} ion.

The $CaCO_3$ and $MgCO_3$ formed are insoluble, so that the Ca^{2+} and Mg^{2+} ions are removed from action, but for the same reason these compounds also accumulate in pipes and boilers. This is a problem in itself because the deposits clog up piping and tubing. Hard water that can be softened by boiling is called **temporary hard water.**

Treatment with alkalis If most of the negative ions are ions other than bicarbonate ion, HCO_3^-, boiling will not help and the hardness is called **permanent.** Adding alkaline materials serves to form insoluble precipitates with Ca^{2+} and Mg^{2+}. Many alkalis are used, especially washing soda, Na_2CO_3, and borax, in which borate ion, $B_4O_7^{2-}$, gives an alkaline reaction in water. If the dissolved salt is $CaSO_4$, the reaction with sodium carbonate again binds the Ca^{2+} ion in insoluble $CaCO_3$:

$$CaSO_4 + Na_2CO_3 \longrightarrow CaCO_3 \downarrow + Na_2SO_4$$

Ion exchange The offending Ca^{2+}, Mg^{2+}, and Fe^{3+} ions are exchanged for other ions that are not troublesome. The exchange takes place in a **zeolite,** a silicate mineral with a sievelike structure. For example, in passing through the zeolite, the hard water comes out without most of its hard ions. Today the natural zeolites have been replaced by synthetic equivalents.

COLLOIDAL DISPERSIONS

The essential feature of colloidal materials is their intermediate size range between the ions and small molecules of true solutions and the coarse particles of suspensions. A colloidal material dispersed in a relatively large volume of water is clear, but can be detected as a colloidal system by the Tyndall effect. However, colloids are by no means restricted to water dispersions, and there are also differences between water dispersions themselves, depending on the total amount of water and that part of it that is bound to the colloidal material, as with organic materials. Colloidal systems can take on many appearances from clear to hazy liquids, nontransparent liquids, fogs, foams, pastes, jellies, and solids, all composed of a colloidal material dispersed in some medium. Any substance that can be brought to, and kept within, the colloidal size range can exist in the **colloidal state.**

THE COLLOIDAL STATE

Surface area Experience shows that granulated sugar dissolves in water faster than an equal amount of cube sugar. Solution

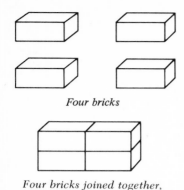

Four bricks

Four bricks joined together, which has more surface area?

takes place at the surface, and since the granules have a total surface area much larger than the cube, solution is faster. Most chemical reactions of solids take place at the surface and go faster if more surface is available to be acted upon. Wheat is a dangerous material to handle or store in large quantities because the cloud of finely dispersed grains presents a huge surface area, and all that is needed to set off an explosively fast combustion is a spark. Colloidal particles are characterized by an especially high ratio of surface area to mass, so that even a small amount of colloid may have a large working surface. This provides increased opportunities for chemical action, for different chemicals to meet and interact. The colloidal nature of protoplasm, the elementary living material, surely contributes to the variety and versatility of its chemical and physical behavior.

Adsorption The large surface area of colloids accounts for their ability to take up and retain on their surfaces foreign particles, gaseous, liquid, or solid. This **adsorption** results because a surface is a region of high energy. A very common adsorbent is finely divided charcoal; its ability to adsorb gases on its surface made it useful for the first gas masks. Charcoal is used today in purifying air and liquids and picking up objectionable odors, particles, coloring matter, and impurities generally. When charcoal adsorbs a gas on its surface, some heat is given off, called the **heat of adsorption,** indicating that surface energy has been reduced. In effect, the surface energy has been put to use. Whereas *ad*sorption releases energy, *de*sorption needs energy, so that the adsorbed matter on charcoal can be removed by heating.

Colloids as catalysts A spontaneous reaction between two substances is often too slow to be useful. In such a case colloidal materials may act as **catalysts,** which speed up slow reactions without appearing as reactants or products. The reacting substances are brought together by adsorption on the colloidal surface and in a sense borrow energy from that surface. Rapid interaction between the reactants is made possible, and when the product or products are formed, they leave the colloid. Since the reaction was spontaneous, energy is released, repaying the catalyst for the energy borrowed from it. The colloid is the same as before the reaction, but the reactants have become products. In industry, colloidally dispersed metals such as platinum and nickel are used as catalysts.

Water-loving and water-hating colloids Whether a colloidal particle is hydrophilic (water-loving) or hydrophobic (water-hating) depends upon whether the colloid has polar features or not.

Being polar, water will attract, and be attracted to, structures that are polar and will reject structures that are not. Generally, colloidally dispersed metals and metal oxides are hydrophobic, whereas organic substances, such as proteins and starches, are hydrophilic. However, organic compounds that lack polarity, such as fats and oils, are hydrophobic and have no affinity for water.

Surface electric charges A special feature of particles in the colloidal state is the presence of electric charges on the surface of the particles. The charge may be positive or negative, but always has the same sign for a given material. For example, the colloidal particles that give the Tyndall effect in the soil-water mixture are likely to include clay particles, whose surface charge is negative. This is also the surface charge of a colloidal dispersion of gold. On the other hand, when a solution of ferric chloride, $FeCl_3$, is slowly added to boiling water, the colloidal particles formed are chemically $Fe(OH)_3$ and have positive surface charges.

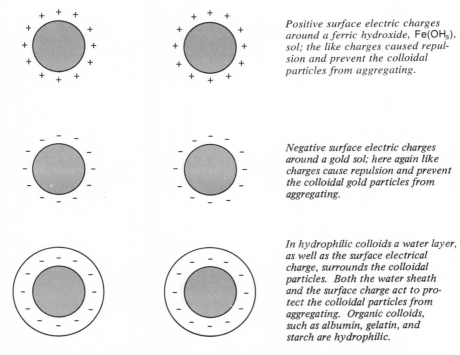

Positive surface electric charges around a ferric hydroxide, $Fe(OH)_3$, sol; the like charges caused repulsion and prevent the colloidal particles from aggregating.

Negative surface electric charges around a gold sol; here again like charges cause repulsion and prevent the colloidal gold particles from aggregating.

In hydrophilic colloids a water layer, as well as the surface electrical charge, surrounds the colloidal particles. Both the water sheath and the surface charge act to protect the colloidal particles from aggregating. Organic colloids, such as albumin, gelatin, and starch are hydrophilic.

FIGURE 6.12 Surface electric charges on colloidal particles. As a whole, each colloidal system is neutral, so that charges equal in number and opposite in sign to those of the surface charges must be distributed throughout the colloidal dispersion.

Since like charges repel, the particles of these dispersions will not aggregate to larger particles and the colloids will be stable (Figure 6.12). A colloidal dispersion is stable so long as its particles remain within the colloidal size range of about 10 to 1000 Å across. Should the particles aggregate and exceed that size, the colloidal state will convert to a suspension of coarse particles. The maintenance of surface charge is therefore necessary for the stability of hydrophobic colloids. Precipitation of the colloid will occur if its surface charge is neutralized by adding ions of opposite charge. Precipitation can also happen if two colloidal dispersions with opposite surface charges are mixed together. In either case, the protective surface charges are neutralized, and the particles aggregate and settle out as a flocculant precipitate. (A flocculant precipitate is a loose, wooly, gelatinous mass. Curdled milk is an example.)

Water as a protective coating Although protected by surface electric charges, hydrophilic colloids also enjoy the further protection of an adsorbed layer of water. The thickness of the water layer and its degree of adhesion vary with the specific colloid and with conditions. For proteins, the most important example, the water layer can be quite thick. This **bound water** can be as much as 4 g of water per 10 g of protein, and to a large extent the stability of hydrophilic colloids depends on this sheath of water. Therefore to precipitate a hydrophilic colloid one must not only neutralize its surface charges but also remove the surrounding layer of adsorbed water.

Colloidal dispersions of solids in water that depend for their stability on the presence of electric surface charges, such as gold and iron colloids, are called **suspensoids.** Hydrophilic colloids, such as those formed by gelatin or starch, whose stability is further supported by interaction with the water medium in the form of bound water are called **emulsoids.** If a hydrophilic colloid, such as gelatin, is added to a dispersion of colloidal gold, which is hydrophobic, the gelatin will surround the gold particles and protect them from being precipitated by ions that would neutralize their surface charge. The gelatin is acting as a **protective colloid** by virtue of its sheath of bound water, which is now also surrounding the gold particles. Of course, the protection has its limits, but the ability of hydrophilic colloids to stabilize hydrophobic colloids is often important. Hydrophobic substances in the blood are kept from precipitating out by hydrophilic proteins.

Stabilization by an emulsifying agent A mixture of mutually insoluble liquids is called an **emulsion.** The most important are an oil-in-water and a water-in-oil combination. Being polar and

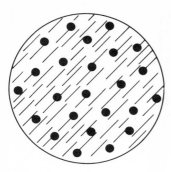

FIGURE 6.13 Dispersed and continuous phases in emulsion. In an oil-in-water system, the oil would be the small drops dispersed in a continuous medium of water. In a water-in-oil emulsion, the water would be the dispersed phase and the oil the continuous phase. An oil in water emulsion is often abbreviated as *o/w*, and a water in oil emulsion as *w/o*.

nonpolar, water and oil refuse to mix, and the addition of a third material is required to permit the oil to become dispersed in the water or the water to become dispersed in the oil. Such a material is called an **emulsifying agent** or **emulsifier.** Soaps and detergents are emulsifiers when added to water, and their cleaning action depends largely on their ability to break up greasy and oily films into droplets of oil dispersed in the water. Emulsifiers act by reducing the surface tension of water so that the water can spread and surround the oil as droplets and by protecting the droplet surface once it is formed. The dispersed phase is the oil; the continuous phase the water (Figure 6.13).

Most emulsifying agents have a common structural feature, namely, a long nonpolar portion and a smaller polar region. In terms of polarity and solubility they are double agents, the nonpolar part being hydrophobic and the polar part being hydrophilic. Emulsions are generally coarser dispersions than most colloidal systems, and the critical factor in many prepared emulsions is their stability. Many foods, both natural and prepared, are largely emulsions (Table 6.2).

Both oil-in-water and water-in-oil emulsions are used in pharmaceuticals and cosmetics; emulsifiers include lanolin, soaps, monoglycerides, and cholesterol. Asphalt is essentially an emulsion, and emulsifiers are used in the preparation of beverages,

TABLE 6.2 Emulsions in food

Food	Emulsion Components	Emulsifier
Milk	Butter fat and water	Casein
Mayonnaise	Oil and water (dilute vinegar, lemon juice)	Egg yolk
Margarine	Water and oil	Lecithin, monoglycerides

cake batters, candies, and many other food products. They are also found in insecticides, herbicides, printing inks, and embalming fluids. Oil-water emulsions are used in metalworking and machining.

COLLOIDAL SYSTEMS

The liquid-liquid system of emulsions is only one of the possible combinations between gases, liquids, and solids (Table 6.3).

An important factor in colloidal systems is the shape of the dispersed particles, generally globular or fibrous. For example, the colloids in the blood are mostly globular proteins, but the colloidal proteins of the connective tissues are fibrous. Fibrous colloids may be quite long, but their colloidal dimension is their thickness. Fibrous, or linear, colloids such as gelatin adsorb much more water than the globular types and are therefore used in forming jellies and semirigid substances. The increased water-holding capacity of linear colloids is explained by their larger surface area, which can adsorb more water. A noodle has more surface area than the same noodle rolled up as a ball.

Sols Of the eight colloidal systems shown in Table 6.3 sols and emulsions are probably the most important and the most common. The sols, which are colloidal solids dispersed in water, are usually divided into hydrophobic suspensoids and hydrophilic emulsoids (discussed earlier) although it should be added that the terminology of colloid chemistry is not firmly fixed. The hydrophilic sols are especially important in living matter and for the foods that maintain life. A particular feature of many hydro-

TABLE 6.3 Types of colloidal systems

Dispersing Medium	Dispersed Phase	Name of System	Examples
Gas	Solid	Aerosol	Dust, smoke
	Liquid	Aerosol	Fog, mist
	Gas	†	
Liquid	Solid	Sol	Gelatin, starch, jellies
	Liquid	Emulsion	Milk, cosmetics
	Gas	Foam	Suds, whipped cream
Solid	Solid	Solid sol	Metal alloys, opal
	Liquid	Solid emulsion	Cheese, butter
	Gas	Solid foam	Pumice, marshmallow

†Cannot form colloidal-sized particles.

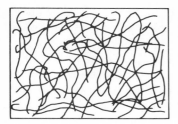

FIGURE 6.14 A tangled network of fibrous colloid holds water two ways: by adsorption as a layer of bound water, and by mechanical trapping of water between the fibers. A colloidal dispersion of this type has been compared to a brush heap.

philic sols is their ability to give a cohesive shape to what is mostly water. A common example is gelatin, a fibrous protein that forms a liquid sol when dispersed in water. As a sol, the liquid is practically clear, and flows like any liquid, but when it is cooled, it sets into a semirigid **gel** that is the base for many desserts. The gel maintains its shape, although it may be 95 to 99 percent water. The gelatin forms a loose but continuous entanglement of fibers in which the water is held (Figure 6.14).

Gelatin can go back and forth between the liquid sol condition and the semirigid gel by cooling and warming. It is also reversible in the sense that if the water is evaporated off, the colloidal condition will be regained by adding water again. The network of gelatin in the gel condition is not entirely stable, and on standing some of the water separates out. This is called **syneresis,** or bleeding, and occurs in other gels as well. Agar-agar, a gel used for bacterial culture which is about 99.8 percent water, also exhibits syneresis, as does a blood clot on standing. The tangle in a blood clot is a protein called fibrin; the clot is about 99.9 percent water, yet the clot seems solid. Proteins and carbohydrates tend to form sols and gels, and materials such as gelatin, starch, pectin, and egg white enter into the making of a great many foods, especially desserts and jellies. Gelatin, for example, not only is a dessert base in its own right but is also used as a protective colloid in making ice cream and in other foods.

Agar-agar is the Malay word for seaweed; the gel is extracted from several varieties.

Smoke and fog The hazy path the light from a motion picture projector makes as it passes through the dust and smoke of a theater is familiar to us all; it is a Tyndall effect on a large scale. When the theater lights go on, the particles that scattered the light cannot be seen. Much dust and smoke is within the colloidal size range, and it is noteworthy that some of these particles carry a surface electric charge. This is the basis of the **Cottrell process,** which precipitates particles from smoke going up industrial chimneys. The air stream passes through a high voltage that serves to impose further charges on the particles, which then settle out on plates that are oppositely charged. Modern systems can remove from 90 to 98 percent of the smoke and dust. This not only keeps the air cleaner but can also be a source of valuable by-products.

Fogs and mists are dispersions of water particles in air, and although they are generally classed as colloids, their particle sizes are variable and often larger than those of true colloids. Smoke and fog are less stable than a dispersion of particles in a liquid or a solid and tend to settle out, or precipitate. As dispersions in a gaseous medium (air), smoke and fogs are called **aerosols.** The same name is given to many products dispensed by pressurized containers using gas as a propellant, but the dispersed particles may or may not be in the colloidal size range.

Foams The gas-in-liquid systems, called **foams,** are also less stable than sols and emulsions, but they are of interest because of their contribution to the taste and texture of foods and because they cannot be made with pure liquids. No matter how long pure water is beaten or stirred it will not support a foam, but add some soap, and you have it. A foam is a mass of gas-filled bubbles, usually air surrounded by water. But water has a high surface tension, and a bubble is mostly surface; the addition of soap lowers the surface tension so that the water can spread around air when it is agitated. The liquid film of the bubble has three layers, two thin outer layers of soap, with a water layer between them; the nonpolar "tail" of each soap molecule is oriented away from the water, while the polar part is toward the water. However, it is obvious that soap is not the only foaming agent, otherwise beer drinking would be more dangerous than it is. Many protein colloids serve as foaming agents. Here, too, shape is a factor, with lamellar and fibrous geometries apparently the most effective. Foams are not stable, but they are pleasant while they last.

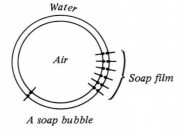

A soap bubble

COLLOID OSMOTIC PRESSURE

Colloids in water exert an osmotic pressure analogous to that of particles in true solution. Their effect is much smaller since there are fewer colloidal particles than particles in an ordinary solution. It has been estimated that a single colloidal particle can consist of between 1000 and 1 billion atoms. If each atom were in solution, together they would cause a much higher osmotic pressure than the single colloidal particle. Nevertheless, colloidal osmotic pressure is significant in maintaining the water balance between the blood and the fluids outside the blood vessels. The protein colloids in the blood, especially the albumins, help safeguard this balance by contributing a small but controlling osmotic pressure.

FIGURE 6.15   The dialyzing process.

When the blood loses albumin due to disease or prolonged malnutrition (particularly if protein intake is minimized), this balance is disturbed.   Water leaves the capillaries and causes swelling (**edema**) in the surrounding tissues.

Dialysis   Under normal conditions the albumins and other protein colloids in the blood are kept there by the walls of the blood vessels.   This is necessary for the body, and the ability of membranes to keep some particles in and other particles out is vital to life.   The same blood-vessel wall that keeps in the larger protein particles also permits the passage in or out of water molecules, as well as of some ions and small molecules.   Membranes that distinguish between colloidal-sized particles and smaller dissolved particles, keeping the larger ones in while allowing the smaller ones to diffuse in or out, are called **dialyzing membranes.** Most body membranes act this way.   Dialyzing membranes are used in the laboratory when a mixture of colloids and particles in solution must be separated from each other.   If the mixture is put into a bag made of a dialyzing material (cellophane or parchment or a bladder from an animal) and suspended in a container through which water is circulating, the smaller dissolved particles will pass slowly out of the bag and be washed away by the water; the larger colloid particles will remain in the bag, so that in time separation can be effected (Figure 6.15).   The same principle is used in the kidney machines that remove poisonous waste products from the blood.

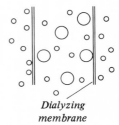

Dialyzing membrane

REVIEW QUESTIONS

1   What two factors in the structure of the water molecule account for its high polarity?
2   Several of the unique properties of water are due to the hydrogen bonding between water molecules.   What are these properties? How does hydrogen bonding account for them?

3 For what reason would you add a surfactant to water?

4 Why is salt sometimes put on icy sidewalks in winter?

5 Carbon tetrachloride, CCl_4, is not soluble in water; why?

6 A freely falling drop of water or any other liquid will assume a spherical shape. Explain why.

7 Why do you add antifreeze to your automobile radiator in the winter? What might happen if you did not?

8 What is meant by saying that two liquids are miscible? Would you expect miscible liquids to be generally similar in their polarity or far apart in their polarity?

9 When a person suffers heat stroke, the skin is dry. What cooling mechanism is not at work?

10 How is it possible for water to act as both an acid and a base and yet be neutral when pure?

11 How can pure water be made acidic? How can it be made basic?

12 Why is pure water a very poor conductor of electric current?

13 Imagine that a sample of soil is added to distilled water, mixed thoroughly, and then filtered. A residue remains on the filter paper, and the filtrate that passes through the filter paper is clear. **a** What is the approximate size of the particles in the residue? **b** Besides filtering, how else could the residue be separated from the liquid filtrate? **c** A narrow beam of light is passed through the clear filtrate, and a positive Tyndall effect is observed. What conclusion would you draw? About what size are the particles causing the Tyndall effect? **d** The filtrate is also tested for its electrical conductivity, and it is found that the liquid is a rather good conductor of current. What sort of chemical substances must be present in the filtrate, and about how large is their particle size? **e** Would you expect the boiling point of the filtrate to be equal to 100°C, above 100°C, or below 100°C?

14 Potassium chloride, KCl, is dissolved in water. **a** What is the solute and what is the solvent in this solution? Explain why the KCl seems to vanish. **b** The solubility of KCl is 35.7 g in 100 ml of water. If 400 g of KCl is added to 1 l of water and mixed thoroughly, how many grams of KCl would remain *undissolved?* **c** If the mixture of 400 g of KCl in 1 l of water is filtered, how much KCl will remain as a residue on the filter paper and how much will pass through with the filtrate? The filtrate is a saturated solution of KCl; what is meant by a saturated solution?

15 Why is it easier to float in seawater than in fresh water?

16 The sketch shows a bag made of a semipermeable membrane wrapped around a glass tube and attached to it. The bag contains a 5% solution of sugar and is immersed in a beaker of pure water, with both sugar solution and water at the same level. What do you expect will happen in time and why?

17 If a solution that is hypotonic with respect to its NaCl concentration is injected into the bloodstream, what will be the effect on the red blood cells?

18 How does a supersaturated solution differ from a saturated solution? Which is the stable condition?

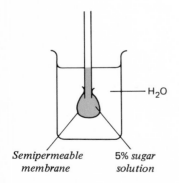

H_2O

Semipermeable membrane *5% sugar solution*

19 What is meant by hard water, and why can it be a problem?

20 What is the difference between temporary and permanent hardness of water? Why are materials like washing soda and borax used as water softeners? Are ion-exchange methods for softening water useful with both temporary and permanent hard water?

21 In general, how would you describe a colloidal system?

22 What are some of the physical forms in which colloidal systems are found?

23 Substances that act as adsorbents are generally colloidal in their particle size; why?

24 Colloidal particles can generally be classified as hydrophilic or hydrophobic. **a** Of the two, which is polar? **b** Which types of organic compounds are hydrophilic and which are hydrophobic? **c** Are colloidal dispersions of metals and metal oxides water-loving or water-hating?

25 How do the surface electric charges on the particles of a colloidal system keep the colloidal state stable? What would happen if these surface electric charges were neutralized by adding ions of opposite charge?

26 In addition to surface electric charges, what additional and very important protection do hydrophilic colloids have that helps keep them stable?

27 What is meant by bound water? Why will fibrous colloids hold more bound water than colloids that have a generally globular shape?

28 How does a hydrophilic colloid, such as gelatin, serve as a protective colloid for a hydrophobic colloid?

29 What is an emulsion?

30 When some oil is mixed or stirred vigorously with a larger volume of water, small oil droplets will form. In time, what happens to these oil droplets and why?

31 Referring to Question 30, how could the dispersion of small oil droplets in water be *stabilized?* In the stabilized system, which is the dispersed phase and which the continuous phase? Is the reverse possible? What is mayonnaise?

32 What essential structural feature must an emulsifying agent have? Why does soap qualify as an emulsifier?

33 Which of the following belong to the class of colloidal systems called a *sol*: beer suds, jelly, smoke, starch pudding, gelatin dessert, cheese? How would you classify the substances that are not sols?

34 A gelatin dessert can be close to 99 percent water, yet when cooled it maintains a semirigid form. How can that be explained? What is the name given to the semirigid form of the gelatin colloid?

35 What is meant by syneresis? Give some examples.

36 What are the three components of a bubble? What is a foam?

37 In advanced malnutrition the blood may lose colloidal proteins called albumins. How does this upset the necessary water balance between the blood and the fluids outside the blood vessels?

38 How does dialysis serve to separate out particles in solution, such as ions and small molecules, from colloidal-sized particles such as proteins, from a water mixture of the two?

REACTION TYPES AND EQUATIONS
Meeting, Interacting, and Changing

In all living matter, simple or complex, there are a multitude of chemical substances, and it is through their interactions that the organism maintains and reproduces itself. Despite the vast diversity of living organisms and the great differences between them, the chemical processes through which they function are remarkably similar.

TYPES OF CHEMICAL REACTIONS

Every chemical reaction results in a change, and every chemical change proceeds by breaking existing bonds and making new ones. Generally, breaking a bond requires an energy imput, whereas bond formation releases energy. The net result is an energy change that is specific to the given reaction. Each chemical reaction is a unique event, where the results obtained depend on the particular materials that interact and the conditions present. At the same time, many chemical reactions have some common distinguishing feature. It is helpful to sort out the major categories. In the list that follows several types of reactions are described in terms of the essential mechanism of the reaction or of the change that is achieved.

Proton transfer The transfer is between a proton donor (or an acid) and a proton acceptor (or a base). Reactions involving proton transfer are acid-base reactions.

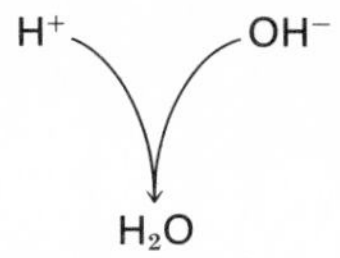

H^+ ⟍ ⟋ OH^-

H_2O

Transfer of a proton, H^+, to OH^- results in water.

Oxidation-reduction An electron is transferred from an electron donor to an electron acceptor. The donor is oxidized and the acceptor reduced, so that electron transfer reactions are called oxidation-reduction reactions.

Oxidation	$Na - 1e^- = Na^+$
Reduction	$Cl + 1e^- = Cl^-$
Oxidation-reduction reaction	$Na + Cl = Na^+Cl^-$ or $NaCl$

Ionic recombinations An exchange of partners takes place between ionic compounds.

Decomposition and hydrolysis reactions A chemical structure breaks up into smaller components. If this breakup occurs by interaction with water, as many organic compounds do, the reaction is called **hydrolysis.**

Condensation reactions Smaller chemical structures are combined into larger structures.

Substitution reactions An atom or group of atoms of a chemical structure is replaced by some other atom or group of atoms.

Addition reactions An atom, or group of atoms, is added to some receptive part of a chemical structure.

Hydration-dehydration reactions A water molecule is added to, or removed from, a chemical material.

Isomerization reactions The atoms of a chemical substance are rearranged so that its structure is altered *without* any change in composition.

Of the reactions listed, proton transfer and electron transfer are the most general, and are often involved in the mechanisms of other types of reactions. Furthermore, although they are of special importance in inorganic chemistry, they are also essential to biological systems.

PROTON TRANSFER

The discussion of proton transfer in Chapter 6 developed these conclusions:

1 The self-ionization of water, which is very slight, proceeds by proton transfer, each transfer resulting in one H^+ and one OH^- ion. With equal numbers of each ion present, water is neutral.

2 Adding a proton donor, or acid, to water increases the number of H^+ ions accordingly, and the solution is acidic.

3 Adding a proton acceptor, or base, to water increases the number of OH^- ions accordingly, and the solution is basic, or alkaline.

4 A group of ionic compounds called hydroxides contain OH^- ions in their structure; when they dissolve in water, the OH^- ions are released by dissociation, rendering the solution alkaline.

That some things are acidic or can become acidic was recognized long before chemists began interpreting what it meant. Many fruits taste acidic, wine sours, and milk turns acid and curdles. On the other hand, some materials are recognized as **antiacid,** which is a good working description of a base. When wine sours, it becomes vinegar. Drop some vinegar on a piece of limestone (an oystershell or a marble chip will do), and there will be fizzing. Some of the acidity of the vinegar will be lost, as well as some of the limestone. Of course, nothing is *really* lost, only changed. The chemist describes it as a **neutralization.** The acid and the base have done each other in.

ACIDS AND BASES

It is convenient to line up acids and bases as chemical opposites against each other as in Table 7.1. Although the differences stand out, there is also some similarity due to the fact that they both yield ions in water.

TABLE 7.1

Acids	*Bases*
1 Sour or tart to the taste, often with a sharp odor	Bitter to the taste and slippery to the touch
2 Turn litmus paper *red*	Turn litmus paper *blue* (base for *blue*)
3 In water solution they yield hydrogen ions, H^+	In water solution they yield *hydroxide* ions, OH^-

4 Both are chemically active; even in small concentrations they may stimulate chemical activity
5 Both are electrolytes in water solution
6 Acids and bases *neutralize* each other by the reaction between the H^+ ions of the acid and the OH^- ions of the base; the products of the reaction are water and a *salt*

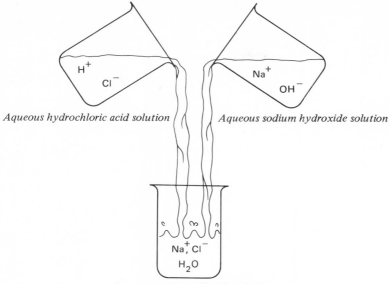

Aqueous hydrochloric acid solution

Aqueous sodium hydroxide solution

Aqueous sodium chloride solution

FIGURE 7.1 Mutual neutralization of acid and base solutions.

The central feature of acid-base reactions is their mutual neutralization. To illustrate this, consider an acid solution in which hydrochloric acid, HCl, is the proton donor and a basic solution in which dissolved sodium hydroxide, NaOH, dissociates to release free hydroxide, OH⁻, ions. It will be assumed that the number of H^+ ions in the acid solution and the number of OH⁻ ions in the alkaline solution are *equal*. When the two solutions are mixed, they neutralize each other as indicated in Figure 7.1. Generally, neutralization reactions, like others, are written as equations. In this instance

$$HCl + NaOH \longrightarrow H_2O + NaCl \qquad (7.1)$$

The neutralization itself is simply the reaction between the H^+ ions of the acid and the OH⁻ ions of the base, resulting in the formation of water molecules. It was pointed out in Chapter 6 that the self-ionization of water is very minute, so that practically all the water molecules formed by the neutralization are *undissociated*. This is also the reason for writing Eq. (7.1) with only one arrow, thus indicating that practically all the H^+ and OH⁻ ions originally present have combined to form water. The second product resulting from an acid-base neutralization is a salt, in this

An aqueous solution of a salt, for example, NaCl, *is sometimes designated* NaCl(aq).

case sodium chloride, NaCl, or table salt. So long as the salt is in water and dissolved, it exists as dissociated ions. If the water is evaporated, the ions will *associate* to form solid NaCl.

ACIDIC MATERIALS

Since an acid is a proton donor, it must contain hydrogen in its structure, so that the general formula of an acid can be taken as HA, where A^- is the negative part of the ionic compound. If acid HA dissociates almost completely, so that we can write

$$HA \longrightarrow H^+ + A^-$$

A single arrow indicates that the strong acid tends to dissociate, completely or almost so.

then HA is a strong acid and just about as many H^+ ions will be formed as HA units were dissolved. A strong acid is therefore a strong proton donor. On the other hand, if HA dissociates only to a limited extent,

$$HA \rightleftharpoons H^+ + A^-$$

A double arrow with a longer arrow in the direction of the undissociated acid indicates a weak acid, i.e., the ions tend to reassociate.

then HA is a poor proton donor and a solution of HA will have many fewer H^+ ions than dissolved units of HA. The strength of an acid is given quantitative expression by the **dissociation constant,** which is essentially a measure of the extent to which the acid dissociates in water. Without discussing how these values are obtained, we give the dissociation constants K_a for a number of acids of general importance and interest in Table 7.2.

Some of the acids listed in Table 7.2 are quite common. Acetic acid (the acid in vinegar) is important in chemical processes, and its carboxylic acid group is characteristic of organic acids. As a

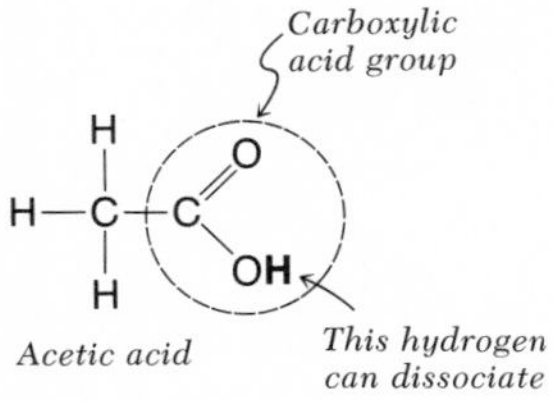

Carboxylic acid group

Acetic acid

This hydrogen can dissociate

TABLE 7.2 The K_a of acids

Acid	Formula	K_a
Strong:		
Hydrochloric	HCl	Very high
Sulfuric	H_2SO_4	Very high
Nitric	HNO_3	>100
Moderately strong:		
Sulfurous	H_2SO_3	1.5×10^{-2}
Phosphoric	H_3PO_4	7.6×10^{-3}
Weak:		
Acetic	CH_3COOH	1.8×10^{-5}
Carbonic	H_2CO_3	4.3×10^{-7}
Ammonium ion	NH_4^+	5.6×10^{-10}
Water	H_2O	1.8×10^{-16}

weak acid it dissociates incompletely:

$$CH_3COOH \longleftrightarrow CH_3COO^- + H^+ \qquad (7.2)$$

Acetic acid   Acetate ion   Hydrogen ion

Acids like H_2SO_4 and H_3PO_4 which have two or three ionizable hydrogens are called **diprotic** and **triprotic acids,** respectively. Their H^+ ions dissociate in sequence, each successive proton donation resulting in a weaker acid:

$$H_2SO_4 \longrightarrow H^+ + HSO_4^- \longleftrightarrow H^+ + SO_4^{2-}$$

Very strong acid   Moderately strong acid

H_2SO_4 is a more effective proton donor and a stronger acid than HSO_4^-.

$$H_3PO_4 \longleftrightarrow H^+ + H_2PO_4^- \longleftrightarrow H^+ + HPO_4^{2-} \longleftrightarrow H^+ + PO_4^{3-}$$

Moderately strong acid   Weak acid, very weak base   Very weak acid, weak base   Strong base

The successive proton donations of H_3PO_4 end in a material that is a rather strong proton acceptor, or base, PO_4^{3-}. The intermediate structures, $H_2PO_4^-$ and HPO_4^{2-}, act as acids if a base is added to the solution and as bases if an acid is added. A single species that can act either as an acid or a base is called **amphoteric.**

Nonmetallic oxides as acids A number of nonmetallic oxides (including carbon dioxide, CO_2; sulfur dioxide, SO_2; and sulfur trioxide, SO_3) have no protons to donate but are nevertheless acidic. Each of the three oxides listed is a gas at room temperature, but if it is bubbled through a basic solution, the solution will be neutralized, indicating that the oxide has acted as an acid. This acidic behavior is due to the oxide's reaction with water to form compounds that are proton donors:

$$SO_2 + H_2O = H_2SO_3 \qquad (7.3a)$$

Sulfur dioxide   Water   Sulfurous acid

$$SO_3 + H_2O = H_2SO_4 \qquad (7.3b)$$

Sulfur trioxide   Water   Sulfuric acid

$$CO_2 + H_2O = H_2CO_3 \qquad (7.3c)$$

Carbon dioxide   Water   Carbonic acid

Sulfur dioxide and sulfur trioxide are pollutants in our atmosphere, largely as a result of burning sulfur-containing fuels and industrial operations involving sulfur or sulfur compounds (Figure 7.2). When rain falls over or near industrial areas where these

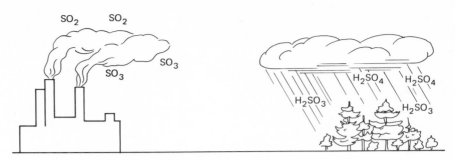

FIGURE 7.2 Industrial pollution often adds sulfur dioxide and sulfur trioxide to the air. Rainfall may cause acids to form and be deposited many miles away.

pollutants are present, the rainwater dissolves some of the SO_2 and SO_3, thereby becoming significantly and even dangerously acidic. When these and similar pollutants react with rainwater, open bodies of water like lakes or rivers, or groundwater, the result is potential damage. Not only the biosphere suffers— plants, aquatic life, animals, and people—but buildings, transportation structures, and even works of art are vulnerable.

An important acidic system: $CO_2 + H_2O$ Equation (7.3c) occurs in both living and nonliving systems. Water is always with us, and carbon dioxide is always being produced wherever carbon-containing compounds are being burned. When such materials as wood, petroleum, or natural gas burn, the flame is obvious and the reaction is rapid and dramatic. We can see the effect as the carbon combines rapidly with the oxygen of the air. Carbon-containing foods ingested by living organisms also burn, but this reaction takes place slowly, quietly, and in a controlled manner to avoid injury to the organism itself. Regardless of the different paths taken, energy and carbon dioxide are produced. Between the burning of fuels and the energy-producing reactions of living matter, enormous quantities of carbon dioxide are constantly being poured into the atmosphere. Fortunately, however, carbon dioxide is also constantly being used up by green plants for synthesizing their structures. Like any other act of construction, this requires energy, and since the energy available to plants is the energy of sunlight, this process is called **photosynthesis.** As a result of these opposing processes, CO_2 production and consumption, there is about 0.03 percent CO_2 in the air. If there were a rise in carbon dioxide output due to large-scale combustion of carbon-containing fuels without a corresponding increase in its consumption by plant life, there would be an increase in atmospheric CO_2. There is evidence that this has, in fact, occurred in

C + O$_2$ $\longrightarrow$ CO$_2$ + thermal energy

Photo means light. Synthesis means building up

Since our atmosphere exerts a pressure, and since CO_2 is part of this atmosphere, it exerts part of that pressure. This is called the partial pressure of CO_2.

FIGURE 7.3 Carbon dioxide is used in some fire extinguishers because CO_2, being heavier than air, cuts off oxygen from the fire.

recent years, and although the increase is apparently small and the long-term effects uncertain, it is one more aspect of pollution.

Since it is the product of an oxidation reaction and already carries all the oxygen atoms it can, CO_2 will not burn. Being heavier than air and therefore able to displace it, carbon dioxide is useful as a fire extinguisher. It acts in effect as a noncombustible blanket that chokes the fire by keeping away the oxygen of the air (Figure 7.3). But in sufficiently large amounts it can also choke off a person's supply of oxygen.

Although it is not classified as a poison (whereas carbon monoxide, CO, is lethal), CO_2 is nevertheless a waste product in the production of energy and must be removed regularly from the cells of the body. The elimination of CO_2 from the tissues, which goes hand in hand with supplying oxygen *to* the tissues, involves a complex series of events, but the net chemical pay off can be understood as a consequence of what happens when carbon dioxide and water are mixed.

The solubility of CO_2 gas in water depends on the pressure, but even at atmospheric pressure it is appreciable. A small fraction of the dissolved CO_2 combines with water to form carbonic acid, a

$$O=C=O \ + \ \overset{H}{\underset{}{\diagdown}}O\overset{H}{\diagup}$$

Dissolved CO_2 *Water*

$$\updownarrow$$

$$O=C\overset{\textstyle OH}{\underset{\textstyle OH}{\diagup}}$$

$$H_2CO_3$$

Carbonic acid

weak acid. Carbonic acid acts more like an in-between compound than a stable material. When gaseous CO_2 enters water, the result is

$$CO_2(g) + H_2O \rightleftharpoons CO_2(aq) + H_2O \rightleftharpoons H_2CO_3 \rightleftharpoons H^+ + HCO_3^- \quad (7.4)$$

Carbonic   Bicarbonate
acid   ion

The longer left-pointing arrows mean that there is less HCO_3^- than H_2CO_3 and less H_2CO_3 than dissolved CO_2. This is the equilibrium condition at 1 atm pressure.

where the notation $CO_2(g)$ means that the CO_2 is present as a gas. Note the three equilibria, all related to each other at the same time. The gaseous CO_2 above the surface of the solution is in equilibrium with the dissolved CO_2 in the solution; the dissolved CO_2 in the water is in equilibrium with the carbonic acid formed; and the carbonic acid is in equilibrium with its dissociation products, H^+ and HCO_3^- ions. A disturbance of any one of these three simultaneous equilibria will alter the other two.

When carbonated drinks are made, carbon dioxide is forced into water under pressure. With increasing pressure, more CO_2 enters the system, disturbing the equilibrium. In effect, the increase in CO_2 tips Eq. (7.4) as indicated in Figure 7.4, so that the reaction flows to the right with a corresponding increase in carbonic acid, and an increase in solution acidity. When enough CO_2 is pushed in, the bottle is capped, and the soft-drink system comes to equilibrium. This differs from the equilibrium represented by Eq. (7.4), where the system is open to the atmosphere. In the soft-drink system the reaction flow to the right, due to higher CO_2 gas pressure, leads to an increase in the number of H^+ ions present in the water and hence an increase in acidity. When the cap is removed, the CO_2 pushed into the water now rushes *out*. If you wait long enough, the system returns to the original equilibrium represented by Eq. (7.4) and the drink is flat.

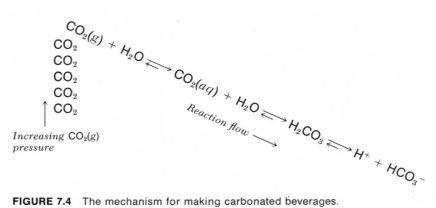

FIGURE 7.4 The mechanism for making carbonated beverages.

$CO_2 + H_2O$ in the bloodstream   What happens in the soft-drink system is analogous to what happens when carbon dioxide is eliminated from the cells of body tissues. Living matter is infinitely more complicated and more subtle than any bottle of soda, and the elimination of CO_2 keeps on going as long as we live. That is, it is a *cyclic* process, and the chemistry of life is built around cyclic processes.

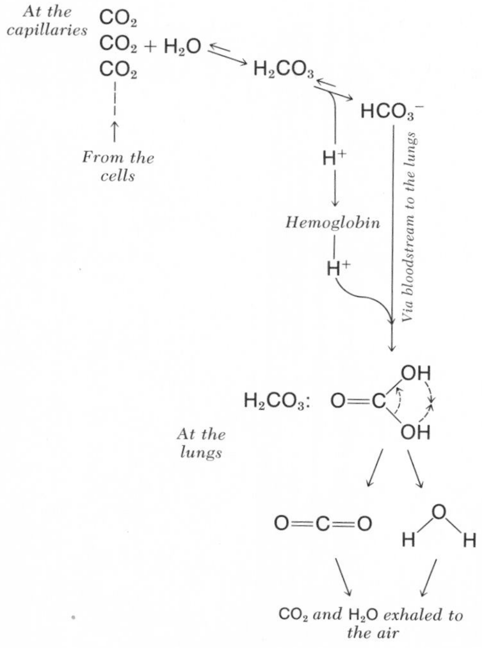

The carbon dioxide produced by the oxidation of food nutrients builds up in the cells, and the pressure it develops makes it diffuse through the cell walls and into the fine capillaries carrying the blood. At this point, many participants enter the picture, including red blood cells, hemoglobin, and an enzyme to speed things up. The net result is to permit much of the CO_2 coming from the cells to enter the liquid plasma of the blood, where it reacts with water to form carbonic acid. In turn, H_2CO_3 gives up a proton to a hemoglobin molecule to become a bicarbonate ion, HCO_3^-. What has happened can again be represented by the tipped equation in Figure 7.4. Excess CO_2 shifts the reaction to the right, increasing the formation of carbonic acid, which by giving up an H^+ ion becomes HCO_3^-. The blood, with its cargo of bicarbonate ions, circulates to the lungs, where the conditions now *reverse* themselves. The hemoglobin molecule obligingly returns the H^+ ion to the bicarbonate ion, which becomes carbonic acid, and as the amount of H_2CO_3 builds up, more of it breaks up into CO_2 and H_2O. With more CO_2 in solution, its escaping tendency, or pressure, also increases, so that at the lungs, where the pressure of incoming air is 1 atmosphere (atm), gaseous CO_2 passes out of the body with the exhaled breath. What began as a push of carbon dioxide at the cells becomes an escape of carbon dioxide at the lungs, thanks to the bloodstream and the equation in Figure 7.4.

Before leaving the H_2O–CO_2 system, it should be noted that Eq. (7.4) is incomplete because the bicarbonate ion, HCO_3^-, can dissociate further,

$$HCO_3^- \rightleftharpoons H^+ + CO_3^{-2}$$

As the relative size of the arrows suggests, this dissociation is extremely small, so that the bicarbonate ion is a better proton acceptor than a proton donor, stronger as a base than an acid. The complete equation representing the water–carbon dioxide system can therefore be written

$$H_2O + CO_2(g) \rightleftharpoons CO_2(aq) + H_2O \rightleftharpoons$$

$$\underset{\substack{\text{Carbonic} \\ \text{acid}}}{H_2CO_3} \rightleftharpoons H^+ + \underset{\substack{\text{Bicarbonate} \\ \text{ion}}}{HCO_3^-} \rightleftharpoons H^+ + \underset{\substack{\text{Carbonate} \\ \text{ion}}}{CO_3^{2-}} \qquad (7.5)$$

BASIC MATERIALS

A base is any chemical entity that will accept protons. Its interaction with water results in the formation of hydroxide ions:

$$H_2O + B \rightleftharpoons BH^+ + OH^-$$

A base is considered strong or weak in accordance with its effectiveness as a proton acceptor. A strong base is a vigorous taker

TABLE 7.3 Common and important bases

Base	Formula	K_b
Hydroxide ion	OH^-	55.6
Phosphate ion	PO_4^{3-}	6.4×10^{-2}
Carbonate ion	CO_3^{2-}	2.1×10^{-4}
Ammonia	NH_3	1.8×10^{-5}
Hydrogen phosphate ion	HPO_4^{2-}	1.6×10^{-7}
Bicarbonate ion	HCO_3^-	2.3×10^{-8}
Acetate ion	CH_3COO^-	5.5×10^{-10}

of protons and in water will increase the number of OH⁻ ions accordingly. A weak base is a reluctant proton acceptor, so that only a small number of OH⁻ ions will be formed by interaction with water molecules. Analogous to the dissociation constant K_a used to indicate the strength of an acid, the basic dissociation constant K_b is the index of the strength of a base. A number of common and important bases are listed in descending order of base strength in Table 7.3; in each case, water is the solvent. The ionic compounds called hydroxides dissociate when dissolved in water, thereby contributing OH⁻ ions directly to the solution. In this respect they differ from the other bases shown, all of which interact with water to produce hydroxide ions. The base takes a proton from a water molecule, which thereby becomes an OH⁻ ion. The stronger bases, whose structures are very attractive to protons, do this more effectively than the weaker bases. If solutions of the bases in Table 7.4 are prepared so that equal volumes of each have the same number of dissolved basic ions or molecules, the relative alkalinity of the solutions will be

$$PO_4^{3-} > CO_3^{2-} > NH_3 > HPO_4^{2-} > HCO_3^- > CH_3COO^-$$

However, an NaOH solution of the same concentration contains more OH⁻ ions per given volume than any of the other bases listed and is more alkaline than any of them. The reason is that

The PO_4^{3-} ion has three negatively charged oxygens with which to attract an H^+ ion.

The carbonate ion, CO_3^{2-}, has two negatively charged oxygens with which to attract H^+ ions.

TABLE 7.4 Interaction between bases and water in order of increasingly weak bases

PO_4^{3-} Phosphate ion	$+ H_2O \rightleftharpoons$	HPO_4^{2-} Hydrogen phosphate ion	$+ OH^-$
CO_3^{2-} Carbonate ion	$+ H_2O \rightleftharpoons$	HCO_3^- Bicarbonate ion	$+ OH^-$
NH_3 Ammonia	$+ H_2O \rightleftharpoons$	NH_4^+ Ammonium ion	$+ OH^-$
HPO_4^{2-} Hydrogen phosphate ion	$+ H_2O \rightleftharpoons$	$H_2PO_4^-$ Dihydrogen phosphate ion	$+ OH^-$
HCO_3^- Bicarbonate ion	$+ H_2O \rightleftharpoons$	H_2CO_3 Carbonic acid	$+ OH^-$
CH_3COO^- Acetate ion	$+ H_2O \rightleftharpoons$	CH_3COOH Acetic acid	$+ OH^-$

only a certain *proportion* of PO_4^{3-}, or CO_3^{2-} ions (both good proton acceptors) succeed in winning a proton from the very stable water molecule, converting it to an OH^- ion. On the other hand, *all* the dissolved NaOH units dissociate, so that if x units of NaOH are dissolved, x OH^- ions will be obtained. If x phosphate ions are dissolved, the number of OH^- ions obtained would be only some *fraction of* x. The common feature of OH^- ions, whether derived from a base-water interaction or from the dissociation of hydroxide, is that *they are themselves proton acceptors.* In accepting a proton, the OH^- becomes a water molecule, so that the formation of water is the characteristic feature of acid-base neutralizations.

Metal oxides and hydroxides The nonmetallic oxides SO_2, SO_3, and CO_2, discussed before, are covalent gases at room temperature and act as acids. On the other hand, a number of **metal oxides** are ionic solids with high melting points and form alkaline solutions in water:

$$Na_2O + H_2O \ = \ 2NaOH$$
Sodium hydroxide

$$K_2O + H_2O \ = \ 2KOH$$
Potassium hydroxide

$$MgO + H_2O \ = \ Mg(OH)_2$$
Magnesium hydroxide

$$CaO + H_2O \ = \ Ca(OH)_2$$
Calcium hydroxide

The common name for CaO is lime. When it is mixed with water, the result is limewater, a solution of Ca(OH)$_2$.

These four common hydroxides, and others, are strong bases because they dissociate completely in solution. However, since $Ca(OH)_2$ is only slightly soluble and $Mg(OH)_2$ *very* slightly soluble, solution of these hydroxides will never have a high concentration of OH^- ions. However, NaOH is quite soluble, and is a very common base, both in the laboratory and industrially. It is produced in large amounts throughout the world and is also known as lye and caustic soda. The dissociation of these four hydroxides is quite simple.

$$NaOH \longrightarrow Na^+ + OH^-$$
$$KOH \longrightarrow K^+ + OH^-$$
$$Ca(OH)_2 \longrightarrow Ca^{2+} + 2OH^-$$
$$Mg(OH)_2 \longrightarrow Mg^{2+} + 2OH^-$$

Although their solubility in water varies from good to negligible, they are all equally vulnerable to neutralization by acids. The neutralization of hydrochloric acid by magnesium hydroxide is

$$2HCl + Mg(OH)_2 \longrightarrow 2H_2O + MgCl_2 \qquad (7.6)$$

The numeral 2 that precedes HCl in the equation is a multiplier, or **coefficient,** and indicates that *two* HCl units are needed to neutralize *one* $Mg(OH)_2$ unit because for each HCl that dissociates *one* H^+ ion is formed, whereas for each $Mg(OH)_2$ that dissociates *two* OH^- ions are formed. For neutralization the two OH^- ions must be matched by two H^+ ions, so that *two* HCl units must be used for each *one* $Mg(OH)_2$ unit:

In acidic solution: $2HCl = 2H^+ + 2Cl^-$

In basic solution: $Mg(OH)_2 = Mg^{2+} + 2OH^-$

The number 2 is a subscript in $Mg(OH)_2$ because the OH^- ions are *part* of the compound and are not free, and a coefficient in $2OH^-$ because the two OH^- ions are *free* and *not* part of a compound. The acidic solution therefore has two HCls for each $Mg(OH)_2$ in the basic solution, and when the two are mixed neutralization results (Figure 7.5).

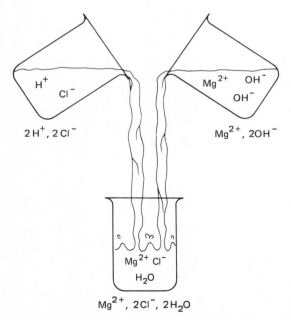

FIGURE 7.5 Neutralization of hydrochloric acid by magnesium hydroxide.

Since the water molecules formed by the acid-base neutralization are almost entirely undissociated, the reaction cannot reverse itself to any significant extent. Also, the Cl^- and Mg^{2+} ions will form an undissociated, solid salt, $MgCl_2$, if the water is evaporated. The overall event is therefore as summarized in the neutralization equation (7.6):

$$2HCl + Mg(OH)_2 \longrightarrow 2H_2O + MgCl_2$$

Hydroxides, such as NaOH and $Mg(OH)_2$, can be formed by the interaction between the corresponding metal oxide and water. If HCl, or any other acid, neutralizes $Mg(OH)_2$, it should also react with MgO, which is $Mg(OH)_2$ less one water molecule; and it does so:

$Na_2O + H_2O = 2NaOH$
$MgO + H_2O = Mg(OH)_2$

$$2HCl + MgO \longrightarrow H_2O + MgCl_2 \qquad (7.7)$$

This is identical with Eq. (7.6) except that one less water molecule is formed, which is the difference between MgO and $Mg(OH)_2$.

Notice that both Eqs (7.6) and (7.7) are **balanced:** the kinds and numbers of atoms on the left side are the same as on the right side. Count the atoms, comparing left to right: how many H atoms, O atoms, Cl atoms, and Mg atoms. Any reaction is essentially a rearrangement of atoms, so that although before and after may not look alike or act alike, the atoms involved must be the *same*, both in kind and number. To tell its story correctly, therefore, every equation should be *balanced*. In balancing an equation keep in mind that only multipliers can be used. Often the multiplier can be seen from the kind of reaction involved. When HCl reacted with $Mg(OH)_2$, the fact that neutralization requires equal numbers of H^+ and OH^- led to the conclusion that *two* HCl units were needed to react with *one* $Mg(OH)_2$, which in turn meant that *two* water molecules were formed for each such reaction.

To further illustrate the action of acids on hydroxides and metal oxides:

$$HCl + NaOH \longrightarrow H_2O + NaCl \qquad (7.8)$$

$$H_2SO_4 + 2KOH \longrightarrow 2H_2O + K_2SO_4 \qquad (7.9)$$

$$2HNO_3 + CaO \longrightarrow H_2O + Ca(NO_3)_2 \qquad (7.10)$$

Fe_2O_3 *is rust. The reaction in Eq. (7.11) is called pickling in the steel industry, where it is used to remove scale from steel surfaces.*

$$6HCl + Fe_2O_3 \longrightarrow 3H_2O + 2FeCl_3 \qquad (7.11)$$

In each instance the reaction with acid produces water, so that these are all acid-base neutralizations, with a corresponding salt also produced in each case. Also, note the balancing of the equations; do they all check out?

Acids versus metals Since acids react with metal oxides, will they also react with the metals themselves? The answer is "yes" and "no"; yes if the metal is active and no if the metal is not active. What active means is discussed under the next reaction type, oxidation-reduction. At this point the action of an acid on an active metal can be represented as analogous to its action on the oxide of that metal. Since magnesium is an active metal, the action of an acid on the metal oxide and on the metal can be compared:

$$2HCl + MgO \longrightarrow H_2O + MgCl_2$$
$$2HCl + Mg \longrightarrow H_2 \uparrow \, + MgCl_2$$

The two are the same except for the (important) difference of an oxygen atom between MgO and Mg and a corresponding difference between H_2O and H_2. In general, therefore, an acid will react with an active metal to produce hydrogen gas and a salt. Since the hydrogen gas escapes, the reaction is irreversible and goes all the way to the right. Chemists sometimes refer to this type of reaction as a *displacement*, since we can think of the magnesium atom, in this case, as pushing out the hydrogen from HCl, and taking its place.

We can imagine that the loss of H_2 gas tips the reaction so that it flows down and to the right.

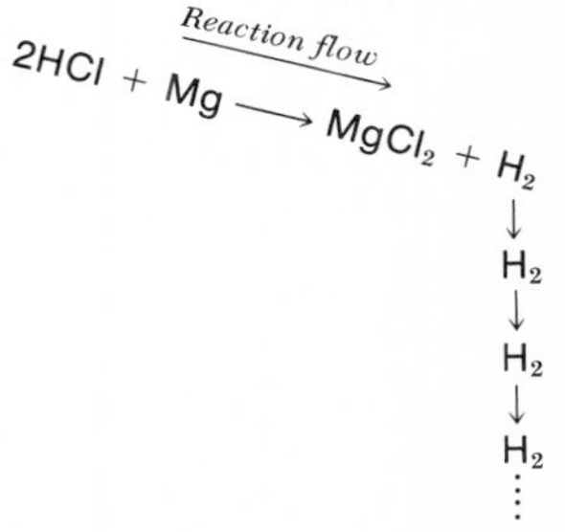

$$2HCl + Mg \xrightarrow{\text{Reaction flow}} MgCl_2 + H_2$$
$$\downarrow$$
$$H_2$$
$$\downarrow$$
$$H_2$$
$$\downarrow$$
$$H_2$$
$$\vdots$$

Acids versus carbonates and bicarbonates We have already mentioned that an acid will attack a carbonate, such as limestone. How it happens can be understood from Eq. (7.5), starting at the right with carbonate ions, $CO_3{}^{2-}$. Suppose that Na_2CO_3 is dissolved in pure water. The dissolved carbonate dissociates into two Na^+ ions and one $CO_3{}^{2-}$ ion for each Na_2CO_3 unit, and the solution is basic since the carbonate ion is a good proton acceptor. This is the equilibrium condition. Now, suppose the system is disturbed by the addition of an acid, say HCl, which will donate protons. The addition of H^+ ions "pushes up" the right side of the system, and the reaction runs downhill to the left (Figure 7.6).

$$Na_2CO_3 \longrightarrow 2Na^+ + CO_3{}^{2-}$$

The reaction will continue to run downhill and to the left so long as $CO_3{}^{2-}$ and H^+ ions are both present. If either or both are used up, the reaction stops. Note that the arrows are in one direction only because the CO_2 gas at the left end of the reaction is escaping into the atmosphere and cannot accumulate in the system to push back and establish a new equilibrium. The reaction is *irreversible* because one end is open and the CO_2 keeps running out. This is an example of a general principle: When a

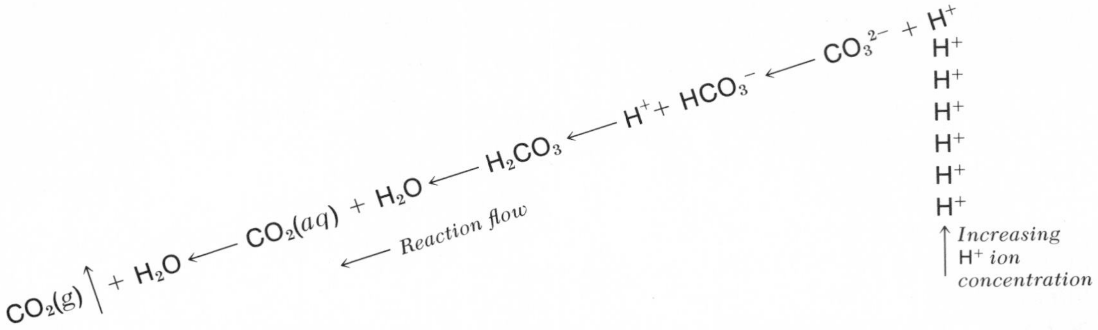

$$CO_2(g) \uparrow + H_2O \longleftarrow CO_2(aq) + H_2O \longleftarrow H_2CO_3 \longleftarrow H^+ + HCO_3^- \longleftarrow CO_3^{2-} + H^+$$

Reaction flow

$$H^+$$
$$H^+$$
$$H^+$$
$$H^+$$
$$H^+$$
$$H^+$$

$\uparrow$ *Increasing H^+ ion concentration*

FIGURE 7.6 Reaction flow for acid attacking a carbonate. The CO_2 gas escapes to the atmosphere, making the reaction irreversible.

product formed by a reaction becomes *unavailable* to the system, whether by escaping as a gas or being locked up as an insoluble or undissociated material, the reaction will proceed in one direction, irreversibly, until the reacting materials are used up.

The equation of Fig. 7.6 can now be read as follows. When an acid is added to a compound containing a carbonate ion, the result is the evolution of CO_2 gas and the formation of water. The carbonate compound can be in solution, or it can be a solid, in which case it starts dissolving and reacting as soon as it hits the acidic solution. If a *bicarbonate*, rather than a carbonate, is used, the same sequence will occur, except that it enters the equation one step to the left of the carbonate. Again, the end products are gaseous CO_2 and water.

Up to now the negative ion of the acid and the positive ion originally associated with the carbonate or bicarbonate have not been shown because actually they were not involved. Nonetheless, they are there, and being opposite in charge, can form a salt together. The equations representing the net result when an acid is added to a carbonate or a bicarbonate are given below. For clarity, the formation of intermediate H_2CO_3 and the presence of the onlooker ions are shown first, with the net result below:

$\uparrow$ means a gas is being evolved and is leaving.

Acid + carbonate:

$$2HCl + Na_2CO_3 \longrightarrow H_2CO_3 + 2Cl^- + 2Na^+$$
$$2HCl + Na_2CO_3 \longrightarrow H_2O + CO_2\uparrow + 2NaCl \qquad (7.12)$$

Acid + bicarbonate:

$$HCl + NaHCO_3 \longrightarrow H_2CO_3 + Cl^- + Na^+$$
$$HCl + NaHCO_3 \longrightarrow H_2O + CO_2\uparrow + NaCl \qquad (7.13)$$

The release of CO_2 when an acid acts on a bicarbonate explains the use of $NaHCO_3$ in baking powders. **Baking soda** is another name for $NaHCO_3$, and when mixed with starch and a weakly acidic material it forms baking powder. When baking powder is added to a wet dough or a batter, the reaction is that of (7.13). Essentially, it is the reaction between the H^+ of the acid and the HCO_3^- ion of the baking soda.

Note that the HCO_3^- ion acts here as a base, neutralizing an acid to yield water and CO_2.

$$H^+ + HCO_3^- \longrightarrow H_2CO_3 \longrightarrow H_2O + CO_2 \uparrow$$

When heat is applied, this reaction goes all the way to the right, and the escaping CO_2 gas lightens and "raises" breads and cakes. It should be noted that at oven temperatures the volume of the CO_2 gas is larger than at room temperatures.

Buffers By referring back again to Eq. (7.5), we see that carbonic acid dissociates in two steps. First H^+ and HCO_3^- ions are formed, and then the HCO_3^- dissociates into H^+ and CO_3^{2-}. The extent of the second dissociation is a small fraction of the first, so that the number of CO_3^{2-} ions formed at the end is very small and a solution of carbon dioxide in water contains much more H_2CO_3 than CO_3^{2-}.

Following the argument used before, it can be seen that the relative amounts of carbonic acid and of bicarbonate ion in the equilibrium

$$H_2CO_3 \rightleftharpoons H^+ + HCO_3^-$$

depend on the concentration of H^+ ions. If the solution is made acidic, we can visualize the right-hand side lifting up due to the increased H^+ ions, and the reaction flowing down to the left, producing an increase in H_2CO_3 and a decrease in HCO_3^-. However, if the solution becomes basic and only a few H^+ ions are present, the reaction flow reverses and there is less H_2CO_3 and more HCO_3^-. In the blood, which always contains carbonic acid and bicarbonate ion since it is always carrying CO_2 from the cells, the conditions are mildly basic, favoring bicarbonate-ion formation, so that the ratio of H_2CO_3 to HCO_3^- is about 1:20. This carbonic acid–bicarbonate ion pair is called a **buffer,** since it acts to resist excessive and dangerous fluctuations of H^+-ion level in the blood. All buffer systems are two-faced, one component acting to neutralize additions of basic materials and the other to neutralize additions of acidic materials; in so doing each becomes the other, as can be seen from the H_2CO_3–HCO_3^- system at work.

When neutralizing added base, H_2CO_3 becomes HCO_3^-:

$$H_2CO_3 + OH^- = HCO_3^- + H_2O$$

When neutralizing added acid, HCO_3^- becomes H_2CO_3:

Reaction flow

$$H_2CO_3 \rightleftharpoons HCO_3^- + H^+$$
$$H^+$$
$$H^+$$
$$H^+$$
$$\uparrow$$

Adding acid

$$H_2CO_3 \rightleftharpoons H^+ + HCO_3^-$$
$$\downarrow$$
$$H^+ + OH^- = H_2O$$
$$\downarrow$$
$$H^+ + OH^- = H_2O$$
$$\downarrow$$
$$H^+ + OH^- = H_2O$$
$$\vdots$$

Pulled out by OH^- ions

Adding base

$$HCO_3^- + H^+ = H_2CO_3$$

Although H_2CO_3–HCO_3^- is the most important buffer in the blood, the phosphate buffer, $H_2PO_4^-$–HPO_4^{2-} is also interesting and important. The monohydrogen phosphate ion, HPO_4^{2-}, is a weak base, and so it helps protect the blood against excess H^+ ions:

$$HPO_4^{2-} \quad + H^+ = \quad H_2PO_4^-$$

Monohydrogen   Dihydrogen
phosphate ion   phosphate ion

The phosphate buffer

$$HPO_4^{2-} \diagup H_2PO_4^-$$

H^+ OH^-

$H_2PO_4^-$ $HPO_4^{2-} + H_2O$

The product of the neutralization is the other half of this buffer system, $H_2PO_4^-$, which acts against OH^- ion additions:

$$H_2PO_4^- + OH^- = HPO_4^- + H_2O$$

In living matter buffers are a critical part of the elaborate system of checks and balances that maintains the stability of the organism. Buffers also have many practical uses in chemical analysis, bacteriological procedures, and industrial operations.

SALTS

Generally, a salt is any ionic compound that does not have an ionizable hydrogen in its structure. In this respect most ionic compounds are salts, which makes the term rather pointless. It is more useful to keep in mind that an acid forms a salt when neutralized by a base, so that HCl, for example, can form a *sodium* salt, NaCl, or a *magnesium* salt, $MgCl_2$, or an *ammonium* salt, NH_4Cl, and so on. Similarly, a hydroxide base such as KOH forms a series of salts with different acids, such as KCl (with HCl) or K_2SO_4 (with H_2SO_4) or K_2CO_3 (with H_2CO_3), and so on.

With relatively few exceptions, all salts dissociate completely in water, so that as a class salts are strong electrolytes. However, there are two main ways in which salts differ from each other:

1 In water solution they may be acidic, basic, or neutral.
2 Their solubility in water may vary greatly.

Neutral, acidic, and basic salts A salt solution will be neutral, acidic, or basic depending on the particular ions it yields in water solution. Let us take NaCl again as an example. Neither the Na^+ ion nor the Cl^- ion has a proton to donate; at the same time, neither ion can accept a proton from water, so that a solution of NaCl is **neutral,** or as neutral as the water in which it was dissolved. A good many salts are neutral (KCl, $NaNO_3$, and $MgCl_2$,

NH_4^+ is acidic because it yields H^+ ions in water to a limited extent, $NH_4^+ \rightleftarrows NH_3 + H^+$

among others), all characterized by the fact that the ions they yield neither donate a proton to water nor accept a proton from water.

Of the acids listed in Table 7.2, the only ion shown is NH_4^+, so that a solution of NH_4Cl is weakly acidic. A number of ammonium salts are used as fertilizer materials because they contribute essential nitrogen to the soil. Among them are ammonium sulfate, $(NH_4)_2SO_4$, ammonium nitrate, NH_4NO_3, and monoammonium phosphate, $NH_4H_2PO_4$, all of which are acidic in reaction.

It was noted earlier that hydroxides are referred to as bases, but they can also be considered as salts that yield OH^- ions in solution. To the extent that they dissolve, the hydroxides dissociate completely and are therefore strong bases. In addition, many salts dissociate to yield basic ions, such as those listed in Table 7.3 and repeated here. Some of these salts merit discussion.

Bases
decreasingly basic →
OH^-
PO_4^{3-}
CO_3^{2-}
NH_3
HPO_4^{3-}
HCO_3^-
CH_3COO^-

Phosphate salts The three phosphate ions are listed in the margin in descending order of basicity; that is, a solution of Na_3PO_4 will be quite strongly basic, a solution of Na_2HPO_4 will be weakly basic, and a solution of NaH_2PO_4 will be faintly *acidic*. Tricalcium phosphate, $Ca_3(PO_4)_2$, which in large deposits is called rock phosphate, is an important fertilizer material and a strong base. However, it is quite insoluble, and to make its phosphorus more available as a soil nutrient it is treated with sulfuric acid, H_2SO_4, the products being $Ca(H_2PO_4)_2$ and a second salt, $CaSO_4 \cdot 2H_2O$. The monocalcium phosphate is both more soluble than the original $Ca_3(PO_4)_2$ and much less basic. The second salt is gypsum, and is a **hydrate** as indicated by the two H_2O molecules it carries. These are not "wet" water molecules but part of the crystalline structure of the compound. The mixture of the two salts, known as **superphosphate,** is a common fertilizer.

PO_4^{3-}
HPO_4^{2-}
$H_2PO_4^-$

$Ca_3(PO_4)_2$ is also a major component of bone structure.

Carbonate and bicarbonate salts A common and cheap basic material is sodium carbonate, Na_2CO_3. It is the washing soda of household use, and industrially is called soda ash. Sodium carbonate is quite soluble in water and ionizes to yield the carbonate ion, so that the solution is quite alkaline. Calcium carbonate also dissociates to yield a CO_3^{2-} ion but is so slightly water-soluble that the solution is only faintly basic. Bicarbonate salts yield the bicarbonate ion, HCO_3^-, which is a very weak base. The sodium salt, $NaHCO_3$, discussed as the essential ingredient in baking powder, is also a common antacid, the "sodium bicarb."

Natural deposits of Na_2CO_3 are not enough to meet demand, so large tonnages are produced industrially by the Solvay process.

$Na_2CO_3 \longrightarrow 2Na^+ + CO_3^{2-}$

$NaHCO_3 \longrightarrow Na^+ + HCO_3^-$

Acetate salts The acetate ion, CH_3COO^-, acts as a very weak base, a very poor proton acceptor. A solution of sodium acetate,

$$H-\overset{\overset{\displaystyle H}{|}}{\underset{\underset{\displaystyle H}{|}}{C}}-\overset{\overset{\displaystyle O}{\|}}{C}-ONa \longrightarrow$$

$$H-\overset{\overset{\displaystyle H}{|}}{\underset{\underset{\displaystyle H}{|}}{C}}-\overset{\overset{\displaystyle O}{\|}}{C}-O^- + Na^+$$

$$NH_3 + H_2O \overset{\longleftarrow}{\rightharpoonup} NH_4^+ + OH^-$$

$$H-\overset{\overset{\displaystyle H}{|}}{\underset{\cdot\cdot}{N}}-H$$
$$\nearrow$$
$$H^+$$

Ammonia

$$H-\overset{\overset{\displaystyle H}{|}}{\underset{\cdot\cdot}{N}}-\overset{\overset{\displaystyle H}{|}}{\underset{\underset{\displaystyle H}{|}}{C}}-H$$
$$\nearrow$$
$$H^+$$

Methylamine

$$\overset{\displaystyle H}{\underset{\underset{\displaystyle H}{|}}{\underset{}{\searrow N \swarrow}}}^{\displaystyle H}$$

*Trivalent nitrogen
in NH_3*

$$H-\overset{\overset{\displaystyle H}{|}}{\underset{\underset{\displaystyle H}{|}}{N^+}}-H$$

*Tetravalent nitrogen
in NH_4^+ ion*

CH_3COONa, is therefore weakly alkaline, as are the K^+ and Ca^{2+} salts of acetic acid. When acetic acid is neutralized by ammonia water, which is basic, the salt obtained is ammonium acetate, CH_3COONH_4. In water the salt dissociates into acetate and ammonium ions; the CH_3COO^- is just about as weak as a base as the NH_4^+ ion is weak as an acid. The opposing effects of the two ions offset each other, with the result that the solution is *neutral*.

Although ammonia gas dissolves readily in water, only a relatively small number of NH_3 molecules react with water and only a small number of OH^- ions are formed. This is shown by the longer left-pointing arrow in the margin equation; the solution is only mildly alkaline. It should be mentioned that this solution is often called ammonium hydroxide, NH_4OH, analogous to $NaOH$ and other hydroxides, in which case its slight alkalinity is explained as the result of its slight dissociation. However, it is more consistent with what is known about ammonia in water and with acid-base theory generally to consider NH_3 as the proton acceptor. It was pointed out in Chapter 5 that NH_3 has two unshared electrons where a hydrogen ion can attach itself.

The same situation would exist if the nitrogen atom were bonded not only to hydrogen atoms but to other atoms as well, as in methylamine. This molecule is also a base, accepting a proton at the unshared electrons of the nitrogen atom, and thus becoming positively charged. Generally, any compound containing *trivalent* nitrogen is a base, because a proton can be accepted by the nitrogen atom, after which this atom has *four* bonds and a plus charge.

Since salts are ionic, and ionic compounds conduct their chemical business through their ions, it is important to recognize the kind and number of ions an ionic compound releases in water solution. The ions may be simple (Na^+) or complex (CO_3^{2-}). A complex ion is a charged *group* of ions that takes part in chemical reactions as a *single entity*. Table 7.5 lists a number of ionic materials and the ions they yield in solution. Follow through each example, and check the names of the complex ions against Table 4.5.

Salt solubility and ionic recombination   Used in reference to living organisms or to foods, "minerals" means the ions contributed by various salts, which in order to dissociate must first dissolve, at least to some extent. Even when an inorganic compound is present as part of a solid structure, such as the $Ca_3(PO_4)_2$ in bones, there are also Ca^{2+}, and PO_4^{3-} ions in plasma, the liquid portion of the blood. The mineral we most often ingest as such is table salt, which dissolves and provides many of the Na^+ ions present in the blood, as well as the Cl^- ions in the stomach juices.

TABLE 7.5

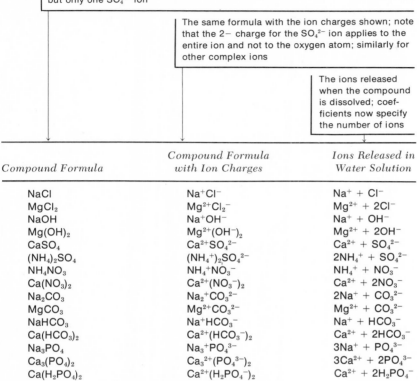

The compound formula with the + ion first and the − ion second; parentheses and a subscript are used to indicate the number of complex ions in a compound; if only one is present, there are no parentheses; for example, $(NH_4)_2SO_4$ contains two NH_4^+ ions, but only one SO_4^{2-} ion

The same formula with the ion charges shown; note that the 2− charge for the SO_4^{2-} ion applies to the entire ion and not to the oxygen atom; similarly for other complex ions

The ions released when the compound is dissolved; coefficients now specify the number of ions

Compound Formula	Compound Formula with Ion Charges	Ions Released in Water Solution
NaCl	Na^+Cl^-	$Na^+ + Cl^-$
$MgCl_2$	$Mg^{2+}Cl_2{}^-$	$Mg^{2+} + 2Cl^-$
NaOH	Na^+OH^-	$Na^+ + OH^-$
$Mg(OH)_2$	$Mg^{2+}(OH^-)_2$	$Mg^{2+} + 2OH^-$
$CaSO_4$	$Ca^{2+}SO_4{}^{2-}$	$Ca^{2+} + SO_4{}^{2-}$
$(NH_4)_2SO_4$	$(NH_4{}^+)_2SO_4{}^{2-}$	$2NH_4{}^+ + SO_4{}^{2-}$
NH_4NO_3	$NH_4{}^+NO_3{}^-$	$NH_4{}^+ + NO_3{}^-$
$Ca(NO_3)_2$	$Ca^{2+}(NO_3{}^-)_2$	$Ca^{2+} + 2NO_3{}^-$
Na_2CO_3	$Na_2{}^+CO_3{}^{2-}$	$2Na^+ + CO_3{}^{2-}$
$MgCO_3$	$Mg^{2+}CO_3{}^{2-}$	$Mg^{2+} + CO_3{}^{2-}$
$NaHCO_3$	$Na^+HCO_3{}^-$	$Na^+ + HCO_3{}^-$
$Ca(HCO_3)_2$	$Ca^{2+}(HCO_3{}^-)_2$	$Ca^{2+} + 2HCO_3{}^-$
Na_3PO_4	$Na_3{}^+PO_4{}^{3-}$	$3Na^+ + PO_4{}^{3-}$
$Ca_3(PO_4)_2$	$Ca_3{}^{2+}(PO_4{}^{3-})_2$	$3Ca^{2+} + 2PO_4{}^{3-}$
$Ca(H_2PO_4)_2$	$Ca^{2+}(H_2PO_4{}^-)_2$	$Ca^{2+} + 2H_2PO_4{}^-$

To reduce tooth decay F^- ions are added to drinking water in the form of soluble sodium fluoride, NaF. The uptake of dissolved minerals in the soil by plant life is the principal path by which animals obtain their mineral requirements.

On the other hand, there are salts that play significant roles because of their *in*solubility. One such salt is calcium carbonate, $CaCO_3$, which has already been discussed as a base of very limited solubility. It is found widely throughout nature and in many different forms. $CaCO_3$ is marble (the colors are minute impurities); it is limestone; it is the shell of an oyster and the pearl in the oyster; it is coral reefs; it is the clear calcite crystals used in optical instruments; it is a geological deposit underlying the island of Manhattan; and it appears as a fine powder in toothpastes. In the mountainous region of northern Italy known as the

Large tonnages of limestone are used as a "flux" in making pig iron. Being basic the $CaCO_3$ combines with acidic impurities in the iron ore, such as silicates.

Dolomites it is found as a **double salt,** $CaCO_3 \cdot MgCO_3$, which is referred to as dolomitic limestone.

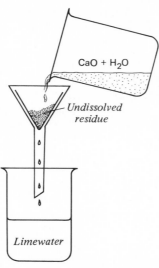

As a salt, $CaCO_3$ is the product of the reaction between acidic H_2CO_3 and basic $Ca(OH)_2$. As mentioned before, a solution of $Ca(OH)_2$ is called limewater, and is prepared by mixing lime and water, then filtering. Since lime, CaO, is only slightly soluble, undissolved CaO remains on the filter paper while the clear liquid passing through is a dilute solution of $Ca(OH)_2$. CO_2 gas passed through the limewater reacts with the water to become H_2CO_3, which is then neutralized by the $Ca(OH)_2$ in solution:

$$CO_2 + H_2O \rightleftharpoons H_2CO_3$$
Carbonic acid

$$H_2CO_3 + Ca(OH)_2 \longrightarrow CaCO_3 \downarrow + 2H_2O \qquad (7.14)$$

As the reaction proceeds, fine particles of insoluble $CaCO_3$ precipitate out, making the limewater turbid. This reaction is a common test for CO_2. If an unknown gas is passed through a clear solution of limewater, the formation of a fine white precipitate is a signal that the gas probably is CO_2. Heating $CaCO_3$ is an example of a decomposition reaction:

$\downarrow$ *means that a precipitate is coming out of the solution.*

$$CaCO_3 \xrightarrow{\text{heat}} CaO + CO_2 \uparrow$$

However, when CaO and CO_2 are brought together in water, they recombine

$$CaO + CO_2 \longrightarrow CaCO_3 \downarrow$$

Although the solubility of a salt can be found in reference handbooks, it is helpful to have some guidelines as a basis for recognizing what happens when different salts are mixed in water, and a rough classification is given here.

1 All Na^+, K^+, and NH_4^+ salts are soluble.
2 All NO_3^- and CH_3COO^- salts are soluble.
3 All Cl^-, Br^-, and I^- salts are soluble, except for AgCl, $PbCl_2$, HgCl.
4 Most OH^-, CO_3^{2-}, and PO_4^{3-} salts are insoluble except where rule 1 applies; for example, NaOH, Na_2CO_3, and Na_3PO_4 are all soluble.
5 SO_4^{2-} salts are generally soluble, although $BaSO_4$ and $CaSO_4$ are important exceptions.

We can now mix and match, and on the basis of the list above see what happens. From rules 1 and 2 both NaCl and $AgNO_3$ are soluble salts, and when they are together in water solution, Na^+, Cl^-, Ag^+, and NO_3^- ions are present. Two of these will "match" to form the insoluble salt, AgCl (see rule 3), which comes down as a fine white precipitate. Adding a drop or two of a silver nitrate solution to about 1 ml of a solution of HCl, KCl, $MgCl_2$, or any chloride salt will similarly result in AgCl precipitating. In other words, this is a *general test* for chlorides. The same idea is used in a test for the presence of sulfate, SO_4^{2-}, ions. From rule 5 we know that $BaSO_4$ is an insoluble salt, so that if Ba^{2+} ions are added to a solution containing SO_4^{2-} ions, the presence of the latter is confirmed by the precipitation of white, insoluble particles of $BaSO_4$. Since $BaCl_2$ is a soluble salt, it is the **reagent** used for testing for sulfate ions in solution; add a drop or two of $BaCl_2$ solution to the unknown solution, and a white turbidity confirms that SO_4^{2-} ions are present. A *reagent* is a chemical *agent* used to bring about a given *re*action.

The insolubility of $BaSO_4$ is also used to advantage in x-ray examination of the gastrointestinal tract. Since these organs are transparent to x-rays whereas barium atoms are relatively opaque to the radiation, the patient has a barium sulfate "cocktail" beforehand. This provides an x-ray picture that outlines the digestive tract so that an abnormality can be seen. If the $BaSO_4$ were not so very insoluble, this would be impossible because Ba^{2+} ions are toxic and cannot be tolerated by the body. But with so very little $BaSO_4$ in solution, the number of Ba^{2+} ions present is negligible.

Many salts, many uses The inorganic salts are many and varied, and some are of vital importance for health, agriculture, and industry. A number have already been mentioned, and several more are listed in Table 7.6 to indicate their variety and range of

NaCl $AgNO_3$
↓   ↓
Na^+ Ag^+
+   ↗↙   +
Cl^- ⤸ NO_3^-

AgCl ↓

This is an example of ionic recombination.

$BaCl_2$
↓
$2Cl^-$
+
SO_4^{2-} ↘ Ba^{2+}

$BaSO_4$ ↓

TABLE 7.6

Salt	Name	Description
$HgCl_2$	Mercuric chloride	Powerful antiseptic; *poisonous*
KNO_3, $NaNO_3$, and NH_4NO_3	Nitrates	Used as fertilizers and in making explosives; NH_4Cl a diuretic
$Bi(OH)_2NO_3$	Bismuth subnitrate	Antacid in stomach disorders
$AgBr$	Silver bromide	Major light-sensitive material in photographic film
$2CaSO_4 \cdot H_2O$	Plaster of paris	Plaster for walls and casts
$AgNO_3$	Silver nitrate, "lunar caustic"	Germicide, formerly added to the eyes of newborn babies; cauterizing agent
$CuSO_4 \cdot 5H_2O$	Cupric sulfate, "blue vitriol"	Fungicide and insecticide
$Al_2(SO_4)_3 \cdot 18H_2O$	Aluminum sulfate	Astringent; contracts tissue and reduces discharges

application. Many salts are found as hydrates, in which the water is part of the crystal structure. Since the water is not held strongly, heating a hydrate drives it off. For example, $CuSO_4 \cdot 5H_2O$ is a deep blue crystalline solid, but when it is heated, the water comes off as steam and white, anhydrous $CuSO_4$ is the result. Since water vapor is always present in the air, hydrates tend to pick up H_2O and revert to the hydrated condition. Any material that can do this effectively can act as a drying agent, or **desiccant.** One of the most widely used drying agents is anhydrous calcium chloride, $CaCl_2$, which forms the dihydrate $CaCl_2 \cdot 2H_2O$ as it removes two water molecules from the air.

Materials that pick up water from the atmosphere are called hygroscopic.

OXIDATION–REDUCTION REACTIONS

Originally, oxidation meant a union with oxygen, and it still does, but only as a special case of a general type of reaction. When a strip of magnesium metal is ignited, it burns in the air with a brilliant light and becomes oxidized to magnesium oxide (magnesia, MgO). In so doing, the magnesium atom gives up its two outer electrons, which are taken by an oxygen atom. The net result is the formation of two oppositely charged ions that unite to form the new chemical entity:

$$Mg - 2e = Mg^{2+}$$
$$O \ + 2e = O^{2-}$$

Adding, we get $Mg + O = MgO$

Since oxygen is present in the air as O_2, the balanced equation for the reaction is

$$2Mg + O_2 = 2MgO$$

which is simply twice the original equation but the same reaction. However, magnesium will also combine directly with chlorine gas, Cl_2, or with fluorine, F_2, and in the same way:

$$Mg - 2e = Mg^{2+}$$
$$2Cl + 2e = 2Cl^-$$

Adding, we get $\quad Mg + Cl_2 = MgCl_2$

Whether combining with oxygen, or with chlorine, the magnesium lost two electrons to its partner. It is the *loss of electrons* that is called **oxidation,** whether the partner is oxygen or any other acceptor of electrons. Looking at it now from the other end, the *gain of electrons* is known as **reduction,** and the total chemical event is called an oxidation-reduction reaction. Oxidation and reduction are simply the two ends of an electron transfer from a donor to an acceptor. Since the magnesium gave its two electrons to oxygen in the first case and to chlorine in the second, causing them to be reduced, magnesium is designated as the **reducing agent.** Note that a reducing agent is itself oxidized. Conversely, since the oxygen and the chlorine were the recipients of the electrons from magnesium, they are termed **oxidizing agents.** Note that an oxidizing agent is itself reduced.

To help the memory: oxidation = electron loss

OXIDATION POTENTIALS

Electron transfer can occur between atoms, molecules, or ions, and oxidation-reduction reactions are important in every aspect of chemistry. When an active metal, such as magnesium, displaces the hydrogen of an acid, such as HCl, hydrogen gas is evolved and a magnesium salt is formed. The reaction may be described as a displacement, but the mechanism is **electron transfer.** Because magnesium holds its outer electrons very loosely, it is a good electron donor; this is what makes it an active metal. On the other hand, the hydrogen ions in the acid solution will gladly accept electrons, thereby becoming hydrogen atoms. The result is a spontaneous passage of electrons from magnesium atoms to hydrogen ions; the hydrogen ions become hydrogen atoms and pair up to escape from the system as H_2 gas, leaving behind Mg^{2+} and Cl^- ions:

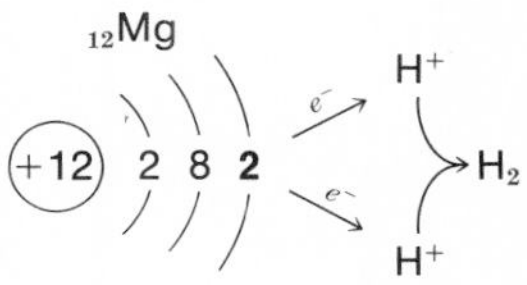

$$Mg - 2e^- \xrightarrow{\text{oxidation}} Mg^{2+}$$

$$2HCl \longrightarrow 2Cl^- + 2H^+ + 2e^- \xrightarrow{\text{reduction}} H_2 \uparrow + 2Cl^-$$

$$Mg + 2HCl \xrightarrow{\hspace{4cm}} H_2 \uparrow + MgCl_2 \qquad (7.15)$$

The better the electron donor and the more eager the electron acceptor, the more vigorous the electron transfer from donor to acceptor will be. The transfer of electrons is an electric current, and the driving force behind a flow of current is its electromotive force, measured in volts (V).

The "pressure" behind an electron transfer from a donor to an acceptor can, therefore, be expressed as a **voltage,** and experimental procedures for determining it have been standardized. When magnesium reacts with HCl, the electrical pressure behind the transfer of electrons from Mg to H^+ is 2.37 V. This value is a measure of the ability of Mg to donate electrons, and is called its standard **oxidation potential** $E°$. In determining the oxidation potentials of materials other than magnesium the *same* electron receiver must be used; otherwise the results would not be comparable. The **couple** $H_2/2H^+$ is the common partner used in determining oxidation potentials, and since it is the point of reference, it is reasonable to fix its own oxidation potential as 0.00 V.

The electron flow from Mg to $2H^+$ is spontaneous, and the oxidation potential of 2.37 V is positive. This also implies that an *equal* and *opposite* voltage, -2.37 V, will reverse the spontaneous electron flow and be a measure of the **reduction potential** of the couple. Using Mg/Mg^{2+} as the couple, this is represented as shown in Table 7.7.

The positive value of the oxidation potential is saying that when Mg loses electrons to H^+ ions, the electrical pressure behind the transfer is $+2.37$ V. The negative value of the reduction potential indicates that a potential of 2.37 V must be *imposed* for Mg^{2+} to receive two electrons from H_2; in other words, the reduction reaction is not spontaneous.

As electron takers, the H^+ ions of the couple are reduced to H_2 gas, as in the reaction with Mg. This means that any material with a positive oxidation potential will exert an electron push on H^+ ions and will displace them from acids. As electron donors,

In a couple the reduced form of the pair is on the left, and the oxidized form is on the right.

TABLE 7.7

Reduction Potential, V	*Couple*	oxidation reaction → ← reduction reaction	*Oxidation Potential,* V
-2.37	$Mg\|Mg^{2+}$	$Mg - 2e = Mg^{2+}$	$+2.37$

TABLE 7.8 Some representative oxidation-reduction potentials

Reduction Potential, V	Couple	Reaction $\frac{oxidation}{reduction}$	Oxidation Potential $E°$, V
−2.92	K|K$^+$	K− e = K$^+$	+2.92
−2.87	Ca|Ca^{2+}	Ca− 2e = Ca^{2+}	+2.87
−2.71	Na|Na$^+$	Na− e = Na$^+$	+2.71
−2.37	Mg|Mg^{2+}	Mg− 2e = Mg^{2+}	+2.37
−0.76	Zn|Zn^{2+}	Zn− 2e = Zn^{2+}	+0.76
−0.44	Fe|Fe^{2+}	Fe− 2e = Fe$^{2−}$	+0.44
0.00	H$_2$|2H$^+$	H$_2$− 2e = 2H$^+$	0.00
+0.34	Cu|Cu^{2+}	Cu− 2e = Cu^{2+}	−0.34
+0.77	Fe^{2+}|Fe^{3+}	Fe^{2+}− e = Fe^{3+}	−0.77
+1.36	2Cl$^-$|Cl$_2$	2Cl$^-$− 2e = Cl$_2$	−1.36
+2.87	2F$^-$|F$_2$	2F$^-$− 2e = F$_2$	−2.87

the H$_2$ molecules are oxidized to H$^+$ ions, so that materials with negative electrode potentials will be reduced by hydrogen gas. A listing of the oxidation potentials for a series of couples arranged in order of decreasing value of $E°$ is very useful as a guide in judging the direction of electron transfer between any two substances on the list. The higher the value of the oxidation potential the more readily it donates electrons and acts as a reducing agent; the lower the value of the oxidation potential the more readily the material accepts electrons and acts as an oxidizing agent. Table 7.8 gives some selected oxidation-reduction potentials.

The higher up a couple is in Table 7.8, the more strongly the indicated oxidation reaction will occur and the more the couple will tend to go from left side to right side. All the metals *above* hydrogen will therefore *displace* hydrogen from an acid by pushing electrons *to* the H$^+$ ion in the same manner as Mg. In fact, K, Ca, and Na will even displace hydrogen from water:

$$2Na + 2H_2O \longrightarrow H_2 \uparrow + 2NaOH$$

The simultaneous formation of a very combustible gas, H$_2$, and a strong alkali, NaOH, plus the heat generated by the reaction is very dangerous, so that sodium metal is usually stored in a nonreacting liquid such as kerosene. Because these metals (K, Ca, Na) are so easily oxidized, they are never found in the earth as the free metal but only as ores, in combination with oxygen, sulfur, chlorine, fluorine, and other atoms, all oxidizing materials. All these active metals will release hydrogen gas from an acid, often too vigorously for safety.

Elementary iron has been found in meteorites, which are from outer space. Iron is also believed to be in inner space—the core of the earth.

Zn   Fe

~~?~~

Oxygen

Zn   Fe

2e⁻   2e⁻

O

ZnO

In generating H_2 gas for laboratory purposes zinc metal is generally used, because the metals above it in the list may react too violently. Zinc will also act as a reducing agent for all the materials listed *below* it. This is the reason why zinc galvanizing protects an iron surface. The zinc is more easily oxidized than the iron. The oxidizing agent in the air is, of course, oxygen, and when oxygen atoms are faced with a choice between Fe and Zn, they will unite with the atom that most readily will *give them* electrons and make them negative. The zinc is the better electron giver, so that the union between zinc and oxygen is the one that prevails, happily for the iron. Magnesium also acts as a sacrificial material and steel pipes and tanks may be protected from oxidation by being attached to magnesium rods that oxidize preferentially. Antioxidants are also used in foods to keep them from going bad. Obviously, metals would not be suitable, but the materials that are used have the same tendency for easy oxidation.

The oxidation potentials listed can also explain what happens when a zinc rod is immersed in a solution that contains Cu^{2+} ions, such as a $CuSO_4$ solution. From the table, zinc will oxidize to Zn^{2+} much more readily than Cu will oxidize to Cu^{2+}, so that the electrons released by the Zn to Zn^{2+} oxidation reaction are transferred to the Cu^{2+} ions in the solution with a push equal to the *difference* between their $E°$ values. This difference is $0.76 - (-0.34)$ V, or 1.10 V. As the Cu^{2+} ions accept the electrons pushed on them from the Zn, they precipitate out as reddish, metallic copper, while the zinc rod gradually disappears as it changes to Zn^{2+} ions. In summary,

$$Zn + CuSO_4 \longrightarrow Cu + ZnSO_4$$

(with $2e^-$ transfer arrow above)

THE OXIDATION OF CARBON AND HYDROGEN

Of the many oxidation-reduction reactions utilized in industry, the most important one is the oxidation of carbon and carbon-containing fuels. This reaction is the major source of energy available to us, whether wood, coal, petroleum products, or natural gas is being burned. The reaction of carbon is simple:

$$C + O_2 = CO_2$$

Since the carbon is oxidized, it must itself be a *reducing* agent, and in fact it is the most important reducing agent used in the extraction of metals from their ores, generally in the form of coke.

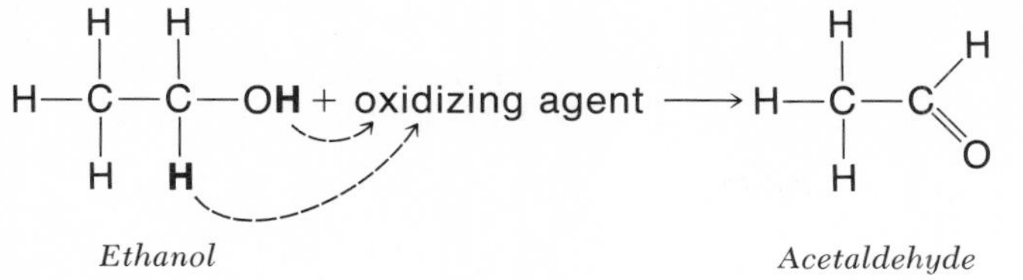

FIGURE 7.7 Ethanol oxidized to acetaldehyde.

Although the oxidation of carbon contributes to the energy supply of living organisms, the oxidation of *hydrogen* is far more important. Again the *net* reaction is very simple:

$$2H_2 + O_2 = 2H_2O$$

Although the reaction as written is actually the *end* result of a sequence of step-by-step oxidations, hydrogen is nonetheless a reducing agent in relation to oxygen. The oxidation of organic compounds, which almost always contain hydrogen atoms, often results in the *loss* of hydrogen to the oxidizing agent. A simple example is what happens when ordinary alcohol, or ethanol, is oxidized; it loses two hydrogen atoms (Figure 7.7). In giving up two hydrogens, the ethanol has, in effect, lost two electrons and two H^+ ions to the oxidizing agent. The loss of electrons is the essential meaning of oxidation, so that the loss of hydrogens by an organic compound is equivalent to an oxidation reaction. That is, *hydrogen loss equals oxidation.* By the same token, gaining hydrogen atoms by an organic compound, each hydrogen contributing its electron *to* the compound, is a *reduction.*

DECOMPOSITION REACTIONS

Although a compound is a stable structure that will not fall apart spontaneously, many compounds can be separated into smaller units or into their elements. Since this involves severing at least some atom-to-atom bonds, decomposition involves some energy input, as heat, electric energy, or energy in some other form. For example, water is decomposed into its elements, H_2 and O_2, by **electrolysis,** the passage through it of an electric current. A good many oxides are vulnerable to decomposition by heating, yielding oxygen gas. A common method of generating oxygen in a laboratory is to heat potassium chlorate, $KClO_3$. The reaction is

$$2KClO_3 \longrightarrow 2KCl + 3O_2 \uparrow$$

Also, metals which are resistant to oxidation (their $E°$ value is low), such as mercury, silver, and platinum, will nevertheless form oxides when treated with strong oxidizing agents. However, the oxides tend to decompose when heated. For example,

$$2HgO \longrightarrow 2Hg + O_2 \uparrow$$

A special and important type of decomposition in biochemical reactions results in the loss of a carbon dioxide molecule. This is called **decarboxylation,** and is undergone by organic acids. Enzymes in the living cell catalyze the reaction, which is part of the energy-producing process.

CONDENSATION REACTIONS

The term **combination** is a very general one in chemistry. It applies to any joining together of atoms or groups of atoms to make larger structures. The oxidation of carbon to carbon dioxide is, of course, a combination, and so is the reaction between the CO_2 and H_2O to form carbonic acid. The phrase "combines with" appears repeatedly when we are talking about chemical reactions. **Condensation,** which has a more restricted meaning, means joining relatively small organic molecules together to produce larger ones. In many cases the joining results in the elimination *at the point where the bond is formed* of a small molecule. The eliminated molecule is usually water, but it may be ammonia or an alcohol. A closer look at condensation reactions must wait until more is said about organic reactions, but it may be mentioned that the class of compounds called **carbohydrates** is largely composed of condensation products. The small units that take part in the condensation are generally six-carbon molecules, or hexoses. Joining these hexoses together, with a loss of a water molecule at each joint, results in the huge chemical structures we call starch and cellulose.

These hexoses are called simple sugars or monosaccharides.

HYDROLYSIS REACTIONS

Lysis is the Greek for loosening, hence hydrolysis and electrolysis.

Hydrolysis can be considered a reverse of condensation, the separation of a condensed structure into its individual units by the action of *water*. For example, the hydrolysis of a starch, if carried to completion, results in only one product, *glucose* molecules, indicating that starch is made up by the successive joining, or condensation, of glucose units. Hydrolysis requires a catalyst (an enzyme in living tissue). In the laboratory small additions of an acid or base serve the purpose. Hydrolysis reactions are essential to a great many biological functions. For example, the entire process of digestion is essentially a series of hydrolysis reactions. The foods we ingest contain chemical structures impossibly large for later passage into the bloodstream and thence to the cells where they are needed. Using a series of enzymes, the digestive organs reduce these huge molecules, by hydrolysis, to smaller units that can be absorbed through the wall of the small intestine. Hydrolysis reactions are also utilized industrially and are one of the most common reactions of organic chemistry.

SUBSTITUTION REACTIONS

A substitution reaction is a replacement of an atom or group of atoms by another atom or group of atoms. The term is generally restricted to organic compounds, and the process is characteristic of them. In fact, one of the reasons why so many organic compounds exist, with new ones being developed constantly, is their ability to undergo substitution. Methane, CH_4, is the simplest of all organic compounds, yet by successive substitution by only *one* kind of atom, say chlorine, it will yield four different substitution products:

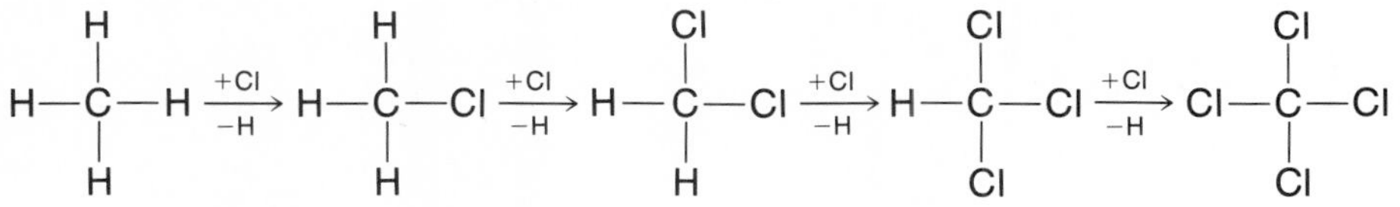

Obviously, the more complex a molecule, the more atoms there are that can be replaced by others and the more substitution products can result. For example, in the benzene molecule, C_6H_6, any one of the six hydrogens can be replaced, after which any one of five can be replaced, etc. The replacing atoms can be Cl, Br, an —OH group, an —NO_2 group, or many others. It is easy to see that substitution products of benzene can proliferate. The carbon atoms serve as the framework for the molecule, and it is the attachments that are being replaced.

ADDITION REACTIONS

Addition reactions, another variety of a combination reaction, are also characteristic of organic compounds. For an additional atom or group to become attached to an organic molecule there

Before

```
    |    |
 —C ∷ C—
    ⌄  ⌄
    H ∶ H
```

After

```
    |   |
 —C ∶ C—
   ∙∙  ∙∙
   H    H
```

The halogens are F, Cl, Br, *and* I.

must be a receptive point or points somewhere in that structure. In Chapter 5, when σ and π bonds were discussed, it was pointed out that a *double* bond between carbon atoms consisted of one σ and one π bond, which are not alike. A carbon compound whose carbon atoms are all held by single bonds is full up, or *saturated*, and cannot accept any additional atoms or groups. The double bond, on the other hand, is a point of *unsaturation*, because the less stable π bond tends to become undone, releasing two electrons that can each form a single σ bond with another atom. This occurs when hydrogen atoms add at a double carbon-to-carbon bond. The more double bonds an organic compound contains, the more unsaturated it is and the more additions it can accept. A variety of atoms or groups can be added, among them hydrogen, halogen atoms, or a water molecule.

HYDRATION-DEHYDRATION REACTIONS

When an unsaturated organic structure takes on a water molecule at a double bond, the reaction is called a **hydration.** If the molecule, at a later time, should *lose* that water, the reaction is a **dehydration.** It will be shown in a later chapter that the addition of a water molecule at a double carbon-to-carbon bond results in the formation of an —OH group, which in carbon chemistry is characteristic of an alcohol.

Hydration and dehydration are of special significance because they are part of the mechanism whereby energy is produced by living matter.

ISOMERIZATION REACTIONS

In one biological reaction a molecule of citric acid loses a molecule of water (dehydration) and in the next step gains a molecule of water (hydration). Does this mean that there is no net change in the citric acid? No, the composition of the citric acid is unchanged, but its arrangement of atoms *has* changed, and this new arrangement makes the next step in the process possible. The rearranged citric acid is called **isocitric acid.** In life chemistry arrangements and rearrangements are of critical importance. The change in the arrangement of atoms of citric acid accomplished by a successive dehydration and hydration is also called an *isomerization* reaction. Two or more compounds are *isomers* if they have the same composition but different arrangements of their component atoms. The different arrangements give different substances with different properties. Isomeric rearrangements occur at several points along the major chemical pathways of living cells, and they are therefore critical reactions.

REVIEW QUESTIONS

1 In general, what always happens in the course of a chemical change, and why does it involve an energy change?

2 Although the reaction between Na and Cl atoms to form NaCl appears to be a simple combination, it is not really. Show how the reaction proceeds due to an electron transfer from one atom to the other and is therefore an oxidation reduction. What are the chemical units that *finally* combine to produce NaCl?

3 Of all the ions, the H^+ ion is the most unusual; explain why.

4 It is obvious that many things are acidic, e.g., vinegar or sour milk, but why can it be said that a piece of limestone is an antacid, or a base?

5 List the differences between acidic and basic materials. Why are both electrolytes?

6 What is meant by acid-base neutralization? Illustrate by means of an example. What are the products obtained as the result of an acid-base neutralization?

7 The concept of proton transfer is used to interpret the nature of acids and bases and their reactions. In terms of proton transfer, what is an acid and what is a base?

8 What ion is always present in a water solution of an acidic material? In a water solution of a basic material? How does this lead to the conclusion that an acid-base neutralization results in the formation of undissociated water molecules plus a salt?

9 What is meant by a strong acid? A weak acid? Give examples.

10 What ions are present in   **a** a solution of HCl,   **b** a solution of H_2SO_4,   **c** a solution of HNO_3?

11 Why is acetic acid referred to as a weak acid?

12 What is meant by an amphoteric material?

13 Why are nonmetallic oxides such as SO_2, SO_3, and CO_2 considered acidic despite the fact that they have no H^+ ions in their molecule?

14 **a** Explain how carbon dioxide is constantly being poured into the atmosphere. **b** Explain how carbon dioxide is constantly being taken *from* the atmosphere. **c** Why must CO_2 be constantly removed from the cells of the body?

15 Water that is exposed to the air will pick up some CO_2 gas from the atmosphere. **a** From a consideration of Eqs. (7.4) and (7.5), what chemical species (molecules, ions) will be present in water due to this interaction with CO_2? Of these, which is present in the largest and which in the least amount? **b** When CO_2 is forced into water under pressure, as in making of soda pop, the solution becomes more acidic. From a consideration of Figure 7.4, explain why the acidity increases. **c** Suppose you were doing an experiment in

which the water must contain as *little* CO_2 as possible; what is a rather simple way of reducing the CO_2 content of the water?

16 **a** In the process whereby CO_2 is eliminated from tissue cells, how and where do the CO_2 molecules enter the bloodstream? **b** What is the chemical response to this increase in CO_2 pressure? **c** When the blood gets to the lungs, how do the hemoglobin molecules assist in converting the HCO_3^- ions to the CO_2 gas and water vapor that is finally exhaled? Do you know of any other function performed by hemoglobin?

17 Why are the hydroxides, such as NaOH and $Ca(OH)_2$, considered bases although they do not react with water to form OH^- ions?

18 The K_b value of the carbonate ion, CO_3^{2-}, is 2.1×10^{-4}, whereas K_b for the bicarbonate ion, HCO_3^{-3}, is 2.3×10^{-8}, so that the carbonate ion is much the stronger base. From a consideration of the structure of the carbonate ion explain its higher basicity.

19 What evidence is there that hydroxide ions, OH^-, whether they result from the dissociation of hydroxides or from the interaction between basic materials and water, are themselves proton acceptors?

20 Why are metallic oxides, such as MgO or CaO, both of which are solids, considered basic in nature? Since a nonmetallic oxide such as CO_2 is acidic and a metallic oxide such as CaO is basic, it would be expected that they would interact, and in fact they do. What compound is formed between them? Does this reaction qualify as an acid-base neutralization even though no water is produced?

21 Each reaction is an acid-base neutralization; complete and balance each equation.
a $H_2SO_4 + NaOH \longrightarrow$ **b** $HNO_3 + Mg(OH)_2 \longrightarrow$
c $H_2SO_4 + Ca(OH)_2 \longrightarrow$ **d** $H_3PO_4 + KOH \longrightarrow$

22 Complete and balance the following:
a $H_2SO_4 + CaO \longrightarrow$ **b** $HNO_3 + Zn \rightarrow$
(Zn is an active metal with a valence of $2+$)
c Of these two reactions, which is an acid-base neutralization?

23 Sodium bicarbonate is a common ingredient in antacids used to relieve acid stomach and that morning-after feeling. From a consideration of Eq. (7.13), how does the bicarbonate ion act to reduce acidity and what products are formed? Note that the excess H^+ ions are *not* entering the sequence as shown in Eq. (7.12); where *can* the H^+ ions be considered as entering the reaction?

24 When $NaHCO_3$ is used as an ingredient of baking powder, the reaction is chemically the same as in Question 23. However, there is a difference in respect to what happens to the CO_2 produced; what purpose does the CO_2 serve when $NaHCO_3$ is used in a baking powder? What must all baking powders include in addition to the bicarbonate?

25 **a** What does a buffer system do? **b** What are the two components of the carbonic acid buffer system in the blood? How does each component contribute a buffering action?

26 Go down the list of salts in Table 7.5 and from a consideration of the ions released by each salt in water solution, indicate whether the solution of that salt is acidic, basic, or neutral.

27 A common test for the presence of CO_2 gas is to pass it through limewater. Describe what happens and *why*.

28 Which of the salts listed below are generally water-soluble, and which are not; give your reason for each answer.

NaCl $Mg(OH)_2$

$CaSO_4$ KOH

Na_2SO_4 $Ca(HCO_3)_2$

CaCO3 $NaHCO_3$

NH_4CO_3 AgCl

29 Look up the formula for epsom salt in your dictionary; is it an anhydride or a hydrate? Why do anhydrides tend to become hydrates in time? Why is $CaCl_2$ used as a desiccant?

30 **a** Which of the following metals will react with an acid, such as HCl, resulting in the production of H_2 gas and the corresponding salt: Fe, Ca, Zn, Cu? **b** Is the reacting metal itself oxidized or reduced? Explain. **c** Are the hydrogen ions of the acid oxidized or reduced?

31 From Table 7.8 the oxidation reaction $Zn - 2e = Zn^{2+}$ is shown as proceeding with a potential $E°$ of $+0.76$ V. However, for this oxidation reaction to proceed there must be an acceptor of the electrons. Relative to the couple $Zn|Zn^{2+}$, which couples will accept these electrons? In so doing, what reaction will these couples undergo?

32 In generating hydrogen gas in a laboratory, why is zinc used rather than calcium metal?

33 Although there are many oxidizing agents in use for specific purposes, the most available of all oxidizing agents is the gaseous oxygen of the atmosphere. Write the equations for the reaction between carbon and oxygen and between hydrogen gas and oxygen. Both these reactions are combustions, rapid reactions that release large quantities of heat in very short time. Can there be slow oxidations? What happens when iron rusts?

34 The simplest alcohol is methanol, and when it is oxidized, it loses two hydrogen atoms to become formaldehyde:

Methanol   Formaldehyde

a Why is the loss of two hydrogens an oxidation reaction? **b** If the formaldehyde were to *gain* two H atoms, it would convert back to methanol; why would this be a reduction?

35 By reference to the several types of reactions discussed toward the end of this chapter, decide which reaction type best describes each of the following. Note that each reaction is designated by a roman numeral.

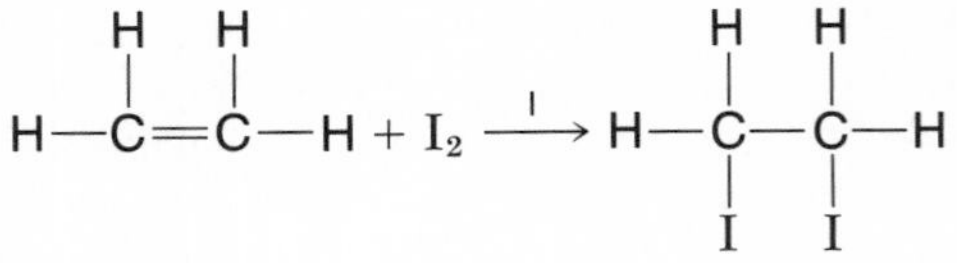

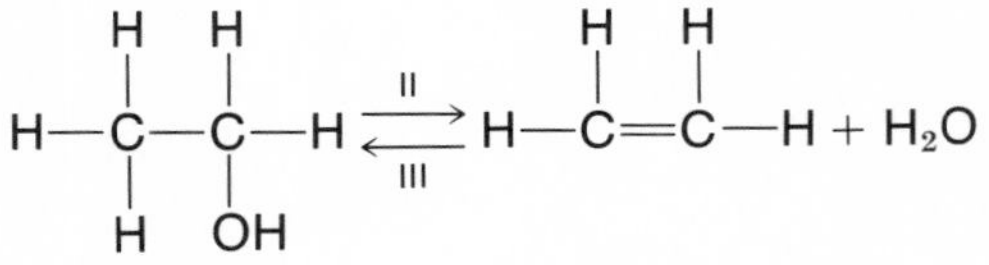

The conversion of methane, CH_4, to carbon tetrachloride, CCl_4 (IV).

$$2KClO_3 \xrightarrow{\text{V}} 2KCl + 3O_2$$

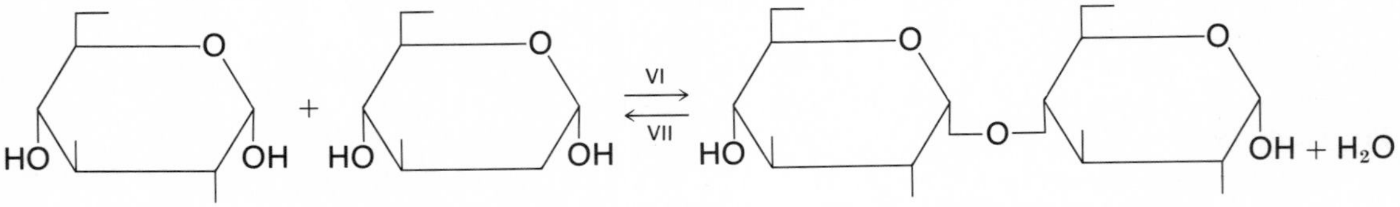

Glucose   Glucose   Maltose

36 The formulas for a number of acidic and basic materials are given below. Identify the substance by name, and indicate whether it is a strong, moderately strong, or weak acid or base. Indicate whether the compound is generally soluble in water or has poor solubility.

HNO_3	H_3PO_4	HCl	$CaCO_3$	H_3BO_3
KOH	$Mg(OH)_2$	$NaOH$	Na_2CO_3	$NaHCO_3$
Na_3PO_4	CH_3COONa	H_2CO_3	NH_4Cl	H_2SO_4

THE MOLE, MOLARITY, NORMALITY, AND pH
Chemical Bookkeeping Among the Honest Atoms

When molecules, atoms, and ions do business with each other, there can be no cheating. A calcium atom cannot be talked into combining with only one chlorine atom; deep in its electron configuration it knows that two are needed. When H^+ and OH^- ions meet, they always unite one to one, forming one water molecule. Obviously, the arithmetical relations reflect the nature of the chemical species that are reacting, and for the most part the arithmetic is simple. Understandably, the more complicated the reaction, and the more variables (temperature, pressure, etc.) that become involved, the more complex the arithmetic; in theoretical studies the treatment also becomes more abstract. However, the essential ideas are not so different. Presented here are a few simple principles for dealing with the quantitative aspects of chemical reactions. Of these, the concept of the **mole** is the most central.

$$Ca + Cl \xrightarrow{\quad} \quad \textit{Not enough}$$
$$Ca + 2Cl \longrightarrow CaCl_2 \quad \textit{Just enough}$$
$$Ca + 3Cl \xrightarrow{\quad} \quad \textit{Too much}$$

THE CONCEPT OF THE MOLE

In the two simple examples given, and in chemical reactions generally, the numbers expressing reactants and products are integers. That is, the relations work out as 1:2, 1:1, 1:3, and so

Ca

Cl $\diagdown$ $\diagup$ Cl

CaCl$_2$

100Ca

100Cl $\diagdown$ $\diagup$ 100Cl

100CaCl$_2$

$$H^+ + OH^- \longrightarrow H_2O$$

$$5000\,H^+ + 5000\,OH^- \longrightarrow 5000\,H_2O$$

$$C + O_2 \longrightarrow CO_2$$

Since the exact *atomic weight of* H = 1.008, 1 *is close enough in most cases.*

on. The ratio is the same no matter how many times the given reaction occurs, so that if 1 Ca atom unites with 2 Cl atoms to form 1 CaCl$_2$ unit, 100 Ca atoms will combine with 200 Cl atoms to produce 100 CaCl$_2$ units. If 1 H$^+$ ion needs 1 OH$^-$ ion to make 1 H$_2$O molecule, then 5000 H$^+$ ions will react with exactly 5000 OH$^-$ ions to make 5000 H$_2$O molecules. So far so good, but suppose you want to obtain a certain amount of CO$_2$ by burning carbon in the air. The reaction tells you that one CO$_2$ molecule is obtained for each carbon atom oxidized. But how would you determine the number of carbon atoms needed to produce, say, 5 g of CO$_2$? All that can be said at this point is that if x CO$_2$ molecules add up to 5 g, an equal number of carbon atoms are needed.

How do you go about counting molecules, atoms, ions, or electrons? Even if it could be done in principle, how long would it take? If carbon atoms were being counted out at the rate of one per second, it would take more than a million billion years to collect 1 g. Obviously, direct counting is out of the question, but a way around it has already been indicated. It would take an impossibly long time to count out the atoms needed to make up 1 g of carbon, but to *weigh* it out is a routine procedure that takes little time or effort. It is not so unusual to *weigh* out a collection of similar articles instead of counting them out. The ordinary carpenter's nails are bought and sold by weight, which is much easier than counting them out singly, even if they must be hammered in one by one. In order to relate some mass of carbon or any other element to the number of atoms in that mass, a conversion factor is required between atom mass and atom population.

What do we know, at this point, of the weights of the hundred or so different atoms of the periodic table? We know their weights relative to each other, their **atomic weights.** For example, each calcium atom, whose atomic weight is 40, is 40 times as heavy as each hydrogen atom, atomic weight = 1. The weight ratio of 40:1 remains the same for all equal numbers of calcium and hydrogen atoms. Using some arbitrary numbers, we can express this as

$$\frac{40}{1} = \frac{\text{wt. of 1 Ca atom}}{\text{wt. of 1 H atom}} = \frac{\text{wt. of 10 Ca atoms}}{\text{wt. of 10 H atoms}}$$
$$= \frac{\text{wt. of 1000 Ca atoms}}{\text{wt. of 1000 H atoms}} = \ldots.$$

In other words, the ratio of the weight of any number of calcium atoms to the weight of the same number of hydrogen atoms is the ratio of their atomic weights, 40:1.

Now imagine that the number of hydrogen atoms keeps increasing, from 1000 to 1 million, to a 100 million, on and on, and with it the weight of all these hydrogen atoms taken together keeps increasing. Eventually, the number of hydrogen atoms will increase to a point where their total weight in grams will just equal the numerical value of the atomic weight of hydrogen; that is, they will weigh 1 g. Symbolize this collection of hydrogen atoms as N, and call it **Avogadro's number.** Now, how much would the same number N of calcium atoms weigh?

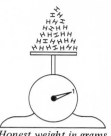

Honest weight in grams

The weights of equal numbers of calcium and hydrogen atoms is always in the same ratio, 40:1, so that if N hydrogen atoms weigh 1 g, N calcium atoms must weigh 40 g. This is simply an extension of the expression above:

$$\frac{40}{1} = \frac{\text{wt. of 1 Ca atom}}{\text{wt. of 1 H atom}} = \frac{\text{wt. of 1000 Ca atoms}}{\text{wt. of 1000 H atoms}}$$

$$= \frac{\text{wt. of } N \text{ Ca atoms}}{\text{wt. of } N \text{ H atoms}} = \frac{40 \text{ g}}{1 \text{ g}}$$

What is true of calcium and hydrogen will also be true for any other atom and hydrogen, the only difference being that the atomic weight would not be 40 but the atomic weight of that atom, so that the weight of N of these atoms would be its atomic weight in grams. That is, the weight of N number of *any* atom is numerically equal to *its atomic weight in grams*. The number N has been experimentally determined as 6.02×10^{23}, a huge number but still a finite collection. If a dozen is the name for a collection of 12 things, a score is the name for a collection of 20 things, and a gross the name for a collection of 144 things, a *mole* is the name for a collection of 6.02×10^{23} things. Unlike these other collections, which are arbitrary and have no special virtue except familiarity, a mole of any chemical species will have a weight, *in grams*, equal to the numerical value of the formula weight of that species, whether it is an atom, a molecule, or an ion.

In the case of atoms, the formula weight is the atomic weight; in the case of molecules, which are composed of atoms, the formula weight is the sum of the atomic weights; in the case of ions, which are also composed of atoms, plus or minus one or more electrons, the formula weight is again the sum of the atomic weights, since the very small electron weights can be neglected. It is also possible to speak of a mole of electrons, that is, a collection of 6.02×10^{23} negative particles whose formula weight is 0.00055 and which together weigh 0.00055 g. Incidentally, this indicates why it is nearly always possible to neglect the weight of electrons since a mole of the lightest atom, hydrogen, weighs a bit over 1 g.

HOW MUCH TO HOW MANY VIA THE MOLE

The mole is really a conversion factor between "how many" by number to "how much" in grams. The "how many" is standardized and fixed at $N = 6.02 \times 10^{23}$, whereas "how much" depends on the formula weight, and therefore differs with the particular chemical species. A mole of any substance therefore has two meanings:

$$N \; (= 6.02 \times 10^{23}) = 1 \text{ mole} = \text{formula weight in grams}$$
$$\longleftarrow \qquad \longrightarrow$$

The two arrows point to the two meanings of "mole," but keep in mind that they are strictly related, by the fact that

N number of any   = the formula weight of that
   chemical species   chemical species in grams

so that N number of H atoms = 1 gram of H atoms

N number of H_2O = 18 grams of H_2O molecules, etc.
   molecules, etc.

The population of each of the species listed is the same, 6.02×10^{23}, or 1 mole. The weight of each mole, however, is generally different, since it depends on the relative weight, or formula, weight of each species. This is like the obvious fact that a

TABLE 8.1 The weight of 1 mole of some molecules, atoms, and ions

Species†	Formula Weight	Calculation	Weight of 1 mole, g
e^-, electron	0.00055		0.00055
H atom	1		1
H^+ ion (proton)	1		1
H_2 molecule	2	2×1	2
C atom	12		12
CH_4 molecule	16	$12 + 4$	16
O atom	16		16
O_2 molecule	32	16×2	32
OH^- ion	17	$16 + 1$	17
H_2O molecule	18	$2 + 16$	18
Na atom	23		23
CO_2 molecule	44	$12 + 32$	44
NaOH unit	40	$23 + 17$	40
$Ca(OH)_2$ unit	74	$40 + (2 \times 17)$	74
$C_6H_{12}O_6$ molecule	180	$6 \times 12 = 72$	180
		$12 \times 1 = 12$	
		$6 \times 16 = \underline{96}$	
		180	

† 6.02×10^{23} of each species makes 1 mole of that species.

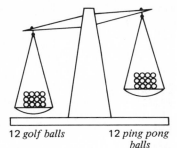

12 *golf balls*   12 *ping pong balls*   **FIGURE 8.1** The population may be the same, but the relative weights vary.

dozen ping-pong balls weigh less than a dozen golf balls, which weigh less than a dozen baseballs. The population is the same in each case, but the relative weights vary (Figure 8.1).

THE MOLE IN CHEMICAL REACTIONS

The usefulness of this concept of the mole is clear. Whereas chemical substances react by number of reacting units in accordance with a specific ratio, what is seen and what can be measured are interactions between weights of reactants to form some weight of product or products. The mole is the conversion factor between the two, so that instead of considering the *numbers* of the reacting species, their *weights* are used. Through the mole relation this becomes the measure of how many units of each reactant participates in the reaction. Let us consider the simple example used before, the burning of carbon in air, which chemically is the oxidation of carbon by oxygen. The chemical event is schematically represented in Figure 8.2, and the equation is

Mole relates weight of product to weight of reactants.

$$C + O_2 = CO_2 \qquad\qquad (8.1)$$

What Figure 8.2 and the equation say is that in this chemical event one carbon atom combines with one molecule of oxygen to

C   + O_2 = CO_2

FIGURE 8.2 Carbon and oxygen combine to form carbon dioxide.

form one molecule of gaseous carbon dioxide. Nothing has been said about how many times this happens. It could be once, in which case it would be practically impossible to know whether it occurred at all, or it could happen $N = 6.02 \times 10^{23}$ times (or some large fraction or multiple of N), which would make it directly observable and measurable. In fact, chemists introduce the mole concept into Eq. (8.1) and all equations by assuming that the chemical event shown takes place $N = 6.02 \times 10^{23}$ times. This does not mean that it *must* take place that many times, but that becomes the point of reference. Remember that N atoms or molecules add up to 1 mole of that atom or molecule and that the weight of that mole adds up to the formula weight of the species in grams. On this basis, Eq. (8.1) takes on a broader meaning, since C now means 1 *mole* of carbon weighing 12 g; O_2 now means 1 *mole* of oxygen molecules weighing 32 g; and CO_2 now means 1 *mole* of carbon dioxide weighing 44 g. The enlarged meaning of Eq. (8.1) can therefore be expressed as

$$C \quad + \quad O_2 \quad = \quad CO_2 \qquad (8.1a)$$
$$\begin{array}{ccc} \text{1 mole, or} & \text{1 mole, or} & \text{1 mole, or} \\ \text{12 g} & \text{32 g} & \text{44 g} \end{array}$$

The sum of the reacting masses, $12 + 32$ g, equals the mass of the product, 44 g. Here, as in all ordinary chemical reactions, mass is conserved. It may be noted that when 1 mole of carbon reacts with 1 mole of oxygen, the thermal energy released is 94,000 cal/mole. In many cases, it is convenient to express the properties of a material or the energy changes resulting from a reaction in terms of a mole of the material or a mole of the reacting substances.

The carbon-oxygen reaction is no more than what happens when charcoal burns in a barbecue. When it occurs, the amount of oxygen is indefinitely large since it is in the air around us. This is referred to as being "in excess," whereas the amount of carbon can be more or less than 1 mole or 12 g.

If, in fact, 60 g of carbon is actually available for burning, how much oxygen will be used up if the burning is complete, and how much carbon dioxide will be produced? The point of departure is what Eq. (8.1a) is saying: 1 mole of carbon reacts with 1 mole of oxygen molecules to form 1 mole of the product, carbon dioxide. Since 12 g of carbon is 1 mole of carbon, 60 g must be 5 moles. The number of moles of a material, usually represented by n, in general equals the mass of the species in grams divided by its formula weight. In this case

$$n = \frac{\text{mass, g}}{\text{formula weight}} = \frac{60}{12} = 5 \text{ moles}$$

Now let us apply Eq. (8.1a):

	C	+	O_2	=	CO_2
Burning 1 mole of carbon:	12 g		32 g		44 g
Burning 5 moles of carbon:	5×12 g $= 60$ g		5×32 g $= 160$ g		5×44 g $= 220$ g

Note again that the sum of the grams of reactants equals the grams of product formed. Also, since the combustion of 1 mole, or 12 g, of carbon yields 94,400 cal, the combustion of 5 moles, or 60 g of carbon, will yield $5 \times 94,400 = 472,000$ cal of heat.

Another element that burns freely in the presence of atmospheric oxygen is hydrogen. In fact, hydrogen is so dangerously flammable that it has been replaced for use in dirigibles and balloons by the heavier but inert gas helium. It is used widely in oxyhydrogen torches where a very hot, concentrated flame is needed, and if you have ever watched one, you may recall water dripping off. The reaction is simple

The last lighter-than-air craft was the dirigible Hindenberg, *which burned in 1937.*

$$H_2 + O_2 = H_2O$$

Although the equation is correct in the sense that it tells us what is happening, it is physically not valid because two oxygen atoms are present at the left, but only one at the right. This can be rectified by multiplying the H_2O at the right by 2:

$$H_2 + O_2 = 2H_2O$$

However, this results in a total of four hydrogen atoms at the right and only two at the left. The final adjustment is therefore a multiplication, by 2, of the H_2 molecule at the left, so that the balanced equation is

$$2H_2 + O_2 = 2H_2O \qquad (8.2)$$

The balanced equation says that *two molecules* of hydrogen are reacting with *one molecule* of oxygen to produce *two molecules* of water. If the event as written above is taken $N = 6.02 \times 10^{23}$ times, there will be *2 moles* of hydrogen molecules reacting with *1 mole* of oxygen molecules to produce *2 moles* of water molecules:

$$2H_2 + O_2 = 2H_2O \qquad (8.2a)$$

2 moles,	1 mole,	2 moles,
or 4 g	or 32 g	or 36 g

The heat evolved in this reaction is close to 58,000 cal per mole of hydrogen gas, or about 29,000 cal per *gram* of hydrogen gas.

Compare this figure with the 94,400 cal obtained when one mole of carbon, weighing 12 g, is oxidized; per *gram* this comes to somewhat less than 8000 cal. In other words, gram for gram, burning hydrogen yields much more thermal energy than burning carbon.

The reaction in Eq. (8.2a) can be used to calculate the masses of reactants and products other than those shown since the *relative* number of moles formed cannot change. Suppose that 288 g of water is to be obtained by the reaction. Since the number of moles equals the number of the grams of substance divided by the formula weight of the substance, $n = 288/18 = 16$ moles; 16 moles of water is 8 times the 2 moles used in the balanced equation (8.2a), so that each mole in that relation must similarly be multiplied by 8:

	$2H_2$	$+$	O_2	$=$	$2H_2O$
For 2 moles of H_2O:	2 moles, or 4 g		1 mole, or 32 g		2 moles, or 36 g
For 16 moles of H_2O:	$8 \times 2 = 16$ moles, or 32 g		$8 \times 1 = 8$ moles, or 256 g		$8 \times 2 = 16$ moles, or 288 g

Although chemical reactions take place by interaction between discrete units of the reacting species in integral ratios, the quantitative measure of the reaction is usually weight. Since the mole connects the number of particles that react with the weight of that number, it is commonly used as the basis for chemical calculations. In fact, Avogadro's number N is the unit of quantity in the SI system of units. A formula such as $CaCl_2$ is taken to mean not only that the compound consists of one atom of calcium and two atoms of chlorine but also represents 1 mole of that compound, that is, 6.02×10^{23} $CaCl_2$ units together weighing $40 + 2(35.5)$ g, or a total of 111 g/mole. Therefore, 0.1 mole of

TABLE 8.2

	Chemical Species		
	H_2O	NaOH	H_2SO_4
Formula weight	$2(1) + 16 = 18$	$23 + 16 + 1 = 40$	$2(1) + 32 + 4(16) = 98$
Weight of 1 mole Units in 1 mole	18 g 6.02×10^{23}	40 g 6.02×10^{23}	98 g 6.02×10^{23}
Weight of 0.1 mole Units in 0.1 mole	1.8 g 6.02×10^{22}	4.0 g 6.02×10^{22}	9.8 g 6.02×10^{22}
Weight of 10 moles Units in 10 moles	180 g 6.02×10^{24}	400 g 6.02×10^{24}	980 g 6.02×10^{24}

CaCl$_2$ would consist of 0.1 N, or 6.02 × 10^{22} CaCl$_2$ units, weighing 0.1 × 111 g, or 11.1 g. Some simple examples are listed in Table 8.2.

SOLUTION CONCENTRATION IN MOLES PER LITER

Since water is the medium for so many reactions, especially those which occur in living matter, it is often important to know how much of a chemical substance is present in some volume of solution. "How much" can be taken to mean the weight of that substance or how many particles of that substance. These two measures of solution concentration are related by the use of the mole. Knowing the concentration of a solution tells us how many units of the solute are available in some volume of solution for reaction with some other material.

One of the materials listed in Table 8.1 is NaOH, which is the base most often used in the laboratory. As indicated, 1 mole of NaOH weighs 40 g and contains Avogadro's number N or 6.02 × 10^{23} NaOH units. Now, suppose that these 40 g are transferred to a 1-l volumetric flask and enough distilled water is added to dissolve it. Then water is added until the solution level just comes to the etched line on the neck of the flask. The solution volume would be 1 l, and it would contain 40 g, or 1 mole, of dissolved NaOH, making it a *1 molar solution*, abbreviated 1 M NaOH (Figure 8.3).

The most common method of designating the concentration of a solution is by its molarity, which is the number of *moles of solute* dissolved in *one liter* of that solution. In the example used, 1 mole of NaOH is dissolved in 1 l of final solution, making it 1 M NaOH. The molarity M of any solution is simply the number of

40 g, *or* 1 mole, *of* NaOH *in a* 1 l *volumetric flask*

Water added to dissolve the NaOH

More water added until the liquid level is just tangent to the etch mark, making the solution volume 1 l

FIGURE 8.3 Preparing a 1 M solution of NaOH.

moles n of the solute divided by the volume V of the solution in liters; that is,

$$M = \frac{n}{V, \text{l}}$$

If the solution containing 1 mole of NaOH has a volume of 2 l, its molarity will be $1/2 = 0.5\,M$; if its volume is 250 ml its molarity will be $1/0.25 = 4\,M$. The examples listed in Table 8.3 will illustrate this further.

Of the four solutes listed in Table 8.3, three dissociate in water and release free ions, and only $C_6H_{12}O_6$, which is the simple sugar glucose, remains undissociated. A solution containing 1 mole of dissolved glucose is therefore water in which N number of glucose molecules are distributed. On the other hand, an ionic solute such as sodium hydroxide dissociates almost completely, so when 1 mole of NaOH is dissolved in some volume of water, what is present in the solution is 1 mole of Na^+ ions, and 1 mole of OH^- ions:

$$NaOH(aq) \longrightarrow Na^+ + OH^-$$

1 mole =	1 mole =	1 mole =
6.02×10^{23} units,	6.02×10^{23} units	6.02×10^{23} units
weighing 40 g	weighing 23 g	weighing 17 g

As discussed in Chapter 6, the presence of OH^- ions is characteristic of water solutions of bases, just as the presence of H^+ ions (more properly, H_3O^+ ions) is characteristic of water solutions of

TABLE 8.3 Calculating solution molarity

Compound	Formula Weight	Proportion Solute, g	Proportion Solution	Moles of Solute n	Volume of Solution V, l	Molarity $M = n/V$, l
NaOH	40	40	1 l	40/40 = 1	1	1/1 = 1
		20	1 l	20/40 = 0.5	1	0.5/1 = 0.5
		20	100 ml	20/40 = 0.5	0.1	0.5/0.1 = 5
$C_6H_{12}O_6$	180	90	1 l	90/180 = 0.5	1	0.5/1 = 0.5
		9	1 l	9/180 = 0.05	1	0.05/1 = 0.05
		45	500 ml	45/180 = 0.25	0.5	0.25/0.5 = 0.5
NaCl	58.5	117	4 l	117/58.5 = 2	4	2/4 = 0.5
		5.85	500 ml	5.85/58.5 = 0.1	0.5	0.1/0.5 = 0.2
		234	2 l	234/58.5 = 4	2	4/2 = 2
H_2SO_4	98	98	1 l	98/98 = 1	1	1/1 = 1
		9.8	1 l	9.8/98 = 0.1	1	0.1/1 = 0.1
		49	100 ml	49/98 = 0.5	0.1	0.5/0.1 = 5

acids. The degree to which a solution is acidic or basic depends on the number of moles of H^+ or OH^- ions present in a liter volume of that solution.

By convention square brackets are used around a chemical species to mean the molarity of that species, that is, its concentration in moles per liter of solution. The notation $[H^+]$ therefore means the molarity of hydrogen ions in a solution, and $[OH^-]$ means the molarity of hydroxide ions. The self-ionization of pure water yields equal concentrations of H^+ and OH^- ions and has been experimentally determined as being 0.0000001 mole/l, or 10^{-7} mole/l, so that $[H^+] = [OH^-] = 10^{-7} M$. When an acidic material such as HCl is added to water, $[H^+]$ increases and $[OH^-]$ decreases proportionately; when a basic material such as NaOH is added to water, $[OH^-]$ increases and $[H^+]$ decreases proportionately. The concentrations of hydrogen and hydroxide ions are related to each other like the opposite sides of a seesaw; as one goes up, the other goes down to the same extent. This inverse relation between $[H^+]$ and $[OH^-]$ derives from the fact that their product is a constant value, which is called the **ion product of water** K_w. Since at neutrality $[H^+] = [OH^-] = 10^{-7}$,

$$[H^+] \times [OH^-] = 10^{-7} \times 10^{-7} = 10^{-14}$$

This value of the ion product remains the same whether the solution is neutral, acidic, or alkaline. If 1 l of solution contains 1 mole of NaOH, the solution is 1 M NaOH and the value of $[OH^-]$ is 1. Substituting this value in the ion-product relation gives

$$K_w = [H^+] \times [OH^-] = 10^{-14}$$
$$[H^+] \times 1 = 10^{-14}$$

so that

$$[H^+] = 10^{-14}$$

Therefore 1 l of a 1 M NaOH solution has a hydroxide-ion population of $1 \times 6.02 \times 10^{23} = 6.02 \times 10^{23}$. The population of H^+ ions in that liter is $10^{-14} \times 6.02 \times 10^{23} = 6.02 \times 10^9$. Even in a strongly basic solution there will be some hydrogen ions, and a strongly acidic solution will also have some hydroxide ions.

Table 8.4 gives the molar concentrations of H^+ and OH^- of a series of increasingly concentrated NaOH solutions and of a series of increasingly concentrated HCl solutions; the reference point and dividing line between acidic and basic is neutral water. Several things about this table should be noted.

The *product* of the molar concentration of the ions is 10^{-14} for all solutions.

TABLE 8.4 [H$^+$] and [OH$^-$] in acidic and basic solutions

	[H$^+$], *or* moles H$^+$/l		[OH$^-$], *or* moles OH$^-$/l	
	10^{-14}	↑	1	1 M
	10^{-13}		10^{-1}	0.1 M
	10^{-12}	increasingly basic	10^{-2}	0.01 M
	10^{-11}		10^{-3}	0.001 M
	10^{-10}		10^{-4}	0.0001 M
	10^{-9}		10^{-5}	0.00001 M
	10^{-8}		10^{-6}	0.000001 M NaOH
	10^{-7} ←	pure water	→ 10^{-7}	
0.000001 M HCl	10^{-6}		10^{-8}	
0.00001 M	10^{-5}		10^{-9}	
0.0001 M	10^{-4}	increasingly acidic	10^{-10}	
0.001 M	10^{-3}		10^{-11}	
0.01 M	10^{-2}		10^{-12}	
0.1 M	10^{-1}		10^{-13}	
1 M	1	↓	10^{-14}	

Each step represents a *tenfold* increase or decrease in ion concentration.

Either [H$^+$] or [OH$^-$] can alone serve as the index of acidity or alkalinity, since when one is known, the other can be calculated.

As a rule, chemists use the value of [H$^+$] for the purpose, but instead of designating the molar concentration of hydrogen ions in a solution as a power of ten, as shown in the table, they use the negative logarithm of [H$^+$]. On this basis a single number, called the pH, becomes the index of solution acidity or alkalinity. Formally, pH $= -$log [H$^+$] and since the logarithm of any number expressed as a power of ten is that power, log $10^{-7} = -7$, and $-$ log $10^{-7} = -(-7) = 7$. That is, a neutral solution has a pH of 7. An acidic solution has a value of [H$^+$] greater than 10^{-7}, such as 10^{-6}, 10^{-5}, 10^{-4}, . . . , and the negative logarithms, or pH values, are respectively 6, 5, 4, ... In other words, acidic solutions have a pH *less* than 7. On the other hand, basic solutions have fewer H$^+$ ions than pure water, so that their [H$^+$] values are less than 10^{-7}, such as 10^{-8}, 10^{-9}, 10^{-10}, . . . , making their pH values 8, 9, 10, ... That is, basic solutions have a pH *greater* than 7. How acidic or alkaline a solution is can be judged by how far its pH is below or above 7, the pH of a neutral solution.

Table 8.5 relates pH to [H$^+$], and also shows the pH of several commonly used liquid materials. Note than an increase or decrease in 1 pH unit means an increase or decrease of 10 times in hydrogen-ion concentration.

Some of the pH values in Table 8.5 are average values. The pH of fluids in living matter has a range of values, although the range

TABLE 8.5 The pH of some common materials

	$\begin{array}{c}-\log[H^+]\\=pH\end{array}$		$[H^+]$, moles/l
	14		10^{-14}
0.1 M NaOH	13		10^{-13}
Washing soda	12		10^{-12}
Ammonia water	11	increasingly basic	10^{-11}
Milk of magnesia	10		10^{-10}
	9		10^{-9}
	8		10^{-8}
Blood			
Pure water	7	neutral	10^{-7}
Urine			
Milk			
	6		10^{-6}
Black coffee	5		10^{-5}
Cola drink	4	increasingly acidic	10^{-4}
Vinegar	3		10^{-3}
Lemon juice	2		10^{-2}
Gastric juice			
0.1 M HCl	1		10^{-1}
	0		$1 = 10^0$

may be quite narrow. Saliva varies in pH from about 6.5 to about 7 depending on stimulation, and the pH of the blood must not deviate more than about one-half pH unit up or down from its average value of 7.4.

Analogous to the pH is the pOH, which is the negative logarithm of $[OH^-]$. The two are related by the fact that their sum is always 14, so that if a solution has a pH of 6 and is acidic, its pOH is 8, indicating that there are 10^{-8} hydroxide ions in a liter of that solution.

$$[H^+] \times [OH^-] = 10^{-14}$$
$$\log[H^+] + \log[OH^-] = -14$$
$$-\log[H^+] - \log[OH^-] = 14$$
$$pH + pOH = 14$$

MEASURING pH

As the quantitative measure of the acidity of a liquid, the pH is probably the most frequently determined chemical value. There are two general methods: (1) using an **indicator** that signals a change in pH by a change in its color or (2) using a **pH meter** that measures the voltage that results when two electrodes are immersed in the solution.

INDICATORS AND pH METERS

Indicators are weak organic acids or bases that change color as they lose or gain a proton, that is, they act as an acid or a base. Different indicators make a color change over a different range of

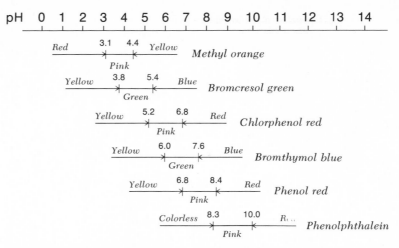

FIGURE 8.4 The pH interval of change and the color changes of several common indicators. In general, an indicator shows three colors: an acidic color at the left, a basic color at the right, and a mixed or intermediate color between the two.

pH values (see Figure 8.4). Indicators have very intense colors, so that generally no more than a drop or two are used in testing a solution.

There are more indicators than those described in Figure 8.4, and the more there are available, the more closely the pH of an unknown solution can be estimated. However, even these few can be used to locate the pH with a fair degree of accuracy. For example, if bromocresol green is added to a solution and a blue color appears, all that can be said is that the pH of the solution is 5.4 or more. However, if adding a drop of bromothymol blue to another small portion of the sample gives yellow, it tells us that the pH of this solution is 6.0 or less. Together the two indicators determine the pH as being between 5.4 and 6.0, which may be good enough for many purposes. With more indicators, even closer results can be obtained. Besides being used for pH determinations, indicators are also used to locate the equivalence point of an acid-base neutralization, that is, they indicate by a color change when equal numbers of H^+ and OH^- ions have been brought together.

Indicator papers, used for many routine pH determinations, are paper strips impregnated with an indicator. When the strip is dipped into the solution, a color change is an index of the solution pH. The best known indicator paper is litmus paper, which simply distinguishes between acid and base, but other papers change color and color intensity with the degree of acidity. Matched against a color code that comes with the strips, the color of the paper can be used to locate the pH of the sample to within a few tenths of a pH unit. Although indicators are simple

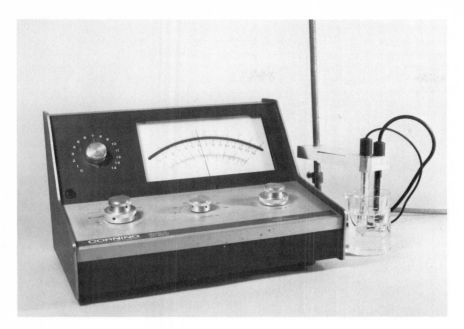

FIGURE 8.5 A pH meter.

and fast and can be used when only a small amount of liquid is available, they are not always as accurate as necessary. Furthermore, the use of color itself introduces some degree of subjective error.

The standard and most accurate method of measuring pH is the pH meter (Figure 8.5). The size, design, and cost of pH meters vary considerably. Some pH meters are powered by batteries, and can be used in the field, and others are so designed that they can be used with only a small amount of solution. However, what they all actually measure is the voltage developed when the electrodes of the instrument are immersed in the solution being tested. This voltage changes directly with changes in the pH of the solution, so that a measure of the voltage becomes a measure of the pH. When the meter has been standardized by adjustment against buffered solutions of known pH, it will give fast and accurate results. The pH meter is an essential piece of equipment in most chemical and biological laboratories.

pH meters are often calibrated in both pH units and millivolts.

ACTIVE AND TOTAL ACIDITY

When a strong acid dissolves in water, it dissociates completely, or nearly so, so that for each unit of acid there is a corresponding number of H^+ ions in the solution. On the other hand, when a weak acid is dissolved, only a fraction of the acid units dissociate,

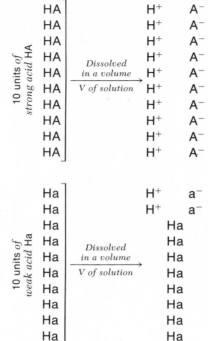

FIGURE 8.6 Solutions of equal concentrations of a strong and weak acid.
Equal concentrations of a strong and a weak acid result in unequal concentrations of free H⁺ ions because the strong acid dissociates completely and the weak acid does not. The two will therefore have different pH values, although both have the same number of available H⁺ ions.

and the number of free H^+ ions in the solution is only a fraction of the total acid units that were added. What the pH measures is the concentration in moles per liter of the free H^+ ions actually present in the solution, those which are on stage, so to speak, and ready to act. The pH is therefore a measure of the *active acidity* of a solution. For weak acids this will differ sharply from the total acidity of the solution, which is the concentration of all the acid units present in the water, whether dissociated or not. Figure 8.6 shows this schematically, with HA as the strong acid and Ha as the weak acid, and each having the same absurdly small number of 10 units in equal volumes of solution.

Although the concentration of the two solutions is the same, the concentrations of their free H^+ ions are quite different and therefore the pH values of the two solutions are different. At the same time, since the total number of *available* H^+ ions, those which are free and those which can be freed, is the same, the total acidity of the two solutions is the same. The total acidity of an acid is its total capacity for neutralizing hydroxide ions. If 1 l each of 0.1 M HC1 and of 0.1 M CH₃COOH is prepared, each solution will contain 0.1 mole of HC1 and CH₃COOH units, respectively. Numerically, this means $0.1 \times 6.02 \times 10^{23} = 6.02 \times 10^{22}$ acid units each. The total acidity of both acids is the same; that is, each can provide 0.1 mole of H^+ ions, which will require 0.1 mole of OH^- ions for their neutralization. HCl offers up all its H^+ at once, so that its pH is 1. On the other hand, only 1 out of about 78 CH_3COOH molecules dissociates to yield H^+ ions, the rest being kept in reserve, so that the pH of 0.1 M acetic acid is only 2.8. When required, the reserves become active, as regulated by the equilibrium between the dissociated and undissociated acid molecules:

$$CH_3COOH \rightleftharpoons CH_3COO^- + H^+ \qquad (8.3)$$

If a base is added to a solution of acetic acid, the hydroxide ions remove the hydrogen ions present in the solution by combining with them to form water, upsetting the equilibrium shown above. To restore the equilibrium some CH_3COOH molecules will dissociate to yield more H^+ ions. As OH^- ions continue to be added, and as CH_3COOH continues to dissociate, a point will be reached when all the acetic acid molecules have dissociated and all the H^+ ions they yielded have been neutralized by an equal number of OH^- ions. This is called the **equivalence point,** when all the ionizable hydrogens of the acid have been just matched by an equivalent number of hydroxide ions. If the base is NaOH, the reaction between the base and each acid is a simple neutralization, expressed quantitatively as

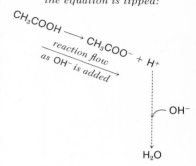

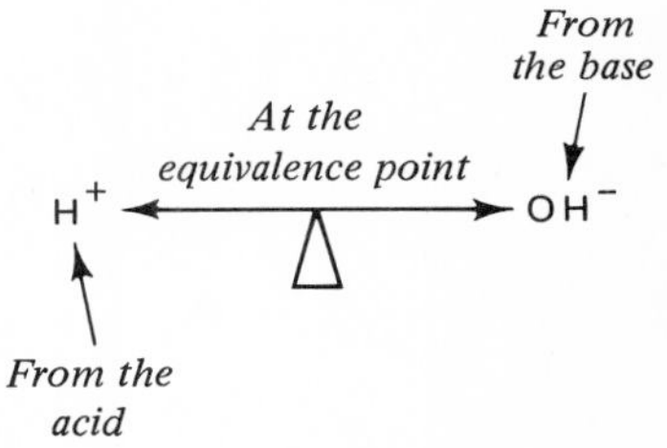

In effect the equation is tipped:

HCl   + NaOH $\longrightarrow$ H_2O + Na^+ + Cl^-
0.1 mole   0.1 mole   0.1 mole   0.1 mole   0.1 mole

CH_3COOH + NaOH $\longrightarrow$ H_2O + Na^+ + CH_3COO^-
0.1 mole   0.1 mole   0.1 mole   0.1 mole   0.1 mole

If n_a is the number of moles of acid and n_b the number of moles of base, then at the equivalence point

$$n_a = n_b \qquad (8.4)$$

This simply restates that at the equivalence point all the ionizable hydrogens provided by the acid have been just neutralized by an equal number of hydroxide ions from the base. The relation between the number of moles of solute n, the molarity of a solution M, and the volume of the solution in liters V has been given by Eq. (8.3), $M = n/V$. If the solute is an acid or a base, respectively, then

$$n_a = M_a V_a \qquad \text{and} \qquad n_b = M_b V_b \qquad (8.5)$$

However, the concentration of a solution of an acid or base, expressed as its molarity, may not in every case be the concentration of the ions that the solution makes available. The concentration of the solution and the concentration of the ions it will provide will be the same only if each acid unit dissociates to yield one H^+ ion or if each base unit dissociates to yield one OH^- ion, as for HCl, CH_3COOH, and NaOH. In general, however, an acid may have more than one ionizable H^+ ion, and a base may yield more than one OH^- ion:

Monoprotic acid: $HCl \longrightarrow H^+ + Cl^-$
Diprotic acid: $H_2SO_4 \longrightarrow 2H^+ + SO_4^{2-}$
Triprotic acid: $H_3PO_4 \longrightarrow 3H^+ + PO_4^{3-}$

Monohydroxy base: $NaOH \longrightarrow Na^+ + OH^-$
Dihydroxy base: $Ca(OH)_2 \longrightarrow Ca^{2+} + 2OH^-$
Trihydroxy base: $Al(OH)_3 \longrightarrow Al^{3+} + 3OH^-$

Suppose that a $1\,M$ solution of each of the three acids is prepared; would 1 l of each of them make available the same number of H^+ ions for reaction with a base such as NaOH? Since any $1\,M$ solution contains 1 mole of solute in 1 l of solution, the three prepared acids each contain 1 mole of HCl, H_2SO_4, and H_3PO_4 units, respectively, in a liter, but the number of moles of H^+ ions would be respectively 1, 2, and 3. This number is called the **normality** N. For $1\,M$ solutions of these three acids, we get

$HCl \longrightarrow$ **1** H^+, so that **1** mole HCl/l contains **1** mole H^+/l = **1** N solution

$H_2SO_4 \longrightarrow$ **2** H^+, so that **1** mole H_2SO_4/l contains **2** moles H^+/l = **2** N solution

$H_3PO_4 \longrightarrow$ **3** H^+, so that **1** mole H_3PO_4/l contains **3** moles H^+/l = **3** N solution

Volume for volume, therefore, a $1\,M$ solution of H_3PO_4 will make available 3 times as many H^+ ions as $1\,M$ HCl, and a $1\,M$ solution of H_2SO_4 will make available twice as many H^+ ions as $1\,M$ HCl.

The relation of normality to the solution molarity M depends on how many H^+ ions the particular acid molecule can make available. As shown above, this is 1 for HCl, 2 for H_2SO_4, and 3 for H_3PO_4. The relation between the normality N and the molarity M of an acid solution is therefore given by the subscript of the ionizable hydrogen in the acid formula. These are the multipliers that relate acid normality and molarity, and some examples are shown, using the acids discussed.

1 M HCl $= 1\ N$ HCl	1 M $H_2SO_4 = 2\ N\ H_2SO_4$	1 M $H_3PO_4 = 3\ N\ H_3PO_4$
0.1 M HCl $= 0.1\ N$ HCl	0.1 M $H_2SO_4 = 0.2\ N\ H_2SO_4$	0.1 M $H_3PO_4 = 0.3\ N\ H_3PO_4$
0.01 M HCl $= 0.01\ N$ HCl	0.01 M $H_2SO_4 = 0.02\ N\ H_2SO_4$	0.01 M $H_3PO_4 = 0.03\ N\ H_3PO_4$

The same discussion can be applied to solutions of bases, in which case the normality is the number of moles per liter of OH^- ions that are available when an alkaline material is dissolved. The multiplier that relates solution molarity to solution normality is now the subscript of the hydroxide group in the formula of the base, which would be 1 for NaOH, 2 for $Ca(OH)_2$, and 3 for $Al(OH)_3$. The conversion from molarity to normality is analogous to the conversion for the acids, so that a 0.1 M solution of NaOH is 0.1 N, a 0.02 M solution of $Ca(OH)_2$ is 0.04 N, and a 0.002 M solution of $Al(OH)_3$ is 0.006 N. The concentration of an acid or a base is usually expressed as its normality, which states unequivocally the number of moles of H^+ or OH^- ions present in 1 l of the solution.

That is, acid normality

$$N_a = \frac{\text{number of moles of } H^+ \text{ ions, } n_{H^+}}{V_a}$$

and base normality

$$N_b = \frac{\text{number of moles of } OH^- \text{ ions, } n_{OH^-}}{V_b}$$

Therefore, the moles of H^+ ions n_{H^+} equal $N_a \times V_a$, and the moles of OH^- ions n_{OH^-} equal $N_b \times V_b$. At the equivalence point of an acid-base neutralization (when equivalent amounts of H^+ ions and OH^- ions have neutralized each other), $n_{H^+} = n_{OH^-}$, and it follows that

$$N_a \times V_a = N_b \times V_b \tag{8.6}$$

This relation is often used in the experimental determination of the concentration of an acid or a base. Since the normality gives the number of moles of H^+ ions that are *available* in an acid, it is a measure of total acidity. Equal volumes of acids of the same normality will therefore neutralize equal volumes of a basic solution. That is, although HCl is a strong acid and CH_3COOH is a weak acid, 1 l of 0.1 N HC1 and 1 l of 0.1 N CH_3COOH have the same number of H^+ ions and can each neutralize 1 l of 0.1 N NaOH solution.

Although normality has been discussed only in relation to acid-base reactions, it is also applicable to other types of reaction where solution molarity may not indicate the number of reacting units. This is especially important in oxidation-reduction reactions, where the number of electrons a chemical species may be able to donate or receive in a reaction must be specified.

STANDARD SOLUTIONS AND TITRATION

Much of the progress in chemistry depends on accurate analysis, determining what chemical species are present and in what amounts. In turn, chemical analysis requires materials whose composition and purity have been so well established that they can be used as references materials, or **standards.** A chemist may sometimes prefer or need to prepare his own standards, but more often it is a saving of time and effort to purchase them from commercial suppliers. In addition, the National Bureau of Standards in Washington provides a number of standard metals and ores, as well as primary standards for use in analytical procedures.

Standard solutions of acids and bases are available from chemical supply houses, often in 1 N or 0.1 N concentrations. The concentration of an acid or base whose normality is not known can then be determined by measuring it against the known concentration of a standard base or acid. In this procedure, called **titration,** the unknown acid (or base) is just neutralized by a standard base (or acid); that is, a measured volume of the unknown solution is taken, and the standard solution is then slowly added to the unknown until the equivalence point is reached. Since the normality of the standard is known and the volumes of acid and base used in the titration were measured, the normality of the unknown solution can be calculated from the other three values.

Although some variations are possible, the usual titration equipment is shown in Figure 8.7. The long, calibrated tubes are **burets,** one for measuring out the volume of the unknown solution to be dispensed and the other for measuring the volume of standard solution used for its neutralization. The stopcock at the bottom of the buret is for regulating the flow, and the accurately graduated marks along the stem can be read to 0.1 ml and estimated to 0.01 ml. After some convenient volume of unknown is dispensed into the Erlenmeyer flask, the milliliters taken are recorded, and a drop or two of indicator are added. The flask is then moved to the second buret containing the standard, which is then added slowly to the unknown, while the flask is swirled. The purpose of the indicator is to signal by a change in color when the addition of the standard is to stop, when the end point is reached. The end point should be as close to the equivalence point as possible, that is when $n_a = n_b$. The most common indicators are phenolphthalein and methyl orange, and although each registers an end point different from the equivalence point, the error is not significant. In performing a titration you should remember that the end point is reached *as soon* as the indicator changes color. If more standard solution is added beyond that, the equivalence point will be passed and the relation $N_aV_a = N_bV_b$ will not apply. For example, if phenolphthalein is the indicator used, it will be colorless in acid and neutral solutions but will turn pink as soon as there is a very slight excess of OH^- ions. Assume that a solution of HC1 of unknown concentration is being titrated with a standard solution of 0.1 N NaOH and that 20.00 ml of the unknown HCl has been dispensed from the buret into the Erlenmeyer flask, to which a drop of phenolphthalein has been added. This volume of acid is V_a. The flask is then transferred to the second buret for adding the NaOH standard solution. The solution will remain colorless until the standard NaOH being added neutralizes all the acid and just passes the theoretical equivalence point by a very small amount. As the NaOH is gradually added and the end point is being approached, a pink color appears and then disappears again. The standard NaOH must now be added drop by drop, swirling the flask to ensure good mixing, until one drop just causes the solution to become pink and remain pink for at least half a minute. That is the end point. If the volume of 0.1 N NaOH used to reach this end point is, say, 18.60 ml, we know that $V_a = 20.00$ ml, $V_b = 18.60$ ml, and $N_b = 0.1\ N$. Only N_a is not known, and that can be calculated from $N_aV_a = N_bV_b$:

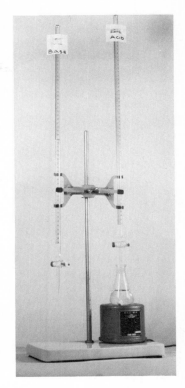

FIGURE 8.7 Titration setup.

$$N_a \times 20.00\ \text{ml} = 0.1\ N \times 18.60\ \text{ml}$$

$$N_a = \frac{0.1 \times 18.60}{20.00} = 0.093\ N \quad Ans.$$

The titration has therefore established that the concentration of the HCl solution is 0.093 N, meaning that it contains 0.093 mole of hydrogen ions in 1 l of solution.

SOLUTION CONCENTRATION

PERCENT

There are many situations where solution molarity or normality is not relevant and concentration expressed as a percentage is simpler and easier to use. This is often true of medications, foods, household materials, and commercial transactions. There are three different ways of determining and expressing the percent of solute present in a solution.

Weight per volume (w/v) In this method the percentage is obtained by taking the ratio of the weight in grams of the solute to the volume of the solution in milliliters and multiplying by 100. If a 50-ml solution of NaCl contains 1 g of NaCl, the w/v percentage is $1/50 \times 100 = 2$ percent. A 2% solution of NaCl therefore contains 2 g of solid NaCl dissolved in a total of 100 ml of solution (Figure 8.8). The formula of the solute is not involved, only the weight, so that in general an $x\%$ solution, expressed in w/v percent, contains x g of the solute in 100 ml of the solution.

An advantage of w/v percent solutions is that the amount of solute in each milliliter of solution is easily obtained. Since a 2% NaCl solution contains 2 g of NaCl in 100 ml of prepared solution, there will be 0.02 g of salt in each milliliter of solution. This can be expressed: 1 ml equals 0.02 g NaCl. Similarly for a 5% glucose solution: 1 ml equals 0.05 g of glucose. The emphasis here is on the amount of solute (salt, glucose, etc.), the water simply being the carrier that delivers the solute. Solutions of this sort are generally made up as w/v percent solutions. A disadvantage of w/v solutions arises because the volume of a liquid changes with temperature, so that the ratio of weight to volume, and hence the percentage, will vary somewhat with temperature. Although not important in most instances, it is a consideration where very exact results are necessary, in which case percent solutions would be prepared by the next method.

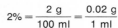

2 g *of NaCl are weighed out and put into a* 100 ml *volumetric flask*

Distilled H_2O *is added to the flask and the NaCl is dissolved*

Distilled H_2O *is added to the mark; the solution volume is* 100 ml *and contains* 2 g *of NaCl*

FIGURE 8.8 Preparing a w/v 2% solution of NaCl. The concentration of the solution will be accurate if each step in its preparation is accurate and there is no accidental loss, gain, or contamination.

Weight per weight (w/w) The percentage here is the ratio of the weight of the solute to the weight of the solution times 100. A 2% solution would be prepared by this method by weighing out 2 g of solute and dissolving it in 98 g of water, so that there would be 2 g of solute per 100 g of solution. Since 98 g of water is just about 98 ml of water, and since the solution of 2 g of solute will not significantly alter the volume of solvent, the solution volume will be quite close to 98 ml. The advantage of w/w solutions is accuracy, although they are more troublesome to prepare.

Volume per volume (v/v) Although liquid-in-liquid solutions can be prepared by the w/w method, it is simpler and often quite acceptable to measure them out by volume. A 10% solution of alcohol in water would be made up by measuring out 10 ml of the alcohol, then adding enough water to make a total volume of 100 ml of solution. In this example, 1 ml of solution would contain 0.1 ml of alcohol.

In describing a percentage solution, the method used for its preparation should be stated. Note that for very dilute solutions, where the quantity of solute is practically negligible compared to the quantity of solvent, all three methods give almost the same results. Practically speaking, most solutions are quite dilute.

PARTS PER MILLION

The term percent means "per hundred," so that a 5% solution means 5 parts of solute per 100 parts of solution. If "part" is taken as being the weight, then there are 5 g of solute per 100 g of solution. But suppose a solution is very dilute, say 0.005%; in this case it is more meaningful to relate the parts of solute to more than 100 parts of solution. This is done by writing 0.005% as a fraction and multiplying both numerator and denominator by 10. Since numerator and denominator are both multiplied by the same number, the value of their ratio will not change. (This has been worked out in the margin.) A 0.005% solution can therefore be expressed as a 50 ppm solution. Eliminating decimals or simplifying them helps us visualize such small concentrations more realistically. If we are told that some pollutant in the water we drink has increased from 5 to 40 ppm, the increase is clearer than if it is expressed as a change in concentration from 0.0005 to 0.004%. Although the "part" in parts per million can be any suitable unit, it is usually taken as the weight in grams. However, since water solutions with concentrations expressed in parts per million have so little solute that the solution weight is really the

$$0.005\% = {}^{0.005}/_{100}$$
$$= {}^{0.05}/_{1,000}$$
$$= {}^{0.5}/_{10,000}$$
$$= {}^{5}/_{100,000}$$
$$= {}^{50}/_{1,000,000}$$
$$= 50 \; parts \; per \; million$$
$$= 50 \; ppm$$

weight of the water, milliliters of solution can be used in place of grams of solution. For example, the solubility of oxygen in water at 25°C is about 8 ppm, meaning 8 g of oxygen per 10^6 ml of water.

REVIEW QUESTIONS

1 When two atoms unite to form a compound, what feature of their atomic structure essentially determines the arithmetical ratio in which the two atoms will combine?

2 Atomic weights are based on the mass of the most common isotope of carbon, $^{12}_6C$ (also written carbon 12), with atomic weight 12. **a** On this basis the atomic weight of helium, He, is 4, and the atomic weight of molybdenum, Mo, is 96. What does this tell us about the weight of *one* He atom and *one* Mo atom compared with the weight of *one* carbon 12 atom? **b** Does it tell us anything about the *actual* weight (really mass) of any one of these atoms? **c** Suppose it is determined experimentally that N (= 6.02×10^{23}) atoms of carbon 12 actually weigh 12 g. What can you now say regarding the weight of N He atoms, and of N Mo atoms? **d** Calculate the actual mass of *one* carbon 12 atom. *Hint:* To simplify matters use 6×10^{23} as the value of N, from which it follows that 6×10^{23} carbon 12 atoms = 12 g.

3 Complete the table by entering the numerical answers called for.

	H_2S	H_2SO_4	CO_3^{2-}	$Mg(OH)_2$	C_2H_5OH
Formula weight	_____	_____	_____	_____	_____
Weight of 1 mole, g	_____	_____	_____	_____	_____
Weight of 0.1 mole, g	_____	_____	_____	_____	_____
Weight of 4 moles, g Number of that chemical species in 4 moles	_____	_____	_____	_____	_____

4 Chapter 7 discussed the reaction between zinc metal and an acid, producing hydrogen gas and a salt. If the acid is HCl,

$$Zn + HCl \longrightarrow H_2 \uparrow + ZnCl_2$$

Atomic weight: 65   1 35.5

a balance the equation. **b** How many moles of acid will be needed to react completely with 1 mole of zinc metal and how many moles of hydrogen gas will be produced? **c** If 260 g of zinc is available and there is an excess of acid, how many grams of hydrogen gas and how many grams of the salt will be produced?

5 In what way is population density (for example, the population den-

sity of Massachusetts is greater than that of Montana) similar to the concentration of a solution? **b** In what ways do they differ?

6 A salt solution is prepared by weighing out 117.0 g of NaCl, which is then transferred quantitatively, that is, without loss or contamination, to a 1-l volumetric flask. The salt is dissolved by adding distilled water to the flask (swirling helps), after which water is further added until the liquid level just becomes tangent to the etch mark on the stem of the flask. **a** How many moles of NaCl are in solution? **b** What is the volume of the solution? **c** What is the molarity of the NaCl solution? **d** How *many* Cl^- ions are there in this liter of solution? **e** It was pointed out in Chapter 7 that mixing Ag^+ and Cl^- ions in water results in the precipitation out of very insoluble AgCl. How many *moles* of Ag^+ ions would you add to this salt solution to just equal the Cl^- ions present?

7 Complete each of the following by entering the suitable answer in the blank space. The relations between grams of solute, number of moles of solute, and solution molarity are recapitulated in the margin. **a** If 10 l of a $Ca(OH)_2$ solution contains 1.48 g of dissolved $Ca(OH)_2$, the solution is ___ M. **b** If 2 l of a KCl solution contains ___ g of dissolved KCl, the solution is 0.1 M. **c** If ___ ml of a $C_6H_{12}O_6$ solution contains 90 g of dissolved $C_6H_{12}O_6$, the solution is 2 M. **d** If 100 ml of a CH_3COOH solution contains 10 g of dissolved CH_3COOH, the solution is ___ M.

$$n = \text{number of moles solute}$$
$$= \frac{\text{grams of solute}}{\text{formula weight of solute}}$$
$$Molarity = \frac{\text{moles of solute}}{\text{liters of solution}}$$

8 The $[H^+]$ of a solution is reported as 10^{-9}. **a** How many moles of H^+ ions are there in 1 l of this solution? **b** How many moles of H^+ are there in 1 l of a *neutral* solution? **c** Is this solution acidic, neutral, or alkaline? **d** How many moles of OH^- ions are there in this solution?

9 The pH of five different solutions is determined experimentally with the results shown in the margin. **a** Which solution is most acidic? **b** Which solution is most basic? **c** Which solution is most nearly neutral?

Solution	pH
1	8.7
2	6.9
3	4.3
4	11.5
5	6.1

10 The pH of a 0.1 M HCl solution is about 1, whereas the pH of a 0.1 M acetic acid, CH_3COOH, solution is close to 3. **a** Which solution has the higher active acidity? Explain. **b** How many moles of H^+ ions are *available* in 1 l of each solution? **c** How many H^+ ions are available in 10 ml of each solution? **d** How many milliliters of 0.1 M NaOH solution will be neutralized by 25 ml of each of the two acid solutions? **e** What is the normality of each acid solution?

11 How would you prepare:
a 2 l of a 0.1 N NaOH solution, **b** 100 ml of an 0.8 N CH_3COOH solution, **c** 500 ml of a 0.001 N $Ca(OH)_2$ solution? *Hint:* What is the *molarity* of a 0.001 N $Ca(OH)_2$ solution?

12 For each of the solutions shown, fill in the blank on the basis of the information provided.

	HNO_3	$Mg(OH)_2$	KOH	H_2SO_4	H_3PO_4
Molarity	6 M	___	___	18 M	___
Normality	___	0.0006 N	0.1 N	___	0.12 M

13 **a** What is the total number of moles of H^+ ions available in 200 ml of a $0.05\,N$ solution of boric acid, H_3BO_3? **b** Is this value the total acidity of this acid solution? **c** How *many* OH^- ions are needed to just neutralize this boric acid solution?

14 In titrating a number of HCl solutions against a standard $0.1\,N$ NaOH solution, an analyst obtains the following data. Calculate the normality of each HCl solution.

	Solution			
	1	*2*	*3*	*4*
HCl used, ml	18.42	12.70	20.02	10.73
$0.1\,N$ NaOH used, ml	9.21	19.05	18.14	16.18
Normality	_____	_____	_____	_____

15 How many milliliters of $0.1\,N$ HNO_3 would be required to neutralize 500 ml of a $0.001\,N$ solution of $Ca(OH)_2$?

16 How would you prepare 200 ml of a 6% w/v solution of NaCl? How much NaCl will be dissolved in each milliliter of this solution?

17 How would you prepare 200 g of a 6% w/w solution of NaCl? Would you expect the solution to have a total volume of 200 ml?

18 The vinegar sold in supermarkets is often labeled 5% acidity, where the acid is acetic acid, CH_3COOH. If we assume this to be 5% w/v, what weight of acetic acid is present in 1 cup of vinegar? (1 cup is about 240 ml.)

19 Although acetone is an organic liquid, the polarity of its molecules makes it water-soluble. How would you prepare 1 l of 4% v/v solution of acetone in water? What volume of acetone would be present in each milliliter of this solution?

20 The concentration of Ca^{2+} ions in a sample of water is reported as 38 ppm. Express this as a percent solution.

21 It has been stated that 1 mg of CN^- ion per kilogram of body weight is a lethal dose for human beings. Express this ratio in ppm and in percent.

ENERGY AND RADIOACTIVITY
Making It Go, and Going It Alone

The idea of **energy** is relatively new in human thought. On the other hand, the idea that things change is very old, supported by all human experience.

Recognition that there is a relation between energy and change began to emerge with the invention that opened our modern technological era, the steam engine. When wood, coal, or any other fuel burns, the fuel is consumed and heat is given off. Before the steam engine the only result of burning fuel was heat. There was one notable exception, the burning of food in living matter, which then, as now, powered all the activities and processes of living systems.

With the steam engine, heat obtained from burning fuel is for the first time purposefully converted into mechanical action, the familiar to-and-fro motion of a piston in a cylinder. This, in turn, can be harnessed to pump water, operate a cotton loom, or turn a ship's propeller, that is, to do *work*. If heat can be converted into mechanical work, the two must have something in common, especially when it is considered that mechanical work can be reconverted back to heat, as by friction. Their common character lies in the fact that they are both forms of *energy*, so that what any

In the first century, Hero of Alexandria, an ingenious inventor, built a simple steam engine, but it was considered a toy.

As the spinning wheel slows down, due to friction between the wheel and the axle, the axle heats up. The mechanical energy (a push of the wheel) has been converted to heat energy.

engine really does is to convert **thermal energy** into **mechanical energy.**

Soon after its invention it was found that a steam engine can rotate the shaft of an electric generator to produce **electric energy,** which by running a motor can also do work. Furthermore, the electric energy produced by a generator can be stored in a chemical battery, whose **chemical energy** can also be converted

FIGURE 9.1 Matter can be converted into energy, but it *cannot* be created or destroyed.

into work, as when an automobile battery starts up the engine. As theory and practice developed, it became apparent that there are many forms of energy, all mutually convertible and all capable of doing work.

The study of the transfer and conversion of energy, called **thermodynamics,** was intensely pursued during the latter part of the eighteenth century and especially the nineteenth century. One result was a general conclusion called the **first law of thermodynamics. It is also called the law of the conservation of energy** since it states that although energy can be transferred from one body to another and can be converted from one form to another, energy itself can neither be created nor destroyed. According to thermodynamics you can't get something for nothing, and you can't reduce something to nothing (Figure 9.1).

INTERNAL ENERGY

Burning a fuel is a chemical reaction (generally an oxidation) and represents a conversion of energy from one form to another. Since energy cannot be created, the heat energy obtained from the reaction can be derived only from some energy contained within the reacting materials, that is, the fuel that burned and the oxygen used for the burning. This contained energy is called the **internal energy** of the substance. Any chemical substance represents not only matter but also an energy content. At a given temperature and pressure, usually taken as 25°C and 1 atm, the internal energy incorporated in a given mass of substance is fixed and specific to that substance. If the substance receives energy, as when water is heated, its internal energy increases accordingly and its temperature rises. When the substance loses energy, as when hot water gives off heat to its surroundings, its internal energy decreases and its temperature drops. Although the gain or loss of thermal energy is the most general way of increasing or decreasing the internal energy of a substance, the same results will be obtained by an input or output of other forms of energy. The internal energy and temperature of a gas will increase if the gas is compressed, since compression requires an input of mechanical energy. On the other hand, if the gas expands by pushing against some smaller external pressure, the internal energy and temperature of the gas will decrease, since this represents an output of mechanical energy.

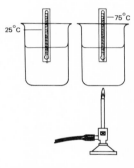

The heat energy from the burner increases the temperature of the water and its internal energy

CHEMICAL CHANGE AND INTERNAL-ENERGY CHANGE

The internal energy in any substance is the consequence of its composition, structure, and bonding. In the course of a chemical reaction the kind and number of atoms in the system remain the

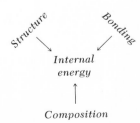

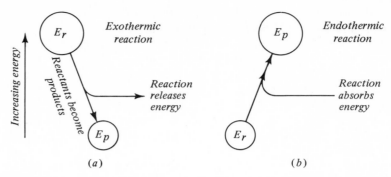

FIGURE 9.2 (*a*) An exothermic reaction is one in which the internal energy of the reactants E_r is greater than the internal energy of the products E_p and consequently the system will release energy in the course of the reaction. The energy change, $\Delta E = E_p - E_r$, is negative, indicating that the energy of the system is reduced as a result of the reaction. (*b*) An endothermic reaction is one in which the internal energy of the reactants is less than the internal energy of the products, and the system must receive energy in order to proceed. The energy absorbed ΔE, equal to $E_p - E_r$, is positive, indicating that the energy of the system increases as a result of the reaction.

$\Delta = change\ in$

same. However, as reactants become products, the atoms rearrange themselves into new combinations of composition, structure, and bonding with internal energy different from that of the original. In general, therefore, a system that undergoes chemical change will also undergo a change in internal energy. Since it is customary to express internal energy as E, the internal energy of the reactants can be written as E_r and that of the products as E_p. Since E_p is the internal energy of the system at the end of the reaction and E_r its value at the beginning, the internal-energy change resulting from the reaction, written ΔE, is $E_p - E_r$.

Since $\Delta E = E_p - E_r$, it follows that if $E_p < E_r$, ΔE is negative and the reaction *releases* heat; if $E_p > E_r$, ΔE is positive and the reaction *absorbs* heat. Figure 9.2 expresses this schematically. An energy-releasing reaction is called **exothermic,** and an energy-absorbing reaction is called **endothermic.**

In discussing the burning of carbon in Chapter 8 it was stated that the oxidation of 1 mole of carbon results in the evolution of 94,400 cal of heat:

$$C \ + \ O_2 \ \longrightarrow \ CO_2 \ \ + 94{,}400\ cal$$

1 mole,	1 mole,	1 mole,	
or 12 g	or 32 g	or 44 g	

E_r = internal energy of reactants $\qquad$ E_p = internal energy of products

The change in internal energy of the system in the course of the reaction is $\Delta E = E_p - E_r = -94{,}400$ cal, where the minus sign means that the internal energy of the system decreases to that extent as 1 mole each of C and O_2 are converted into 1 mole of CO_2. The 94,400 cal that leave the system are lost to the system but gained by its surroundings, which is a way of saying that the reaction is exothermic.

FROM HEAT TO WORK

The fact that all forms of energy are interconvertible is used by our technological society in many ways. There is, however, one important restriction. All the forms of energy so far mentioned (including mechanical, electric, and chemical, and other varieties such as radiant and magnetic) can be converted into thermal energy without any loss. This is an advantage, for instance, when electric energy is put into the heating element of a toaster, where it is converted into heat energy plus some radiant energy in the form of light. Since radiant energy in turn results in heat, the net result is the complete conversion of the electric energy into thermal energy. The reverse is not possible; heat cannot be *fully* converted into electric energy or mechanical energy or any other form of energy, because in the process of conversion some of the heat escapes. This heat is not destroyed (that would violate the first law of thermodynamics), but it becomes *unavailable* for doing work.

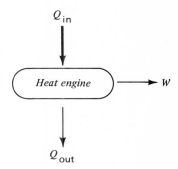

The fact that no heat engine can completely convert the heat energy received into mechanical energy is a matter of experience. A schematic view of what happens is shown in the margin. A quantity of thermal energy, symbolized Q_{in}, is *received* by the engine; some smaller quantity of thermal energy Q_{out} *leaves* the engine. The *difference* between heat energy in and heat energy out is the mechanical work W delivered by the engine:

$$Q_{in} - Q_{out} = W \qquad (9.1)$$

If it were possible to reduce Q_{out} to zero, Q_{in} would equal W and all the heat received would be converted into mechanical energy. But it can't be done. Even if the engine were "ideal," suffering no energy loss from friction or leakages of any kind, some quantity of heat would still leave the engine, reducing the work performed to that extent. It must therefore be some feature of the conversion process itself that accounts for the heat loss from the engine. The effort to analyze this built-in inefficiency led to a fundamental principle of science, the second law of thermodynamics.

THE SECOND LAW OF THERMODYNAMICS

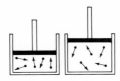

An ordinary deck has 52 cards. If it had 520 cards, the number of possible disorderly arrangements would increase astronomically.

The theoretical analysis of the conversion of thermal energy into mechanical energy is rather lengthy and involved, but the essential feature explaining the inability of a heat engine to convert heat completely into work is not difficult to visualize. Work is done in a heat engine when the heat it receives acts to expand a gas; the expanding gas pushes against a piston, and work is performed. However, any gas is a disorderly collection of particles moving at random with varying speeds and in varying directions. When a gas expands at a given temperature, it has even more room in which to be *more disorderly.* The larger the portion of space over which the gas particles are scattered the more random their motion. However, if the engine is to keep on working, the expanded gas must return to its original smaller volume and its original more orderly state. But now we encounter a fundamental fact of nature: *a disorderly condition does not become more orderly of its own accord;* in fact, in all natural, spontaneous occurrences, the reverse is true. If a fresh deck of cards, perfectly arranged in four suits with each suit further arranged in ascending order from ace to king, is opened and shuffled for the first time, the orderly arrangement is gone. If the cards are reshuffled again and again, the original ordered arrangement will not return. *Possibly* it may, but when you consider how very many disorderly arrangements there are compared with the one orderly arrangement being sought, would you care to wait for order to come out of disorder? Of course, the cards could be turned face up, and sorted into their original sequence, but that would require effort, which is energy.

The added disorder due to the expansion of the gas will not undo itself, but for the engine to continue doing work the gas must regain its original more orderly state. An increase in disorder is spontaneous, but its reduction requires energy, which in this instance is the energy Q_{out} that leaves the engine. This thermal energy is not lost since it is gained by the surroundings, but it *is* lost to the engine for doing work. Essentially what happens is a trade-off, a *decrease* in disorder as the gas returns to its original state against an *increase* in disorder as heat Q_{out} flows out of the engine. As we shall shortly see, the spontaneous flow of heat is equivalent to an increase in disorder.

The fact that heat energy cannot be completely converted into mechanical energy is one way of stating the second law of thermodynamics. Apart from the practical limitations imposed on the conversion of heat to work, this statement of the second law indicates a *directional* factor in the way the physical world operates. Other forms of energy (mechanical, electric, chemical,

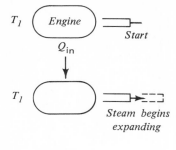

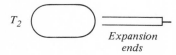

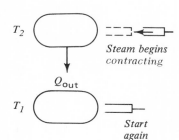

FIGURE 9.3 Steam engine cycle, schematic and simplified. The net result of the cycle is that the work W is done in an amount $W = Q_{in} - Q_{out}$.

and so on) can all be transformed in full into thermal energy, but the reverse is not possible because the *tendency to disorder* intervenes. This tendency is itself directional, since once the disorder of a system increases (the shuffled cards, the expanded gas), the disorder will not repair itself; the system will not return of its own accord to a more orderly state.

The tendency toward disorder is so commonplace that we overlook its significance. When you stop to think of it, there are many one-way events that are accompanied by a decrease in order and an increase in randomness. A vibrating string will stop vibrating, and its scattered energy does not reassemble to set it vibrating again. A drop of ink in water will diffuse outward, and its particles will not recollect to re-form the drop. If you stir water in a beaker, the whirling pattern dies out, not to return. Barely touch a soap bubble, and its regular geometry collapses. These events are examples of an ordered condition spontaneously changing to a less orderly state.

The most general and important of the one-way events that result in increasing disorder is the flow of heat from a warmer to a cooler region, sometimes expressed "heat runs downhill." If one

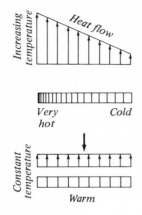

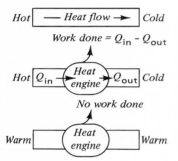

Warm

The bar is uniformly warm and there is no heat flow, and no work done.

Gone

end of a metal bar is very hot and the other end is cold, heat will flow from the hot to the cold end, so that in time the temperature of the bar becomes uniform. What happens is the spontaneous change of an orderly temperature sequence (very hot, hot, quite hot, warm, lukewarm, cool, cold) to an undifferentiated sequence of warm, warm, warm, warm, warm, warm, warm. It might be imagined that as heat flows down the bar, the rapidly moving particles originally at the hot end and the slowly moving particles at the cold end mix together to give a uniform average temperature. There is no change in the energy of the bar due to the mixing, simply a net uniform distribution of energy in place of a high-to-low distribution. In principle, therefore, an *unmixing* could take place which would return the bar to its original high-to-low temperature distribution, and since there would be neither a gain nor a loss of energy, it would be permitted by the first law of thermodynamics. But it never happens; nature prefers the mixed-up condition, and the more mixed up things get, the greater the disorder.

The increase in disorder that accompanies the flow of heat from warmer to colder can also be recognized by imagining a heat engine inserted between the hot and the cold ends of the bar. So long as there is a temperature drop, Q_{in} heat will enter the engine, Q_{out} heat will leave, and work $W = Q_{in} - Q_{out}$ will be obtained. But if the bar is at a uniform temperature because of the flow of heat from hot to cold, the insertion of the imaginary heat engine will yield no work, since there can be no Q_{in} and no Q_{out}. The flow of heat from hot to cold, which is a spontaneous one-way event, represents a decrease in order and an increase in randomness, since the directed, orderly work that *could* have been done is lost; the potential energy of the heat at the higher temperature has been dissipated. Furthermore, when heat flows from any warm object, such as the bar, to the cooler atmosphere, the heat transferred similarly becomes unavailable for doing work.

Although heat flows spontaneously from hot to cold, we all know that a refrigerator or an air conditioner moves heat *uphill* from a cooler to a warmer region. But this is an *arranged* process, arranged by man, in the course of which *more* heat runs downhill spontaneously than is forced uphill.

In the course of all natural, spontaneous events, whether it is the damping out of a vibrating string or the flow of heat downhill, there is an increase in disorder. This is the constant one-way event that accompanies all change, whether natural or arranged. When the disorder of a particular system is decreased, as when heat is moved uphill, this is more than offset by the increase in disorder due to the effort required. The degree of disorder in a given substance or system is called its **entropy,** and another way of stating the second law of thermodynamics is to say that the entrophy of the universe is always increasing.

ENTROPY

Entropy is considered an intrinsic property of a material and is therefore regarded as a state function. The entropy of many substances has been calculated from experimental data. The entropy of a number of chemical materials, all at 25°C, is given in Table 9.1. The differences are revealing. Hard diamond has the lowest value, and we can visualize that the carbon atoms have very little opportunity for vibrating widely in their tightly held tetrahedral arrangement. Liquid water has a much higher entropy, reflecting its greater randomness of thermal motion, while gaseous oxygen has the highest entropy in the table since its molecules are free to zip around in all directions and with a wide range of velocities.

ENTROPY AND EQUILIBRIUM

Since all change, including chemical change, is accompanied by entropy increase, where there is *no change* there will be *no entropy increase.* In other words, if increasing entropy is a criterion of change, unchanging entropy is a criterion of equilibrium. In applying this criterion to some system, it is the entropy of the universe (the system and its surroundings) that is involved.

If it were possible to build an entropy meter, analogous to a speedometer, it could be used to follow the increasing entropy of a chemical system and its surroundings as the reactants become products and move toward equilibrium. If the entropy meter were read like a speedometer, we could imagine the pointer moving rapidly at the outset of the reaction, slowing down as it continues, and finally coming to a stop when the reaction stops and equilibrium is reached. Unfortunately, there are no entropy meters, and in practice it is quite difficult, if not impossible, to determine the entropy change of both a system and its surroundings experimentally. Furthermore, chemists are generally more concerned with the system that is changing than with the surroundings and would prefer a criterion of equilibrium expressed in some property of the system itself.

This property is **free energy.** Since entropy is the underlying intrinsic property, or state property, involved in change, free energy must be related to entropy and must itself also be a state property that is inherent in a material.

TABLE 9.1 Standard entropy at 25°C

Material	cal/K
Diamond (C)	0.54
Iron	6.49
Copper	7.97
Liquid water	16.72
Ethyl alcohol	38.40
Oxygen gas	49.00

The state functions (or state properties) of a system (or of a material) specify its condition. In addition to entropy, some important state properties are internal energy, temperature, pressure, and volume.

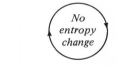

No entropy change

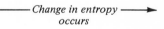

Change in entropy occurs

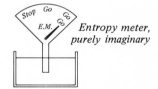

Entropy meter, purely imaginary

Reaction starts

Reaction reaches equilibrium

FREE ENERGY

When a chemical reaction reaches equilibrium and entropy increase comes to an end, the system is stable and unchanging; because it is unchanging, it can do no work. On the other hand, the path from the start of the reaction to equilibrium, along which entropy is always increasing, is also the path along which work can be done. Even if no work was actually performed because there was no mechanism for doing it, it *could* have been done in principle.

It was mentioned earlier in this chapter that a chemical reaction often results in the liberation of heat. For example, if metallic zinc is added to a solution of copper sulfate, the more electropositive zinc displaces the copper and the products are zinc sulfate and metallic copper plus some heat that is lost to the atmosphere. No work is done, and all the heat is wasted. However, the same reaction can be used to operate an electric battery that can do electrical work. Furthermore, by controlling the conditions, the reaction can be made reversible, so that the work done will be the maximum work possible. The fact that some fixed, maximum quantity of work can be extracted, at least in principle, in the course of a reaction led to the conclusion that there is a potential, or capacity, for work, called a **work function,** that is inherent in all the chemical species involved in the reaction. This state property is the free energy G, so that the free-energy change ΔG of a reaction is the free energy of the products less the free energy of the reactants. In the example given, the combined free energy of Zn and $CuSO_4$ is greater than the combined free energy of Cu and $ZnSO_4$, the difference being the maximum work obtainable from this reaction. Note that the free-energy change in the reaction is negative, since $\Delta G = G_p - G_r$ and $G_r > G_p$.

The absolute values of the free energy of chemical substances cannot be determined, but this presents no problem since what is significant in a reaction is the free-energy *change* of the reaction. Experimental data and methods are available for determining this change. However, the free-energy change of a reaction depends on the concentration of the reactants and on the temperature, so that standard conditions must be agreed upon if the results are to be useful. These conditions are a concentration of reactants of 1 mole/l and a fixed temperature, generally 25°C. When so determined, the free-energy change of the reaction is called the **standard-free energy change** $\Delta G°$.

The heat liberated in the course of an exothermic reaction is the **enthalpy change** ΔH. It is analogous to the internal energy change ΔE, discussed before, the difference being the work of

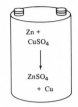

Zn +
$CuSO_4$

↓

$ZnSO_4$
+ Cu

Electrical energy from a chemical fraction

G is for J. Willard Gibbs, an outstanding American scientist of the late nineteenth century, who was a pioneer in thermodynamics.

expansion that a gas would perform if it were involved in the reaction. In the reactions that occur in living matter, the components are mostly liquids or solids, so that ΔH and ΔE have practically the same value and the two terms can be used interchangeably.

The free-energy change ΔG of a reaction is related to the internal-energy change ΔE and the entropy change ΔS by the equation

$$\Delta G = \Delta E - T \, \Delta S \qquad (9.2)$$

Note that all the terms in the right-hand side of the equation are state functions, so that free energy must also be a state property.

All the changes refer to the system itself, not to the surroundings. This equation is obtained by combining the first and second laws of thermodynamics, and its derivation is available in standard texts on the subject. For our purpose it suffices to see how the relation contributes to our understanding of chemical reactions. In the course of a spontaneous reaction, the work potential, or free energy, of the system is reduced, reaching a minimum when equilibrium is achieved. So long as the system remains at equilibrium there is no further change in its free energy. If a free-energy meter were available (like our imaginary entropy meter), it could be used to monitor the free-energy changes of a reaction from start to finish. At the very outset the meter would show a rapid drop in free energy as the reaction proceeds rapidly down the "reaction hill"; as the pace slackens, the meter records a slowing decrease of free energy; and when the bottom is reached and the system achieves equilibrium, the meter records zero change. From start to finish, the cumulative free-energy change ΔG is *negative* because the free energy of the products is below that of the original reactants. Therefore we now have a criterion for a spontaneous reaction: *its free-energy change is negative.* By the same token, if thermodynamic calculations show that a given reaction results in a positive value of ΔG, then it cannot occur spontaneously. To complete the picture, when the spontaneous reaction loses its spontaneity and slows down to equilibrium and to zero net change, the free-energy change also becomes zero. In summary;

1 For a spontaneous, irreversible reaction, ΔG is negative.
2 For a nonspontaneous, arranged reaction, ΔG is positive.
3 At equilibrium, where free energy is minimum, ΔG is zero.

There is a general parallel between the entropy increase of the system plus the surroundings and the free-energy decrease of the

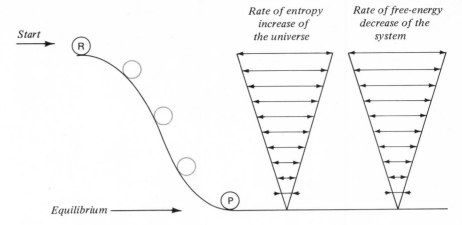

Start

R

Equilibrium ⟶

P

*Rate of entropy
increase of
the universe*

*Rate of free-energy
decrease of the
system*

*A spontaneous reaction moving
toward equilibrium and reaching
it.*

FIGURE 9.4 Entropy increase of the universe and free energy decrease of the system for a spontaneous reaction. At the left the reactants are seen ready to roll down the "reaction hill" toward equilibrium. As the reaction proceeds, the entropy of the system plus the surroundings increases, while the free energy of the system decreases. When equilibrium is reached, there is no further increase in entropy and no further decrease in free energy. The height of the hill represents the net decrease in free energy of the system that is driving it to equilibrium. The horizontal lines have no quantitative values and simply indicate that the driving force behind the reaction diminishes in intensity the closer the system is to equilibrium.

system itself as a spontaneous reaction moves down a reaction hill. This is illustrated in a schematic and nonquantitative fashion in Figure 9.4.

The farther away a system is from its equilibrium condition, that is, its stable condition, the more unstable the system is. An explosion is a prime example of an unstable system that reaches equilibrium in a very short time. Also, the farther away a system is from equilibrium, the greater the decrease in free energy when equilibrium is reached and the more negative the ΔG of the reaction. The decrease in free energy is therefore thought of as the driving force behind a reaction, moving it on toward its stable, equilibrium condition. It is the drop in free energy that makes a reaction go. But as can be seen from $\Delta G = \Delta E - T\,\Delta S$, the value of ΔG is made up of two terms. Individually they can be plus, minus, or even zero, but together they must add up to a negative value if the reaction is spontaneous. Furthermore, the more negative this sum is, the greater the push behind the reaction. If ΔE is negative and the reaction releases heat, this will contribute to the negative value ΔG. Although this is often the case, it can also happen that ΔE is substantially zero or even positive in value, yet the reaction is spontaneous. This will occur if the reaction leads to an increase in the disorder of the system, in which case the entropy change ΔS is positive and the term $-T\,\Delta S$ takes on a negative value. If this negative value of ΔS is large enough to more than cancel out the positive value of ΔE, the sum of the two terms will be negative, so that ΔG will be negative, in which case the reaction will go spontaneously.

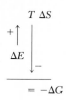

An example from the literature of thermodynamics is of biochemical interest. Proteins are very large molecules that arrange themselves into specific and complex three-dimensional structures. This spatial conformation is necessary to their biological function, and the loss of that structure, called **denaturation,** reduces or destroys the usefulness of the protein. The most general way of denaturing a protein is to raise the temperature. When a protein is exposed to a higher temperature, it denatures spontaneously although the change is accompanied by a very substantial increase in internal energy; that is, ΔE for the denaturation is positive. This implies that the change also results in a value of $-T\,\Delta S$ sufficiently large to more than offset the positive value of ΔE and so make ΔG for the reaction negative. The high negative value of $-T\,\Delta S$ derives first from the increase in disorder as the organized structure of the protein becomes undone, thereby sharply increasing its entropy, and second, because the value of T, which multiplies ΔS, has been increased. Despite increased energy, denaturation is spontaneous. This result is interesting not only in its own right but also because it points to the general conclusion that increasing temperatures favor those reactions and structures which result in an increase in disorder. Quiet, orderly conditions are favored by lower temperatures.

Orderly helical conformation

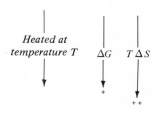

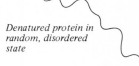

Denatured protein in random, disordered state

$$\Delta G = \Delta E - T\,\Delta S \qquad (negative)$$

FREE ENERGY AND CHANGE OF PHASE

A change of **phase** can be considered a change in the physical state of a material, of which melting and boiling are the most familiar. It is obvious that the most orderly form of water is ice. On melting, ice becomes the less orderly liquid water, which on being heated to boiling changes to its most disordered and random state—steam. Each phase change is accompanied by an intake of heat energy at a constant temperature (0°C and 100°C, respectively), called a **transition temperature.** In summary:

At 0°C: 1 g of ice $\xrightarrow[\text{melting}]{+80\text{ cal}}$ 1 g of liquid water

At 100°C: 1 g of liquid water $\xrightarrow[\text{boiling}]{+540\text{ cal}}$ 1 g of steam

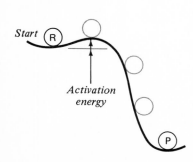

Note that melting 1 g of ice at 0°C requires an *intake* of 80 cal. Suppose that only 20 cal was received. What would happen? Instead of 1 g, only 20/80, or ¼ g, of the ice would melt, so that ¼ g of water would coexist quietly with the ¾ g of ice remaining, and if heat neither entered nor left, everything would remain that way. This is a description of equilibrium; in fact, if it were *not* equilibrium something would be going on spontaneously, and it isn't. For any change to occur, say the formation of more water, some more heat must enter. If more ice is to re-form from the water, some heat must leave. If neither happens, nothing happens. The melting of ice into water and (by the same reasoning) the vaporization of water into steam are equilibrium processes. The reverse processes, steam condensing into water and water freezing into ice, are also equilibrium processes. If the entering or leaving of heat were *stopped* at any point in these phase transformations, the system would remain stationary.

For a system in equilibrium, $\Delta G = 0$, so that for these phase changes $\Delta G = \Delta E - T\,\Delta S = 0$, from which $\Delta E = T\,\Delta S$. This says that the 80 cal which is taken up by the gram of ice in becoming water, which represents an increase in internal energy, is all used up in loosening the bonds that hold the ice particles together, allowing them to exist as more disorderly liquid water. The temperature does not change because all the heat energy received goes into this changeover, with an accompanying increase in entropy. The transition from liquid water to steam needs more thermal energy than the ice-to-water change because whereas in going from ice to liquid water looser bonds are replacing tighter ones the formation of steam requires the complete separation of all the molecules and the undoing of all bonds between them. In turn, this expresses itself in increased disorder and increased entropy of steam over the same quantity of water.

ACTIVATION ENERGY

The fact that the free-energy change for a reaction is negative *permits* the reaction to go, but *how fast* it will go is quite another matter. In a great many cases, and especially in the chemical reactions that occur in living matter, the rate at which a reaction occurs without help is so slow as to be useless. The reason is that the reacting materials are generally not at the peak of the reaction hill, but in the hollow, as sketched in the margin. In order for the reaction to take advantage of its negative ΔG, the reactants must be boosted over this energy barrier, after which they can interact and proceed spontaneously to equilibrium. The energy required to overcome this barrier is called the **energy of**

activation of the reaction. If it is high, few of the reacting molecules will be energetic enough to make the jump, with the result that for all practical purposes the reaction does not proceed. In living matter the height of this energy barrier is reduced by the use of enzymes, which therefore occupy a critical role in all life chemistry.

FREE ENERGY AND THE POINT OF EQUILIBRIUM

If negative free-energy change tells us that a reaction *may* go, and if furthermore it does go at a useful rate, there is still the question: How *far* will it go? That is, if we represent the reacting materials as R, and the products formed by the reaction as P, how much of each will be present when the system reaches its point of equilibrium? Of course, if P is gaseous and escapes from the system, no equilibrium can be reached and the reaction proceeds until all the R is used up. The same happens when ionic reactants form products that are un-ionized or insoluble, since once formed, they cannot participate in any reverse reaction and in that sense are removed from the system.

More generally, however, especially in organic and biochemical reactions, the tendency for R to become P is to some extent offset by the reverse tendency of P to become R, and this is represented by P → R. Generally speaking, therefore, the point of equilibrium will be somewhere between 100 percent P and 0 percent R, and 0 percent P and 100 percent R. If the tendency R → P is greater than the reverse tendency R ← P, there will be more P than R at equilibrium; if, however, R → P proceeds with less drive than R ← P, there will be less P obtained at the equilibrium condition and more R. But the driving force behind a reaction is the resulting free-energy change ΔG, obtained when the system reaches equilibrium. The system can be considered either the reactants R, or the products P; in either case the system will fall to the same equilibrium level of free energy, and the same ratio of R to P will be obtained. Remember that the equilibrium condition is always the minimum-free-energy condition. Whether more reactants R or more products P are found in the equilibrium composition depends on which loses more free energy in coming to equilibrium. If the free-energy drop when R arrives at equilibrium exceeds the free-energy drop as P achieves equilibrium, R → P has more drive behind it than R ← P. The result will therefore be the existence of more P than R at equilibrium. Should the resulting free-energy loss be the reverse and R ← P have more push behind it, less P than R will be present in the equilibrium mixture.

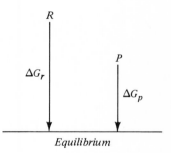

ΔG_r is more negative than ΔG_p and the equilibrium mixture is richer in products than in reactants.

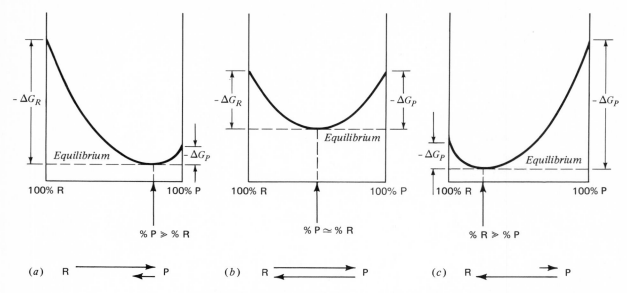

FIGURE 9.5 Equilibrium: (*a*) the equilibrium is to the right, giving more products than reactants; (*b*) the equilibrium is about at midpoint, giving approximately equal amounts of products and reactants; (*c*) the equilibrium is to the left, giving more reactants than products.

This is represented schematically in Figure 9.5. In each case the minimum energy is the lowest point of the imaginary string joining the free-energy level of 100 percent R and 100 percent P. Since the equilibrium condition is that of minimum free energy, this low point of the string also identifies the relative amounts of products and reactants.

100% R = *reactants only*
100% P = *products only*

ISOTOPES AND RADIOISOTOPES

So far spontaneous reactions between different chemical materials have been considered as *the* method whereby chemical energy is converted into other forms, especially heat energy. Not only is energy released by these reactions, but the entropy of the universe is increased and the free energy of the reacting system is minimized. The great bulk of the energy obtained today still comes from these rather ordinary and readily observable reactions, such as the combustion of coal, petroleum, natural gas, and wood.

All these chemical reactions involve interactions between the outer electrons of the reacting atoms; in these reactions the atomic nuclei do not participate directly. Of course, the number of protons in the nucleus of each atom is also the number of electrons of the atom and in that respect determines its chemical behavior, but apart from acting as a center of positive charge, the nucleus does not enter into the reaction and undergoes no change.

There is a class of energy-producing processes that involve no chemical changes in the usual sense but derive from changes in the atomic nucleus. The spontaneous change of an atomic nucleus is called **radioactivity,** and the substance involved is called radioactive. The fact that the nucleus of a radioactive atom is spontaneously changing means that the nucleus is unstable and that its change is moving it to a more stable condition. A closer look at the atomic nucleus is, therefore, in order.

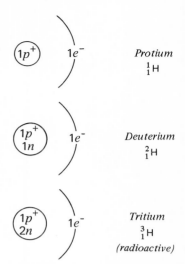

FIGURE 9.6 The isotopes of hydrogen.

ISOTOPES

Although there are 92 naturally occurring elements, from hydrogen to uranium, plus 13 more synthetic elements by last count, there are many more different *atomic nuclei*. If two atoms have the same atomic number Z, they have the same number of outer electrons and the same type of chemical behavior and are members of the same atomic species. However, they may still differ in the number of neutrons in their nucleus, and since a neutron contributes mass but no charge, they will therefore differ in their atomic *weight*. Such atoms are called **isotopes,** and even the simplest atom, hydrogen, exists in nature as three different isotopes, known respectively as protium, deuterium, and tritium (Figure 9.6). There are a number of points to keep in mind concerning isotopes.

The isotopes of an element differ only in the number of neutrons in their nuclei; their atomic weights will differ accordingly.

The atomic weight of an element is the average value of the different atomic weights of its isotopes. For example, the two major isotopes of chlorine are $^{35}_{17}Cl$, present as 75.5 percent of a normal sample of chlorine, and $^{35}_{17}Cl$, present as 24.3 percent; the result is that the atomic weight of chlorine averages out to 35.45.

The difference in the number of neutrons between two isotopes produces a slightly heavier and a slightly lighter version of the *same* element, *not* two different elements.

Having the same atomic numbers, and therefore the same electron configurations, isotopes of a given element have similar chemical properties and reactions. A sample of water, generally represented as H_2O, may contain P_2O, D_2O, and T_2O since all the three hydrogen isotopes react with oxygen in the same way. However, since deuterium and tritium are present in only very

$^{35}_{17}Cl$ *has* 18 *neutrons in its nucleus and therefore an atomic weight of* 35. $^{37}_{17}Cl$ *has* 20 *neutrons in its nucleus.*

P = *protium*
D = *deuterium*
T = *tritium*

small percentages, water is largely P_2O. Water containing the heavier isotopes is called **heavy water.**

Isotopes may be stable, or they may be radioactive. A radioactive isotope is called a **radioisotope.** The fact that tritium, with one proton and two neutrons, is a radioisotope whereas deuterium, with one proton and one neutron, is stable suggests that an atomic nucleus becomes unstable when there is a sufficient imbalance between the number of neutrons and protons. When that happens, the nucleus of the atom may relieve its instability by firing out, at high speed, one of three different particles, often accompanied by a high-energy radiation called **gamma rays.** Whenever the atomic nucleus loses one of these particles, its atomic number changes, so that the atom **transforms** into some other element; its original identity is lost, and a new identity is gained. This atom disintegration, or atom decay, is entirely spontaneous. The rate at which it occurs is specific for each radioisotope and is expressed as its **half-life,** meaning the time it takes for half of any sample of the isotope to disintegrate.

THE PARTICLES EJECTED

Aluminum 25 is $^{25}_{13}$Al

Magnesium 25 is $^{25}_{12}$Mg

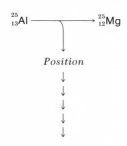

The several particles that are emitted by radioisotopes and the consequent changes in atomic number are as follows.

Beta particles These can have either a negative charge or a positive charge. A **positive beta particle,** called a **positron,** has the same mass as an electron; in fact, it can be called a positive electron. When a positron is ejected from an atom nucleus, the atomic number is reduced by 1. Loss of positrons is relatively infrequent and is restricted to isotopes of the lighter elements. For example, aluminum 25 emits a positron to become magnesium 25.

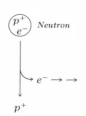

A **negative beta particle** is simply a negative electron, and its loss from the nucleus increases the atomic number by 1, since the loss of a negative electron is equivalent to a gain of a positive proton. This electron does not come from the electrons around the nucleus but *from* the nucleus. The question arises: How did it get there? It has been suggested that the neutral neutron is made up of a positive proton and a negative electron. When the restless nucleus ejects the electron, the proton remains, increasing the atomic number. Although easy to visualize this is an oversimplification, and a more subtle process of energy emission is probably at work.

Alpha particles These particles are identical to the nucleus of the helium atom and consist of two protons and two neutrons.

When an atom nucleus loses an alpha particle, its atomic number is reduced by 2 and its atomic mass is decreased by 4.

A high-velocity stream of beta or alpha particles may also be called beta rays and alpha rays, but these are particles and so differ from gamma rays.

An alpha particle

Gamma rays In many cases atomic disintegration also results in the emission of **gamma rays.** Since they are radiant energy (just as visible light is), gamma rays have no charge or mass, so that their loss has no effect on atomic number or atomic mass. The specific feature of gamma rays compared with other radiation is their high energy. Radiant energy arises from the vibration of a charged particle, somewhat like snapping the end of a rope up and down to give it a wave motion. In the case of radiant energy the energy expended in generating these vibrations is communicated by the resulting wave, which travels outward at the same speed for all radiation, the speed of light. However, the energy of the radiation can vary widely, from the soft radio waves used for broadcasting to the hard x-rays and gamma rays. Soft and hard refer to their ability to penetrate matter. Soft radiation oscillates slowly; hard radiation oscillates very fast.

c = speed of light
  = 300,000 km/s

The energy of radiation is often expressed in **electronvolts** (eV) (Figure 9.7). This unit can be visualized as the energy a negative electron attains as it falls toward a positive electron under the electric push, or potential, of 1 V. When it is considered that 1 eV results when a single electron falls down a single rung of the electrical ladder, it can be appreciated that it represents a very small amount of energy. Larger multiples are therefore often used, including the kiloelectronvolt (keV), which is 1000 eV, the megaelectronvolt (MeV), which is 10^6 eV, and the gigaelectronvolt (GeV), which is 10^9 eV.

NATURAL AND ARTIFICIAL RADIOACTIVITY

Although Wilhelm Röntgen discovered x-rays in 1895, there was no awareness of radioactivity as a spontaneous physical process. In 1896 the French physicist Henri Becquerel observed that uranium can fog a photographic plate, and further experimentation led to the conclusion that this was due to some kind of rays emanating from the uranium. In 1898 thorium was found to have the same effect, and in 1902 Marie Curie isolated a small amount of radium in the form of radium chloride, $RaCl_2$. Since then radioactivity and radioactive processes have had an enormous and lasting impact.

The most dramatic application of radioactivity is, of course, the generation of massive amounts of energy released in an uncon-

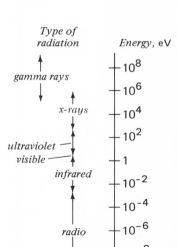

FIGURE 9.7 The energy of radiation.

trolled manner as atomic and hydrogen bombs. Similar nuclear reactions are used to generate energy in a controlled manner in nuclear reactors. In both instances radioactive fallout and radioactive contamination are dangers. It is obvious that radioactivity is a double-edged sword, to be used judiciously and sheathed carefully. Apart from the mechanically destructive effects of a nuclear blast, living matter is sensitive to radioactive matter and can be damaged by radiation. At the same time, it must be recognized that living matter has always been exposed to some radioactivity since life began. In addition to the spontaneous radiation coming from radioisotopes found in our soil, water, and atmosphere, cosmic rays come from space. These naturally occurring radiations, called **background radiation,** are believed to have been instrumental in the evolution of plant and animal life by altering the hereditary material in living cells. These alterations must have been of a random nature, so that some led to changes that could not be tolerated, resulting in death, whereas others resulted in changes that were to the advantage of the organism. This background radiation is still with us, but now we also have man-made radiation, either unwanted, as in the form of fallout and contamination, or as purposefully used in therapy, research, and the generation of power.

Three major factors enter into the action of radioactive materials on living organisms. These are half-life, degree of penetration, and ionizing effect.

HALF-LIFE

The half-life of a radioisotope, which is the time required for the disintegration of half of its atoms, varies enormously. A few examples are listed in Table 9.2.

It is obvious that organisms exposed to radioisotopes with a long half-life will suffer the cumulative effects of the radiation emitted. Also, the source of radiation can be passed on from plant to animal organisms. An example is strontium 90, found in the fallout from an atomic bomb, which has a half-life of 25 years. It tends to deposit in bone tissue because of its chemical similar-

TABLE 9.2 Radioisotopes

Isotope	Name	Radiation Emitted	Half-Life
$^{238}_{92}$U	Uranium 238	Alpha, gamma	4.5×10^9 years
$^{226}_{88}$Ra	Radium 226	Alpha, gamma	1590 years
$^{222}_{86}$Rn	Radon 222	Alpha	3.8 days
$^{214}_{84}$Po	Polonium 214	Alpha	1.6×10^{-4}s

TABLE 9.3   Shielding ability of lead, copper and aluminum

Isotope	Energy of Gamma Radiation, MeV	Half-Value Thickness,† cm		
		Lead	Copper	Aluminum
Radium 226	0.188	0.035	0.38	1.92
Cobalt 60	1.17, 1.35	0.86	1.36	4.42

† That thickness which will reduce the intensity of the given gamma radiation by one-half.

ity to calcium. From the fodder of a cow it can appear later in the milk a child drinks, still radiating beta rays.

PENETRATION

If radiation cannot penetrate tissue, its capacity for damage is reduced; similarly, a penetrating radiation will go a long way and may do great damage. The greatest penetration results from gamma rays. In cancer therapy this is turned to advantage by directing the rays toward the tumor in order to destroy its cells. However, this must be done in a very controlled way, for healthy tissue is also damaged. In using gamma rays and x-rays shielding is a matter of great practical importance. Table 9.3 gives data on the relative shielding ability of different materials. Lead is clearly the best. It should be noted that the higher the energy of the radiation the more shielding is required.

Beta particles, plus or minus, have relatively little ability to penetrate; even the more energetic ones penetrate soft human tissue or water only to a depth of about ¼ mm. Alpha particles are stopped even more readily, a single sheet of ordinary paper being sufficient.

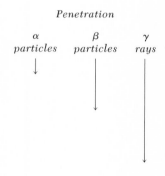

IONIZATION

The ionization caused by radiation is not the transfer of one or more electrons from one atom to another, as in ordinary chemical reactions. Ionization by alpha particles (the most effective in this respect) represents damage to molecules and therefore damage to the functioning of the tissue containing those molecules. A good deal depends on the rate at which the damage is done. Small amounts of damage slowly incurred may be repaired by the cells themselves, whereas a large amount of damage, especially over a short time, can be dangerous or even lethal. The damage causes a swelling of the cell and its nucleus and an increase in viscosity of the cell fluid. The cell wall functions improperly, and

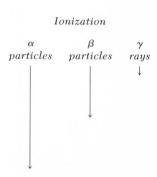

the chromosomes that carry the hereditary genes may be altered. Since alpha particles have very little penetrating power, they do little damage if the radioactive isotope is outside the body. However, if the radioactive material is swallowed, as in food or liquids, or enters through surface cuts, it can do a great deal of damage, because of the high ionizing ability of alpha particles.

Beta particles have less ionizing capability than alpha particles, but under some conditions they can cause serious damage. Beta particles are sufficiently penetrating to cause skin lesions. Internally, they are less ionizing than alpha particles but more penetrating. Although gamma radiation causes little ionization, it can cause much damage as it penetrates living matter, depending upon the extent to which the particular tissue absorbs the energy being transmitted to it and the intensity and duration of the radiation. The more intense the gamma radiation and the longer it acts the more damage results.

RADIOISOTOPES IN HEALTH

Despite their capacity for damage to living matter, and sometimes because of it, radioisotopes are widely used in medicine. Gamma radiation from radium and cobalt 60 are used in radiation therapy against cancer. Iodine 131 has a half-life of 8 days. When it is taken into the body as a water solution of an iodine salt, it becomes concentrated largely in the thyroid. If it is picked up too quickly or too slowly, some malfunction of the thyroid is indicated. Iodine 131 has also been used against cancer of the thyroid, yet excessive dosage could lead to cancer. Other radioisotopes have been used to study the circulation of the blood, to sterilize foods, to destroy harmful insects, to irradiate plants to lead to more productive varieties through mutations, and in many other ways.

RADIOISOTOPES IN RESEARCH

Radioactive isotopes have been a boon to research in medicine, chemistry, and industry. Their great value lies in the constant announcement of their presence through the radiation they emit. In whatever form they are used—as elements or in combination with other atoms in compounds, as solid metals or in water solution—they can be detected and followed by metering instruments like the geiger counter. In this way, the path of a chemical reaction or the details of some process can be followed with the radioisotope acting as the **tracer.** Sodium 24 has been used as the tracer in studying the circulation of the blood. Organic compounds have been tagged by incorporating radioactive carbon 14 in their structure and used to study the selection of food by plants, the destination of chemical materials introduced into plant and animal organisms and how they get there, the behavior of fats and oils in the body, the movement of blood cells, and many other normal processes. All this is possible because the radioactive carbon atoms behave like all carbon atoms except that they can be followed around. The mechanisms of many chemical reactions have been clarified in this way.

REVIEW QUESTIONS

1 Before the invention of the steam engine two natural forces (apart from the muscle power of men and animals) were used for doing mechanical work. What were those natural forces, and are they still in use?

2 Explain in your own words what is meant by the law of the conservation of energy.

3 The simple act of striking a match and making it burn involves two different kinds of energy, each of which is converted into thermal energy, or heat. What are those two different kinds of energy?

4 When NaOH pellets are dissolved in water, the temperature of the water increases. Which would you say has a higher internal energy, 1 g of solid NaOH or 1 g of dissolved NaOH? Is the solution of NaOH in water an exothermic or endothermic reaction?

5 Heating $CaCO_3$ causes its decomposition into CaO and CO_2.

$$CaCO_3 \xrightarrow{\Delta} CaO + CO_2 \uparrow$$

Is this reaction exothermic or endothermic? Do the reactants or the products have the higher internal energy?

6 Why does good insulation help keep a house warmer in the winter and cooler in the summer? What fundamental natural tendency is being slowed down in both cases?

7 In poker a straight flush beats a straight. **a** Which is the more orderly arrangement of cards, the straight or the straight flush? **b** Which arrangement would you expect to occur less frequently?

8 In Iceland the warm water from natural hot springs is used to heat greenhouses in which vegetables, fruits, and flowers are grown. Would you say that this heat is doing work? Is a tomato or a head of lettuce a more orderly arrangement than a heap of soil with a seed somewhere in its midst? If this hot water were *not* piped in from the springs, it would simply cool off in the air, since heat runs downhill; would that represent an increase in disorder?

9 One way of stating the second law of thermodynamics is to say that no mechanism or engine can completely convert the heat energy it receives into mechanical energy. What happens to the heat energy that is *not* converted into work?

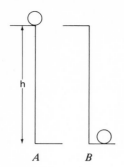

A B

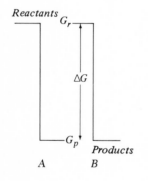

Reactants G_r

ΔG

G_p *Products*

A B

10 Another, equivalent, statement of the second law is to say that *all change* is accompanied by an increase in the disorder of the universe, that is, of the system and its surroundings. Does that mean that *portions* of the universe cannot be made *more* orderly? Can you think of some examples? Does that negate the second law?

11 What is meant by entropy, and what symbol is used to represent it?

12 Would you expect a gram of liquid water to have more or less entropy than a gram of ice?

13 A weight is shown in two positions, A and B, separated by a vertical distance h. **a** Can work be done as the weight goes from position A to position B? **b** How does the amount of work that can be done vary with the value of h? **c** Suppose the weight simply falls from A to B, without doing *any* work in lifting or moving anything. Since the potential energy of the weight at A is not used for work, into what other form of energy will it be converted when it lands at B? **d** Can the weight do any work at position B? **e** Is B the equilibrium position of the system?

14 Instead of a weight, as in Question 13, consider a chemical system in which the reactants have a free energy G_r at position A and the reaction products have a free energy G_p at position B. The free-energy difference ΔG is analogous to the height h in Question 13. **a** What is the maximum amount of work that can be done by the system as it goes from position A to position B? **b** Suppose that there is no mechanism available whereby this free energy can be converted into work (such as an electric battery); since no work is done, into what other form of energy will this free energy be converted? **c** Can the system do any work at position B? **d** Once having arrived at position B, will there be any further change in free energy? Is position B the equilibrium condition?

15 Using the criterion of free-energy change, distinguish between a spontaneous, irreversible reaction, an arranged, nonspontaneous reaction, and a system that has reached equilibrium.

16 Identify each term in the expression

$$\Delta G = \Delta E - T \, \Delta S$$

a If ΔE is negative, meaning that heat is *released* in the course of the reaction, and if the system becomes *more* disordered as it comes to equilibrium, what will happen to the value of ΔG? Will the reaction be spontaneous? **b** In some cases ΔE is positive, meaning that the reaction *absorbs* heat as it proceeds, yet the reaction is spontaneous and the value of ΔG is therefore negative. For ΔG to be negative while ΔE is positive, what must the value of $T \, \Delta S$ be?

17 The sketch shows a piece of ice floating in a volume of water, with the entire system so well insulated that heat can neither leave nor enter the system. **a** What is the temperature of the water? Of the ice? **b** Is the system undergoing any change in free energy? **c** If the piece of ice weighs 12 g and the insulation is removed, how many calories will the system absorb from the warmer air around it in order to just melt the ice? Will the entropy increase as the ice melts? **d** Why is the heat of vaporization of a gram of water greater than the heat of melting of a gram of ice?

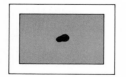

18 In the sketch, by reference to the letters A, B, and C, which is the heat of activation? Which is the net energy released by the reaction?

19 In striking a match, what provides the heat of activation? Would it be uncomfortable if matches did *not* need a heat of activation?

20 Three different sets of reactants R and products P are shown. The free-energy loss of each reactant and of each product when they reach equilibrium is represented by the height of the vertical line. At equilibrium, which system will consist mostly of reactants? Which mostly of products? Which of about equal amounts of both?

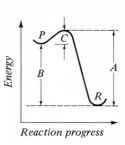

Reaction progress

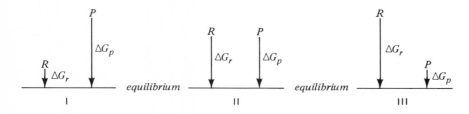

21 In what essential respect does a radioactive material differ from a nonradioactive material? Is radioactivity a spontaneous process?

22 How do the three isotopes of hydrogen differ from each other?

23 Are all isotopes radioactive?

24 Describe what happens when an atom disintegrates, or decays.

25 What is meant by the half-life of a radioisotope?

26 What is the mass of and the charge on a positron? What is the mass of and the charge on a negative beta particle?

27 What are alpha particles? What is the charge on an alpha particle? How does the mass of an alpha particle compare with that of a positron, or negative beta particle?

28 What are beta rays and alpha rays?

29 Does the emission of a gamma ray from a radioisotope alter the atomic weight or the atomic number of the isotope? Explain why.

30 Why are gamma rays considered very hard radiation? How do they compare with x-rays in that respect?

31 What is meant by background radiation?

32 What are the three major features of radioactive materials that affect their action on living organisms?

33 Why can a radioisotope with a long half-life represent a danger to living matter and to human life?

34 Which type of radiation emitted by radioisotopes has the greatest penetrating power? Which radiation has the smallest penetration?

35 What type of radiation is most effective in causing ionization? What is the effect on the cells resulting from ionization caused by radioactivity?

36 How does the damage done by radioactivity relate to duration, intensity, and rate?

37 What are some ways in which radioisotopes are used to combat disease?

38 Why are radioisotopes so useful in medical and industrial research?

10

INTRODUCTION TO ORGANIC CHEMISTRY
In the Land of the Living, Chemically Speaking

Organic chemistry derives its name from an early belief that organic compounds could be made only by living organisms. This belief was fostered by a theory known as the "vital force" principle. Before the advent of modern organic chemistry, the only sources of organic compounds were seeds, leaves, roots, animals, and microorganisms.

In 1829 Wohler made urea, previously obtained only from urine, from an inorganic substance. When he evaporated a solution of ammonium cyanate, the compound that was left was identified as urea:

$$NH_4OCN \longrightarrow (NH_2)_2CO$$

Ammonium cyanate,   Urea, an organic compound
an inorganic compound

Biochemistry—the chemistry of biological systems
Medicinal chemistry—the science of "designing" and evaluating drugs
Polymer chemistry—the science of modern synthetics and polymers

After this breakthrough it was realized that organic substances can be made, or synthesized, from inorganic substances. Today chemists even synthesize compounds that never occur in nature.

During 1858 to 1870 the structure of organic compounds was elucidated. The theories and postulates of early organic chemists laid the foundations of modern organic chemistry, which today generally means the chemistry of carbon and carbon compounds.

BONDS FORMED BY CARBON

Carbon forms four covalent bonds. In a compound these bonds can be present in several different ways:

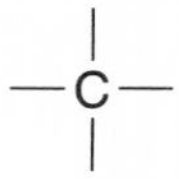

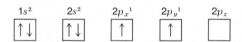

Four single bonds 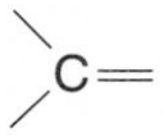   Two single and one double bond

$$-C\equiv$$

One single and one triple bond

The fact that carbon has *four* bonds is important. From the periodic table we know that the atomic number of carbon is 6, which means that it has six protons and six electrons. The electrons in carbon are distributed in their orbitals as follows:

$1s^2$ $2s^2$ $2p_x{}^1$ $2p_y{}^1$ $2p_z$

| ↑↓ | ↑↓ | ↑ | ↑ | |

The bonds of covalent compounds are each formed by sharing a pair of electrons, where generally each of the two atoms bonded contributes one electron. But, as can be seen, the carbon atom has only two unpaired electrons and presumably could form only two covalent bonds. In fact, however, carbon forms four bonds, so that a change in the electron distribution of carbon must occur when it forms a compound. From physical organic chemistry and quantum mechanics, we know that carbon when forming a compound "promotes" one of the electrons from the $2s$ level to the empty $2p_z$ level by hybridization:

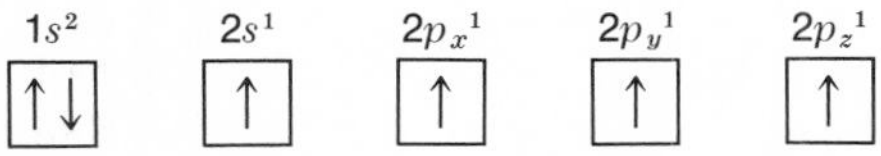

$1s^2$ $2s^1$ $2p_x{}^1$ $2p_y{}^1$ $2p_z{}^1$

| ↑↓ | ↑ | ↑ | ↑ | ↑ |

The carbon atom has now four unpaired electrons in four different orbitals and can share these four electrons with atoms of other elements to form four covalent bonds. Of the four electrons that carbon will now contribute to the formation of four covalent bonds, one comes from the $2s$ level and the other three from the $2p$ levels. This is somewhat like making a drink from one

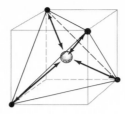

FIGURE 10.1 Tetrahedral structure.

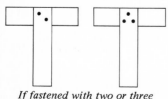

part of vermouth and three parts of whiskey. The result is something new, neither exactly whiskey nor exactly vermouth but a Manhattan. The carbon atom will have now four bonds, called sp^3 hybrids, that are unique in being identical. Another unique property of the carbon atom and its four bonds is that they form a regular tetrahedron. The bonds from carbon point to the corners of the tetrahedron, and the carbon atom itself is in the center of that tetrahedral structure. From the geometry of a tetrahedron it follows that the angles between the bonds must be 109.50° (Figure 10.1). In this arrangement of the four single bonds of carbon lines drawn to connect the atoms attached to a carbon atom form a three-dimensional structure, the tetrahedron. Since each attached atom is held by a single bond, it can rotate freely around the bond.

Carbon can also satisfy the four-bond rule by forming two single bonds and one double bond. These bonds are in a planar arrangement, called a **trigonal configuration,** and the bond angles are 120°. A carbon atom with such an arrangement of bonds has a planar structure, and the bonds are sp^2 hybridized.

The third combination of bonds for carbon is one single and one triple bond. Carbon again has four bonds, but the geometry is planar and linear, with bond angles of 180°. In such an arrangement, the bonds are sp hybrids.

Single bonds permit free rotation around the bond. You can visualize this by taking two pieces of wood and driving a nail through the two pieces. It is possible to rotate the two pieces of wood relative to each other around the nail. If you take a second nail and drive it through the two pieces of wood, you have a rigid structure. Rotation around the nails is difficult, to say the least. This is analogous to the situation in a carbon compound with a double bond. The free rotation around the double bond is restricted, and the result is a fixed trigonal planar geometry. By driving a third nail into our two pieces of wood, we make the structure even more rigid. A triple bond is like having three nails connecting two pieces of wood.

STRUCTURE OF CARBON COMPOUNDS

The early organic chemists also found that carbon has an unusual property which few other elements have; carbon atoms link to other carbon atoms to form carbon chains. These chains can be straight or branched

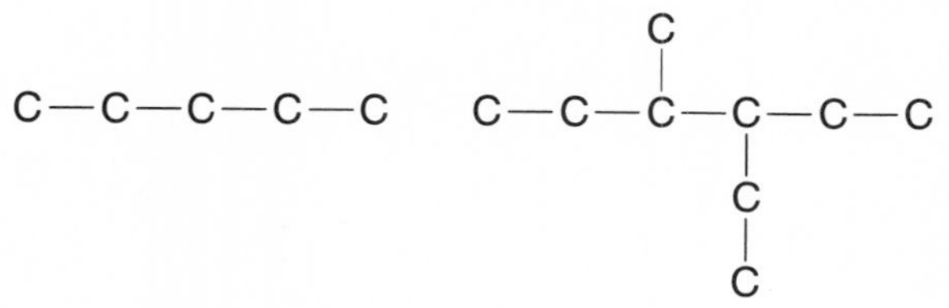

or the chains can be cyclic:

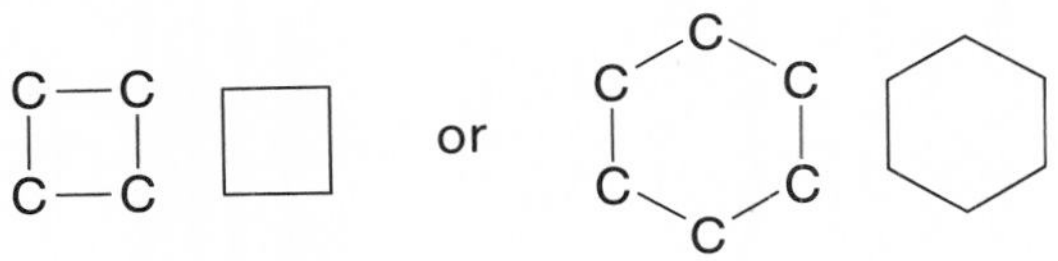

All the structures shown have carbon-to-carbon single bonds, but the other atoms attached to carbon to complete its four bonds are omitted. The atom most commonly attached is hydrogen. When we fill in the missing attachments with hydrogen atoms, the structures become

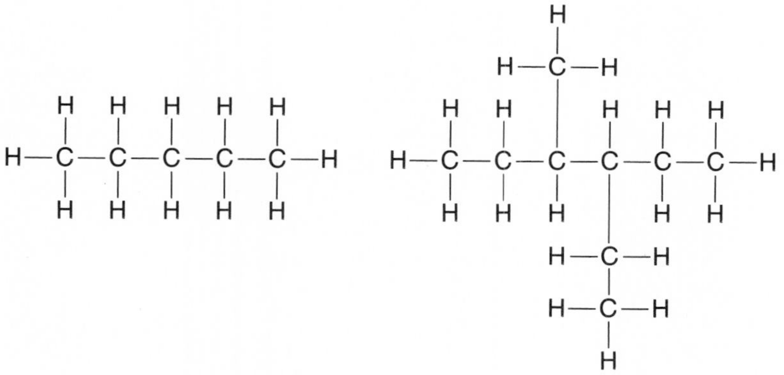

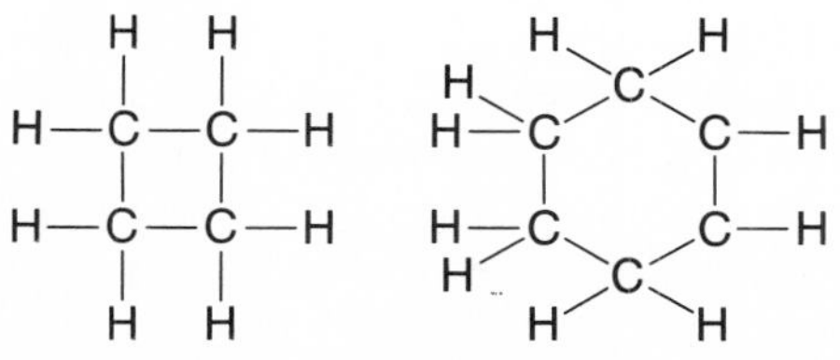

A third way of writing the structures is

$$CH_3CH_2CH_2CH_2CH_3 \qquad CH_3CH_2CH(CH_3)CH(CH_2CH_3)CH_2CH_3$$

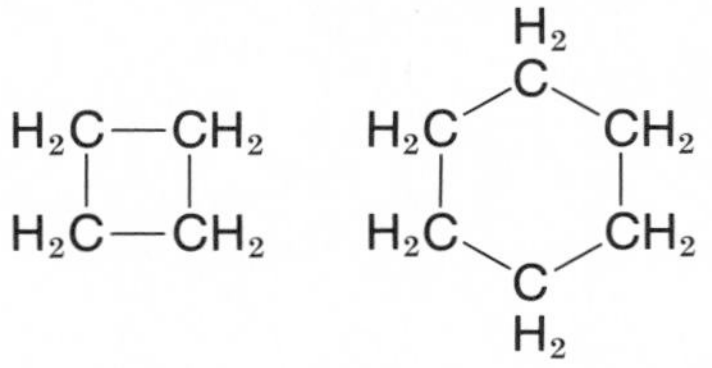

These representations tell you nothing about the geometry of the molecule. A better way of showing the carbon chains might be

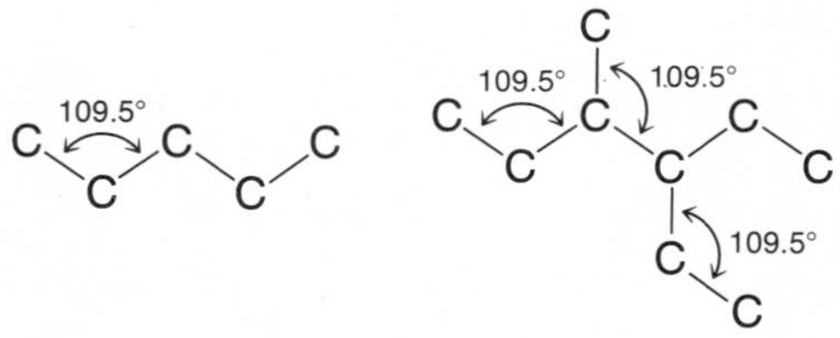

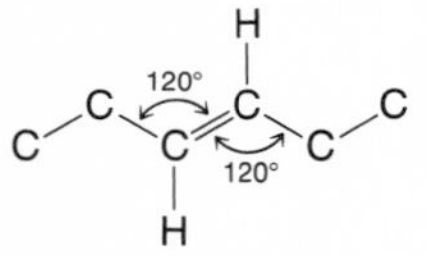

Structure I (trans)

Structure II (cis)

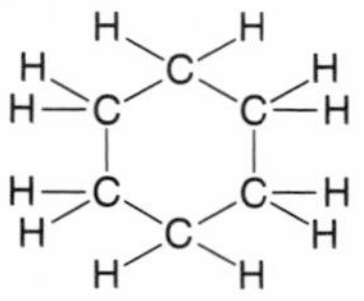

No rotation around the carbon-carbon bonds is possible.

These structures are more logical but clumsier than linear representations because the carbons in a compound are in the tetrahedral configuration with bond angles of 109.5°. A zigzag chain shows this better than a straight chain. A second important feature of these single-bonded structures is that the atom attached to each carbon can rotate freely around the single bond. Different arrangements of the same structure are called conformations or **conformers** of the compound.

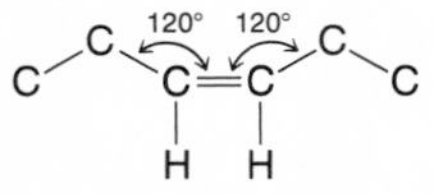

Whereas sp^3-hybrid bonds indicate a single-bonded carbon structure in a tetrahedral configuration, sp^2-hybrid bonds indicate a double bond. Because of the four-bond rule, the carbons sharing the double bond hold one hydrogen each. Since these carbons cannot rotate because of the double bond between them, the position of the hydrogen atom each holds is fixed. In structure I (margin) these hydrogens are in the trans configuration, that is, on opposite sides of the double bond. In structure II these hydrogens of the double-bonded carbons are in the cis configuration, which is on the same side of the double bond.

The introduction of the double bond into a carbon compound results in a rigid structure at the double bond—no more free rotation. This means that attachments are fixed in a trans configuration or a cis configuration at the carbons at the double bond. It also introduces into the compound the geometry of sp^2 hybridization, that is, a trigonal planar geometry with bond angles of 120° Another case in which free rotation of carbon atoms is prohibited occurs when they are arranged into a cyclic structure.

If a compound has sp-hybrid bonds, it contains a triple bond between two carbons. This also means restricted rotation, and the four central atoms shown are in a linear arrangement, with bond angles of 180° for the triple-bonded carbons.

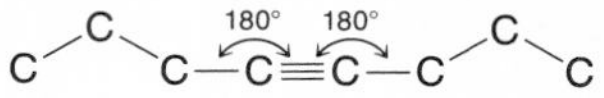

CLASSIFICATION OF ORGANIC COMPOUNDS

Three major divisions of organic compounds are generally recognized. The first division consists of **aliphatic** compounds, of which the simplest and "parent" compound is methane, CH_4. The aliphatic compounds may have their carbons arranged in straight- or branched-chain sequence or in a cyclic (ring) form. In the second major division are the **aromatic** compounds, for which the parent compound is benzene, C_6H_6. The structure of benzene is cyclic, with six carbon atoms in the ring, where they may be thought of as being joined by alternating single and double bonds. The third division is made up of the **heterocyclic** compounds, which are cyclic but contain in their rings not only carbon atoms but also atoms of other elements. The atoms most often found with carbon in the ring of heterocyclic compounds are nitrogen, oxygen, and sulfur. Examples of heterocyclic compounds are given in Figure 10.2.

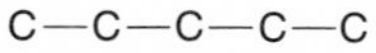

Organic compounds are further classified within these major divisions according to composition, structure, and functional groups. Composition in this case simply means the number and kind of atoms present in the compound. The composition of a compound can be summed up by a molecular formula:

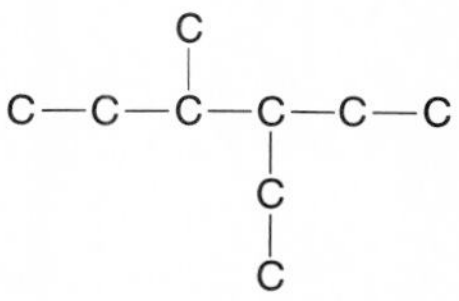

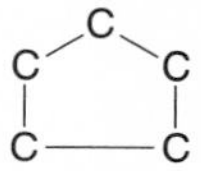

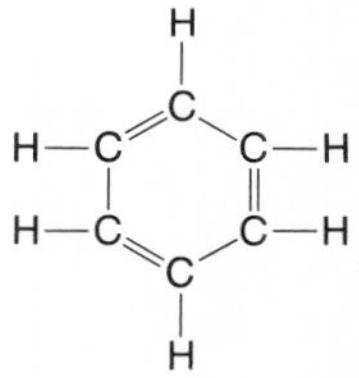

$$C_6H_{12}O_6 \qquad C_5H_9NO_2 \qquad C_6H_5Br$$

Structural formulas for these compounds are:

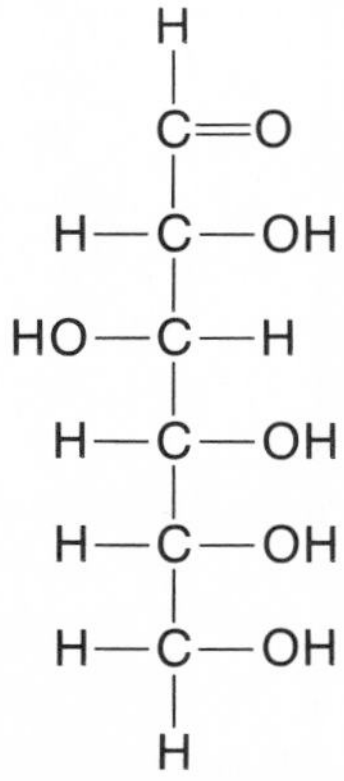

Straight

Branched

Cyclic

simplified as

or

Glucose

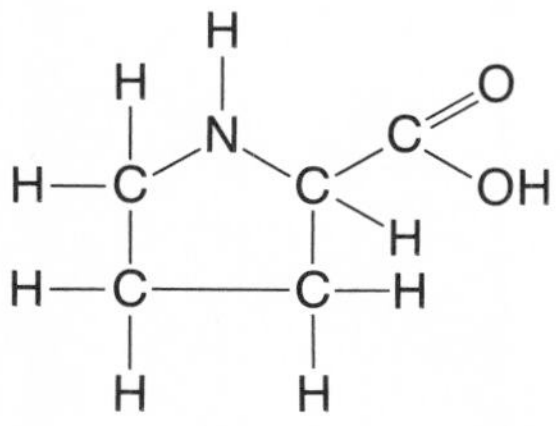

Proline

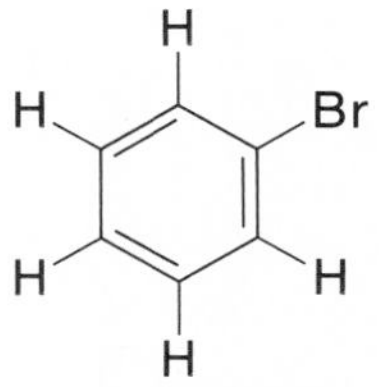

Bromobenzene

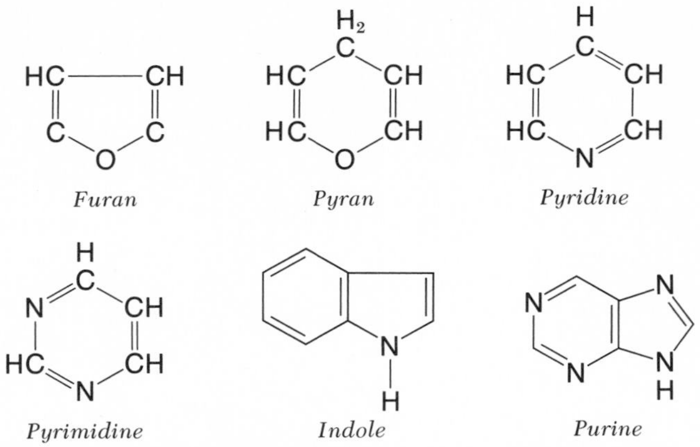

FIGURE 10.2 Examples of heterocyclic compounds.

TABLE 10.1 Elements common in organic compounds and their covalent number

Symbol	Name	Covalent Number
C	Carbon	4
H	Hydrogen	1
O	Oxygen	2
N	Nitrogen	3
S	Sulfur	2
P	Phosphorus	3
I	Iodine	1
Br	Bromine	1
Cl	Chlorine	1
F	Fluorine	1

Molecular formulas are written so that the carbons are given first, then the hydrogens, and then all other atoms in alphabetical order. This can be seen in the second compound. The most common elements are listed in Table 10.1.

Since molecular formulas say nothing about the arrangement of atoms in compounds, structural formulas, such as the ones shown above, are in general use. They tell us whether a compound is aliphatic, aromatic, or heterocyclic. They show the sequence of linkage of atoms and whether there is a straight or branched chain, a cyclic structure, or a combination of these. The structural formula also shows the type and number of functional groups present and where they are located. Table 10.2 illustrates these points.

FUNCTIONAL GROUPS

Chemists classify organic compounds by their **functional groups.** Each class of compounds has its own functional group, which imposes physical and chemical properties on the compounds.

A functional group is an arrangement of atoms within an organic structure, or attached to it. Names and structures of the major functional groups are given in Table 10.3. Special attention should be given to their structure and *specific* arrangements of the atoms forming them. With a few exceptions the functional group name remains the same whether the major structure is aliphatic, aromatic, or heterocyclic.

In Table 10.3 *one* functional group is defined at a time, but two or more, alike or different, may be present in a compound. Several examples are given in Table 10-4.

TABLE 10.2 Examples of organic compounds

Structural Formula and Name	Type of Compound	Functional Group and Name
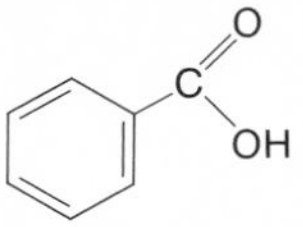 Benzoic acid	Aromatic	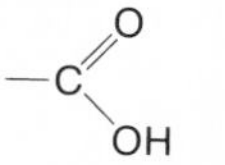 Carboxyl group
$CH_3CH_2CH_2CH_2OH$ 1-Butanol	Aliphatic	—OH Alcohol
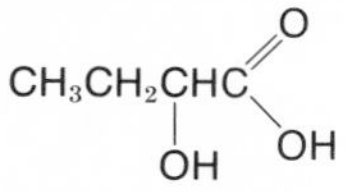 CH_3CH_2CHC 2-Hydroxybutanoic acid	Aliphatic	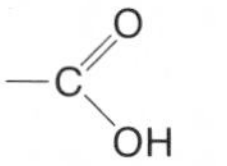 Carboxyl group —OH Alcohol
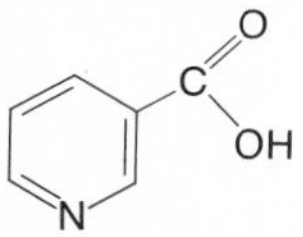 Nicotinic acid	Heterocyclic	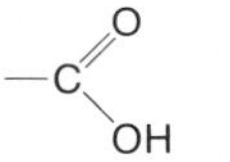 Carboxyl group
CH_3CH_2—O—CH_2CH_3 Diethyl ether	Aliphatic	—C—O—C— Ether
CH_3CH_2CH=CH_2 1-Butene	Aliphatic	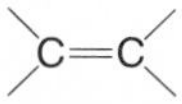 Alkene, carbon-carbon double bond
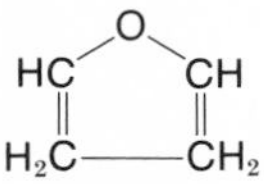 Furan	Heterocyclic	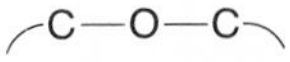 Cyclic ether

Let us take a closer look at some organic structures to see how the functional group behaves and how it affects the compound. The compound shown here is an alkane, an aliphatic compound containing only the elements carbon and hydrogen (and thus a hydrocarbon). The compound has only single bonds between carbon and carbon and between carbon and hydrogen:

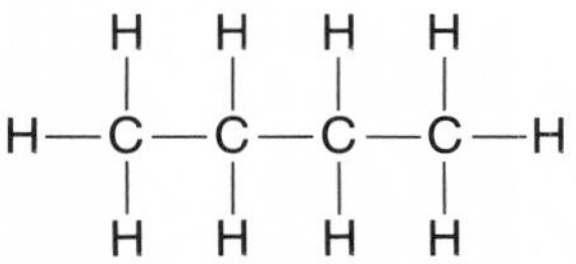

TABLE 10.3 Functional groups

Class Name	Functional Group	Example of Compound Structure
Alkane		C—C—C—C—C
Alkene	Double bond	C—C—C=C—C
Diene	Two double bonds	C=C—C=C—C
Alkyne	Triple bond	C—C—C—C≡C
Ether	Carbon-oxygen-carbon bridge	C—C—O—C—C—C
Alcohol	—OH, hydroxyl group	C—C—C—C—OH
Aldehyde	$-\overset{\displaystyle O}{\overset{\|}{C}}-H$, terminal carbonyl group	$C-C-C-\overset{\displaystyle O}{\overset{\|}{C}}-H$
Ketone	$-\overset{\displaystyle O}{\overset{\|}{C}}-$, carbonyl group	$C-C-\overset{\displaystyle O}{\overset{\|}{C}}-C-C$
Organic acid	$-\overset{\displaystyle O}{\overset{\|}{C}}-OH$, carboxyl group	$C-C-C-\overset{\displaystyle O}{\overset{\|}{C}}-OH$
Ester	$-\overset{\displaystyle O}{\overset{\|}{C}}-O-C$, ester linkage	$C-C-\overset{\displaystyle O}{\overset{\|}{C}}-O-C$
Amine	—NH$_2$, amino group	C—C—C—C—NH$_2$
Amide	$-\overset{\displaystyle O}{\overset{\|}{C}}-NH_2$, amide linkage	$C-C-\overset{\displaystyle O}{\overset{\|}{C}}-NH_2$
Thiol	—SH, thiol group	C—C—C—SH
Acid anhydride	$-\overset{}{\underset{}{C}}\overset{O}{\diagup}$ $\overset{}{\diagdown}$ O , anhydride linkage $-\overset{}{\underset{}{C}}\overset{}{\diagup}$ $\underset{O}{}$	$C-C\overset{O}{\diagup}$ $\diagdown$ O $C-C\overset{}{\diagup}$ $\underset{O}{}$
Alkyl halide	—X, halide	C—C—C—C—X†
Acid halide	$-\overset{\displaystyle O}{\overset{\|}{C}}-X$, acyl halide	$C-C-\overset{\displaystyle O}{\overset{\|}{C}}-X$†

† X = F, Cl, Br, I.
Note: H atoms attached to carbons have been omitted for clarity.

TABLE 10.4 Polyfunctional organic compounds

Structure	*Functional Groups*	*Class of Compound*
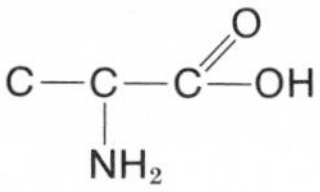	$-\overset{O}{\overset{\|}{C}}-OH$ Carboxyl group; $-NH_2$ Amino group	Amino acid
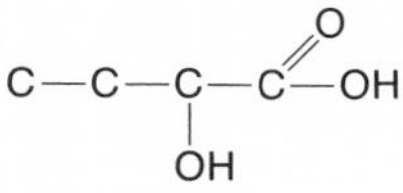	$-\overset{O}{\overset{\|}{C}}-OH$ Carboxyl group; $-OH$ Hydroxyl group	Hydroxy acid
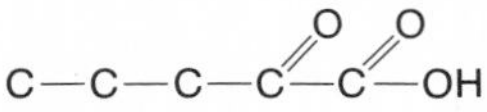	$-\overset{O}{\overset{\|}{C}}-OH$ Carboxyl group; $-\overset{O}{\overset{\|}{C}}-$ Carbonyl group	Keto acid
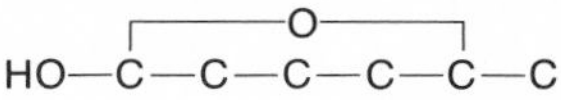	C—O—C Ether linkage; $-OH$ Hydroxyl group	Hemiacetal
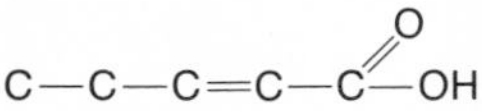	—C=C— Double bond; $-\overset{O}{\overset{\|}{C}}-OH$ Carboxylic acid	Unsaturated acid

TABLE 10.5 Electronegativity values
F > O > Cl, N > Br > C, H

Element	*Value*
F	4.0
O	3.5
Cl	3.0
N	3.0
Br	2.8
C	2.5
H	2.1

From Table 10.5 we see that carbon and hydrogen differ very little in their electronegativity value. This means that in the compound above, electrons are almost equally shared in the whole structure. A compound in which bonding electrons are equally shared between the atoms is nonpolar. If we take an aliphatic hydrocarbon and remove one hydrogen from each of two adjacent carbon atoms, a new compound results. This new compound is an **alkene,** with a double bond in its structure. The new compound now has an uneven electron distribution. The carbons connected by the double bond share more electrons

C : C: :C : C

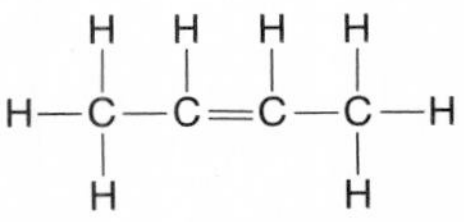

An alkene

between them than any other atoms in the compound. This creates a higher electron density in the region of the double bond. Since this higher electron density attracts certain reagents, some reactions can take place that could not occur when a single bond was present. In effect, the double bond, with its increased electron density, has become a functional group.

By placing an —OH, or hydroxyl, group at the end of our model compound, we arrive at the structure

$$H-\underset{\underset{H}{|}}{\overset{\overset{H}{|}}{C}}-\underset{\underset{H}{|}}{\overset{\overset{H}{|}}{C}}-\underset{\underset{H}{|}}{\overset{\overset{H}{|}}{C}}-\underset{\underset{H}{|}}{\overset{\overset{H}{|}}{C}}-OH$$

From Table 10.5 we see that oxygen is more electronegative than carbon. This means that oxygen attracts electrons more strongly. In our new compound with the —OH group, electrons shared between carbon and oxygen are pulled *closer* to the oxygen atom. As a result, the oxygen atom becomes more negative and the carbon to which oxygen is attached becomes less negative. This uneven sharing of electrons introduces **polarity** into this compound. We usually express this by labeling oxygen partial minus (δ^-) and labeling the carbon partial plus (δ^+). Remember the discussion on the polarity of the water molecule in Chapter 6. The polarity of the organic compound shown was introduced into the molecule by the functional —OH group. Here again, some reactions are promoted by this polarization arising from the functional group.

$$\overset{\delta^-}{\overset{\cdot\cdot}{\underset{..}{O}}}$$
$$\underset{\delta^+}{H}\qquad\underset{\delta^+}{H}$$

Polarity

ISOMERISM

The formula C_4H_{10} tells us that the compound contains four carbons and ten hydrogens. In writing a structure for this compound, we could come up with $CH_3CH_2CH_2CH_3$, which satisfies the requirement of four carbons and ten hydrogens. Another structure, $CH_3CH(CH_3)CH_3$, contains the same number of carbons and hydrogens. These two structures thus differ in the arrangements of the same atoms. The first structure is a straight-chain compound and the second structure a branched-chain compound. Whenever two or more compounds have the same number and kind of atoms, that is, the same *molecular* formula, but a different atom arrangement, they are said to be isomeric to each other and are called **isomers.**

Another case of isomerism can be shown with the molecular

C—C—C—C

C—C—C
|
C

Two possible structures for C_4H_{10}

formula C_2H_6O. The atoms of this formula can be arranged in two ways:

$$CH_3CH_2OH \qquad CH_3 - O - CH_3$$

An alcohol An ether

Both compounds contain the same number of carbons, hydrogens, and oxygen with different structural arrangements. Furthermore, there are two different functional groups, which makes this a case of **functional-group isomerism.**

Another case of functional isomerism can be shown with the summation formula C_3H_6O. The atoms can be arranged to give different structures:

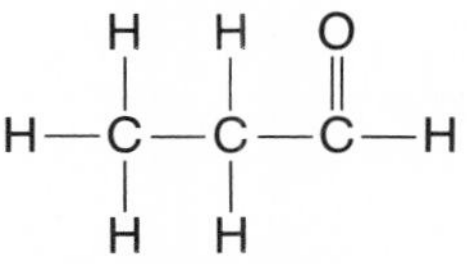

An aldehyde 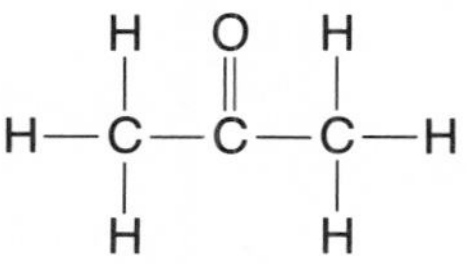 A ketone

The number of individual atoms is the same, but the structures and the functional groups differ.

Another type of isomerism can be shown with the molecular formula $C_4H_{10}O$. For this formula we can write four structures, each representing a different alcohol:

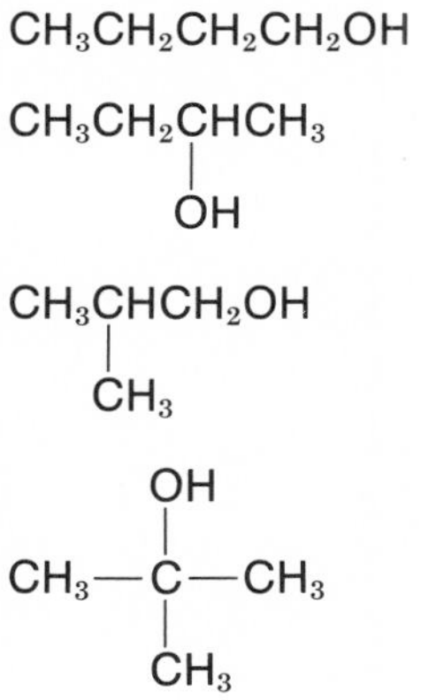

$$CH_3CH_2CH_2CH_2OH$$

$$CH_3CH_2\underset{\underset{OH}{|}}{C}HCH_3$$

$$CH_3\underset{\underset{CH_3}{|}}{C}HCH_2OH$$

$$CH_3 - \underset{\underset{CH_3}{\overset{\overset{OH}{|}}{C}}}{} - CH_3$$

In this case we find the hydroxyl functional group at different positions in the molecules. This type of isomerism, called **position isomerism,** can involve any functional group.

Another type of position isomerism is represented by the molecular formula C_4H_8. Two structures can be written:

$$CH_3 - CH_2 - CH = CH_2 \qquad CH_3 - CH = CH - CH_3$$

Several ethers could also be composed of these same atoms; for example
$CH_3 - CH_2 - O - CH_2 - CH_3$.

These compounds show position isomerism because the functional group, the double bond, can be in different positions. Furthermore, the second compound represents another type of isomerism which is called **cis-trans** or **geometric isomerism,** a type of **stereoisomerism** (*stero*, space).

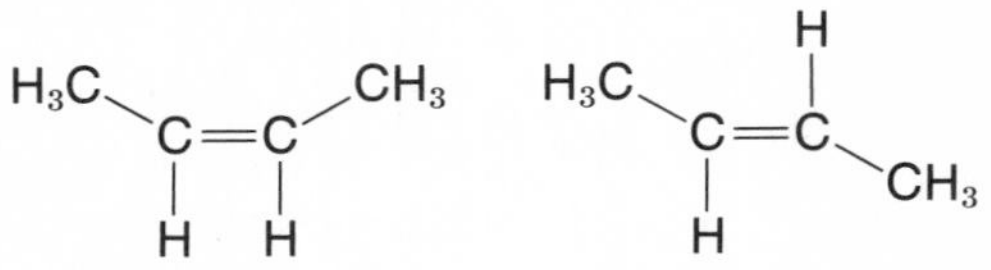

$$H_3C \diagdown \underset{\underset{H}{|}}{C} = \underset{\underset{H}{|}}{C} \diagup CH_3 \qquad H_3C \diagdown \underset{\underset{H}{|}}{C} = \underset{\underset{H}{|}}{C} \overset{\overset{H}{|}}{\diagup} CH_3$$

Cis   *Trans*

Cis-trans (geometric) isomerism refers to the location of the atoms or groups on the carbons with the double bond. This type of isomerism occurs chiefly at double bonds within a carbon chain. If the hydrogens attached to the double-bonded carbons are both on the same side of the double bond, the compound is the cis isomer. If the hydrogens are on the opposite sides of the double bond, the compound is the trans isomer. This type of isomerism is very important because it fixes the relationship of groups as well as the geometry of the molecule. The biological activity of molecules often depends on the suitable geometry of such double-bonded compounds.

Finally, another kind of stereoisomerism is exhibited by organic compounds, containing an **asymmetric carbon atom,** that is, one which has four different attachments, such as

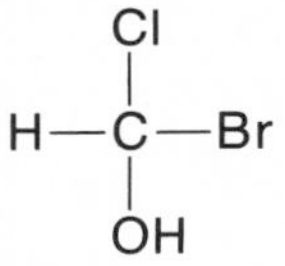

$$H - \underset{\underset{OH}{|}}{\overset{\overset{Cl}{|}}{C}} - Br$$

The two compounds in Figure 10.3 are mirror images and cannot be superimposed on each other. The carbon atom in each compound is asymmetric because, considered as a geometric unit, it

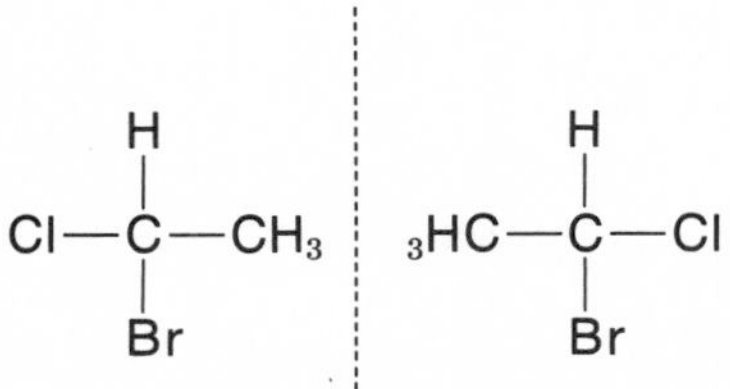

FIGURE 10.3 Mirror images.

$$Cl - \underset{\underset{Br}{|}}{\overset{\overset{H}{|}}{C}} - CH_3 \qquad {}_3HC - \underset{\underset{Br}{|}}{\overset{\overset{H}{|}}{C}} - Cl$$

Compound I   *Compound II*

Mirror Images

has no plane of symmetry; no hypothetical plane can cut it into identical values. Compounds I and II are completely identical from the standpoint of physical properties and chemical reactivity, except for one distinguishing physical characteristic. One of these compounds rotates polarized light to the left, and the other compound rotates polarized light to the right. More on this **optical isomerism** will be found in Chapter 13.

ORGANIC REACTIONS

Organic compounds undergo reactions specific to the skeletal structure and functional groups they contain. These reactions are influenced by the type of functional group and their location within a given molecule. Furthermore, skeleton and functional groups influence each other and thus help modify reactivity. Organic reactions can be classified into four general categories:

1 Displacement, or substitution, reactions
2 Elimination reactions
3 Addition reactions
4 Rearrangement reactions

In general, inorganic reactions proceed rather quickly and often take place under conditions that permit them to go to completion. Organic reactions are somewhat different:

1 Organic reactions are generally slow compared with inorganic reactions.
2 Most organic reactions are reversible.
3 Most organic reactions do not go to completion.

The extent to which they proceed is governed by an equilibrium condition, which we can express as

$$A + B \rightleftharpoons C + D$$
$$\text{Reactants} \qquad \text{Products}$$

This equation says that A and B are reacting to form C and D. The two arrows indicate that the reaction is reversible. When a certain amount of A has reacted with B and a certain amount of C and D has been formed, no further net conversion occurs; the further reaction of A and B to form more C and D is exactly balanced by reaction C and D to re-form A and B. This situation is known as an equilibrium. Organic reactions often attain equilibrium, and it is the task of the organic chemist to shift such an equilibrium in favor of the desired products. We discuss organic reactions in more detail in Chapter 11.

REVIEW QUESTIONS

1 What was the vital force principle?

2 Write the structure of the first organic compound synthesized from inorganic compounds.

3 Name some modern branches of chemistry that can be traced to organic chemistry.

4 The most important concept for the element carbon as it occurs in organic structures is hybridization, which enables carbon to have four covalent bonds. Describe hybridization in your own words.

5 List the types of bonds and the types of carbon structures that can be formed by carbon atoms.

6 Make sketches of the different bond arrangements that can occur between carbon atoms and describe their geometry and bond angles.

7 Describe free rotation and restricted rotation in an organic compound and show how it comes about.

8 What are the major divisions of organic compounds?

9 Define **a** heterocyclic compound, **b** branched structure, **c** aromatic compound, **d** cyclic structure, **e** electronegativity.

10 Make a list of the functional groups listed in this chapter. Write the structural formulas of the compounds and identify the functional groups by using a different color.

11 Look through Chapters 11 to 19 and pick out organic compounds with more than one functional group.

12 Use structures to show **a** structural isomers, **b** position isomers, **c** stereoisomers.

13 How many different organic structures of organic compounds can you write from the summation formula C_3H_6O? Identify them by the general group they belong to. *Hint:* There are four.

ORGANIC NOMENCLATURE
ORGANIC CLASSES
The Name of the Game and of the Players

11

Of all the atoms, carbon is the most proficient builder of chemical structures. The architecture of the molecule is the skeleton of carbon atoms to which other atoms, most often hydrogen or oxygen, bond covalently to fill out the structure and make a stable compound. The variety of possible combinations is infinite; as many as 2 million carbon compounds are known today, with more forthcoming each year.

As the number of known compounds proliferated during the nineteenth century, it became apparent that a systematic approach to their naming was needed. In 1892 an international organization of chemists, now known as the International Union of Pure and Applied Chemistry (IUPAC), called a conference in Geneva for that purpose. The task of bringing order into organic nomenclature was difficult not only because of the sheer number and varieties of compounds but also because a common nomenclature had already sprung up, but without plan, often inexact or ambiguous. Many names had originated with workers in medicine and biology who by habit and sentiment preferred their own terms to the somewhat more cumbersome but more exact names proposed by the Geneva Convention.

Nonetheless, the 1892 conference was a turning point in the development of a rational organic nomenclature. With continuing revisions and extensions from time to time, the IUPAC system is the framework for the naming of the carbon compounds. The system strives to provide names that will *unambiguously* identify the compound as well as convey its structure. The older, less exact names that persist are lumped together under the heading of **common and derived names** for organic compounds. The industrial and commercial aspects of chemistry still generate complications in the naming of compounds. A widely sold taste enhancer is marketed as Accent but its chemical name is monosodium glutamate, the sodium salt of glutamic acid.

The IUPAC system will be used here, although where necessary the common and derived names will also be indicated. We shall present only those essential aspects of the system which concern us; the complexities of the system we leave to the professional chemist.

IUPAC name: monosodium glutamate
Common abbreviation: MSG
Trade or commercial names: Accent, Ajinomoto, Vetsin, Zest

HYDROCARBONS

ALIPHATIC ALKANES

The simplest organic compounds consist only of carbon and hydrogen atoms. This general class of organic structures, called **hydrocarbons,** has several subclasses. The simplest subclass, called **alkanes,** is characterized by the fact that its members are made up of chains of carbon atoms and the carbon-to-carbon bonding is always a *single* bond. Structures of this kind are called **saturated;** they cannot take any more atoms into their structure, as all the bonds are fully occupied. All the alkanes fit the general molecular formula C_nH_{2n+2}, where n is the number of carbons in the compound. For example, if n is 3, the formula of that three-carbon alkane is C_3H_8. Any compound whose formula is compatible with the general summation formula is an alkane. For example, $C_{12}H_{26}$ must be an alkane. Table 11.1 gives the names of the first 10 straight-chain alkanes; you will see that all end with -*ane*.

We speak of compounds, with one, two, or three carbons as having one possible isomer since only one structure is possible. However, when we come to a four-carbon alkane, C_4H_{10}, two structural formulas can be written, as shown in Table 11.2. Note that only the carbon skeleton is used in the skeletal structural formula. This method will be used generally with the *clear and express reminder* that carbon really has *four* bonds, all of which must be satisfied either by carbon-to-carbon bonds or by attachment to hydrogen atoms.

TABLE 11.1 The first 10 straight-chair alkanes

Molecular Formula and Name	Number of Isomers
CH_4 Methane	1
C_2H_6 Ethane	1
C_3H_8 Propane	1
C_4H_{10} Butane	2
C_5H_{12} Pentane	3
C_6H_{14} Hexane	5
C_7H_{16} Heptane	9
C_8H_{18} Octane	18
C_9H_{20} Nonane	35
$C_{10}H_{22}$ Decane	75

TABLE 11.2   Isomers of a four-carbon alkane

Name	Molecular Formula	Condensed Structural Formula	Skeletal Structural Formula
Butane	C_4H_{10}	$CH_3CH_2CH_2CH_3$	C—C—C—C
Name?	C_4H_{10}	$CH_3CH(CH_3)CH_3$	C—C—C | C

The first structure shown for C_4H_{10} is a linear structure with its four carbons in sequence, called butane. But what is the name of the structure below it, also C_4H_{10}?

Let us apply the rules of the IUPAC system and name this C_4H_{10} isomer. These rules are:

1 Determine the longest continuous carbon chain. This will be called the **parent structure** and will be named according to the number of carbons it contains. In the isomer we are naming the longest sequence in the chain contains three carbons, and so the parent name is propane.

2 Number the atoms of this chain from one end to the other, starting at the end nearer a branch. In this case you can see that various ways of numbering are possible, but they all turn out to be equivalent.

3 Identify all branching groups and the location on the parent chain where the branching occurs. In our example the attachment coming off is a methyl group, CH_3—, at carbon atom 2 of the parent chain. (In this example, 2 is the only number possible, since you get it no matter how the numbering is done.)

4 Build up the name as follows:

   a Write the position number of the branching chain, in this case 2.

   b Write the name of the branching group, in this case methyl.

   c Write the name of the parent structure, in this case propane.

   d Put them together: 2-methylpropane.

The next compounds in the aliphatic series of the alkanes have the molecular formula C_5H_{12}; three isomers are possible. The longest continuous carbon chain contains five carbons, and the parent name is therefore pentane. The first branched structure has the parent name butane. The location of the branch is established by numbering the carbons in the parent chain. Counting from right to left, the branch occurs at carbon 2,

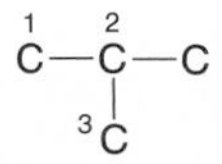

Branched isomer of C_4H_{10}:

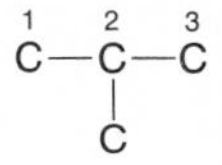

C—C—C
    |
    C

or

C—C—C
    |
    C

or

C—C—C
    |
    C

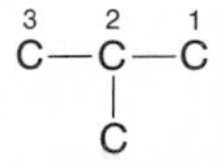

Each represents the same isomer; only the numbering sequence is different.

Isomers of C_5H_{10} with IUPAC names:

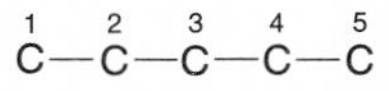

C—C—C—C—C

Pentane

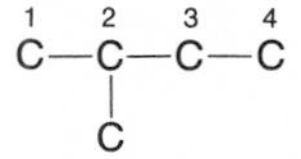

C—C—C—C
    |
    C

2-Methylbutane

    C
    |
C—C—C
    |
    C

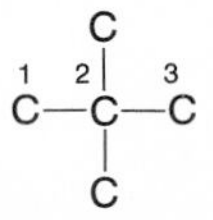

2,2-Dimethylpropane

Number of Methyl Groups	Name
1	Methyl
2	Dimethyl
3	Trimethyl
4	Tetramethyl
5	Pentamethyl
6	Hexamethyl

whereas counting from left to right, the branch is at carbon 3. In this instance, the lower number is 2, counting from right to left. The complete name for the isomer is therefore 2-methylbutane; 3-methylbutane would be incorrect. The third isomer of C_5H_{12} is the branched structure with the longest continuous sequence of three carbons, so that the parent name is propane. The branching here consists of *two* methyl groups, which must be called *di*methyl, since there are two. Both branches are at carbon 2 of the parent chain, and the number is repeated in the name. The complete systematic name for the third isomer is therefore

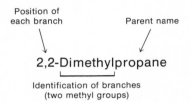

Position of each branch

Parent name

2,2-Dimethylpropane

Identification of branches (two methyl groups)

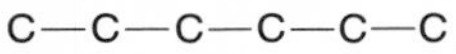

C—C—C—C—C—C
Hexane

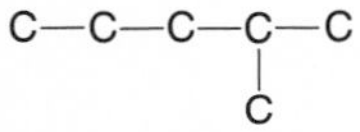

C—C—C—C—C
      |
      C
2-Methylpentane

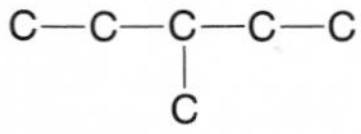

C—C—C—C—C
    |
    C
3-Methylpentane

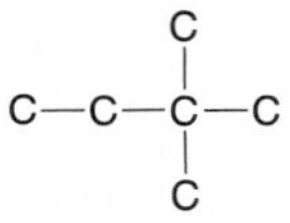

     C
     |
C—C—C—C
     |
     C
2,2-Dimethylbutane

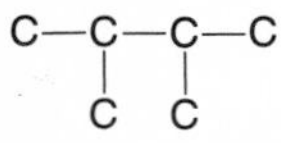

C—C—C—C
  | |
  C C
2,3-Dimethylbutane

The next higher group of alkanes contains six carbons; the molecular formula is C_6H_{14}, and five isomers are possible.

In all the compounds considered so far the branches were methyl groups, but there can be other groups as well. These groups are generally called alkyl groups, represented as —R. They are conceptually derived from parent alkanes by the removal of one hydrogen and named by changing the ending in the parent name to —*yl*. Common alkyl groups are listed in Table 11.3 with their parent names and structures. The letter R in a structure means that some unspecified alkyl is present (methyl, ethyl, propyl, or some other). A few explanations are in order for some of the names in Table 11.3. So far we have used only the proper IUPAC names, but many common names are also in use. One common naming practice is to place the letter *n*-, which stands for **normal**, before the name for straight-chain compounds, so that butane is *n*-butane, pentane is *n*-pentane, etc. In the common system, the alkanes with a methyl group attached to their next-to-last carbon are called **isoalkanes**, so that structure I has the common name **isobutane** and structure II the common name **isopentane**:

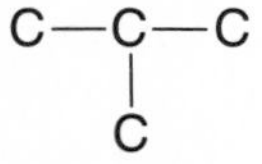

C—C—C
   |
   C

Structure I

$\overset{4}{C}$—$\overset{3}{C}$—$\overset{2}{C}$—$\overset{1}{C}$
        |
        $\overset{5}{C}$

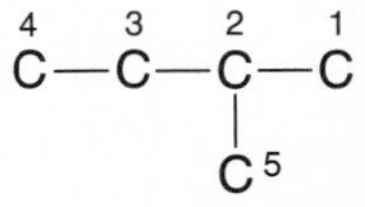

Structure II

A carbon atom in a given structure can also be described or classified according to how many *other* carbons it is attached to.

TABLE 11.3   Common alkyl groups

Alkane	Alkyl Name	Structure
Methane	Methyl	CH_3-
Ethane	Ethyl	CH_3CH_2-
Propane	Propyl	$CH_3CH_2CH_2-$
	Isopropyl	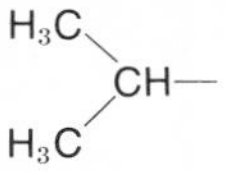 $\begin{array}{c} H_3C \\ \diagdown \\ CH- \\ \diagup \\ H_3C \end{array}$
Butane	Butyl	$CH_3CH_2CH_2CH_2-$
	Secondary butyl, *sec*-butyl	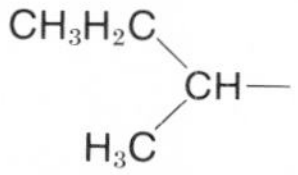 $\begin{array}{c} CH_3H_2C \\ \diagdown \\ CH- \\ \diagup \\ H_3C \end{array}$
	Isobutyl	$\begin{array}{c} CH_3CHCH_2- \\ | \\ CH_3 \end{array}$
	Tertiary butyl, *tert*-butyl	$\begin{array}{c} CH_3 \\ | \\ H_3C-C- \\ | \\ CH_3 \end{array}$

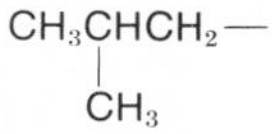

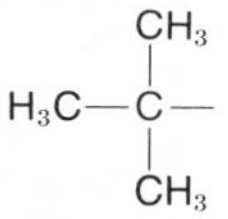

Consider structure II again: carbons 1, 4, and 5 are attached to only one other carbon. Such carbons are called **primary carbons atoms,** and the hydrogens attached to them are called primary hydrogen atoms. If, say, an —OH group is attached to a primary carbon, the compound will be a primary alcohol. Carbon 3 in structure II is attached to two other carbons and is classified for this reason as a **secondary carbon atom.** Carbon 2 is attached to three other carbons and is therefore a **tertiary carbon atom.** Compounds in which functional groups are attached to these carbons are referred to as **primary, secondary,** and **tertiary compounds,** respectively. The name secondary alcohol tells us that the structure has an —OH group bonded to a carbon which is attached to two other carbons.

CYCLOALKANES

The last group of the saturated hydrocarbons is the **cycloalkanes.** One can visualize their formation by removing a hydrogen from each end of a carbon chain and then connecting the two end carbons (Table 11.4).

TABLE 11.4 Examples of cycloalkanes

Carbon Atoms Present	Structure		Name
3	$\begin{array}{c} CH_2 \\ H_2C\!-\!\!-\!CH_2 \end{array}$	△	Cyclopropane
4	$\begin{array}{c} H_2C\!-\!CH_2 \\ H_2C\!-\!CH_2 \end{array}$	☐	Cyclobutane
5	$\begin{array}{c} H_2 \\ C \\ H_2C \qquad CH_2 \\ H_2C\!-\!\!-\!CH_2 \end{array}$	⬠	Cyclopentane
6	$\begin{array}{c} H_2 \\ C \\ H_2C \qquad CH_2 \\ H_2C \qquad CH_2 \\ C \\ H_2 \end{array}$	⬡	Cyclohexane

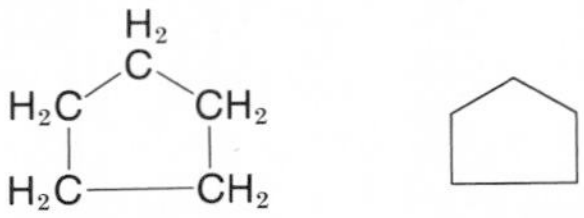

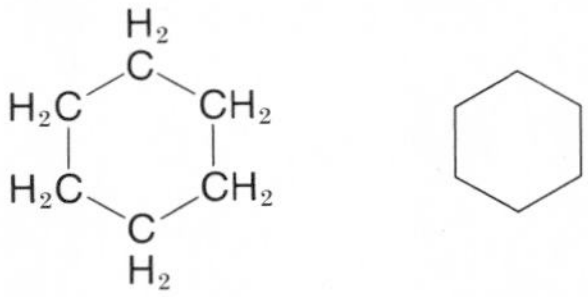

ALKENES

The single bonded alkanes are also called paraffins; the alkenes have a carbon-to-carbon double bond and are also called olefins.

The next hydrocarbon subclass is that of the **olefins,** unsaturated hydrocarbons with one carbon-to-carbon double bond. They can be considered as derived from corresponding alkanes by removing one hydrogen each from neighboring carbons, then forming a double bond between them. The region around a double bond has a higher energy and is more reactive, so that the double bond is classified as a functional group. The general molecular formula for the alkenes is C_nH_{2n}, and the name ending in the IUPAC system is *-ene.* The common names for the alkenes end in *-ylene,* as in ethylene, propylene, and butylene (Table 11.5).

The IUPAC system rules apply as before, except that the longest carbon chain must contain the double bond and the number given to the position of the double bond (the number of the first of the two carbons involved) must be the lowest possible. Because the double bond is a functional group, it takes precedence over branches for numbering purposes. For example, in the formula in the margin the longest carbon chain containing the double bond has six carbons; therefore, the parent name is hexene. The double bond is located between carbon 2 and carbon 3, or between carbon 4 and carbon 5, depending on the direction of counting. Using the lower number for locating the bond, we get 2-hexene. Since there are two methyl groups at positions 4 and 5 (counting right to left), the complete name is 4,5-dimethyl-2-hexene.

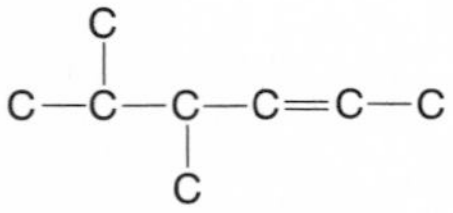

$$\begin{array}{c} C \\ | \\ C\!-\!C\!-\!C\!-\!C\!=\!C\!-\!C \\ | \\ C \end{array}$$

4,5-Dimethyl-2-hexene

TABLE 11.5 **Examples of alkenes**

Carbon Atoms Present	Molecular Formula	Structure	Carbon Skeleton	IUPAC Name	Common Name
2	C_2H_4	$CH_2\!\!=\!\!CH_2$	C=C	Ethene	Ethylene
3	C_3H_6	$CH_3CH\!\!=\!\!CH_2$	C—C=C	Propene	Propylene
4	C_4H_8	$CH_3CH_2CH\!\!=\!\!CH_2$	C—C—C=C	1-Butene	α-Butylene†
	C_4H_8	$CH_3CH\!\!=\!\!CHCH_3$	C—C=C—C	2-Butene	β-Butylene†

	C_4H_8	$CH_3C(CH_3)\!\!=\!\!CH_2$	C—C=C
			   |
			   C
				2-Methyl-propene	Isobutylene

† Obsolete.

One of the important features of the alkenes is that they may possess position isomerism in relation to the functional group. As can be seen from Table 11.5, 1-butene and 2-butene differ only in the position of the double bond. Furthermore, some of the alkenes exhibit cis-trans (geometric) isomerism. Because of hindered (restricted) rotation around the double bond, the carbons connected by the double bond and their attachments are fixed in space. 2-Butene provides a good example. Geometric isomerism is exhibited by compounds with double bonds where neither of the double-bonded carbons is attached to two identical atoms or groups of atoms. 1-Butene does not show this isomerism because carbon 1 has the same two attachments, two hydrogens.

trans-2-Butene *cis-2-Butene*

1-Butene

ALKYNES

A third subclass of hydrocarbons contains a triple bond. Two hydrogens each are considered to have been removed from adjacent carbons, so that a triple bond between the two carbons results. The molecular formula for this class is C_nH_{2n-2}, and in the IUPAC system the parent name ends in *-yne* (Table 11.6). As the functional group, the triple bond again takes precedence over branching and must receive the lower number available. The last compound in Table 11.6 locates the triple bond (counting from left to right) at carbon 1 and the methyl at carbon 3.

Unsaturated hydrocarbons discussed up to this point have contained only one double or one triple bond. It is possible for a molecule to have more than one double or triple bond or both. Compounds with two double bonds are called **dienes;** with three double bonds they are called **trienes,** and with more double bonds **polyenes.** When the double bonds are separated by single bonds, so that single and double bonds alternate, the arrangement is called a **conjugated system.**

1,3-Butadiene

2-Methyl-1,3-butadiene
(isoprene)

1,3,5-Hexatriene

2,4-Dimethyl-1,3,5-hexatriene

TABLE 11.6 Examples of alkynes

Carbon Atoms Present	Molecular Formula	Structure	Carbon Skeleton	IUPAC Name	Common Name
2	C_2H_2	H—C≡C—H	C≡C	Ethyne	Acetylene
3	C_3H_4	CH_3—C≡C—H	C—C≡C	Propyne	Methylacetylene
4	C_4H_6	CH_3—CH_2C≡C—H	C—C—C≡C	1-Butyne	Ethyl acetylene
		CH_3C≡C—CH_3	C—C≡C—C	2-Butyne	Dimethyl acetylene

5	C_5H_8	H—C≡C—CH—CH_3
		             CH_3
			C≡C—C—C
			       C
				3-Methyl-1-butyne	Isopropyl acetylene

Unsaturated hydrocarbons can also appear in cyclic skeletons:

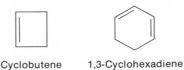

Cyclobutene   1,3-Cyclohexadiene

The several subclasses (alkanes, alkenes, alkynes, cycloalkanes, etc.) together constitute the **aliphatic hydrocarbons.**

ALIPHATIC COMPOUNDS CONTAINING OTHER ELEMENTS BESIDES CARBON AND HYDROGEN

Organic compounds in the aliphatic system often contain other elements besides carbon and hydrogen.

ALKYL HALIDES

The first group contains an element of the halogen family, Group VII: F, Cl, Br, or I. The monohalogen compounds are known as **alkyl halides** or monohalogenated hydrocarbons. In the IUPAC system they are named as substitution products of the parent alkanes. In the common naming, one identifies the alkyl group and designates the halogen by giving it the ending -*ide*, as in naming inorganic salts (Table 11.7). Common names are mainly used with short-chain compounds.

By substituting more than one halogen we get dihalogen, trihalogen, and polyhalogen compounds, which are mostly named by the IUPAC system.

ETHERS

R and R' are two different alkyl groups.

The first representatives of compounds containing carbon, hydrogen, and oxygen are the **ethers,** with the general formula

TABLE 11.7 Examples of monohalogen compounds

Carbon Atoms Present	Parent	Structure	Carbon Skeleton	IUPAC Name	Common Name
1	Methane	CH_3Cl	C—Cl	Chloromethane	Methyl chloride†
2	Ethane	CH_3CH_2Cl	C—C—Cl	Chloroethane	Ethyl chloride
3	Propane	$CH_3CH_2CH_2Cl$	C—C—C—Cl	1-Chloropropane	n-Propyl chloride
		$CH_3CH(Cl)CH_3$	C—C—C, Cl	2-Chloropropane	Isopropyl chloride
4	Butane	$CH_3CH_2CH_2CH_2Cl$	C—C—C—C—Cl	1-Chlorobutane	n-Butyl chloride
		$CH_3CH_2CH(Cl)CH_3$	C—C—C—C, Cl	2-Chlorobutane	sec-Butyl chloride
		$CH_3CH(CH_3)CH_2Cl$ propane	C—C—C—Cl, C	1-Chloro-2-methyl-propane	Isobutyl chloride
		$CH_3C(Cl)(CH_3)CH_3$	Cl, C—C—C, C	2-Chloro-2-methyl-propane	tert-Butyl chloride

† This could also be Br, F, or I to become, in the common system, bromide, fluoride, or iodide, respectively.

R—O—R′. As we can see, oxygen acts here as a bridge between two alkyl groups. The common names, which are more in use for the shorter-chain compounds, identify the two alkyl groups and then add the word **ether.** In the IUPAC system the smaller alkyl group is combined with the oxygen and called an **alkoxy group** and then substituted into the name of the longer chain (Table 11.9). Consider an example in the IUPAC system:

$$CH_3—O—CH_2CH_2CH_2CH_3$$

Methoxy   Butane

The shorter alkyl group CH_3— plus oxygn gives CH_3—O—, the **methoxy group.** Using this as prefix to the longer carbon chain, whose parent name is **butane,** gives us the IUPAC name for the ether, 1-**methoxybutane.**

ALCOHOLS

Compounds containing carbon, hydrogen, and oxygen which are isomeric to ethers are the monohydroxy compounds R—OH, al-

TABLE 11.8   Examples of higher-substituted halogen compounds

Carbon Atoms Present	Structure	Carbon Skeleton	IUPAC Name	Common Name
1	CH_2Cl_2	Cl │ C—Cl	Dichloromethane	Methylene chloride
	$CHCl_3$	Cl—C—Cl │ Cl	Trichloromethane	Chloroform
	CCl_4	Cl │ Cl—C—Cl │ Cl	Tetrachloromethane	Carbon tetra-chloride
	CCl_2F_2	F │ Cl—C—Cl │ F	Dichlorodifluoro-methane	Freon
2	CH_3CHCl_2	C—C—Cl │ Cl	1,1-Dichloroethane	Ethylidene chloride
	CH_2ClCH_2Cl	Cl—C—C—Cl	1,2-Dichloroethane	Ethylene chloride
	CH_3CCl_3	Cl │ C—C—Cl │ Cl	1,1,1-Trichloro-ethane	Methyl-chloroform

cohols (Table 11.10).   R again stands for any alkyl chain.   The IUPAC system name uses the parent alkane name; the ending is changed to *-ol* to designate an **alcohol.**   The position of the —OH group is indicated by a number which must be the lowest possible.   The common names are formed by naming the alkyl chain and adding the word **alcohol.**

TABLE 11.9   Common ethers

Molecular Formula	Structure	IUPAC Name	Common Name
C_2H_6O	CH_3—O—CH_3	Methoxymethane	Dimethyl ether
C_3H_8O	CH_3CH_2—O—CH_3	Methoxyethane	Methyl ethyl ether
C_4H_{10}	CH_3CH_2—O—CH_2CH_3	Ethoxyethane	Diethyl ether
	$CH_3CH_2CH_2$—O—CH_3	Methoxypropane	Methyl n-propyl ether
	$(CH_3)_2CH$—O—CH_3	2-Methoxypropane	Methyl isopropyl ether

TABLE 11.10 Common alcohols

Parent	Structure	IUPAC Name	Common Name
CH_4	C—OH	Methanol	Methyl alcohol
C_2H_6	C—C—OH	Ethanol	Ethyl alcohol
C_3H_8	C—C—C—OH	1-Propanol	*n*-Propyl alcohol
	C—C—C with OH on middle C	2-Propanol	Isopropyl alcohol
C_4H_{10}	C—C—C—C—OH	1-Butanol	*n*-Butyl alcohol
	C—C—C—C with OH on third C	2-Butanol	*sec*-Butyl alcohol
	C—C—C—OH with C below second C	2-Methyl-1-propanol	Isobutyl alcohol
	C—C—C with OH above and C below central C	2-Methyl-2-propanol	*tert*-Butyl alcohol

TABLE 11.11 Common diols, triols, and polyols

Structure	IUPAC Name	Common Name
C—C with OH OH	1,2-Ethanediol	Ethylene glycol†
C—C—C with OH OH	1,2-Propanediol	Propylene glycol
C—C—C with OH OH OH	1,2,3-Propanetriol	Glycerin, glycerol
C—C—C—C—C with OH on second and fourth C	1,3-Pentanediol	
C—C—C—C—C—C with OH OH OH OH OH OH	1,2,3,4,5,6-Hexahydroxy-hexol	Sorbitol, etc.

† Glycols are organic compounds that contain two hydroxyl groups.

Structures with more than one —OH group are named **diols** if two —OH groups are present, **triols** with three —OH groups, and **polyols** with more than three. Some common names of importance, especially for those containing two —OH groups, are listed in Table 11.11. Cyclic alcohols receive the prefix *cyclo-*, as in the cycloalkanes.

ALDEHYDES

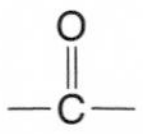

Several classes of organic compounds have one oxygen attached by a *double* bond to carbon. **Aldehydes** have the general formula RCHO, where R = H or any alkyl group. The CO group called the **carbonyl group** is not exclusive with aldehydes, as we shall see. The aldehyde functional group, —CHO, can occur only at the end of a carbon chain. The IUPAC system establishes the name by using the parent alkane name and changing the ending to *-al*. The common names are derived from the common names of organic acids with the ending changed to *-aldehyde*. The name *form*aldehyde comes from *form*ic acid (Table 11.12).

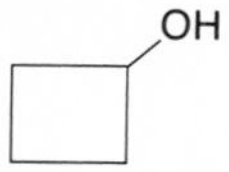

Carbonyl group

Cyclobutanol

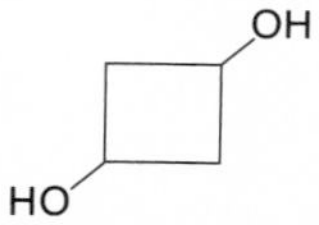

1,3-Cyclobutanediol

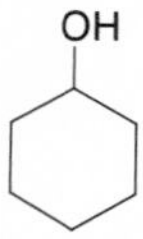

Cyclohexanol

TABLE 11.12   Examples of aldehydes

Parent	Structure	IUPAC Name	Common Name
Methane	H—C=O (H)	Methanal	Formaldehyde
Ethane	CH₃C=O (H)	Ethanal	Acetaldehyde
Propane	CH₃CH₂C=O (H)	Propanal	Propionaldehyde
	CH₂=CHC=O (H)	2-Propenal	Acrolein
Butane	CH₃CH₂CH₂C=O (H)	Butanal	Butyraldehyde
	C—C—C=O (C,H)	2-Methylpropanal	Isobutyraldehyde
Pentane	C—C—C—C—C=O (H)	Pentanal	Valeraldehyde

KETONES

Ketones have the general formula R—C(=O)—R', where R stands for any alkyl group and R' indicates that two different alkyl groups may be present, for example, methyl and ethyl. Ketones also contain the carbonyl group and are isomeric with aldehydes:

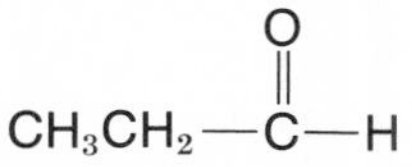

$$CH_3CH_2 - \overset{\overset{\displaystyle O}{\|}}{C} - H$$

An aldehyde

C_3H_6O:

$$CH_3\overset{\overset{\displaystyle O}{\|}}{C} - CH_3$$

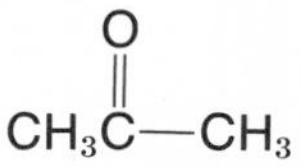

A ketone

Isomeric (same number of carbon, hydrogen, and oxygen atoms)

The IUPAC system name uses the parent alkane name and changes the ending to *-one*, indicating the ketone. The location of the keto (or oxy) group is identified within the chain by giving it the lower number. The C=O in ketones is located between two carbon atoms, which places it *within* the carbon chain and never at the end. The common names designate the groups attached to the carbonyl group and add the word **ketone** (Table 11.13).

ORGANIC ACIDS

The next class of compounds containing the carbonyl group is the **organic acids.** To the carbonyl carbon is added an —OH group, to form a new functional group called the **carboxyl group,** —COOH. Monoprotic (one H^+ ion from one carboxyl group) aliphatic acids are known as **fatty acids;** they are true acids in the sense that they liberate hydrogen ions, are sour to taste (when soluble enough), and behave as respectable acids should. In the IUPAC naming system the carboxyl group must be contained in the longest carbon chain, and the carboxyl carbon becomes carbon 1 in the chain. Carboxyl groups are located at the end of carbon chains. The parent alkane, with its ending changed to *-oic acid*, gives the name to the acid. In the common system, the names for the acids were usually derived from the natural products from which they were originally isolated. The general formula for monoprotic acids is RCOOH, where R could be H or any alkyl group (Table 11.14). Organic acids can be di-, tri-, or even polyprotic acids. This means they have two, three, or more carboxyl groups, as illustrated in Table 11.15.

Two systems for specifying the location of the carbons in the parent chains of carboxylic acids are in use. The first (used in the IUPAC system) assigns the carboxylic acid carbon the

TABLE 11.13 Examples of ketones

Carbon Atoms Present	Structure	IUPAC Name	Common Name
3	$C-\overset{\overset{\textstyle O}{\|}}{C}-C$	Propanone	Dimethyl ketone, acetone
4	$C-\overset{\overset{\textstyle O}{\|}}{C}-C-C$	Butanone	Methyl ethyl ketone
5	$C-C-C-\overset{\overset{\textstyle O}{\|}}{C}-C$	2-Pentanone	Methyl *n*-propyl ketone
5	$C-C-\overset{\overset{\textstyle O}{\|}}{C}-C-C$	3-Pentanone	Diethyl ketone
5	$C-\overset{\overset{\textstyle O}{\|}}{C}-\underset{\underset{\textstyle C}{\|}}{C}-C$	3-Methyl-2-butanone	Methyl isopropyl ketone
8	$C-C-C-\overset{\overset{\textstyle O}{\|}}{\underset{\underset{\textstyle C}{\|}}{C}}-\overset{\overset{\textstyle C}{\|}}{C}-C$	2,5-Dimethyl-3-hexanone	Isopropyl isobutyl ketone
5	(cyclopentanone structure)	Cyclopentanone	
7	(3-methylcyclohexanone structure)	3-Methylcyclohexanone	

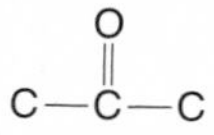

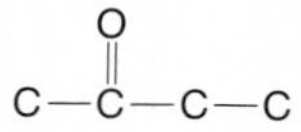

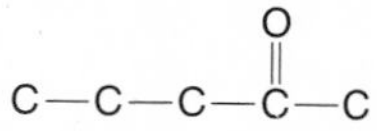

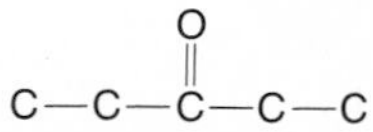

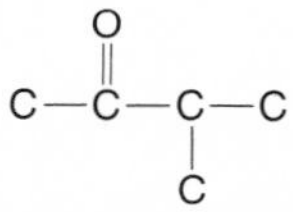

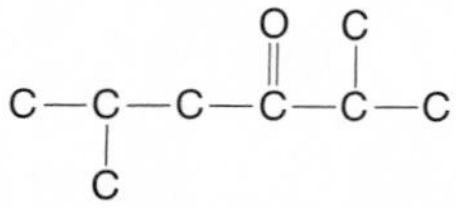

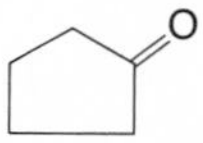

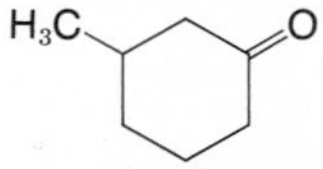

$$H_3C-\overset{4}{C}-\overset{3}{C}-\overset{2}{C}-\overset{\overset{\textstyle O}{\|}}{\underset{1}{C}}-OH$$

$$\overset{4}{\underset{\gamma}{C}}-\overset{3}{\underset{\beta}{C}}-\overset{2}{\underset{\alpha}{C}}-\overset{1}{C}$$

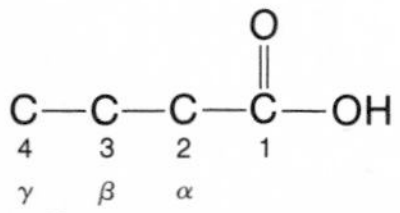

Numbers are used with IUPAC names; Greek letters may be used with common names.

number 1, with all other carbons in the chain numbered thence. In the common naming system Greek letters may be used. The carboxylic carbon is identified by the word acid, and the successive carbons thence in the chain are designated as alpha (α), beta (β), gamma (γ), delta (δ), etc. In the structure shown in the margin, carbon 2 is α, carbon 3 is β, and so on.

AMINES

Another class of organic compounds consists of those containing nitrogen besides carbon and hydrogen. These compounds often

TABLE 11.14 Common organic acids

Carbon Atoms Present	Structure	IUPAC Name	Common Name
1	H—C=O | OH	Methanoic acid	Formic acid
2	CH_3—C=O | OH	Ethanoic acid	Acetic acid
3	CH_3CH_2—C=O | OH	Propanoic acid	Propionic acid
4	$CH_3CH_2CH_2$—C=O | OH	Butanoic acid	Butyric acid
	CH_3CH—C=O | | CH_3 OH	2-Methylpropanoic acid	Isobutyric acid
5	$CH_3CH_2CH_2CH_2$—C=O | OH	Pentanoic acid	Valeric acid
12	$CH_3(CH_2)_{10}$—C=O | OH	Dodecanoic acid	Lauric acid

contain the $-NH_2$ group, called the **amino group** in the IUPAC system. The general formula for primary **amines** is RNH_2; for secondary amines it is R_2NH, and for tertiary amines it is R_3N.

Replace one hydrogen from NH_3 by an R group ⟶ RNH_2

A primary amine

Replace two hydrogens from NH_3 by R groups ⟶ R_2NH

A secondary amine

Replace three hydrogens from NH_3 by R groups ⟶ R_3N

A tertiary amine

The common naming system identifies the alkyl groups and then adds the ending -*amine*. The IUPAC system identifies the alkane and the position of the amino group and names the compound as an alkanamine (Table 11.16). The prefix *N*- denotes a substituent on the N atom replacing a hydrogen atom. For the shorter-chain amines the common names are more used; they are accepted by the IUPAC system.

TABLE 11.15 Examples of polyprotic acids

Carbon Atoms Present	Structure	IUPAC Name	Common Name
2	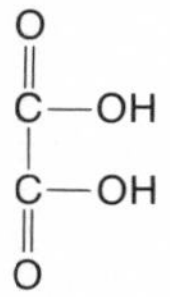	Ethanedioic acid	Oxalic acid
3	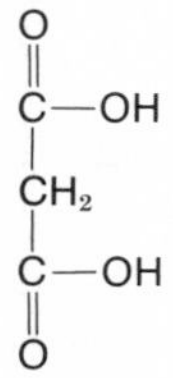	Propanedioic acid	Malonic acid
4	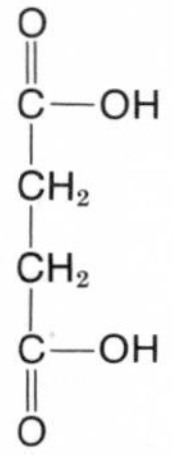	Butanedioic acid	Succinic acid
	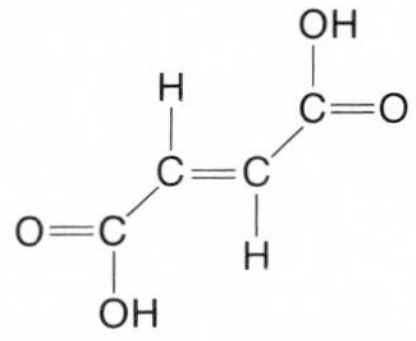	*trans*-Butenedioic acid	Fumaric acid
	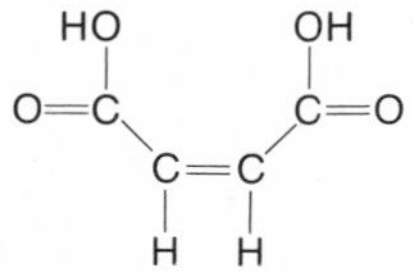	*cis*-Butenedioic acid	Maleic acid
5	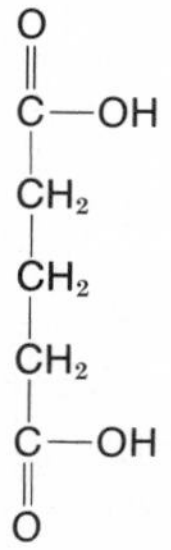	Pentanedioic acid	Glutaric acid

Table 11.15 (continued)

Carbon Atoms Present	Structure	IUPAC Name	Common Name
6	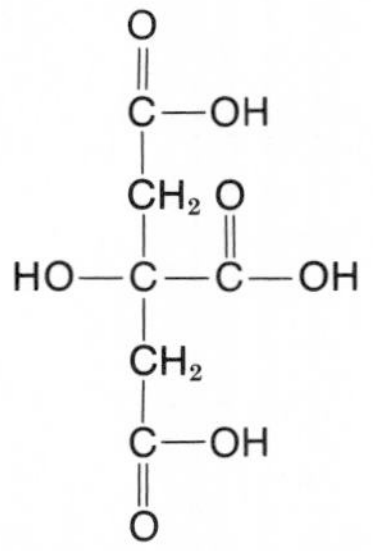 (structure of citric acid)	2-Hydroxy-1,2,3-propanetricar-boxylic acid	Citric acid

ACID DERIVATIVES

Several classes of organic compounds can be considered as derivatives of organic acids.

Salts By replacing the hydrogen in the —OH of the carboxyl group by positive ions we get **carboxylic salts.** They are named like inorganic salts by citing the positive ion first and then the organic acid part. In the IUPAC system and in the common names the ending of the acid name is changed to -ate (Table 11.17). By replacing the whole —OH group of the carboxyl group we can get esters, amides, acid halides, and acid anhydrides.

TABLE 11.16 Common amines

Carbon Atoms Present	Structure	IUPAC Name	Common Name
1	CH_3NH_2	Methanamine	Methylamine
2	$CH_3CH_2NH_2$	Ethanamine	Ethylamine
3	$CH_3CH_2CH_2NH_2$	1-Propanamine	n-Propylamine
	CH_3CHCH_3 NH_2	2-Propanamine	Isopropylamine
	CH_3CH_2—NH CH_3	N-Methylethanamine	Methylethylamine
	CH_3CH_2—N—CH_3 CH_3	N,N-Dimethylethanamine	Dimethylethylamine

TABLE 11.17 Examples of acid derivatives

Structure	IUPAC Name	Common Name
Carboxylic Salts		
$CH_3\overset{O}{\overset{\|}{C}}-OK$	Potassium ethanoate	Potassium acetate
$CH_3\overset{O}{\overset{\|}{C}}H\overset{}{C}-ONH_4$, with CH_3 below	Ammonium 2-methyl-propanoate	Ammonium isobutyrate
Esters		
$CH_3-\overset{O}{\overset{\|}{C}}-O-CH_2CH_3$	Ethyl ethanoate	Ethyl acetate
$H-\overset{O}{\overset{\|}{C}}-O-CH$ with CH_3-CH_2 up and CH_3 down	sec-Butyl methanoate	sec-Butyl formate
Amides		
$CH_3-\overset{O}{\overset{\|}{C}}-NH_2$	Ethanamide	Acetamide
$CH_3CH_2CH_2-\overset{O}{\overset{\|}{C}}-NHCH_3$	N-Methylbutanamide	N-Methylbutyramide
Acyl Halides		
$CH_3CH_2-\overset{O}{\overset{\|}{C}}-Cl$	Propanoyl chloride	Propionyl chloride
$CH_3CHCH_2-\overset{O}{\overset{\|}{C}}-Cl$, with OH below	3-Hydroxybutanoyl chloride	β-Hydroxybutyryl chloride
Acid Anhydrides		
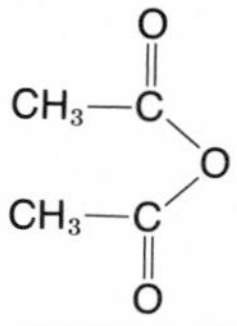 $CH_3-\overset{O}{C}$ and $CH_3-\overset{}{C}$ joined by O, with O below	Ethanoic anhydride	Acetic anhydride

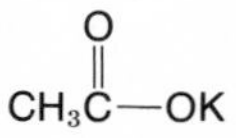

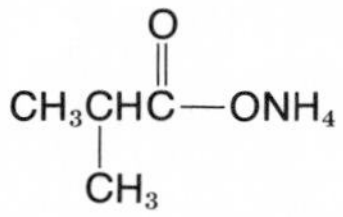

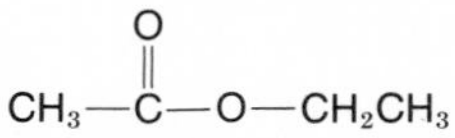

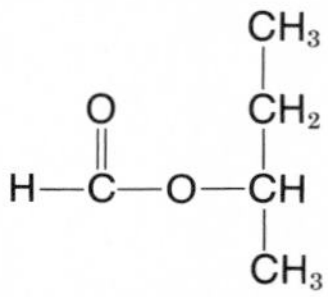

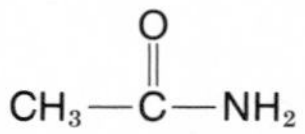

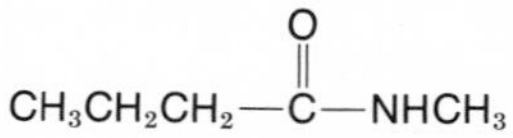

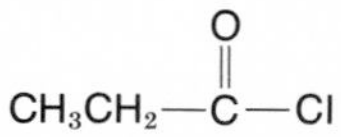

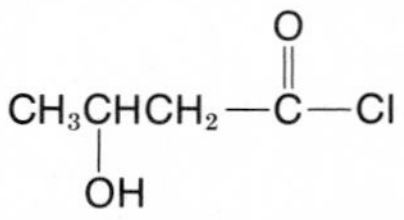

Esters These compounds are formed by replacing the hydroxyl group with an alkoxy group R—O—; the result is a functional group consisting of a carbonyl carbon connected by an oxygen bridge to another carbon. The naming is basically the same as for the salts except that the alkyl group is specified instead of a positive ion (Table 11.17).

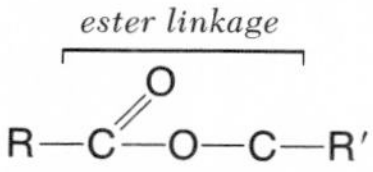

ester linkage

$$R-C-O-C-R'$$

General formula

Amides When the acid —OH group is replaced with an amino group, the resulting functional group consists of a carbonyl carbon having a nitrogen attachment. The general structure of simple amides is $RCONH_2$. The IUPAC system and common names both use the parent acid name and add the ending *-amide* (Table 11.17).

Acyl halides This group is formed by substituting a halogen for the —OH of the carboxyl group. The general formula for the acyl halides (also called acid halides) is RCOX, where X = F, Cl, Br, or I. The naming in both systems uses the parent acid name and replaces the ending *-ic acid* by *-yl*, with the halide name added as a separate word (Table 11.17).

Anhydrides Acid anhydrides are the product when one molecule of water is eliminated from two acids. The general anhydride structure is shown in Table 11.17. The name in both systems comes from that of the specific parent acid, to which the word **anhydride** is added.

SULFUR ANALOGS

In looking back at some of the functional groups we can pick out structures that have some similarities. An example is

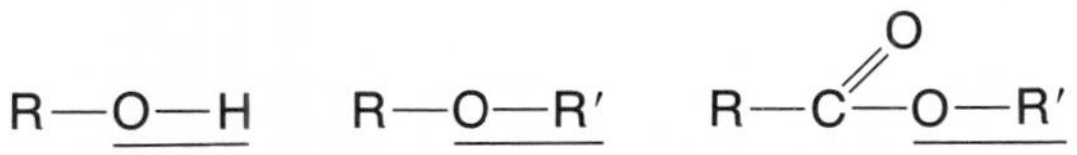

$$R-O-H \qquad R-O-R' \qquad R-C-O-R'$$

Alcohol   Ether   Ester

These three groups of organic compounds all contain an oxygen bridge. If we replace the oxygen atom by a sulfur atom, we obtain the sulfur analogs:

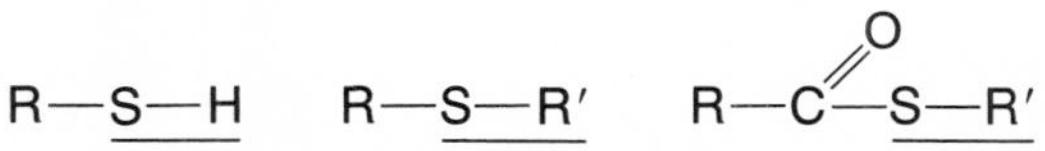

$$R-S-H \qquad R-S-R' \qquad R-C-S-R'$$

Thiol   Thioether   Thioester

The S—H group as found in the thiols is also called a **mercapto** group. In common usage these compounds are called **mercaptans** (Table 11.18).

TABLE 11.18 Examples of mercaptans

Structure	IUPAC Name	Common Name
CH_3CH_2SH	Ethanethiol	Ethylmercaptan
CH_3CH_2—S—CH_3	Methylthioethane	Methylethylthioether
CH_3—$\overset{\displaystyle O}{\overset{\|}{C}}$—S—$CH_2CH_2CH_3$	Propylthioethanoate	n-Propylthioacetate

AROMATIC COMPOUNDS

The aliphatics include alkanes, alkenes, alkynes, and cycloalkanes.

The second major hydrocarbon class (the aliphatics were the first) consists of the **aromatic** compounds. The parent compound for most aromatic compounds is C_6H_6, benzene. The designation as aromatic came from early days of organic chemistry, when benzene and benzene-containing compounds were noted for their aromatic odors. Today, the idea of aromatic structures relates to the benzene ring, or rings, present. Most cyclic structures with alternating single and double bonds exhibit aromatic characteristics. Aromatic compounds may contain a benzene ring or a similar cyclic structure, more than one benzene ring, or benzene rings with one or more sides in common, called **fused rings.** As we see from the examples in the margin, we can have pure aromatic compounds as well as blends of aromatic and aliphatic structures. It is very important that we be aware of these differences since the naming and reactivity of a compound depend on its structure. Let us look first at some aromatic structures containing only carbon and hydrogen.

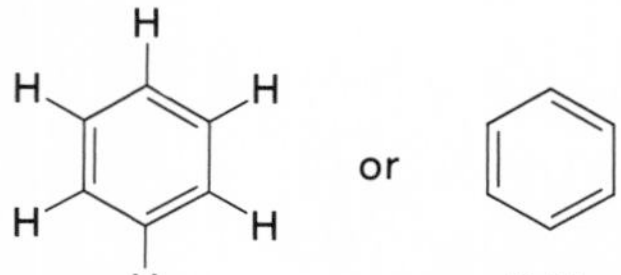

Molecular formulas for C_6H_6

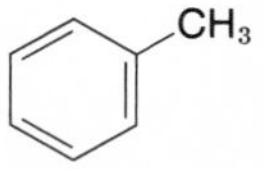

Methylbenzene (toluene)

Benzene C_6H_6

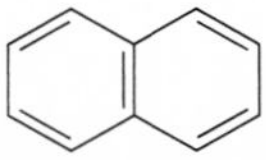

Naphthalene, two fused benzene rings

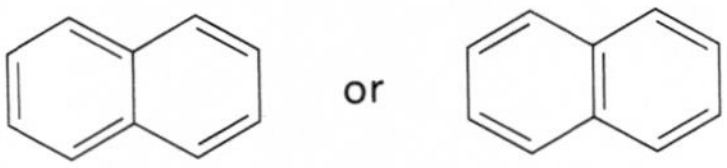

Naphthalene $C_{10}H_8$

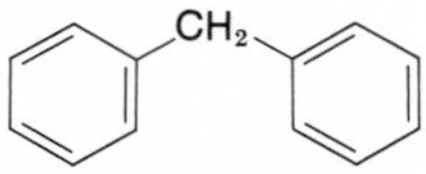

Diphenylmethane

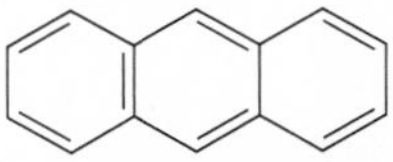

Anthracene, $C_{14}H_{10}$

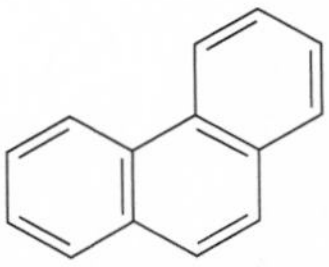

Phenanthrene, $C_{14}H_{10}$

AROMATIC HYDROCARBONS (ARENES)

When we remove one hydrogen from benzene or from nephtha-
lene, we have the phenyl, C_6H_5—, or the naphthyl group, $C_{10}H_7$—.
These correspond to alkyl groups in the aliphatic system but are
called **aryl** groups. They are abbreviated with the symbol Ar,
which comes from **arenes,** another name for aromatic hydrocarbons. To create new classes of aromatic compounds we
replace hydrogens by other atoms or groups of atoms on the aromatic ring. Replacing one hydrogen of a benzene ring gives a
monosubstituted benzene.

Only one monosubstituted compound is possible since the
group W can replace any one of the six hydrogens and still give
the same compound. When a second hydrogen is substituted,
we have several possibilities for the position of the second substituted group (call it Z) (see Fig. 11.1).

The ortho-meta-para designating system goes well with
common names, and it is much used for simple compounds. The
abbreviations o-, m-, and p- are used in compound names. For
more complicated systems the IUPAC numbering system is preferable.

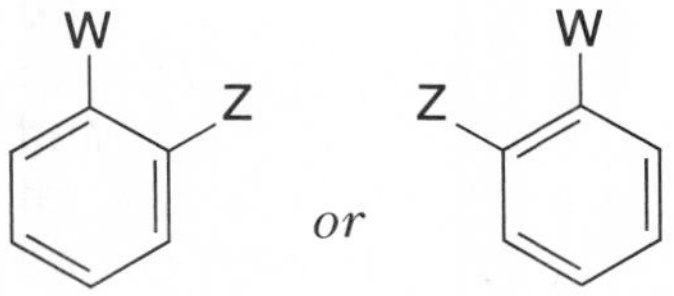

1,2-, *or ortho, substitution*
(substitution on adjacent carbons in the ring)

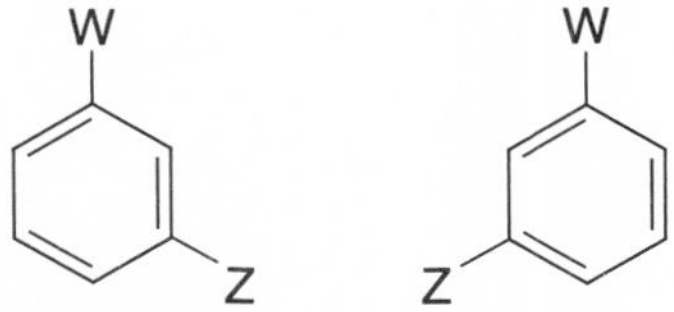

1,3-, *or meta, substitution*
*(substituents one carbon removed from each
other in the ring)*

1,4, *or para, substitution*
*(substituents two carbons removed from each
other on the ring)*

FIGURE 11.1 Ortho, meta, and para substitution around a benzene ring.

ALKYL BENZENES

Benzenes with substituted alkyl groups are called alkyl benzenes (Table 11.19).

AROMATIC HALIDES

Aromatic halogen compounds are named in the IUPAC system by giving the position and the name of the halogen and then following with the parent name of the aromatic compound. The common system uses the aryl group name followed by the name of the halide (Table 11.20).

TABLE 11.19 Examples of alkyl benzenes

Structure	IUPAC Name	Common Name
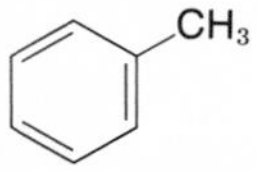 CH_3	Methylbenzene	Toluene
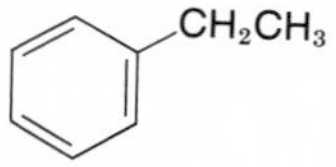 CH_2CH_3	Ethylbenzene	Phenylethane
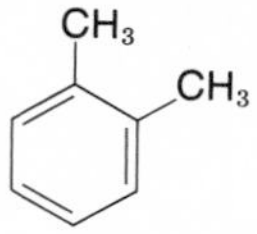 CH_3 CH_3	1,2-Dimethylbenzene	*o*-Xylene
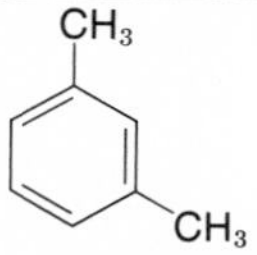 CH_3 CH_3	1,3-Dimethylbenzene	*m*-Xylene
 CH_3 CH_3	1,4-Dimethylbenzene	*p*-Xylene
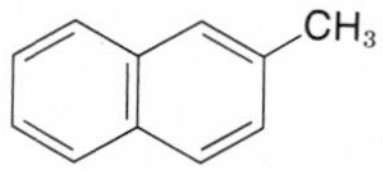 CH_3	2-Methylnaphthalene	β-Methylnaphthalene
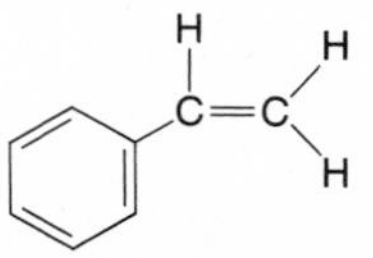	Phenylethene	Styrene

TABLE 11.20 Examples of aromatic halides

Structure	IUPAC Name	Common Name
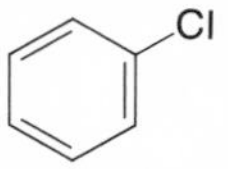	Chlorobenzene	Phenyl chloride
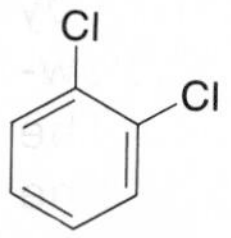	1,2-Dichlorobenzene	o-Dichlorobenzene
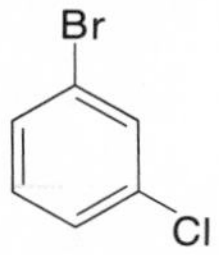	1-Bromo-3-chlorobenzene	m-Bromochlorobenzene
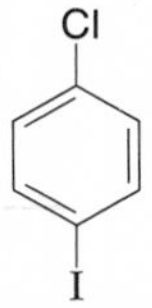	1-Chloro-4-iodobenzene	p-Chloroiodobenzene
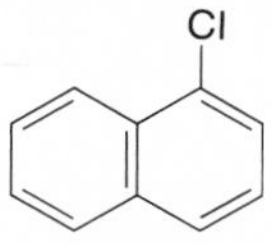	1-Chloronaphthalene	α-Chloronaphthalene
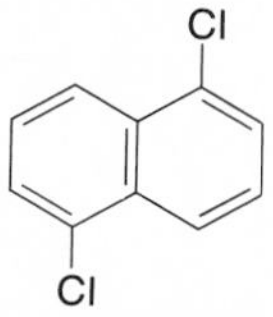	1,5-Dichloronaphthalene	
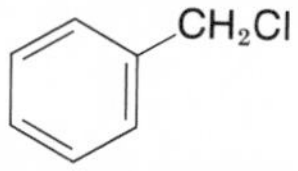	Chloromethylbenzene†	Benzyl chloride

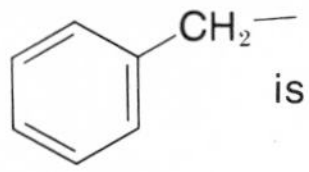

† (structure) is the benzyl group.

AROMATIC ETHERS AND ALCOHOLS

The simple aromatic and mixed aromatic and aliphatic ethers are named as indicated for the aliphatic system (Table 11.21). Structures obtained by replacement of one or more hydrogens on an aromatic ring by —OH groups(s) are generally known as **phenols** (Table 11.21).

TABLE 11.21 Examples of aromatic ethers and alcohols (phenols)

Structure	IUPAC Name	Common Name
Ethers		
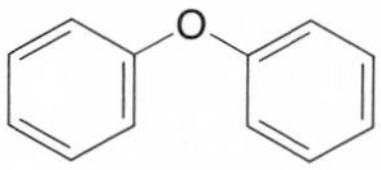	Phenoxybenzene	Diphenyl ether
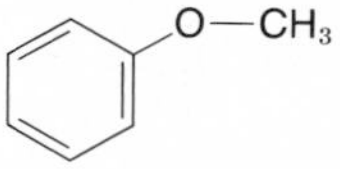	Methoxybenzene	Methyl phenyl ether
Phenols		
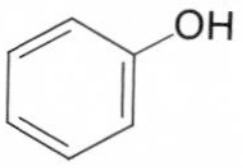	Phenol	Phenol
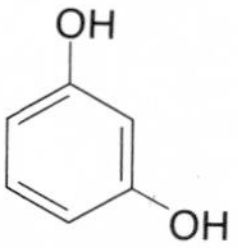	1,3-Benzenediol	Resorcinol
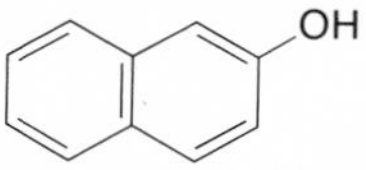	2-Naphthalenol	β-Naphthol
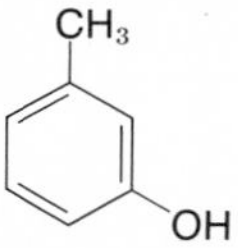	3-Hydroxymethylbenzene	*m*-Cresol
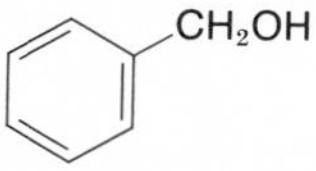	Hydroxymethylbenzene	Benzyl alcohol

AROMATIC ALDEHYDES AND KETONES

In the aromatic aldehydes the —CHO (aldehyde) group is attached to the benzene ring. In this case the IUPAC system signals the presence of an aldehyde group with the suffix *-carbaldehyde* and indicates what carbon the group is attached to. The common name is derived from the parent acid with the ending changed to *-aldehyde* (Table 11.22).

For ketones which have one or two aryl groups attached to the carbonyl group the common naming system uses the designation *phenyl* for the benzene ring and adds the word *ketone*. In the

TABLE 11.22 Examples of aromatic aldehydes and ketones

Structure	IUPAC Name	Common Name
Aldehydes		
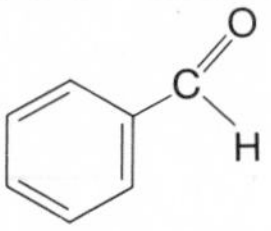	Benzenecarbaldehyde	Benzaldehyde
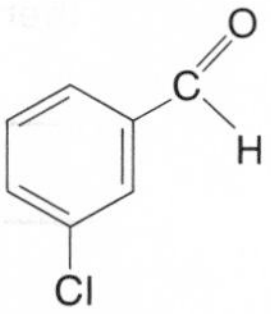	3-Chlorobenzenecarbaldehyde	*m*-Chlorobenzaldehyde
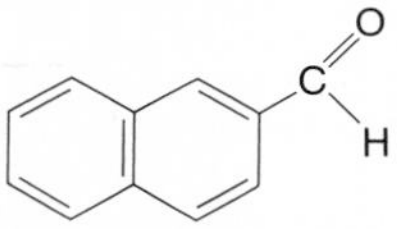	2-Naphthalenecarbaldehyde	*β*-Naphthaldehyde
Ketones		
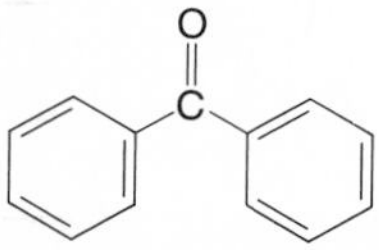	Benzophenone (from benzoic acid)	Diphenyl ketone
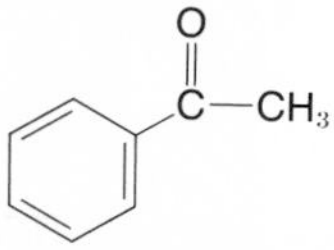	Acetophenone (from acetic acid)	Methyl phenyl ketone
Quinones		
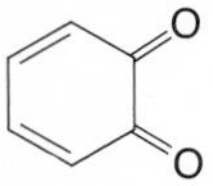	1,2-Benzoquinone	*o*-Benzoquinone
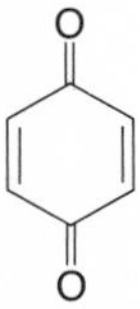	1,4-Benzoquinone	*p*-Benzoquinone
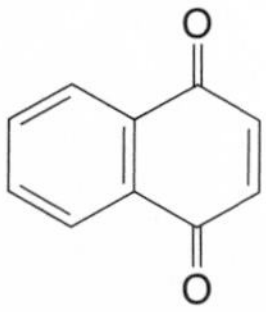	1,4-Naphthoquinone	*p*-Naphthoquinone

IUPAC system such ketones are named as the substitution products of the corresponding acids by removal of the ending *-ic acid* from the acid name and use of the ending *-phenone* when the attachment is a phenyl group (Table 11.22).

Cyclic diketones In aromatic ring structures the possibility exists for the formation of cyclic diketones. These compounds are known as **quinones** and exist as ortho and para diketones (Table 11.22).

AROMATIC AMINES

These amines mostly follow the nomenclature outlined for the aliphatic amines. Some common names for simple aromatic amines are also used in the IUPAC system (Table 11.23).

AROMATIC ACIDS AND ACID DERIVATIVES

The naming of aromatic acids cannot follow the rules given for aliphatic acids since there is no end of a chain that can be regarded as changed to a —COOH group. The IUPAC names are therefore formed by citing the name of the parent arene and adding the suffix *-carboxylic acid, dicarboxylic acid*, etc., as appropriate with numbers to show positions (Table 11.24).
  The acid derivatives discussed in the aliphatic naming system

TABLE 11.23 **Examples of aromatic amines**

Structure	*IUPAC Name*	*Common Name*
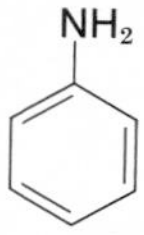 NH$_2$ (benzene ring)	Benzenamine, aniline	Aniline
 CH$_3$ (benzene ring) NH$_2$	4,Methylbenzenamine, *p*-toluidine	*p*-Toluidine
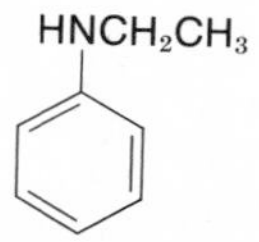 HNCH$_2$CH$_3$ (benzene ring)	*N*-ethylaniline	

TABLE 11.24   Examples of aromatic acids

Structure	IUPAC Name	Common Name
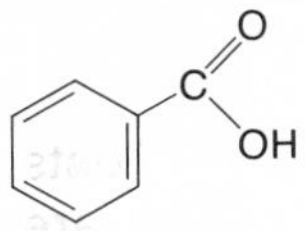	Benzenecarboxylic acid	Benzoic acid
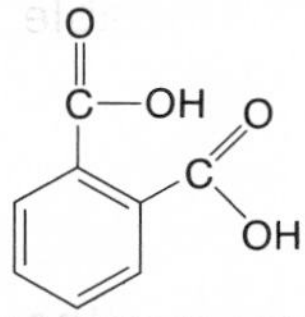	1,2-Benzenedicarboxylic acid	o-Phthalic acid
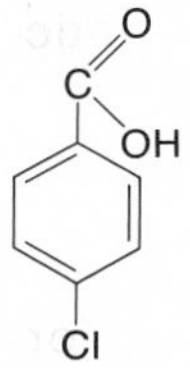	4-Chlorobenzenecarboxylic acid	p-Chlorobenzoic acid
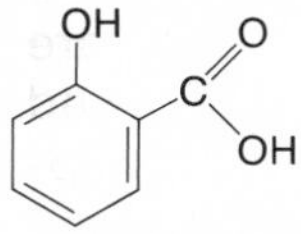	2-Hydroxybenzenecarboxylic acid	Salicylic acid
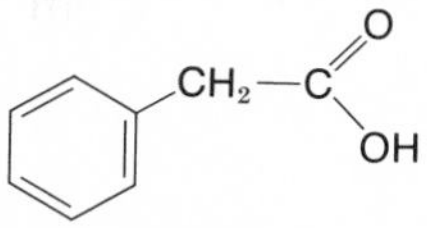	2-Phenylethanoic acid	Phenylacetic acid

are known as well for the aromatic compounds. Their nomenclature can be readily understood from the examples shown in Table 11.25.

HETEROCYCLIC COMPOUNDS

The last division of organic compounds, the **heterocyclic compounds,** are very important as they exist in many variations in natural products, being found in such diverse types as vitamins, alkaloids, natural and synthetic drugs, nucleic acids, and proteins. They have cyclic structures which contain carbon as well as one or more other atoms, the hetero atoms, in their ring structures. The hetero atoms may appear once, twice, three, or more

TABLE 11.25   Examples of derivatives of aromatic acids

Type	Structure	Name
Salt	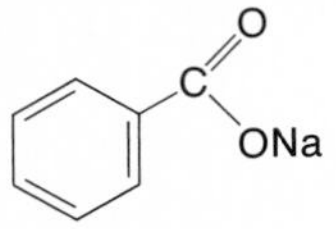	Sodium benzoate
Ester	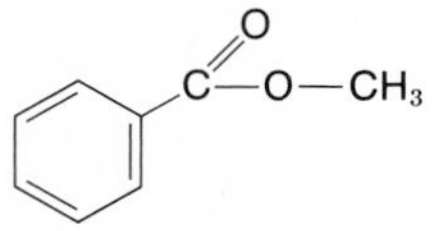	Methyl benzoate
	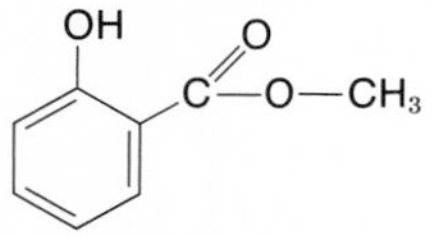	Methyl salicylate
Amide	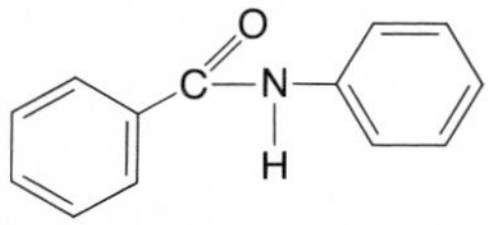	Benzanilide
	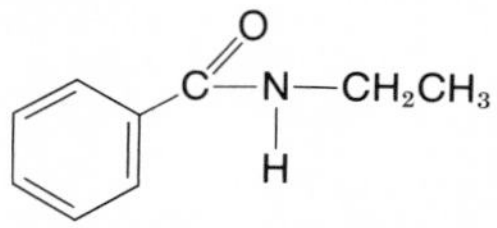	N-Ethylbenzamide
Acid halide	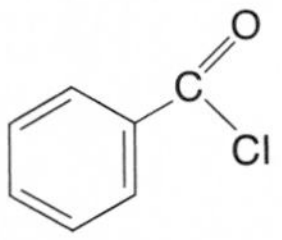	Benzoyl chloride
Acid anhydride	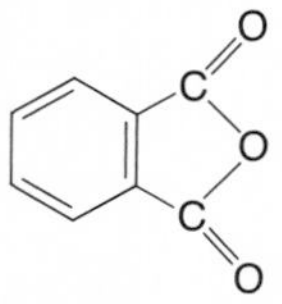	Phthalic anhydride

times within a ring, and a compound may have one, two, three, or more rings in its structure. For each such arrangement a different name must be assigned, so that the names are legion and complex. Many of the parent compounds are known by their original name, that is, the name derived from the natural product from which the compound was first isolated. Lately the IUPAC system has caught up with assigning names for these com-

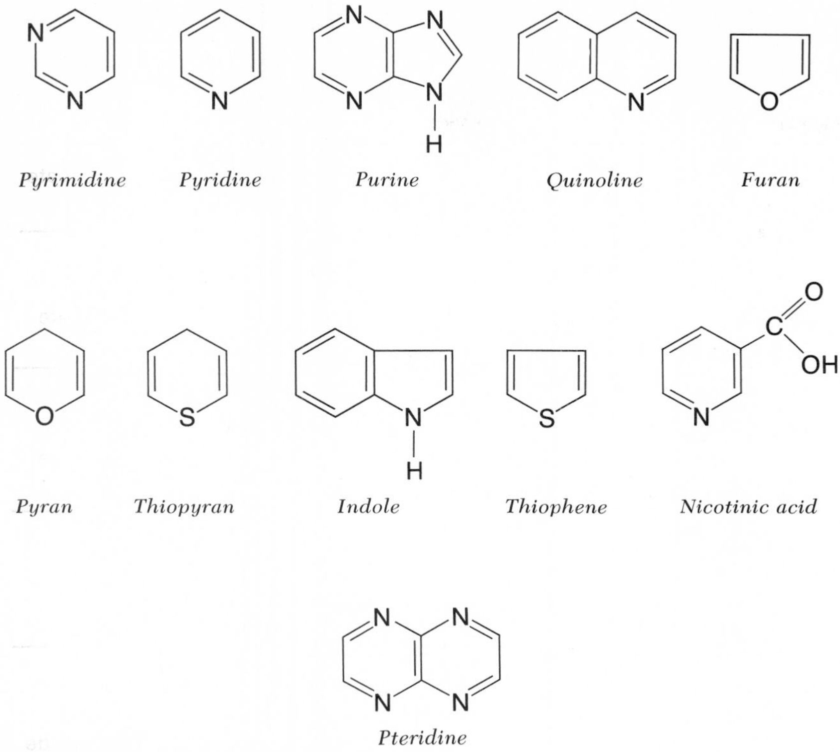

Pyrimidine   *Pyridine*   *Purine*   *Quinoline*   *Furan*

Pyran   *Thiopyran*   *Indole*   *Thiophene*   *Nicotinic acid*

Pteridine

FIGURE 11.2 Selected heterocyclic systems.

pounds. At this point we introduce only a few structures and names of heterocyclic compounds, primarily those which will appear in further discussions (Figure 11.2).

Most of these structures will appear again attached to some aliphatic or aromatic structure. The numbering system for the heterocyclic structures will not be discussed until they arise in later discussions.

POLYFUNCTIONAL COMPOUNDS

In many instances organic structures contain more than one functional group. These groups are generally included in naming the compound, since all functional groups must be identified. Examples of polyfunctional compounds and their names are shown in Table 11.26.

TABLE 11.26 Examples of polyfunctional compounds

Structure	Name	Functional groups
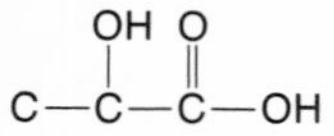 $\overset{\text{OH}}{\underset{}{C}}\overset{\text{O}}{\underset{}{}}$ C—C—C—OH	2-Hydroxypropanoic acid, α-hydroxypropionic acid	—OH, hydroxyl; —COOH, acid
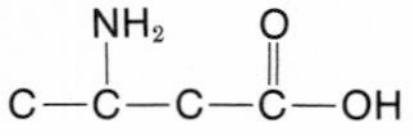 $\overset{\text{NH}_2}{}\overset{\text{O}}{}$ C—C—C—C—OH	3-Aminobutanoic acid, β-aminobutyric acid	—NH₂, amino; —COOH, acid
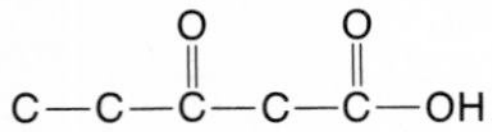 $\overset{\text{O}}{}\overset{\text{O}}{}$ C—C—C—C—C—OH	3-Ketopentanoic acid, β-ketovaleric acid	C=O, keto; —COOH, acid
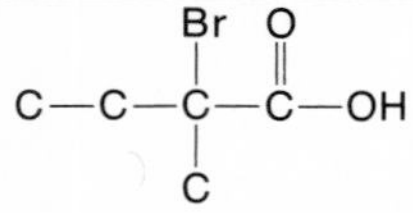 $\overset{\text{Br}}{}\overset{\text{O}}{}$ C—C—C—C—OH with C below	2-Bromo-2-methylbutanoic acid, α-bromo-α-methylbutyric acid	—Br, halogen; —COOH, acid
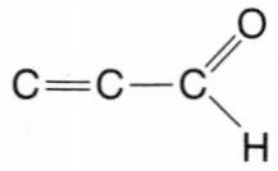 C=C—C with O double bond and H	2-Propenal, acrolein	C=C, double bond; C=O, aldehyde
H₂N—C—C—OH	2-Aminoethanol, ethanolamine	—NH₂, amino; —OH, hydroxyl
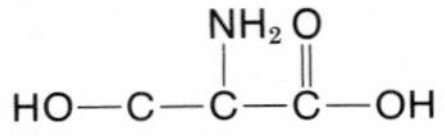 $\overset{\text{NH}_2}{}\overset{\text{O}}{}$ HO—C—C—C—OH	2-Amino-3-hydroxypropanoic acid, α-amino-β-hydroxypropionic acid, serine	—NH₂, amino; —OH, hydroxyl; —COOH, acid
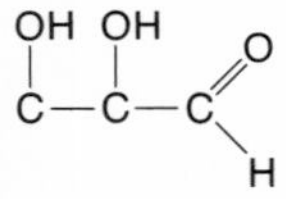 $\overset{\text{OH}}{}\overset{\text{OH}}{}$ C—C—C with O and H	2,3-Dihydroxypropanal, glyceraldehyde	—OH, hydroxyl; C=O, aldehyde
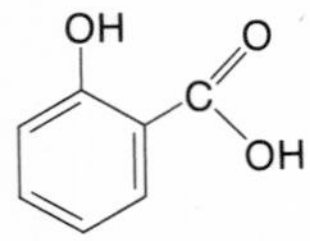 OH on benzene ring with C=O and OH	2-Hydroxybenzenecarboxylic acid, salicylic acid, O-hydroxybenzoic acid	—OH, hydroxyl; —COOH, acid
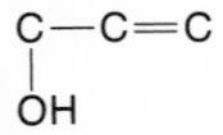 C—C=C with OH below	2-Propen-1-ol	C=C, double bond; —OH, hydroxyl
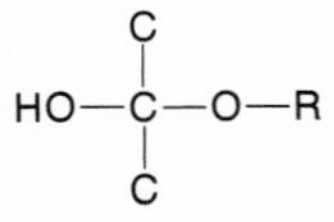 HO—C—O—R with C above and C below	A hemiacetal	—OH, hydroxyl, and C—O—R, ether, both on the same carbon
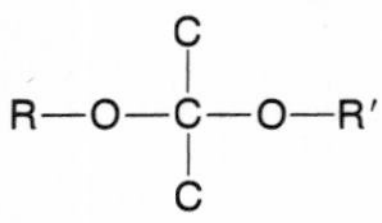 R—O—C—O—R' with C above and C below	An acetal	Diether both ether groups on the same carbon

TABLE 11.26 (continued)

Structure	*Name*	*Functional groups*
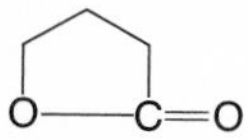 O—C=O	A lactone (γ-butyrolactone)	Cyclic ester
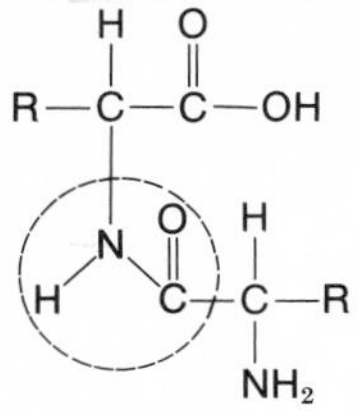		A peptide linkage; joining of two amino acids through an amide linkage

REVIEW QUESTIONS

1 Name these compounds:

a
$$CH_3\overset{\overset{\displaystyle CH_3}{|}}{\underset{\underset{\displaystyle CH_3}{|}}{C}}CH_3$$

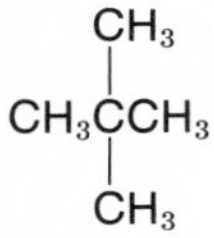

b
$$CH_3CH_2\overset{\overset{\displaystyle CH_3}{|}}{\underset{\underset{\displaystyle CH_3}{|}}{C}}CH_3$$

c
$$CH_3CH_2\overset{\overset{\displaystyle H}{|}}{\underset{\underset{\displaystyle CH_3}{|}}{C}}CH_2CH_3$$

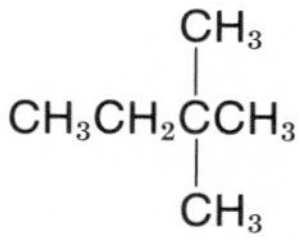

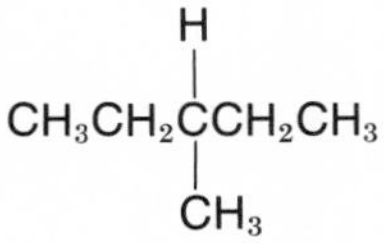

d
$$CH_3-CH_2-\underset{\underset{\displaystyle CH_3}{|}}{CH}-\underset{\underset{\displaystyle CH_3}{|}}{CH}-\underset{\underset{\displaystyle CH_3}{|}}{CH}-CH_2-CH_3$$

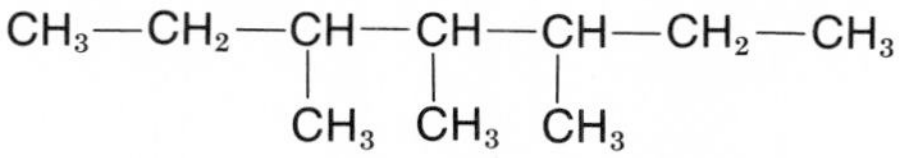

2 Write the structures of **a** 2,4-dimethylhexane, **b** 3-ethylpentane, **c** 3,3,5-trimethylnonane, **d** n-hexane, **e** isopentane.

3 Write all structures for **a** C_5H_{12}, **b** C_6H_{14}, **c** C_7H_{16}, **d** C_8H_{18}.

4 What is meant by a primary, a secondary, a tertiary carbon?

5 Write the following structures: **a** 1,3-dimethylcyclohexane, **b** 1,1-dimethylcyclobutane.

6 Name the compounds:

a —CH_3 b —$CH_2CH_2CH_3$

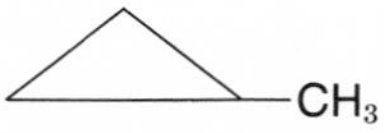

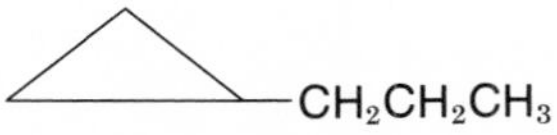

c 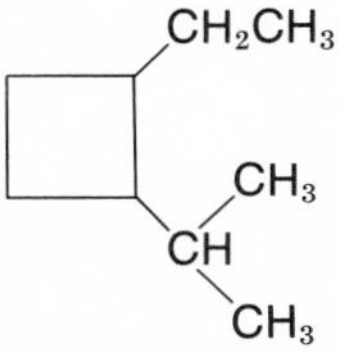   d

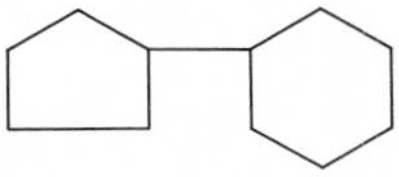

7 Name these compounds:

a CH_2=CHCH$_3$

b $CH_3\overset{\displaystyle H}{\underset{\displaystyle CH_3}{C}}CH$=CHCH$_2CH_3$

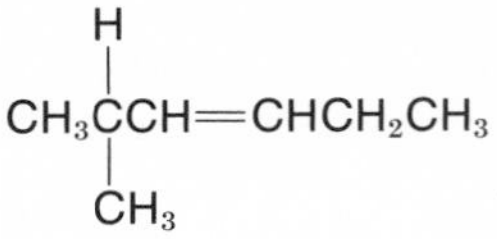

c CH$_3$CH$_2$C≡CCH$_3$

d $CH_3\overset{\displaystyle H}{\underset{\displaystyle \underset{\displaystyle CH_3}{CH_2}}{C}}C$≡CCH$_2CH_3$

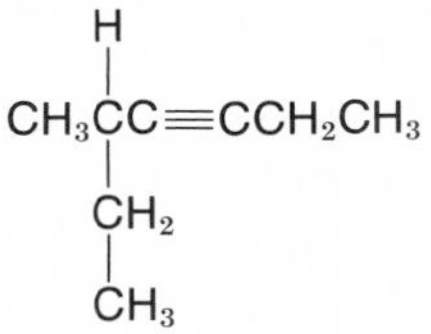

8 Write the following structures: **a** 2,3-dimethyl-1-butene, **b** 3,4-dimethyl-1-pentyne, **c** trans-2-pentene, **d** 2-methyl-1,3-butadiene **e** 2,5-dimethyl-1,3,5-heptatriene.

9 Name the compounds:

a CH$_2$ClBr

b $CH_3CH_2\overset{\displaystyle Cl}{\underset{\displaystyle Cl}{C}}CH_3$

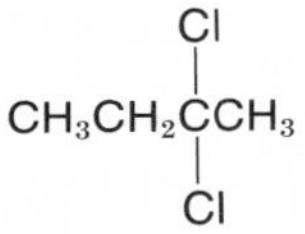

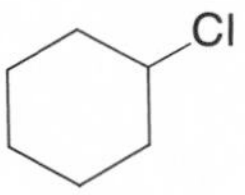

c (cyclohexane with Cl)

d CHCl$_2$CHCl$_2$

e $CH_3\overset{\displaystyle Cl}{\underset{\displaystyle CH_3}{C}}CH_2CH_3$

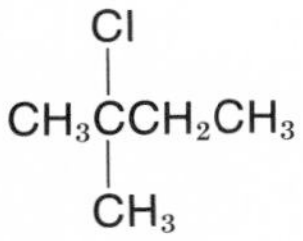

10 Write the following structures: **a** 2,2-dimethyl-1-chlorobutane, **b** 2,4-dibromo-2-pentene, **c** chloroform, **d** 1,1,2-trichloropropane.

11 Name the compounds:

a CH$_3$CH$_2$—O—CH$_2$CH$_2$CH$_3$

b $CH_3CH_2\overset{\displaystyle H}{\underset{\displaystyle OH}{C}}CH_2CH_3$

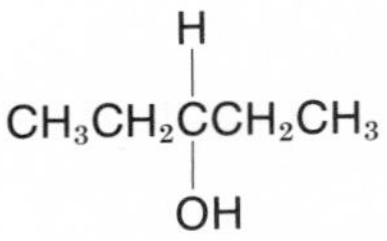

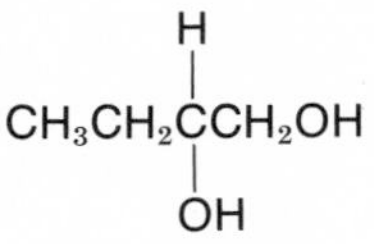

c $CH_3CH_2\overset{\displaystyle H}{\underset{\displaystyle OH}{C}}CH_2OH$

d $HO\overset{\displaystyle H}{\underset{\displaystyle HOH}{C}}CHCH_2CH_2CH_3$

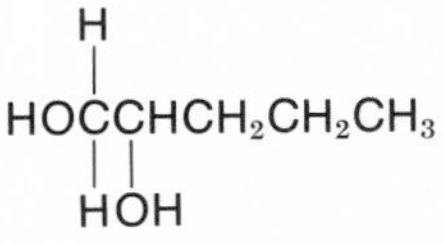

e CH$_2$=CHCH$_2$CH$_2$OH

12 Write the following structures: **a** diisopropyl ether, **b** 2,3-butanediol, **c** *tert*-butyl alcohol, **d** 3-octanol, **e** 1,2,3-cyclohexanetriol.

13 Name the compounds:

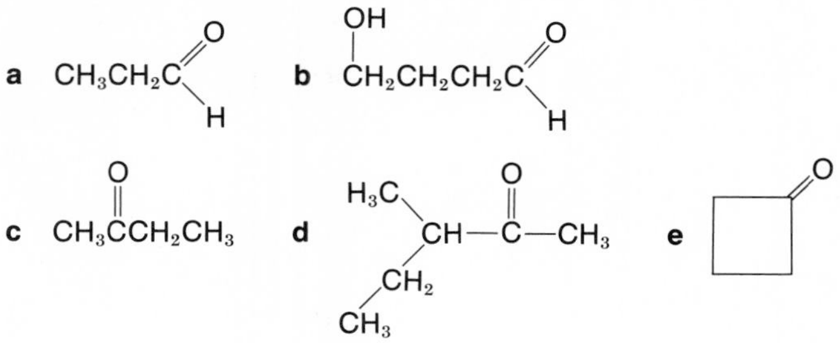

a $CH_3CH_2C\!\!\diagdown\!\!\overset{O}{\underset{H}{}}$ **b** $\overset{OH}{\underset{}{CH_2CH_2CH_2C}}\!\!\diagdown\!\!\overset{O}{\underset{H}{}}$

c $CH_3\overset{O}{\overset{\|}{C}}CH_2CH_3$ **d** $\overset{H_3C}{\underset{\underset{CH_3}{CH_2}}{CH}}\!\!-\!\!\overset{O}{\overset{\|}{C}}\!\!-\!\!CH_3$ **e** (structure of cyclobutanone)

14 Write the following structures. **a** 2-propenal, **b** ethyl isopropyl ketone, **c** heptanal, **d** 2-methyl-3-hexanone, **e** cyclohexanone.

15 Name the compounds:

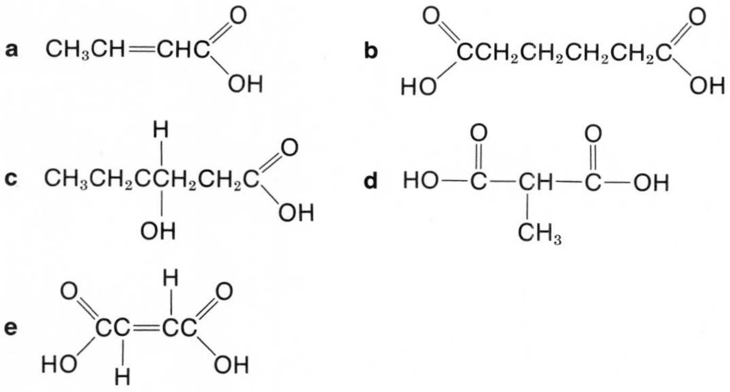

a $CH_3CH\!=\!CHC\!\!\diagdown\!\!\overset{O}{\underset{OH}{}}$ **b** $HO\!\!\diagup\!\!\overset{O}{\overset{\|}{C}}CH_2CH_2CH_2CH_2C\!\!\diagdown\!\!\overset{O}{\underset{OH}{}}$

c $CH_3CH_2\overset{H}{\underset{OH}{C}}CH_2CH_2C\!\!\diagdown\!\!\overset{O}{\underset{OH}{}}$ **d** $HO\!\!-\!\!\overset{O}{\overset{\|}{C}}\!\!-\!\!\overset{}{\underset{CH_3}{CH}}\!\!-\!\!\overset{O}{\overset{\|}{C}}\!\!-\!\!OH$

e $\overset{O}{\overset{\|}{C}}C\!=\!\overset{H}{\underset{}{C}}C\overset{O}{\underset{OH}{}}$ $\underset{HO \quad H}{}$

16 Write the following structures: **a** oxalic acid, **b** hexanedioic acid, **c** 3-pentenoic acid, **d** 2-hydroxybutanoic acid.

17 Name the compounds:

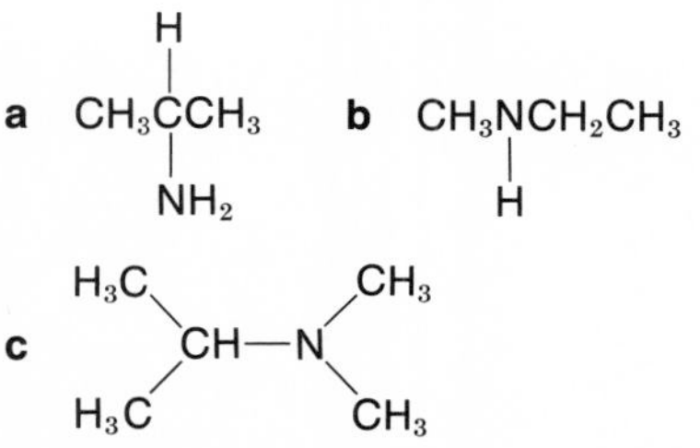

a $CH_3\overset{H}{\underset{NH_2}{C}}CH_3$ **b** $CH_3NCH_2CH_3$ $\underset{H}{}$

c $\overset{H_3C}{\underset{H_3C}{}}CH\!\!-\!\!N\overset{CH_3}{\underset{CH_3}{}}$

18 Write the following structures: **a** trimethylamine, **b** 2,3-diaminobutane, **c** *N,N*-diethylaminopropane.

19 Name the compounds:

a CH_3CH_2C (=O)Br

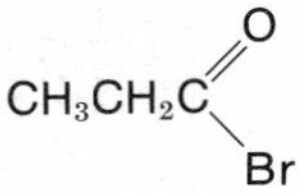

b $CH_3CH_2CH_2C$(=O)$—OCH_2CH_3$

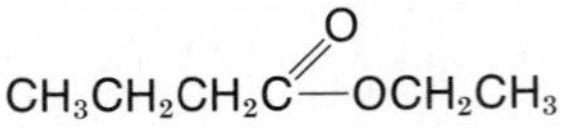

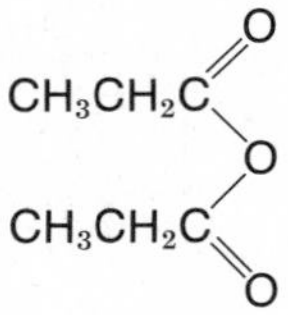

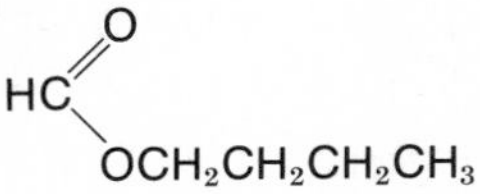

c CH_3CH_2C(=O)$—O—$(O=)CCH_2CH_3

d HC(=O)$—OCH_2CH_2CH_2CH_3$

e CH_3C(=O)$—N$(CH_3)(CH_3)

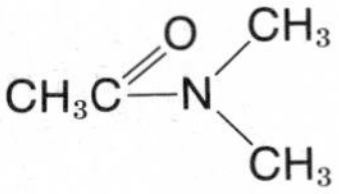

20 Write the following structures: **a** N-ethylbutanamide, **b** isopropyl-butanoate, **c** propylformate, **d** pentanoyl bromide, **e** hexanoyl chloride, **f** propanoic anhydride.

21 Name the compounds:

a (naphthalene)

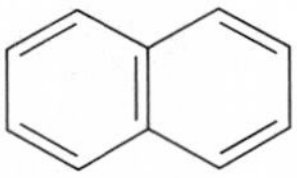

b (benzene ring with CH_2CH_3)

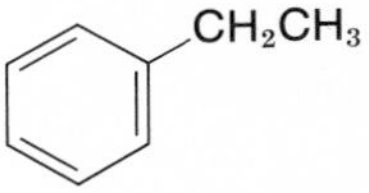

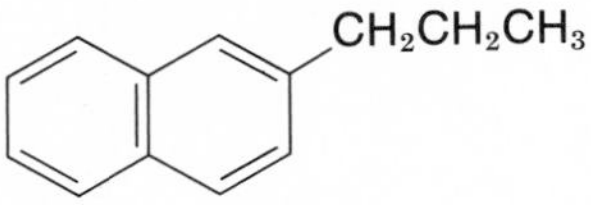

c (naphthalene with $CH_2CH_2CH_3$)

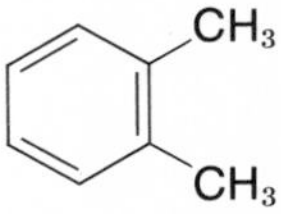

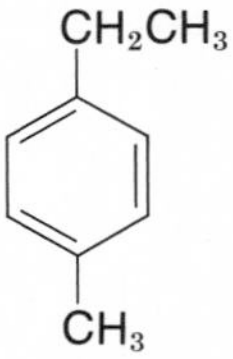

d (benzene ring with CH_3 and CH_3, ortho)

e (benzene ring with CH_2CH_3 and CH_3)

f (benzene ring with Br)

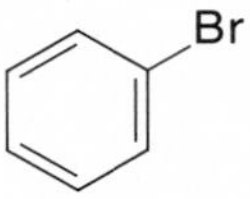

22 Write the following structures: **a** 1,2-dichlorobenzene, **b** 2-methylnaphthalene, **c** m-dibromobenzene, **d** α-chloronaphthalene, **e** isopropylbenzene, **f** 3-phenyl-2-butene, **g** N-methylaniline, **h** 3-ethylaminobenzene.

23 Name the compounds:

a

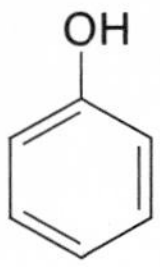

b

c

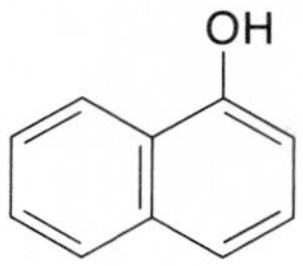

d

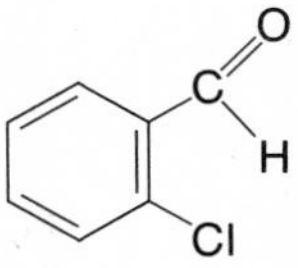

e

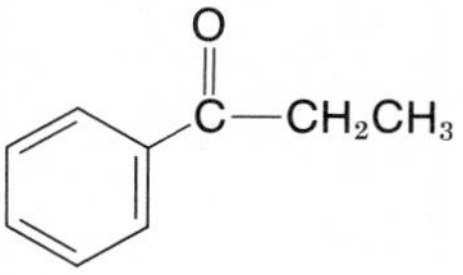

f

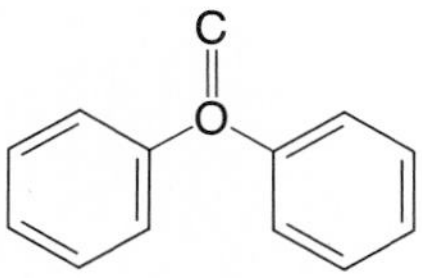

g

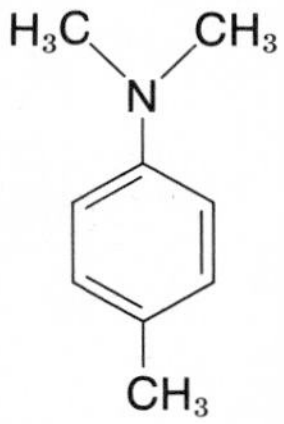

24 Name the compounds:

a

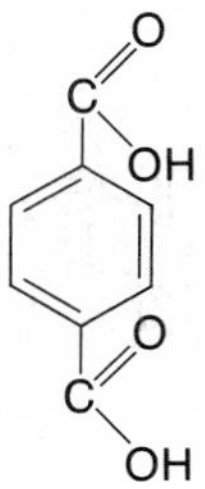

b

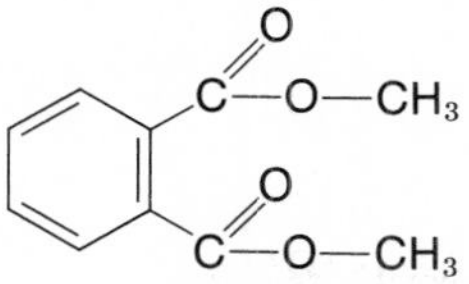

c

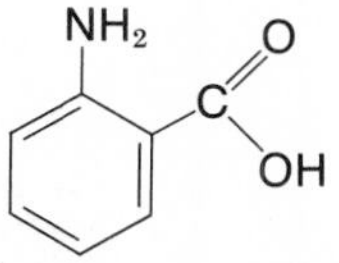

d

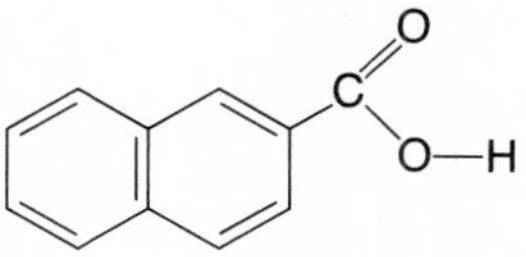

e

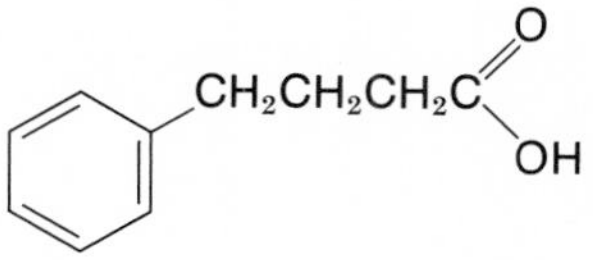

f

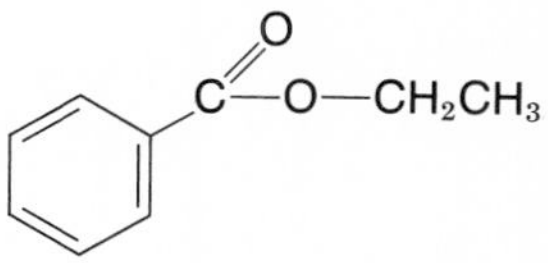

g

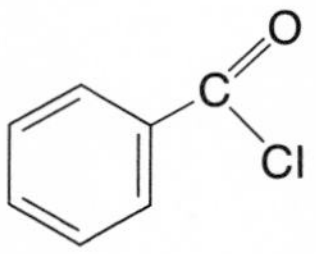

h

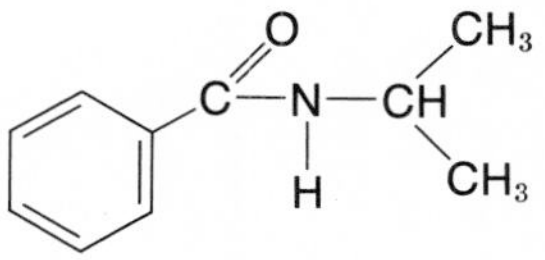

25 Write the following structures: **a** *m*-phthalic acid, **b** isopropylbenzoate, **c** 4-aminobenzoic acid, **d** phthalic anhydride, **e** *N*-methylbenzamide, **f** 2-naphthylpropionic acid, **g** salicylic acid, **h** naphthoyl chloride.

26 Name these compounds:

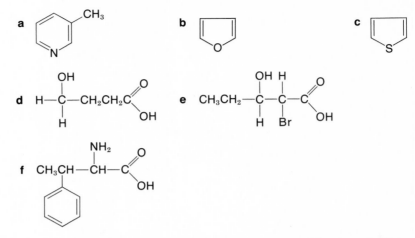

27 Write the following structures: **a** nicotinic acid, **b** quinoline, **c** thiopyran, **d** 3-ketohexanoic acid, **e** α-hydroxy-β-phenylpropionic acid, **f** β-(p-hydroxyphenyl)-α-aminopropionic acid.

ORGANIC REACTIONS
How the Game Is Played

In this chapter we shall look closely at the functional groups present in organic compounds, since it is at these groups that reactions occur. The preparation of new compounds involves breaking existing bonds and making new bonds, generally requiring some expenditure of energy. It is often difficult to make one compound directly from another; many steps may be needed. There are several reasons for this, for example, excessive energy requirements or competing reactions. Many organic reactions are made possible only by the use of a catalyst, which lowers the energy required to start the reaction, the **activation energy,** thereby greatly facilitating the reaction (Figure 12.1). Some catalysts are useful for only specific reactions, whereas others speed up many different reactions. Catalysts that promote biochemical processes are called **enzymes.** Enzymes may be very specific or more general in their action.

An organic compound generally undergoes those reactions which are characteristic of the functional group or groups contained in its structure. Nevertheless, some reactions are common to a large number of compounds, the most obvious being **combustion,** or burning. Nearly all organic compounds will burn by

Functional group =

$$-OH, \quad \underset{}{\text{C}}{=}O, \quad -NH_2, \quad -\overset{\text{O}}{\underset{}{\text{CH}}},$$

etc.

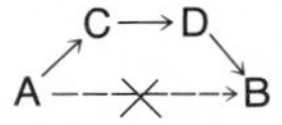

$$
\begin{array}{c}
\text{C} \longrightarrow \text{D} \\
\text{A} \text{---} \times \text{---} \!\!\rightarrow \text{B}
\end{array}
$$

When the direct A *to* B *path is blocked, it may be necessary to take a less direct path.*

Combustion = oxidation

283

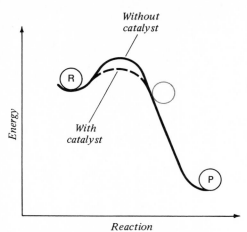

FIGURE 12.1 A catalyst helps a reaction to get going by reducing the height of the "hump" over which the reaction must pass.

reacting with oxygen, thereby releasing energy, carbon dioxide, and water. In some cases the composition of the organic compound is such that its burning yields undesirable or dangerous chemical products that pollute the air; the burning of some plastics produces a lethal gas.

Organic reactions have certain general characteristics:

1 Reactions take place at or near functional groups.
2 If a molecule contains more than one functional group, the characteristic reactions at each group may be affected.
3 The structure of an organic molecule influences its reactions and may serve to favor or impede the formation of a specific product.

In discussing organic reactions, we begin by considering the functional groups and their reactions.

REACTIONS OF ALKANES

General alkane formula $= C_nH_{2n+2}$:
CH_4
C_2H_6
C_3H_8
etc.

Alkanes are not very reactive. However, they are the starting material for many important compounds.

COMBUSTION

In excess oxygen, after being ignited, alkanes burn completely to carbon dioxide, water, and energy:

$$C_5H_{12} + 8O_2 \longrightarrow 5CO_2 + 6H_2O + energy$$

This rapid oxidation of alkanes, which we call combustion, makes them important as fuels. With lesser amounts of oxygen, an incomplete reaction results with the formation of carbon monoxide, CO, a toxic gas.

PARTIAL OXIDATION

At elevated temperatures, and with a limited supply of oxygen, alkanes may be partially oxidized to yield a mixture of oxygen-containing compounds, including alcohols, aldehydes, and carboxylic acids. This process is utilized extensively by the petroleum industry to manufacture increasingly large amounts of the petrochemicals.

$$C_nH_{2n+2} \xrightarrow[O_2]{200-500°C} \begin{cases} RCH_2OH \\ \textit{An alcohol} \\[1ex] RC\!\!{\overset{O}{\underset{H}{}}} \\ \textit{An aldehyde} \\[1ex] RC\!\!{\overset{O}{\underset{OH}{}}} \\ \textit{A carboxylic acid} \end{cases}$$

Petrochemicals are organic chemicals made from petroleum or natural gas.

CRACKING; PYROLYSIS

At elevated temperatures and in the *absence* of oxygen, alkanes undergo a thermal decomposition called **cracking.** This process is used in the petroleum industry to break up the long-chain alkanes found in crude oil into smaller molecules. The long-chain alkanes, which are high-boiling oils or even solids, are converted by this process into such useful commodities as gasoline, diesel fuels, and propane and butane gas:

$$C_{18}H_{38} \xrightarrow[N]{450-500°C} \begin{cases} C_8H_{16} + C_6H_{12} + C_4H_{10} \\ C_8H_{18} + 2C_4H_8 + C_2H_4 \\ C_6H_{14} + C_6H_{12} + C_4H_8 + C_2H_4 \\ C_6H_{14} + C_6H_{12} + C_4H_8 + 2C + 2H_2 \end{cases}$$

REPLACEMENT; SUBSTITUTION REACTION

A very important reaction of alkanes is the **substitution** of one or more hydrogens by a halogen. F_2, Cl_2, and Br_2 (but not I_2) react with alkanes in the vapor phase to produce halogen-substituted alkanes. These compounds, also known as alkyl halides, are important as intermediates for organic preparations and as solvents for the chemical industry.

R—X
X=F, Cl, *or* Br
Alkylhalide

Before we go any further, let us look for some explanations. We begin with the concept of a **free radical.** A covalent bond can break in two different ways. One is for each of the two fragments to take with it a single electron. This happens when the Cl_2 molecule is split into two chlorine free radicals. This is called a **homolytic cleavage** because each fragment carries one electron from the covalent bond.

$$:\!\overset{..}{Cl}\!:\overset{xx}{\underset{xx}{Cl}}\!x \longrightarrow \overset{..}{\underset{..}{Cl}}\!\cdot \quad Cl\cdot \\ x\overset{xx}{\underset{xx}{Cl}}x \quad Cl\cdot$$

Homolytic cleavage

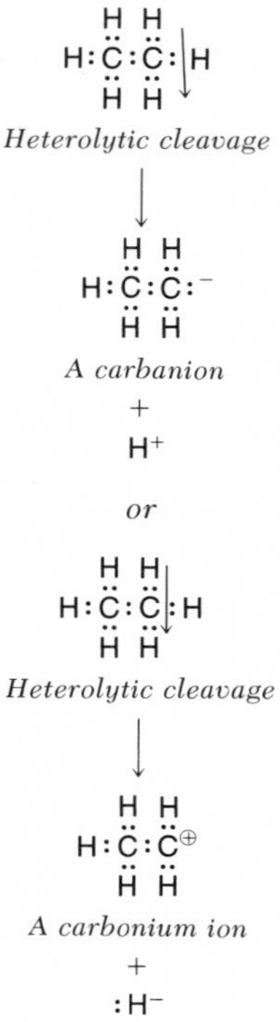

H H
H:C:C:|H
H H
Heterolytic cleavage

↓

H H
H:C:C:⁻
H H
A carbanion
+
H⁺

or

H H
H:C:C:|H
H H
Heterolytic cleavage

↓

H H
H:C:C⊕
H H
A carbonium ion
+
:H⁻

In the second way of breaking a covalent bond both electrons are carried away by one of the fragments; this is called **heterolytic cleavage** and is discussed more fully later.

Free radicals are particles that *lack* one electron to achieve an octet configuration. They are very reactive and are constantly on the search for the missing electron which will give them lower energy and more stability. The higher the energy of a free radical, the more different products are obtained in a reaction in which it is involved.

Cracking and partial oxidation of alkanes are high-energy free-radical reactions which result in many products. Since these chain reactions do not proceed very selectively, the halogen-substitution reactions of alkanes also give mixtures of products. The reaction proceeds in several distinct steps:

Step 1: $\quad Cl_2 \xrightarrow[\text{ultraviolet light}]{\Delta \text{ or}} 2Cl\cdot$

Chlorine atoms, free radicals

Step 2: $\quad Cl\cdot + CH_4 \longrightarrow HCl + CH_3\cdot$

Methyl free radical

Step 3: $\quad CH_3\cdot + Cl_2 \longrightarrow CH_3Cl + Cl\cdot$

An alkyl halide   Free radical

Step 1 is called the **chain-initiating reaction,** and steps 2 and 3 are called the **chain-propagating reactions,** since a Cl· from step 1 or step 3 goes to step 2 and the CH₃· from step 2 goes on to step 3, over and over again. This chain reaction can be stopped only when a free radical combines with another free radical. It is estimated that several thousands of these alternating-step reactions occur before such a reaction is terminated:

$$CH_4 + Cl_2 \longrightarrow CH_3Cl + HCl$$
$$CH_3Cl + Cl_2 \longrightarrow CH_2Cl_2 + HCl$$
$$CH_2Cl_2 + Cl_2 \longrightarrow CHCl_3 + HCl$$
$$CHCl_3 + Cl_2 \longrightarrow CCl_4 + HCl$$

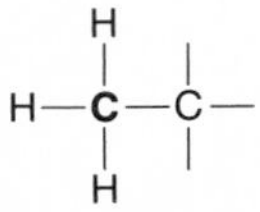

A slightly different picture appears in the reaction of longer-chain or even branched alkanes with halogens. A new factor is acting here.

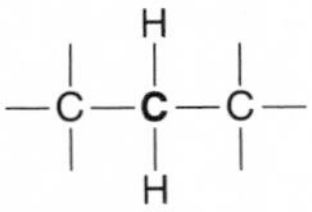

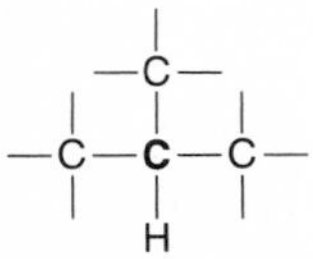

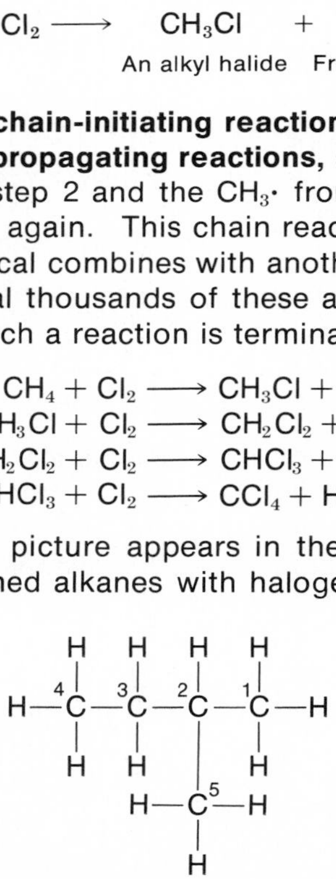

H
|
H—C—C—
|
H
A primary carbon

H
|
—C—C—C—
|
H
A secondary carbon

—C—
|
—C—C—C—
|
H
A tertiary carbon

2-Methylbutane

In examining this compound we can identify three different kinds of carbons. Carbons 1, 4, 5, primary carbons; carbon 3 is a secondary carbon; and carbon 2 is a tertiary carbon. This classification of carbon atoms is used to identify substitution products in which the hydrogens are replaced by other atoms or groups of atoms. For instance, we have primary, secondary, and tertiary alkyl halides and primary, secondary, and tertiary alcohols. Associated with this classification of the carbons is the strength of their bonds and their reactivity. Tertiary carbons are more reactive sites than secondary carbons, which, in turn, are more reactive than primary carbons, as a rule. This is because of the differing *strength* of the bonds that primary, secondary, and tertiary carbons form with their attachments, which also influence their reactions. Since bond breaking is the necessary step in the production of free radicals, we can say that breaking of primary carbon-hydrogen bonds is the most difficult, and the breaking of a tertiary carbon-hydrogen bond the easiest to achieve.

Ethanol, a primary alcohol

2-Propanol, a secondary alcohol

2-Methyl-2-chloropropane, a tertiary alkyl halide

Difficulty of bond breaking $\propto$ strength of carbon-hydrogen bonds

Primary > secondary > tertiary

This important fact should be remembered.

Since the Cl· free radical is quite energetic, it succeeds in breaking primary, secondary, and tertiary C—H bonds, resulting in the formation of primary, secondary, and tertiary alkyl chlorides. However, the Br· free radical is less energetic, and for the most part can break only tertiary C—H bonds (which are the easiest to break), so that tertiary alkyl bromide is the predominant product formed.

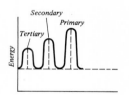

2-Methyl-2-bromobutane, a tertiary alkyl halide

Understanding reaction mechanisms and the effects of structure, reactants, and other influences on the outcome of a reaction enables the chemist to design successful and economical experiments and production methods.

REACTIONS OF ALKENES

The olefins, or alkenes, have the functional group C = C, the carbon-carbon double bond. This double bond is relatively easy to break, and several reagents can be added to it. Thus there are

General alkene formula =
C_nH_{2n}: C_2H_4
$\qquad C_3H_6$
$\qquad C_4H_8$
etc.

several types of **addition reactions** to the double bond, namely, hydrogenation, halogenation, hydrohalogenation, hydration, polymerization and oxidation (combustion).

HYDROGENATION

Hydrogenation is the addition of H_2, hydrogen gas, to the double bond to produce alkanes. It requires a catalyst (usually nickel, palladium, or platinum as a finely divided powder) and usually elevated temperatures and pressures.

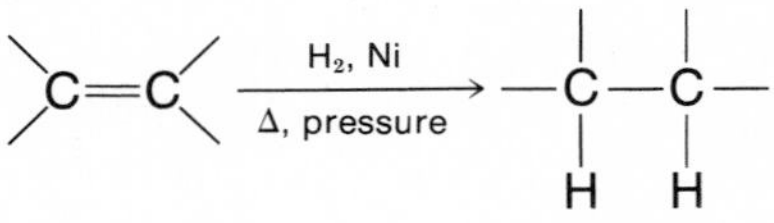

Alkene   Alkane

HALOGENATION

Halogenation is the addition of a halogen, Cl_2 or Br_2, to the double bond.

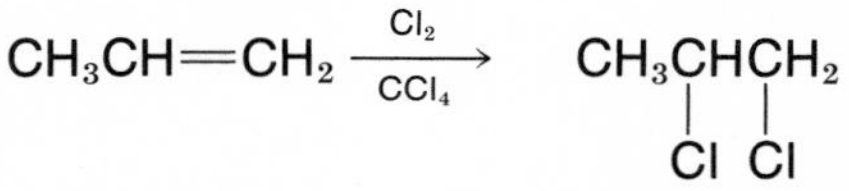

$$CH_3CH{=}CH_2 \xrightarrow[CCl_4]{Cl_2} \quad CH_3CHCH_2$$
$$\qquad\qquad\qquad\qquad\qquad\qquad | \quad |$$
$$\qquad\qquad\qquad\qquad\qquad\qquad Cl \quad Cl$$

Propene   1,2-Dichloropropane

This reaction proceeds without a catalyst at room temperature in an inert solvent. The reaction actually involves a two-step mechanism:

Step 1:

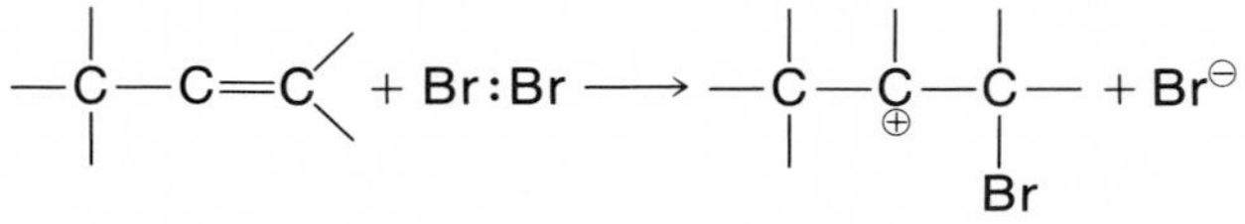

Step 2:

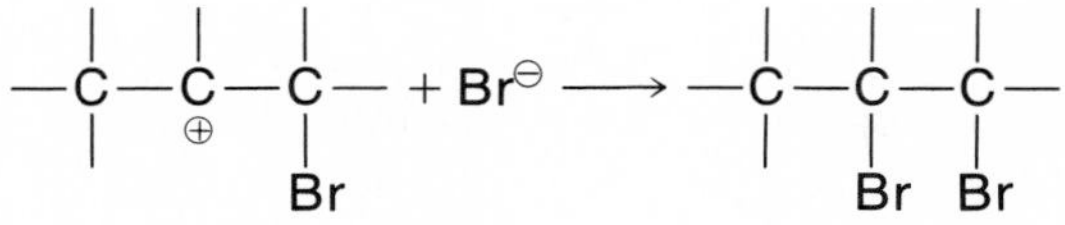

In step 1 the Br—Br bond becomes polarized, and the positive end of the polarized bond attaches to a carbon of the double bond:

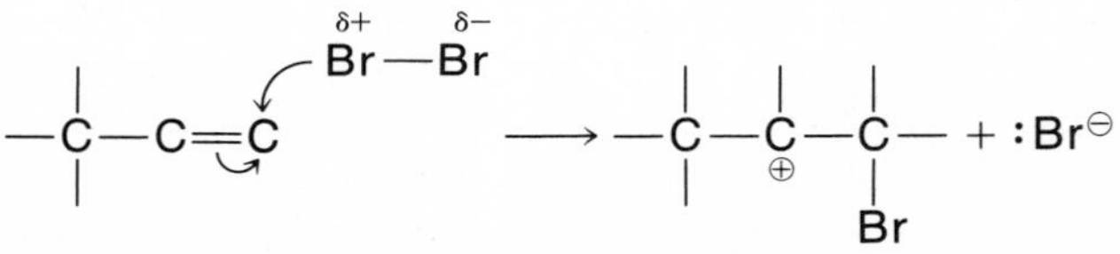

The result is the formation of a bromide ion, Br⁻, and a **carbonium ion,** —C⁺, which combine in step 2 to form the final product, the dihalide:

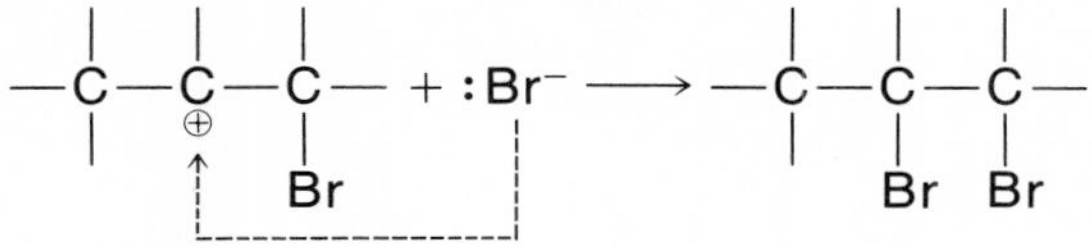

$$-\overset{|}{\underset{\oplus}{C}}-\overset{|}{\underset{Br}{C}}-\overset{|}{C}- \ + \ :Br^- \longrightarrow \ -\overset{|}{\underset{|}{C}}-\overset{|}{\underset{Br}{C}}-\overset{|}{\underset{Br}{C}}-$$

HYDROHALOGENATION

Hydrohalogenation is the addition of HX, where HX = HCl, HBr, or HI, to a double bond with the formation of an alkyl halide:

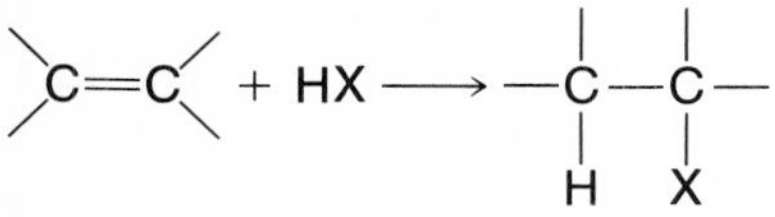

$$\underset{}{\overset{}{>}}C=C\overset{}{\underset{}{<}} + HX \longrightarrow -\overset{|}{\underset{H}{C}}-\overset{|}{\underset{X}{C}}-$$

The reaction proceeds via a two-step mechanism involving the polarization of the double bond with the addition of the H⁺ ion from the acid to the partially negative carbon of the double bond:

Step 1:

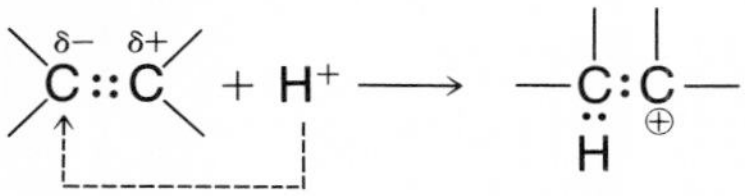

$$\overset{\delta-}{\underset{}{}}C::C\overset{\delta+}{\underset{}{}} + H^+ \longrightarrow -\overset{|}{\underset{H}{C}}:\overset{|}{\underset{\oplus}{C}}-$$

Carbonium ion

The resulting intermediate is a carbonium ion. In the second step the halide X⁻ becomes attached to the carbonium ion to form the alkyl halide:

Step 2:

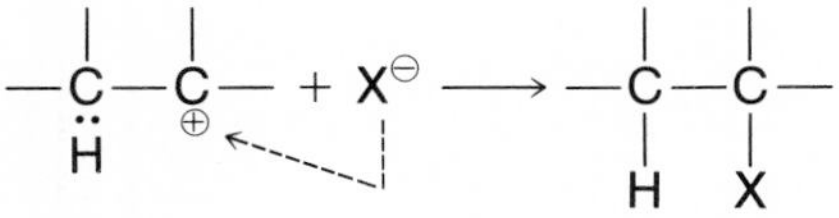

$$-\overset{|}{\underset{H}{C}}-\overset{|}{\underset{\oplus}{C}}- + X^\ominus \longrightarrow -\overset{|}{\underset{H}{C}}-\overset{|}{\underset{X}{C}}-$$

Alkyl halide

This addition reaction follows **Markovnikov's Rule,** which states that in the addition of an acid to a carbon-carbon double bond the hydrogen ion of the acid becomes attached predominantly to that carbon of the double bond which already bears the *greater* number of hydrogens. For example,

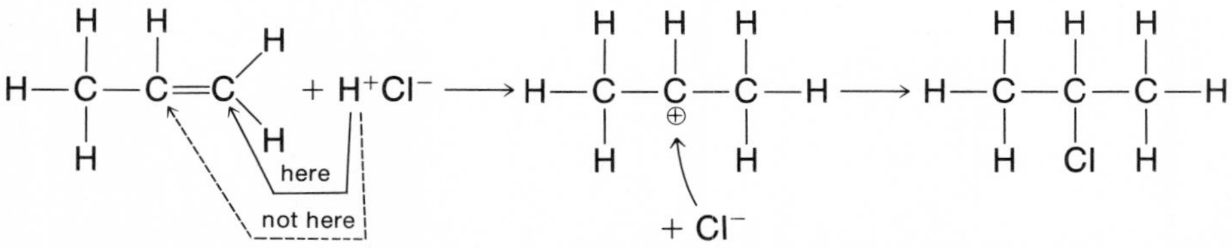

Propene + H⁺Cl⁻ → A secondary carbonium ion + Cl⁻ → 2-Chloropropane

or

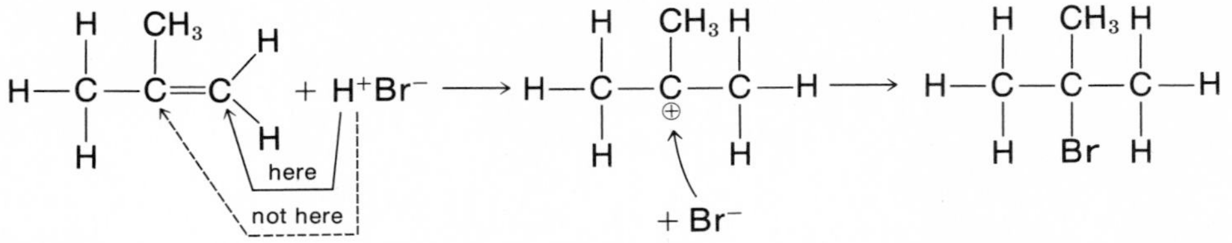

H   CH₃   H
| | /
H—C—C＝C   + H⁺Br⁻ ——→
| | \
H   H
              here
         not here

H   CH₃ H
| | |
H—C—C—C—H ——→
| ⊕ |
H   H
         ↑
      + Br⁻

H   CH₃ H
| | |
H—C—C—C—H
| | |
H   Br   H

2-Methylpropene   A tertiary   2-Bromo-2-methylpropane
                                    carbonium ion

From the above reaction sequence we see that the intermediate carbonium ion formed is the secondary or the tertiary carbonium ion rather than the primary carbonium ion. From investigations of such addition reactions we have learned that the orientation in this reaction sequence is controlled by how readily the intermediate carbonium ion can be formed. The formation of carbonium ions follows the sequence tertiary, secondary, primary, which also indicates their relative stability. The addition of sulfuric acid to an olefin follows the same mechanism and Markovnikov's rule:

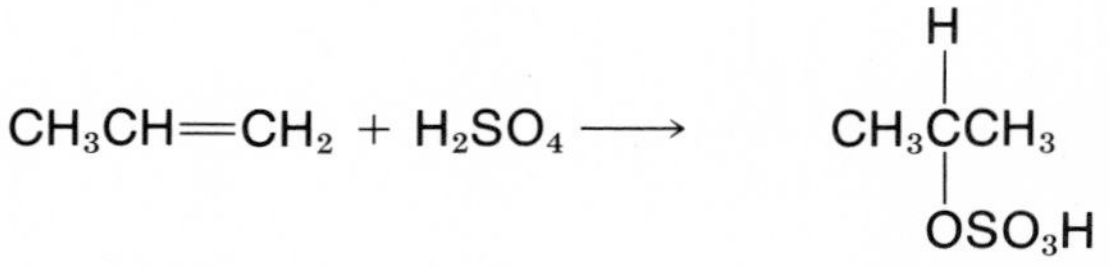

$$CH_3CH{=}CH_2 + H_2SO_4 \longrightarrow$$
$$\begin{array}{c} H \\ | \\ CH_3CCH_3 \\ | \\ OSO_3H \end{array}$$

Propene   Isopropyl hydrogen sulfate

HYDRATION

Hydration is the addition of water across the double bond, to give an alcohol, and is the common industrial process for making of alcohols. It requires the presence of an acid (usually sulfuric) as a catalyst, and the reaction follows Markovnikov's rule:

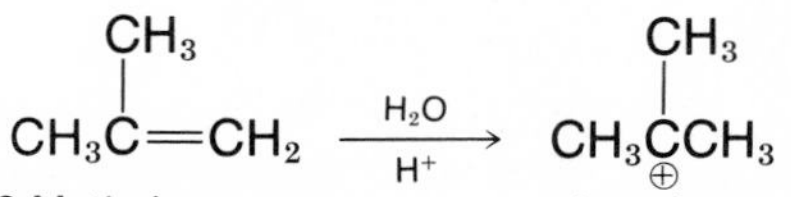

$$\begin{array}{c} CH_3 \\ | \\ CH_3C{=}CH_2 \end{array} \xrightarrow[H^+]{H_2O} \begin{array}{c} CH_3 \\ | \\ CH_3CCH_3 \\ \oplus \end{array}$$

2-Methylpropene   A tertiary
                                  carbonium ion

$$\begin{array}{c} CH_3 \\ | \\ CH_3CCH_3 \\ \oplus \end{array} + H_2O \longrightarrow \begin{array}{c} CH_3 \\ | \\ CH_3CCH_3 \\ | \\ OH \end{array} + H^+$$

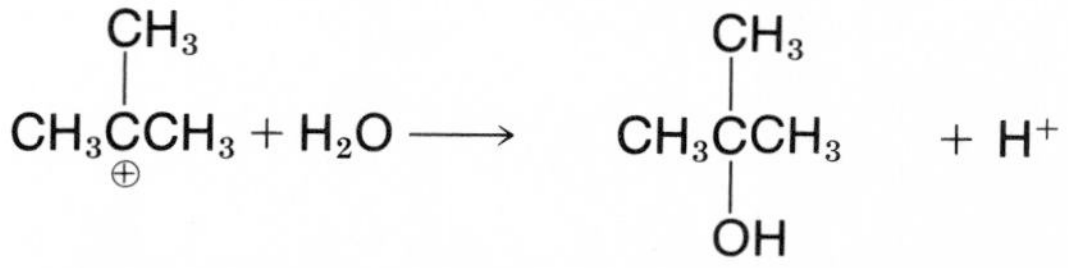

2-Methyl-2-propanol

POLYMERIZATION

Polymerization is the process by which a large number of small molecules are joined together to form a large molecule, the polymer; for example, polyethylene, polypropylene, and polyvinylchloride. You are familiar with these materials, which are known as **plastics** and widely used in films, pipes, furniture, wall coverings, table tops, electrical insulation, casings of radios and television sets, nylon, Orlon, and many other fabrics. There are two general reactions for making polymers. One is by **addition polymerization,** usually involving a free-radical mechanism:

Initiating reaction: $R—O:O—R \longrightarrow 2R—O\cdot$

A peroxide   Free radicals

Propagating reactions:

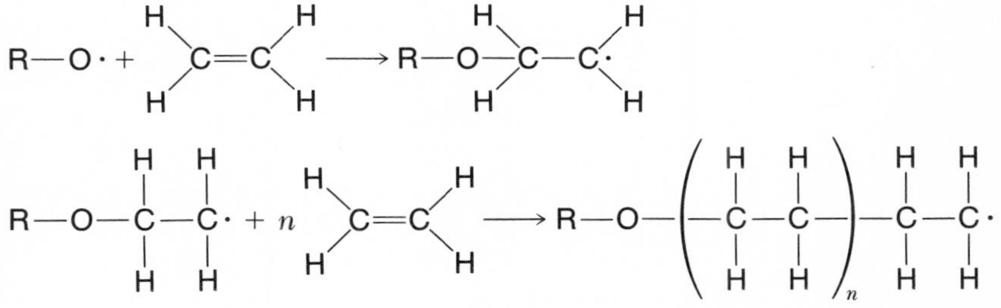

Polymer, in this case polyethylene

The second type of reaction is a **condensation reaction,** in which the polymer is formed by joining (condensing) a large number of molecules, and at each juncture eliminating some small molecule such as water, as in making nylon:

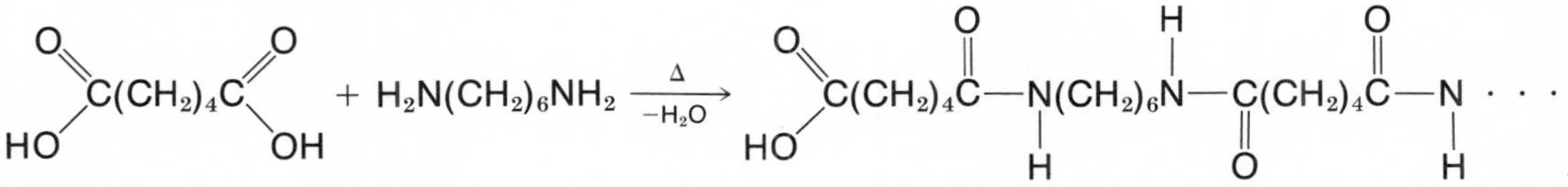

Adipic acid   Hexamethylene diamine   Nylon

OXIDATION; COMBUSTION

Partial oxidation of the carbon-carbon double bond gives glycols, which are compounds containing two hydroxyl groups:

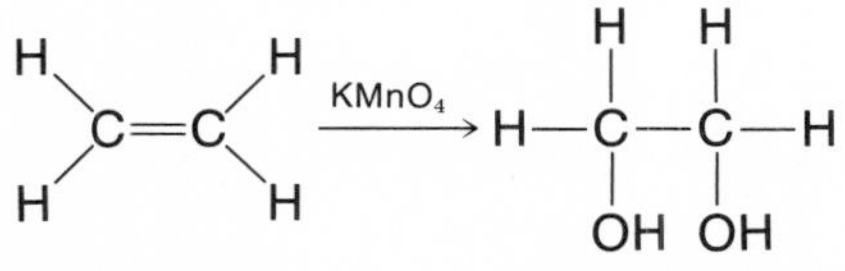

Ethene (ethylene)   1,2-Ethanediol (ethylene glycol)

Ethylene glycol and propylene glycol are the principal components in permanent antifreeze.

Vigorous oxidation causes the **cleavage** of the carbon-carbon double bond with the formation of aldehydes, ketones, and carboxylic acids. In the presence of sufficient amounts of O_2 the olefins will burn when ignited, producing CO_2, H_2O, and energy.

SUBSTITUTION REACTIONS

Substitution reactions of the alkenes are less common. They are illustrated by halogenation, which proceeds by the same free-radical mechanism as the alkanes. The reaction proceeds in this way only at *elevated* temperatures:

$$CH_3CH = CH_2 + \xrightarrow[600°C]{Cl_2} \qquad ClCH_2CH = CH_2$$

Propene   3-Chloro-1-propene (allyl chloride)

The intermediate formed in this reaction is the free radical ·$CH_2CH = CH_2$, called an **allyl free radical.** Hydrogens from carbons adjacent to the double bond are most readily substituted, and hydrogens from the carbons bearing the double bond are the least readily substituted. These hydrogens are known as **allylic** and **vinylic** hydrogens, respectively:

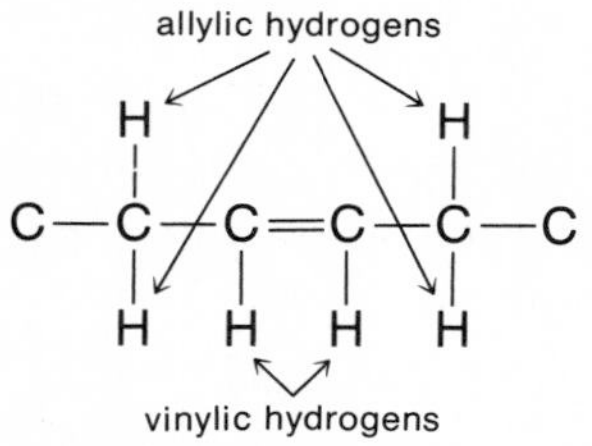

allylic hydrogens

vinylic hydrogens

Thus the relative ease with which hydrogens can be replaced can be extended:

allylic > tertiary > secondary > primary > vinylic

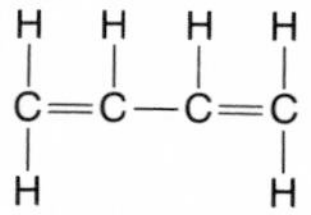

This sequence represents the ease of formation of free radicals as well as the relative stability of the free radicals formed.

The reactions of dienes and polyenes are similar to those of the olefins, as are also the reactions of alkynes. A summary of alkane and alkene reactions is shown in Figure 12.2.

Butadiene, a diene

REACTIONS OF AROMATIC HYDROCARBONS

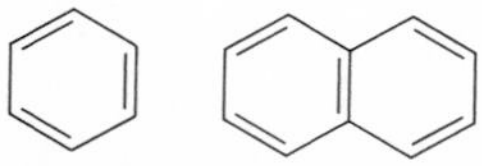

Aromatic hydrocarbons

The most common reaction of the aromatic hydrocarbons is substitution, the replacement of one or more hydrogens of the aromatic ring structure by other atoms or group of atoms. The main

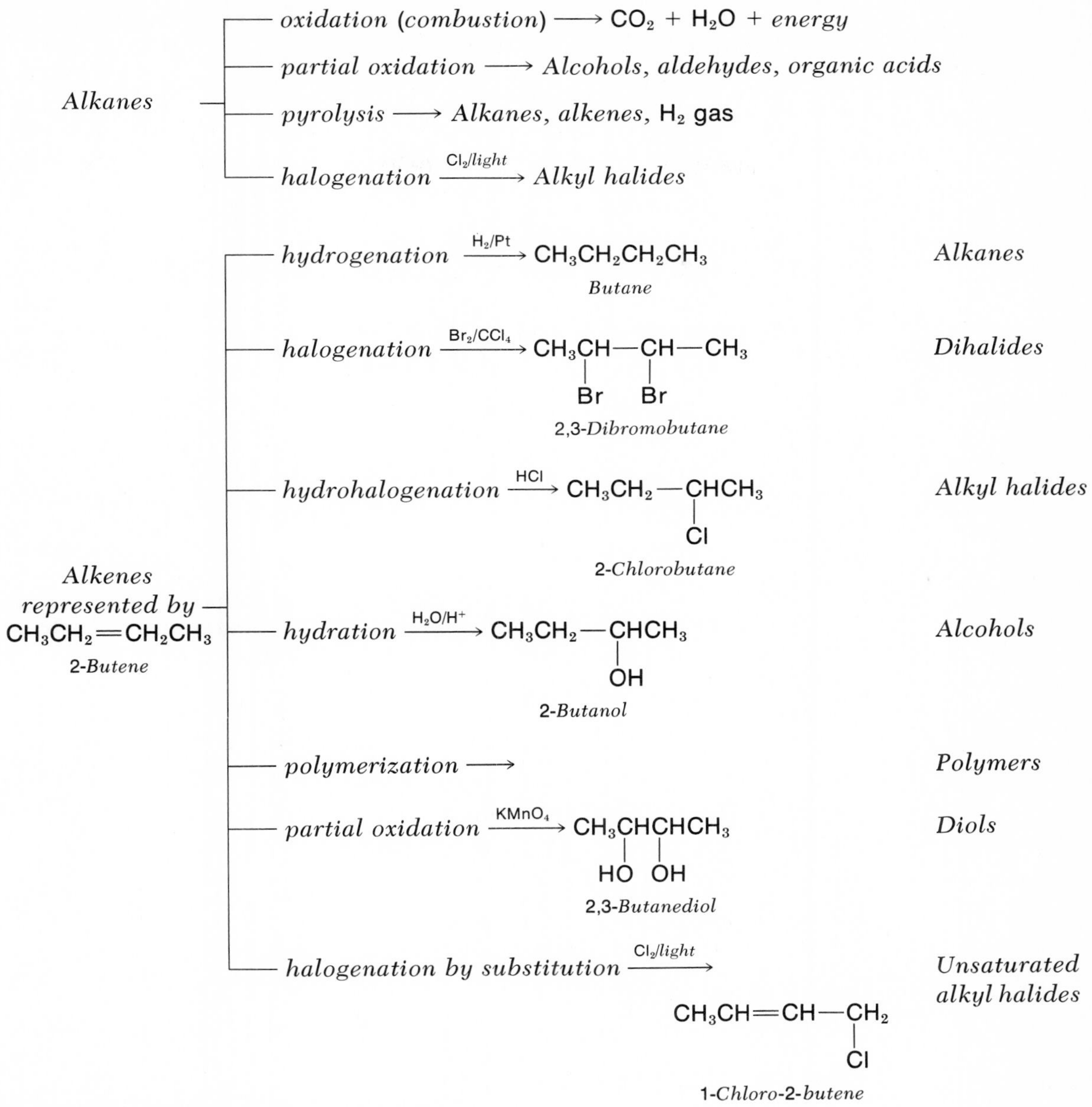

Alkanes
- oxidation (combustion) ⟶ CO_2 + H_2O + energy
- partial oxidation ⟶ Alcohols, aldehydes, organic acids
- pyrolysis ⟶ Alkanes, alkenes, H_2 gas
- halogenation $\xrightarrow{Cl_2/light}$ Alkyl halides

Alkenes represented by
$CH_3CH_2{=}CH_2CH_3$
2-Butene

- hydrogenation $\xrightarrow{H_2/Pt}$ $CH_3CH_2CH_2CH_3$ ..... Alkanes
  Butane

- halogenation $\xrightarrow{Br_2/CCl_4}$ $CH_3CH{-}CH{-}CH_3$ ..... Dihalides
  $\quad\quad\quad\quad\quad\quad$ Br $\quad$ Br
  2,3-Dibromobutane

- hydrohalogenation $\xrightarrow{HCl}$ $CH_3CH_2{-}CHCH_3$ ..... Alkyl halides
  $\quad\quad\quad\quad\quad\quad\quad$ Cl
  2-Chlorobutane

- hydration $\xrightarrow{H_2O/H^+}$ $CH_3CH_2{-}CHCH_3$ ..... Alcohols
  $\quad\quad\quad\quad\quad\quad\quad$ OH
  2-Butanol

- polymerization ⟶ ..... Polymers

- partial oxidation $\xrightarrow{KMnO_4}$ $CH_3CHCHCH_3$ ..... Diols
  $\quad\quad\quad\quad\quad\quad\quad$ HO $\;$ OH
  2,3-Butanediol

- halogenation by substitution $\xrightarrow{Cl_2/light}$ ..... Unsaturated alkyl halides
  $CH_3CH{=}CH{-}CH_2$
  $\quad\quad\quad\quad\quad\quad\;$ Cl
  1-Chloro-2-butene

FIGURE 12.2   Reactions of alkanes and alkenes.

substitution reactions are:

1   Halogenation: substituting —F, —Cl, —Br, or —I for H
2   Nitration: substituting —NO_2 for H
3   Sulfonation: substituting —SO_3H for H
4   Alkylation: substituting —R groups for H

The reaction can be classified as an **electrophilic aromatic substitution.**

NITRATION OF BENZENE

The mechanism of this reaction consists of three individual steps:

Step 1: $HNO_3 + 2H_2SO_4 \longrightarrow H_3O^+ + 2HSO_4^- + {}^\oplus NO_2$

In this step we have the formation of a **nitronium ion,** an electron-seeking ion, or **electrophile,** which in the next step attacks the benzene ring:

Step 2:

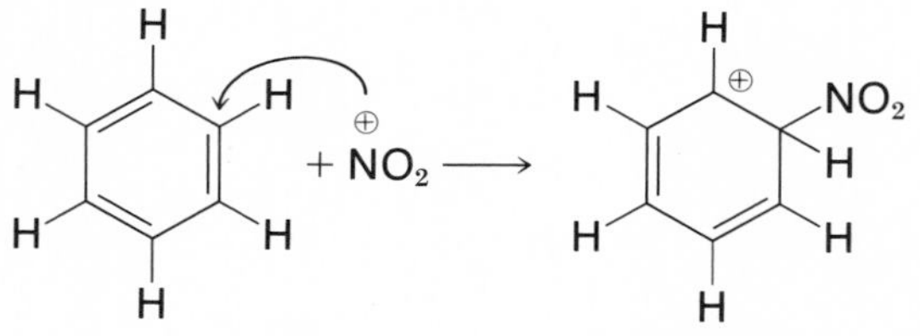

A special carbonium ion

The nitronium ion attaches itself to the benzene ring because the benzene ring is electron-rich and the nitronium ion is electron-poor. In other words, they attract each other.

Step 3:

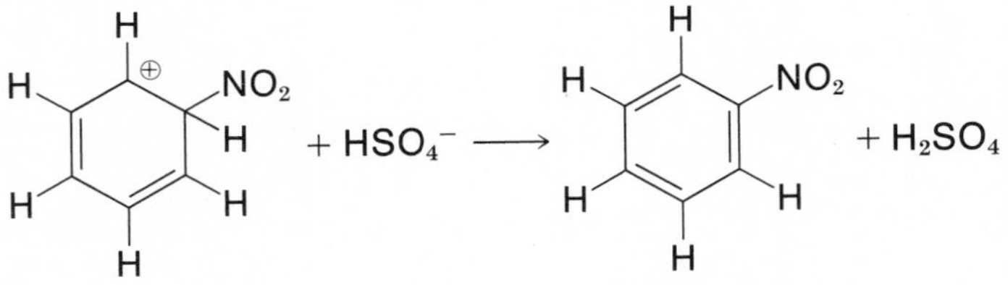

Nitrobenzene

In this part of the mechanism the hydrogensulfate ion removes a hydrogen ion, so that nitrobenzene is produced and the aromatic benzene ring is restored. The overall reaction can be written

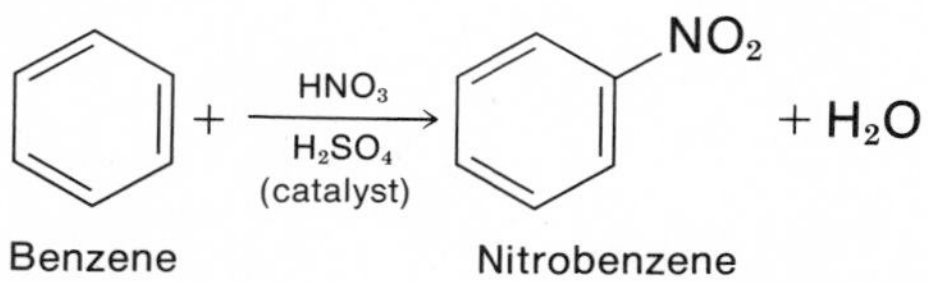

Benzene   Nitrobenzene

OTHER SUBSTITUTION REACTIONS

We can represent the other substitution reactions of benzene similarly, keeping in mind that the complete reaction follows the ionic nitration mechanism.

Halogenation:

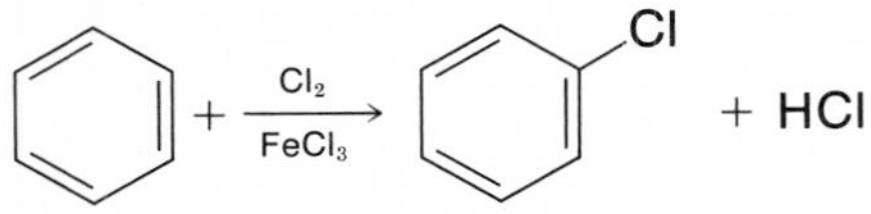

Chlorobenzene

Sulfonation:

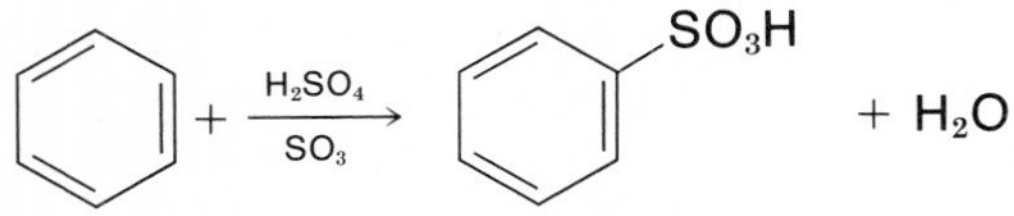

Benzenesulfonic acid

Alkylation (Friedel-Crafts alkylation)

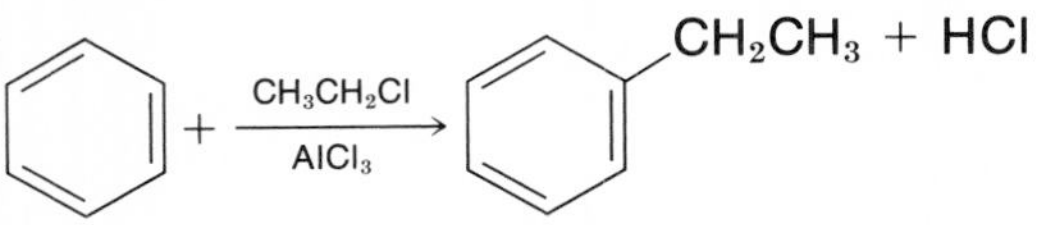

Ethylbenzene

If further substitution occurs, the group or groups originally present will influence the reaction in two ways. They may *slow down* or *speed up* the further substitution in comparison with the first substitution. This is referred to as **activation** (speed-up) and **deactivation** (slow-down) of the aromatic ring system. The second influence is that these groups *direct* the new incoming group to certain positions. We can identify two different kinds of groups: **ortho-para-directing groups** and **meta-directing groups** (Table 12.1). The ortho-para-directing groups are usually activating groups, and the meta-directing groups are deactivating groups. For instance, further nitration of nitrobenzene will yield m-dinitrobenzene (1,3-dinitrobenzene):

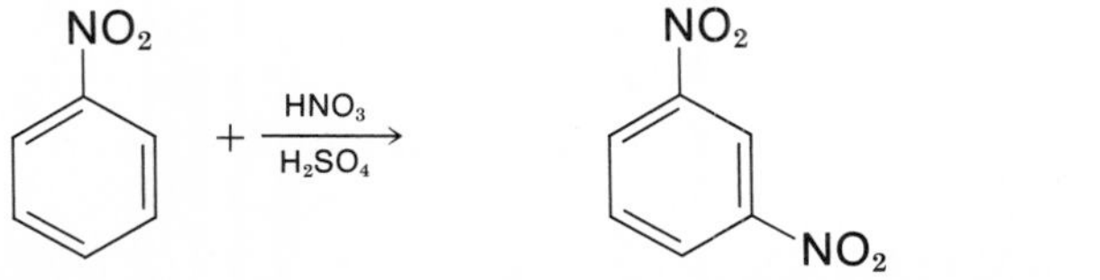

Nitrobenzene m-Dinitrobenzene (1,3-dinitrobenzene)

The second substitution took place at the carbon meta to the original nitro group in the benzene ring. The nitro group is therefore a meta-directing group. The nitration of nitrobenzene proceeds much more slowly than the nitration of benzene. Thus the nitro group is a deactivating group and retards the second reaction. Knowing the effects of the substituted groups helps the chemist plan and carry out organic syntheses.

The nitration of toluene (methylbenzene) yields a mixture of products, almost entirely o- and p-nitrotoluene:

TABLE 12.1 Ortho-para- and meta-directing groups

Ortho-Para-directing Groups (Activating)	Meta-directing Groups (Deactivating)
—NH₂	—N→O
	—C≡N
—OH	—S—OH
—OR	
—Cl	—C—OH
	—C—OR
—Br	—C—H
—R	—C—R

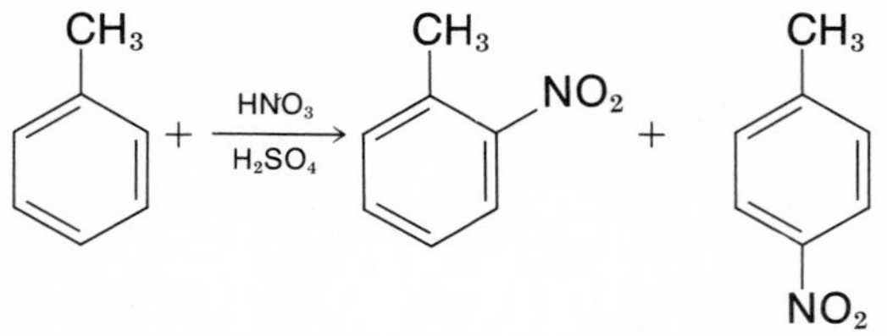

Toluene   o-Nitrotoluene   p-Nitrotoluene

As you can see, the methyl group has an ortho-para-directing influence on further substitution. The nitration of toluene proceeds faster than the nitration of benzene, indicating that the methyl group is an activating group.

REACTIONS OF ALKYL BENZENES

Alkyl benzenes can undergo reactions involving the aliphatic alkyl group; these include halogenation and oxidation.

HALOGENATION

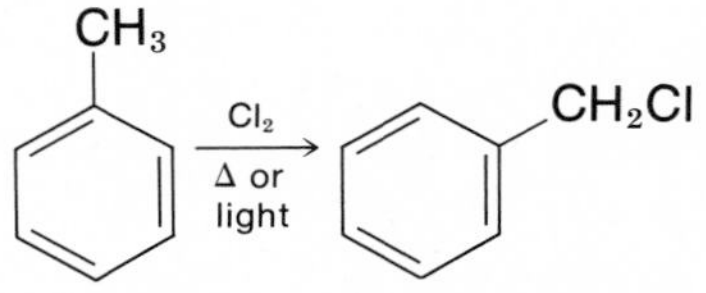

Toluene   Benzyl chloride

OXIDATION

KMnO₄, potassium permanganate, is a widely used oxidizing agent.

The combination of aromatic and aliphatic components in alkyl benzenes makes the alkyl side chain reactive enough to be oxidized by hot aqueous $KMnO_4$. The oxidation proceeds along the chain up to its attachment to the ring, and *only* one carbon of the side chain remains in the form of a carboxyl group:

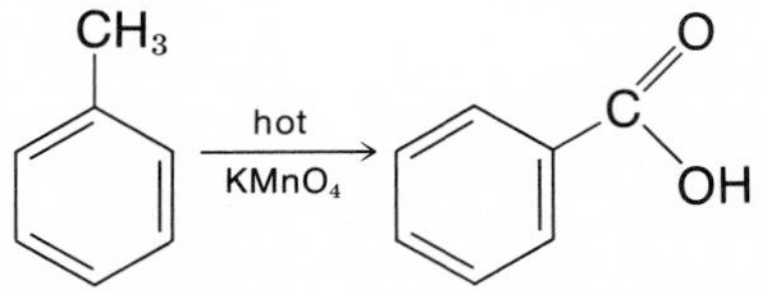

Toluene   Benzoic acid

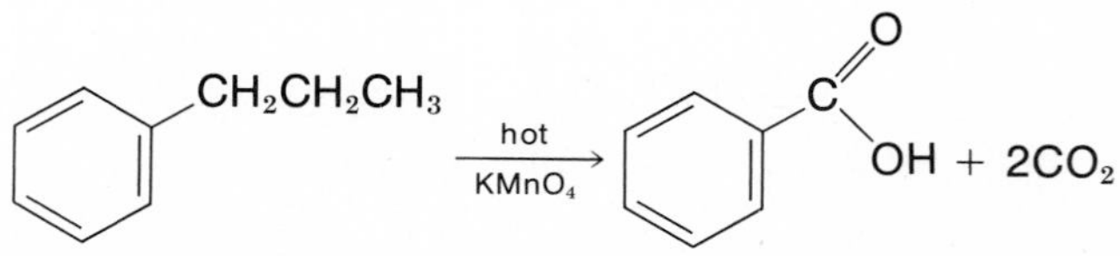

n-Propylbenzene   Benzoic acid

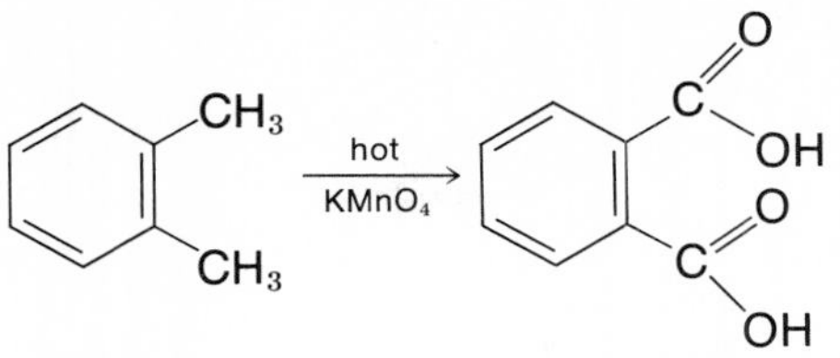

o-Xylene   Phthalic acid

The alkyl benzenes also undergo reactions of the aromatic ring; the alkyl groups influence the reaction as expected for ortho-para-directors:

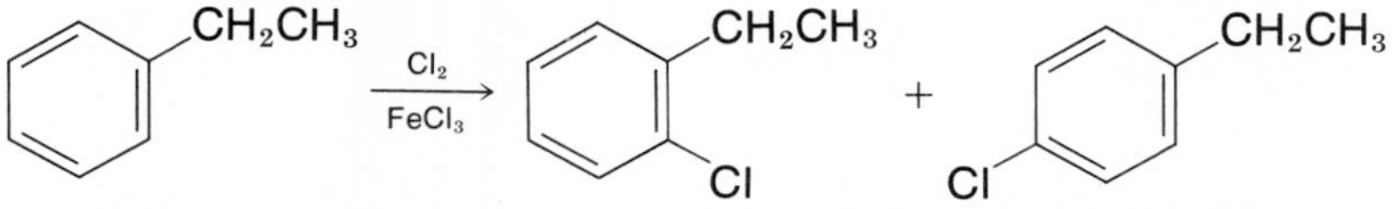

Ethylbenzene   o-Chloroethylbenzene   p-Chloroethylbenzene

Figure 12.3 summarizes the reactions of benzene and alkyl benzenes.

REACTIONS OF ALKYL HALIDES

Alkyl halides, like aryl halides and other halogenated hydrocarbons, are important organic compounds. They are used in the manufacture of plastics such as polyvinylchloride (PVC) and Teflon, which is a fluorohydrocarbon polymer. They find use as solvents and as propellants in aerosol cans. Many of the most effective germicides, herbicides, and insecticides are halogenated hydrocarbons such as DDT (*d*ichloro*d*iphenyl*t*richloroethane), an insecticide; 2,4-D(2,4-dichlorophenoxyacetic acid), a herbicide; and hexachlorophene, a potent germicide:

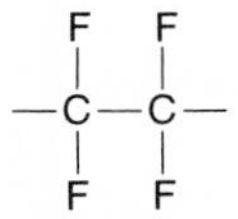

The Teflon monomer

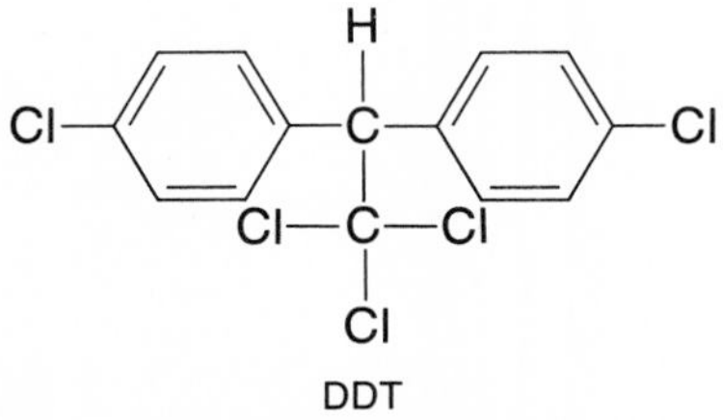

DDT 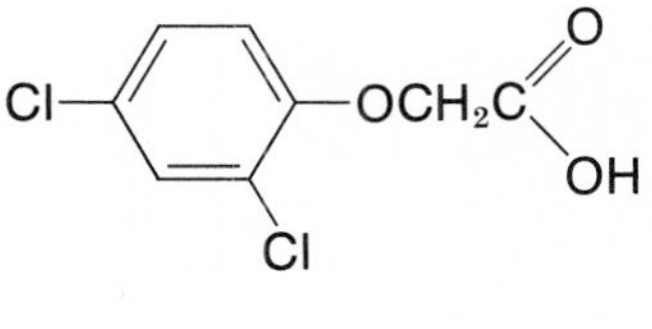   2,4-D 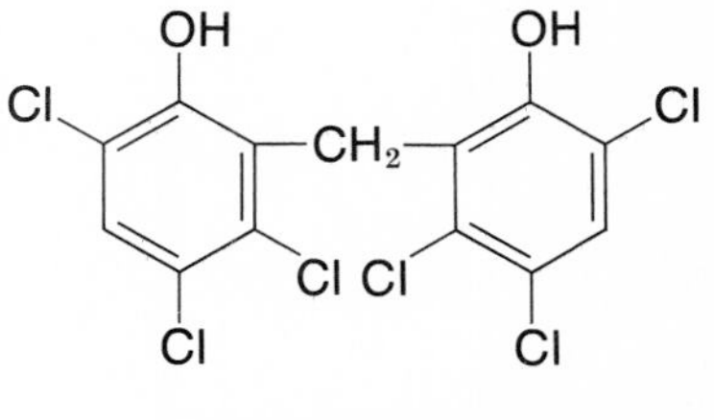   Hexachlorophene

To the organic chemist, alkyl chlorides are important intermediates in the synthesis of other organic compounds. Their most important reactions are **nucleophilic substitution, elimination,** and reaction with magnesium to prepare the important **Grignard reagent.**

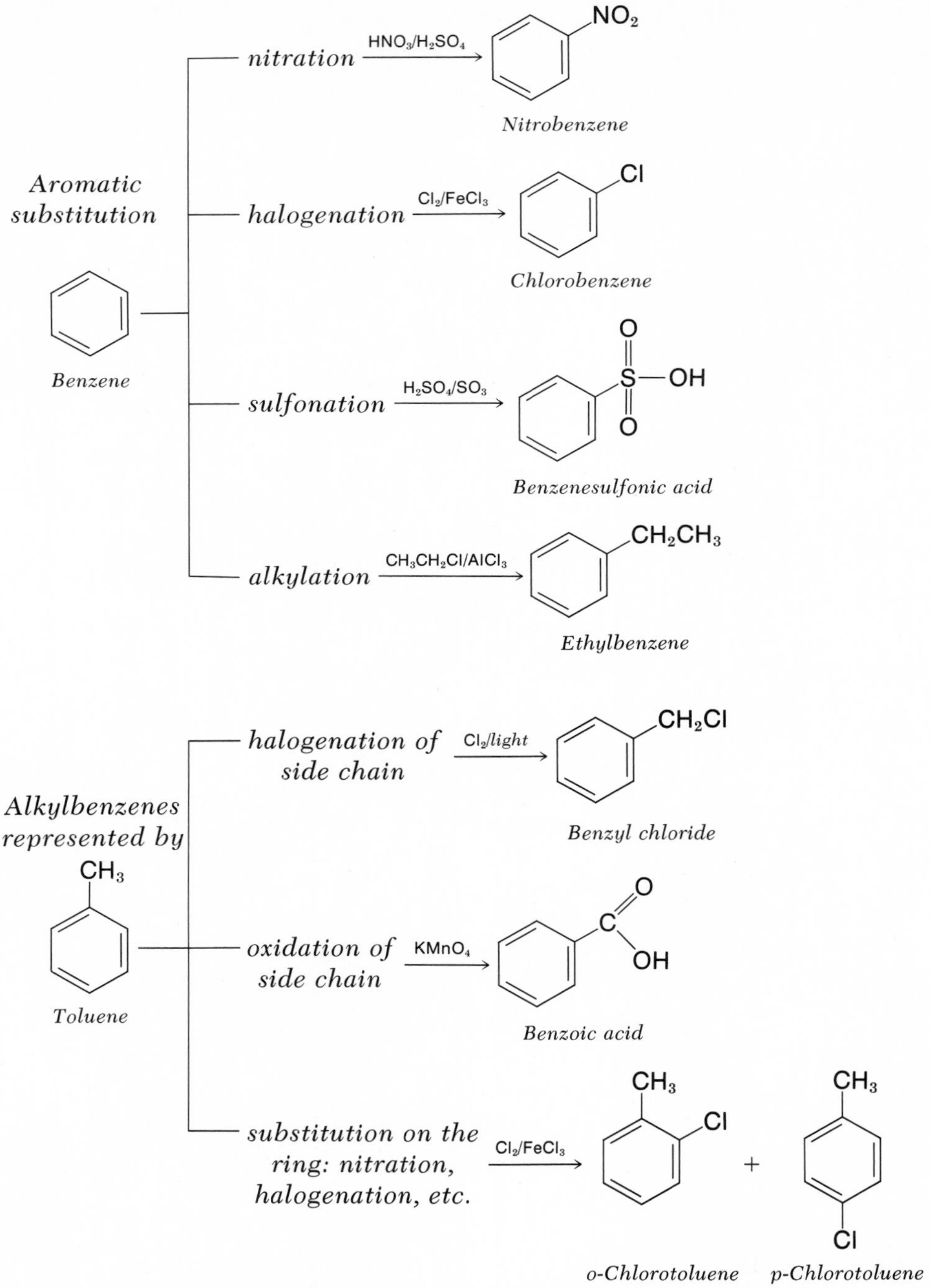

FIGURE 12.3 Reactions of benzene and alkyl benzenes.

NUCLEOPHILIC SUBSTITUTION

Nucleophilic substitution is the displacement of a halogen by a nucleophilic, or nucleus-loving reagent (the opposite of electrophilic). The nucleophilic reagents are electron-rich and are often negative ions. Let us look at the reaction mechanisms and the products that can be formed. Tertiary alkyl halides usually react by a mechanism called S_N1, which means substitution, nucleophilic, unimolecular. It is a two-step mechanism, and the first step is the *ionization* of the alkyl halide:

S_N1 mechanism step 1:

$$H_3C\!-\!\overset{\displaystyle CH_3}{\underset{\displaystyle CH_3}{C}}\!-\!Br \xrightarrow{\text{slow}} H_3C\!-\!\overset{\displaystyle CH_3}{\underset{\displaystyle CH_3}{C}}\!\oplus \quad + \quad Br^-$$

A tertiary alkyl halide   A tertiary carbonium ion

In the ionization step, a carbonium ion is formed. This step is a *slow* reaction and involves only the alkyl halide. Since the rate of the reaction is determined by this step and only one type of molecule is involved, the mechanism is designated as **unimolecular.** The tertiary carbonium ion formed reacts in step 2 with the nucleophilic reagent:

S_N1 mechanism step 2:

$$H_3C\!-\!\overset{\displaystyle CH_3}{\underset{\displaystyle CH_3}{C}}\!\oplus \;+\; OH^- \xrightarrow{\text{fast}} H_3C\!-\!\overset{\displaystyle CH_3}{\underset{\displaystyle CH_3}{C}}\!-\!OH$$

2-Methyl-2-propanol, an alcohol

The electron-*rich* hydroxyl group combines with the electron-*poor* nucleus of the carbonium ion to form the product.

Primary alkyl halides follow a different reaction mechanism, called S_N2, which means substitution, nucleophilic, bimolecular. In this mechanism there is only one step, but the reaction goes through an energy-rich stage which is called the **transition state.** The formation of the transition state involves both reagents, and the rate of the reaction depends on the concentration of the alkyl halide as well as the concentration of the nucleophilic reagent. For this reason the reaction is called **bimolecular.**

S_N2 mechanism:

$$OH^- + \underset{CH_3CH_2H_2C}{\overset{H}{C}}\!\!-\!Br \longrightarrow \left[\underset{CH_3CH_2H_2C}{\overset{H}{HO\cdots C\cdots Br}} \right] \longrightarrow \underset{CH_3CH_2CH_2}{\overset{H}{HO\!-\!C\!-\!H}} + Br^-$$

n-Butyl bromide (1-bromobutane)   Transition state   1-Butanol, a primary alcohol

In this reaction mechanism, the nucleophilic reagent, the OH^-, attacks the carbon nucleus *opposite* the point at which the halogen is attached. A transition state is formed in which the carbon atom has five bonds. We could say that the carbon atom has three covalent bonds and two covalent half bonds expressed by the dotted lines. The —OH group then pushes the halogen out so that the primary alcohol is produced.

ELIMINATION

A second kind of reaction competes with S_N1 and S_N2 reactions, namely, elimination reactions known as E_1 and E_2: elimination, unimolecular and elimination, bimolecular. The first step of the S_N1 mechanism consists of the formation of the carbonium ion. At this point an *elimination* of a hydrogen ion from the carbonium ion can take place, forming an alkene rather than an alcohol. This is then an E_1 rather than an S_N1 reaction.

E_1 mechanism step 2:

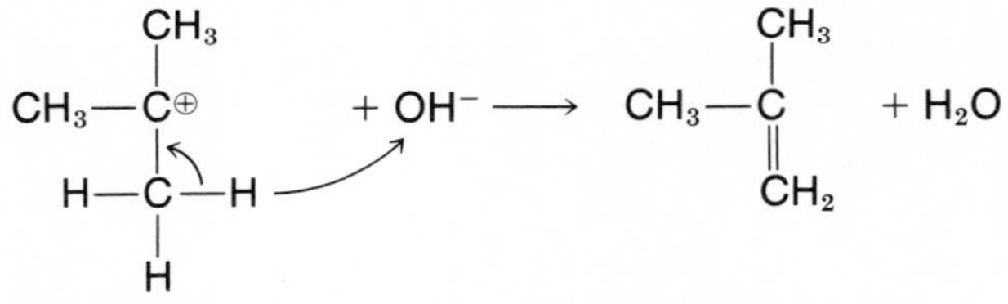

A tertiary carbonium ion   2-Methylpropene, an alkene

The E_2 reaction *competes* with the S_N2 mechanism in the following way:

E_2 mechanism:

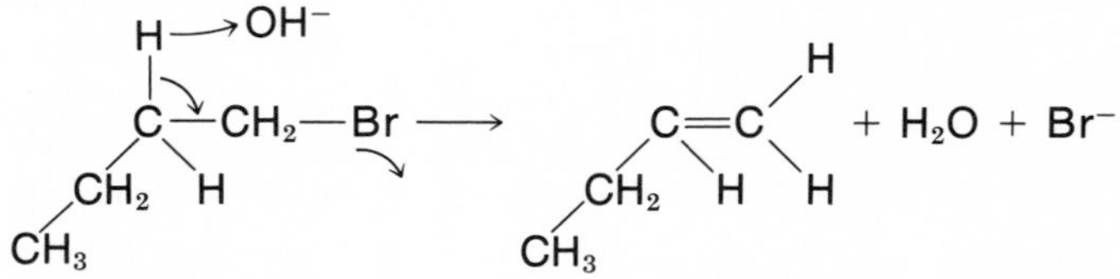

1-Bromobutane, a primary alkyl halide   1-Butene, an alkene

Here the nucleophilic agent abstracts a hydrogen atom on the carbon atom next to the carbon bearing the halogen, and, perhaps simultaneously, the halogen leaves as halide ion. A double bond can thus be formed. Substitution is the preferred pathway of reaction of primary alkyl halides, whereas elimination is the preferred path with tertiary alkyl halides. Secondary alkyl halides are somewhere in the middle.

REACTION WITH MAGNESIUM

Alkyl halides react with magnesium metal in anhydrous ether to form alkylmagnesium halides, RMgX:

$$RX + Mg \xrightarrow{\text{dry ether}} RMgX$$

The product formed is known as the **Grignard reagent** in honor of the French chemist, Victor Grignard, who discovered it. The reagent can be added to carbonyl compounds such as aldehydes, ketones, and carbon dioxide to produce alcohols and carboxylic acids:

Step 1:

R′—C(=O)(H) + RMgX ⟶ R′—C(H)(R)—O—MgX

| Aldehyde | Grignard reagent | | Addition product |

In this first step the R group of the Grignard reagent adds to the carbonyl carbon of the aldehyde, and the MgX adds to the carbonyl oxygen. This is followed by formation of the product, in step 2, by the addition of water or dilute acid. Figure 12.4 outlines some of the reactions and products obtained with the Grignard reagent.

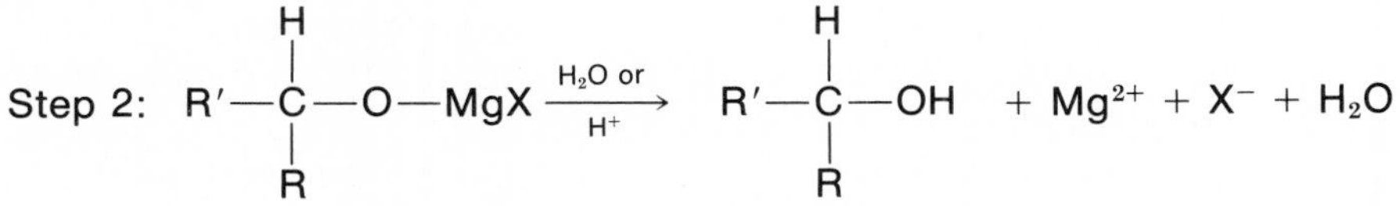

Step 2:

R′—C(H)(R)—O—MgX $\xrightarrow[\text{H}^+]{\text{H}_2\text{O or}}$ R′—C(H)(R)—OH + Mg^{2+} + X^- + H_2O

A secondary alcohol

REACTIONS OF ALCOHOLS AND PHENOLS

The functional group of the alcohols and phenols is the —OH group. Aliphatic —OH compounds are called alcohols and aromatic —OH compounds (with —OH attached to the ring) are known as phenols. They differ somewhat in their characteristics, but for the most part undergo the same reactions, which are mainly the reactions of the —OH group. These are important compounds, used as solvents in the paint and chemical industries and as intermediates in the chemical industry. Phenols are also important starting materials for plastics, dyes, and pharmaceuticals, as well as for the preparation of herbicides, fungicides, and

CH_3CH_2OH
An alcohol

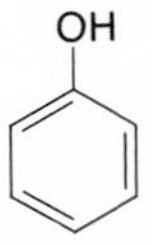

OH
A phenol

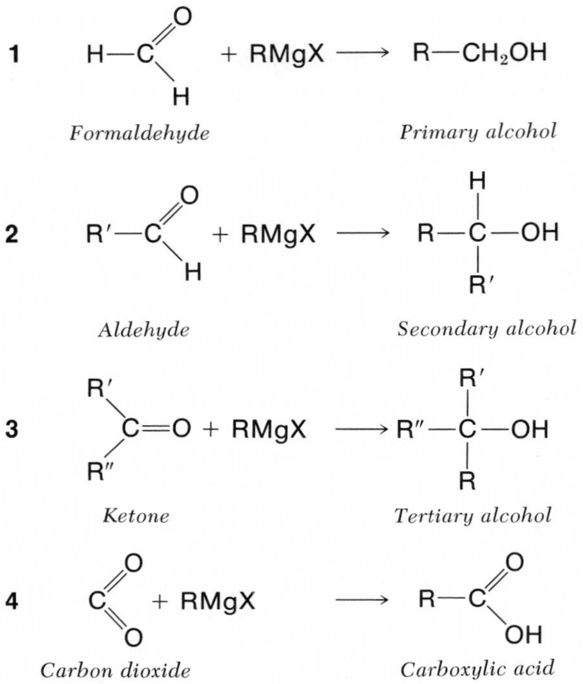

1 $H—C\overset{\displaystyle O}{\underset{\displaystyle H}{\Big\lVert}}$ + RMgX ⟶ R—CH₂OH

Formaldehyde   *Primary alcohol*

2 $R'—C\overset{\displaystyle O}{\underset{\displaystyle H}{\Big\lVert}}$ + RMgX ⟶ $R—\overset{\displaystyle H}{\underset{\displaystyle R'}{\overset{|}{\underset{|}{C}}}}—OH$

Aldehyde   *Secondary alcohol*

3 $\overset{\displaystyle R'}{\underset{\displaystyle R''}{C}}{=}O$ + RMgX ⟶ $R''—\overset{\displaystyle R'}{\underset{\displaystyle R}{\overset{|}{\underset{|}{C}}}}—OH$

Ketone   *Tertiary alcohol*

4 $\overset{\displaystyle O}{\underset{\displaystyle O}{C}}$ + RMgX ⟶ $R—C\overset{\displaystyle O}{\underset{\displaystyle OH}{\Big\lVert}}$

Carbon dioxide   *Carboxylic acid*

FIGURE 12.4 Grignard reactions with carbonyl compounds.

germicides. In biological systems, alcohols are found as flavor and odor components and vitamins but primarily as polyhydroxy components of carbohydrates and lipids.

In the reactions of the alcohols, we can distinguish between (1) the removal of the hydrogen from the hydroxyl group and (2) the removal of the entire —OH group from the carbon holding it:

$$R—O \!\!\mid\!\! H \qquad R \!\!\mid\!\! O—H$$

Oxidation of alcohols is different from both of these.

DISPLACEMENT OF THE HYDROXYL HYDROGEN

Reaction with active metals The most reactive metals such as sodium react with alcohols to form alkoxides (salts) and hydrogen gas:

$$2ROH + 2Na \longrightarrow \quad 2RO^-Na^+ \quad + H_2 \uparrow$$

Sodium alkoxide,
a salt

The alkoxides are used with alkyl halides in the Williamson synthesis to produce ethers. They also find use as basic catalysts in a number of organic reactions.

Ester formation Esters are formed in the reaction of alcohols with organic acids, acid halides, and acid anhydrides. The general reaction is

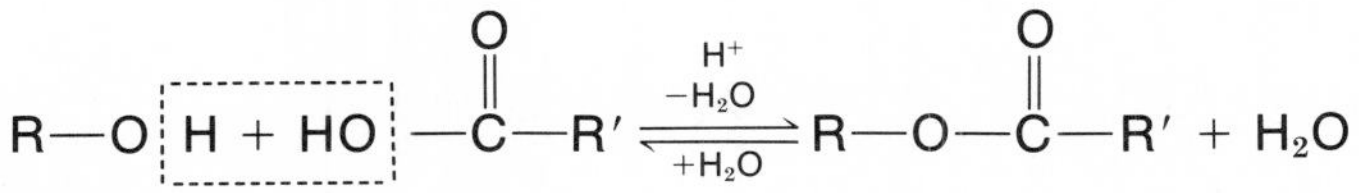

$$R—O\!:\!H + HO\!:\!—\overset{\displaystyle O}{\overset{\|}{C}}—R' \underset{+H_2O}{\overset{H^+ \; -H_2O}{\rightleftharpoons}} R—O—\overset{\displaystyle O}{\overset{\|}{C}}—R' + H_2O$$

An alcohol   An ester   An acid

We shall discuss the esters further as acid derivatives. Inorganic acids, such as sulfuric acid and nitric acid, also react with alcohols to form esters. Indeed, esters of nitric acid are among the main components of explosives like smokeless gunpowder and dynamite.

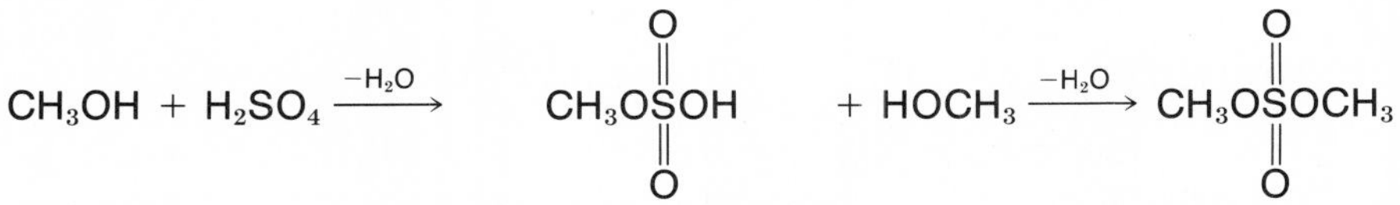

$$CH_3OH + H_2SO_4 \xrightarrow{-H_2O} \quad CH_3O\overset{\displaystyle O}{\underset{\displaystyle O}{\overset{\|}{\underset{\|}{S}}}}OH \quad + HOCH_3 \xrightarrow{-H_2O} CH_3O\overset{\displaystyle O}{\underset{\displaystyle O}{\overset{\|}{\underset{\|}{S}}}}OCH_3$$

Methanol   Methyl hydrogen sulfate   Dimethyl sulfate

DISPLACEMENT OF THE OH GROUP

Formation of alkyl halides Reagents such as halogen acids, in the presence of a zinc halide, will produce an alkyl halide:

$$ROH \xrightarrow{HCl/ZnCl_2} RCl + H_2O$$

Phosphorus trichloride, PCl_3, and thionylchloride, $SOCl_2$, can be used as well.

The reaction proceeds by the formation of a carbonium ion, and the more stable the carbonium ion the faster the reaction. The relative reactivity of the alcohols (tertiary > secondary > primary) can be used to differentiate between the different types.

Reaction with sulfuric acid The reaction of alcohols with sulfuric acid can result in two different kinds of products. At lower temperatures and an excess of alcohol, ether formation is favored, whereas at higher temperatures and an excess of sulfuric acid, dehydration of the alcohol is favored, resulting in alkene formation:

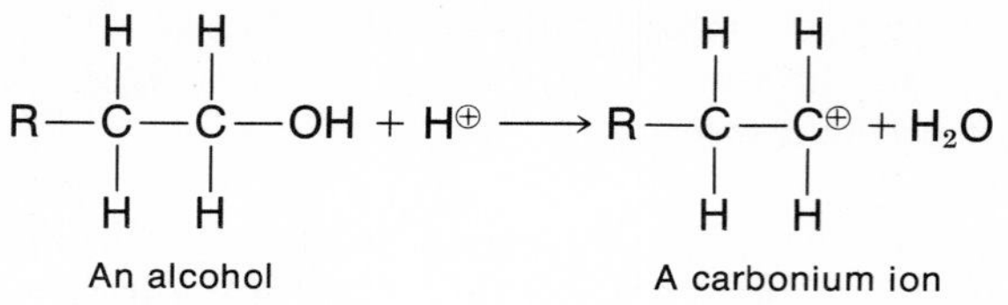

$$R-\overset{\overset{\displaystyle H}{|}}{\underset{\underset{\displaystyle H}{|}}{C}}-\overset{\overset{\displaystyle H}{|}}{\underset{\underset{\displaystyle H}{|}}{C}}-OH + H^{\oplus} \longrightarrow R-\overset{\overset{\displaystyle H}{|}}{\underset{\underset{\displaystyle H}{|}}{C}}-\overset{\overset{\displaystyle H}{|}}{\underset{\underset{\displaystyle H}{|}}{C^{\oplus}}} + H_2O \qquad (12.1)$$

An alcohol   A carbonium ion

$$R-\overset{\overset{\displaystyle H}{|}}{\underset{\underset{\displaystyle H}{|}}{C}}-\overset{\overset{\displaystyle H}{|}}{C^{\oplus}} + \overset{}{\underset{\underset{\displaystyle H}{}}{O}}-CH_2-CH_2-R \longrightarrow R-\overset{\overset{\displaystyle H}{|}}{\underset{\underset{\displaystyle H}{|}}{C}}-\overset{\overset{\displaystyle H}{|}}{\underset{\underset{\displaystyle H}{|}}{C}}-O-\overset{\overset{\displaystyle H}{|}}{\underset{\underset{\displaystyle H}{|}}{C}}-\overset{\overset{\displaystyle H}{|}}{\underset{\underset{\displaystyle H}{|}}{C}}-R + H^+ \quad (12.2a)$$

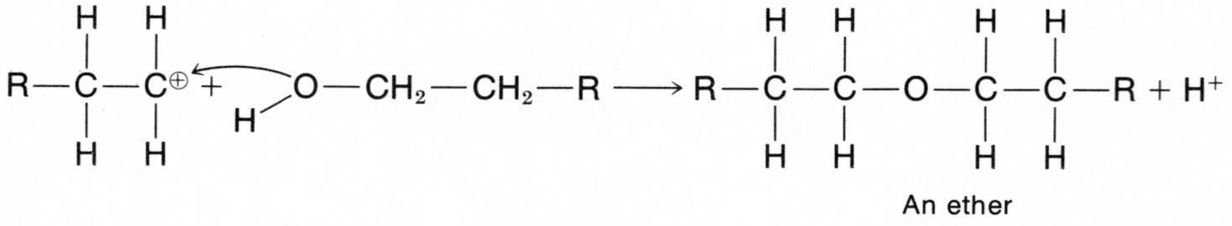

An ether

or

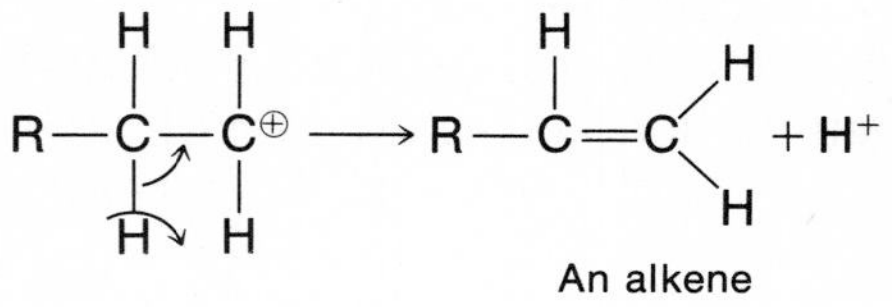

$$R-\overset{\overset{\displaystyle H}{|}}{\underset{\underset{\displaystyle H}{}}{C}}-\overset{}{C^{\oplus}} \longrightarrow R-\overset{\overset{\displaystyle H}{|}}{C}=\overset{\overset{\displaystyle H}{}}{\underset{\underset{\displaystyle H}{}}{C}} + H^+ \qquad (12.2b)$$

An alkene

OXIDATION

The product obtained by the oxidation of an alcohol depends on whether the alcohol is primary, secondary, or tertiary and on the conditions of the reaction:

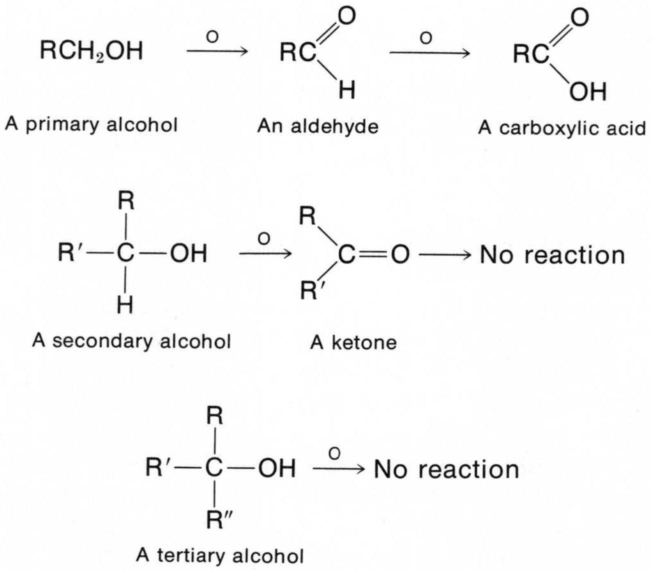

$$RCH_2OH \xrightarrow{O} RC\overset{\displaystyle O}{\underset{\displaystyle H}{\diagdown}} \xrightarrow{O} RC\overset{\displaystyle O}{\underset{\displaystyle OH}{\diagdown}}$$

A primary alcohol   An aldehyde   A carboxylic acid

$$R'-\overset{\overset{\displaystyle R}{|}}{\underset{\underset{\displaystyle H}{|}}{C}}-OH \xrightarrow{O} \overset{\displaystyle R}{\underset{\displaystyle R'}{\diagup}}C=O \longrightarrow \text{No reaction}$$

A secondary alcohol   A ketone

$$R'-\overset{\overset{\displaystyle R}{|}}{\underset{\underset{\displaystyle R''}{|}}{C}}-OH \xrightarrow{O} \text{No reaction}$$

A tertiary alcohol

The reactions of the alcohols are summarized in Figure 12.6.

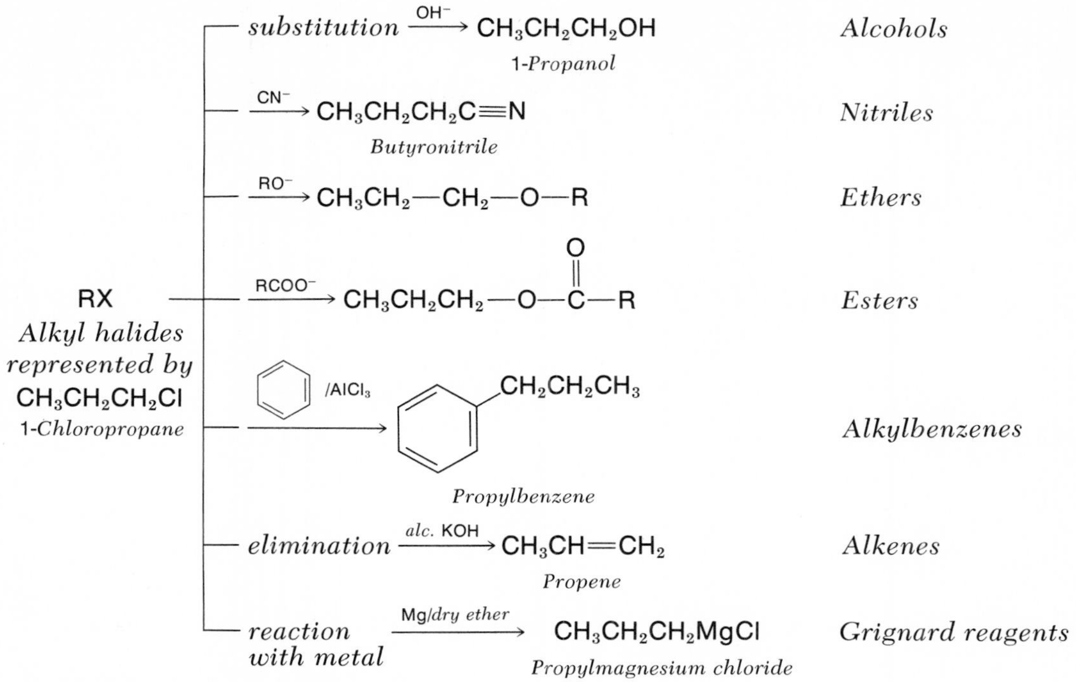

substitution $\xrightarrow{\text{OH}^-}$ CH$_3$CH$_2$CH$_2$OH *Alcohols*

1-*Propanol*

$\xrightarrow{\text{CN}^-}$ CH$_3$CH$_2$CH$_2$C≡N *Nitriles*

Butyronitrile

$\xrightarrow{\text{RO}^-}$ CH$_3$CH$_2$—CH$_2$—O—R *Ethers*

$\xrightarrow{\text{RCOO}^-}$ CH$_3$CH$_2$CH$_2$—O—C(=O)—R *Esters*

RX
*Alkyl halides
represented by*
CH$_3$CH$_2$CH$_2$Cl
1-*Chloropropane*

/AlCl$_3$ → (ring)—CH$_2$CH$_2$CH$_3$ *Alkylbenzenes*

Propylbenzene

elimination $\xrightarrow{\text{alc. KOH}}$ CH$_3$CH=CH$_2$ *Alkenes*

Propene

reaction
with metal $\xrightarrow{\text{Mg/dry ether}}$ CH$_3$CH$_2$CH$_2$MgCl *Grignard reagents*

Propylmagnesium chloride

FIGURE 12.5 Reactions of alkyl halides.

REACTIONS OF ALDEHYDES AND KETONES

The aldehydes and ketones are the first group of compounds we have studied that contain the carbonyl group $\diagdown$C=O as the functional group. Aldehydes and ketones are compounds that generally have pleasant odors; they are the principal sources of flavor and aroma in perfumes and spices. Aldehyde and ketone groups are important in such biological compounds as carbohydrates, steroids, and the intermediates formed in metabolism. In the chemical industry the shorter-chain ketones, such as acetone, methyl ethyl ketone, and diethyl ketone, find extensive use as solvents. Formaldehyde, the first member of the aldehyde series, is used as a disinfectant, as a preservative for biological specimens, in embalming fluid, and in the manufacture of plastics.

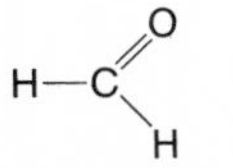

H—C(=O)—H

Formaldehyde

  The reactions of the aldehydes and ketones are oxidation, addition to the carbonyl group, and replacement of α hydrogens (helped by the presence of the carbonyl group):

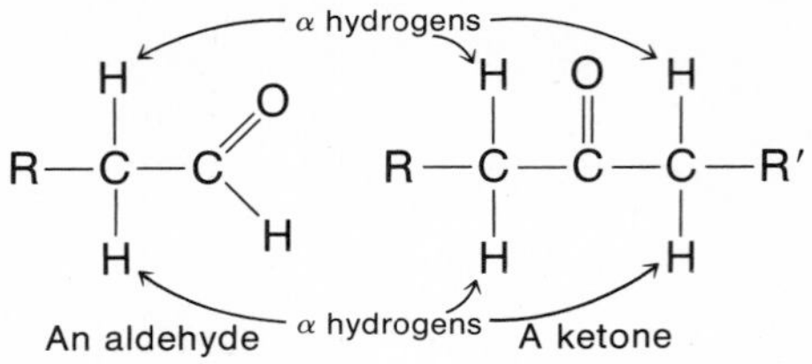

An aldehyde α hydrogens A ketone

OXIDATION

Mild oxidizing reagents like Benedict's solution readily convert aldehydes into acids, while the bright blue cupric complex of the reagent [which may be regarded as $Cu(OH)_2$ for simplicity] is reduced to orange-red cuprous oxide, Cu_2O:

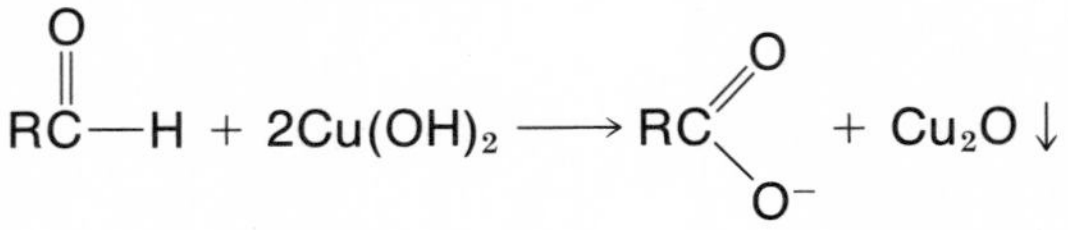

(Blue)   (Red)

Since ketones will not react with such a mild reagent, this reaction is used to differentiate between aldehydes and ketone.

ADDITION REACTIONS AT THE CARBONYL GROUP

This reaction, typical of the aldehydes and ketones, represents **nucleophilic addition.** Since the carbonyl carbon is deficient in electrons, it is there that the attack of the electron-rich nucleophilic reagent takes place:

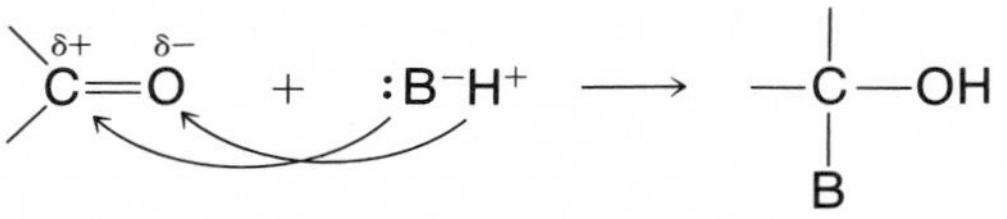

Carbonyl group   Nucleophilic   Addition product
of aldehydes   reagent
and ketones

REACTION WITH ALCOHOLS

The addition of an alcohol to an aldehyde leads to the formation of a hemiacetal:

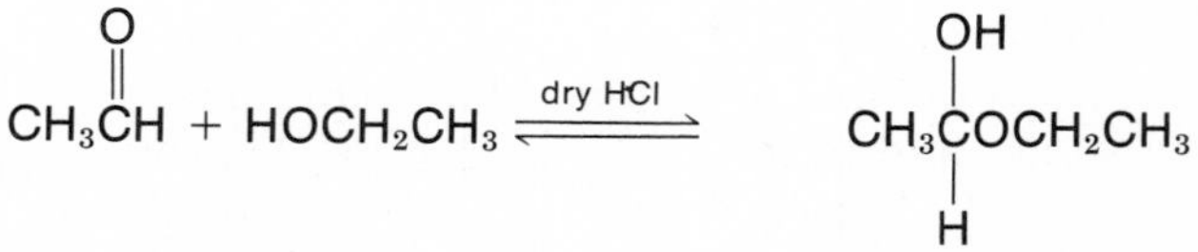

A hemiacetal, simultaneously an
ether and an alcohol

Hemiacetals are usually unstable and can react further with the alcohol to form the acetal:

$$CH_3\underset{\underset{H}{|}}{\overset{\overset{OH}{|}}{C}}OCH_2CH_3 + HOCH_2CH_3 \xrightarrow[\text{HCl}]{\text{dry}} CH_3\underset{H}{\overset{OCH_2CH_3}{C}}OCH_2CH_3$$

An acetal, a diether

The hemiacetal system is a characteristic part of the cyclic struc-ture of the monosaccharides. Ketones can form ketals but require different and more exacting reaction conditions.

$$R\underset{R'}{\overset{O—R''}{\underset{O—R''}{C}}}$$

A ketal

ADDITION OF GRIGNARD REAGENT

This reaction has already been discussed under the reactions of the alkyl halides; see the summary in Figure 12.4.

ADDITION OF HYDROGEN

The addition of hydrogen to aldehydes and ketones constitutes a **reduction.** In this process the hydrogens are added to the car-

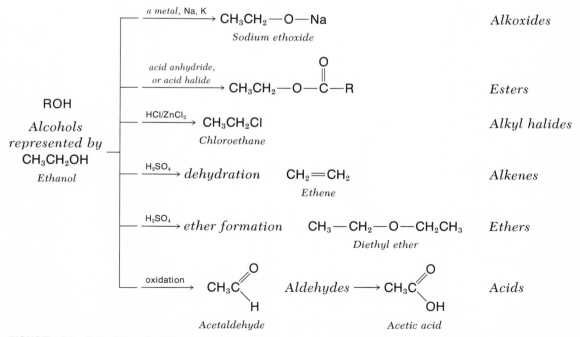

ROH

Alcohols represented by
CH_3CH_2OH
Ethanol

a metal, Na, K → $CH_3CH_2—O—Na$ *Sodium ethoxide*		*Alkoxides*
acid anhydride, or acid halide → $CH_3CH_2—O—\overset{\overset{O}{\|\|}}{C}—R$		*Esters*
HCl/ZnCl$_2$ → CH_3CH_2Cl *Chloroethane*		*Alkyl halides*
H$_2$SO$_4$ → *dehydration* $CH_2{=}CH_2$ *Ethene*		*Alkenes*
H$_2$SO$_4$ → *ether formation* $CH_3—CH_2—O—CH_2CH_3$ *Diethyl ether*		*Ethers*
oxidation → $CH_3\overset{\overset{O}{\nearrow}}{C}_{\searrow H}$ *Acetaldehyde*	*Aldehydes* → $CH_3\overset{\overset{O}{\nearrow}}{C}_{\searrow OH}$ *Acetic acid*	*Acids*

FIGURE 12.6 Reactions of alcohols.

bonyl carbon and the oxygen to produce primary and secondary alcohols, respectively. This is done by catalytic hydrogenation with platinum, palladium, or nickel catalysts, or the hydrogen is provided by chemical reducing agents such as lithium aluminum hydride, $LiAlH_4$:

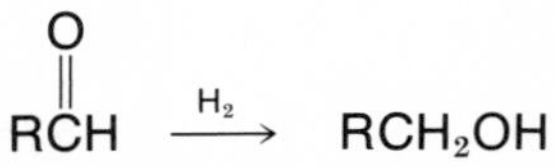

$$\underset{\text{Aldehyde}}{R\overset{\overset{\textstyle O}{\|}}{C}H} \xrightarrow{\text{ }H_2\text{ }} \underset{\text{Primary alcohol}}{RCH_2OH}$$

$$\underset{\text{Ketone}}{\overset{\textstyle R}{\underset{\textstyle R'}{\diagdown}}C=O} \xrightarrow{\text{ }H_2\text{ }} \underset{\text{Secondary alcohol}}{\overset{\textstyle R}{\underset{\textstyle R'}{\diagdown}}C\overset{\textstyle H}{\underset{\textstyle OH}{\diagup}}}$$

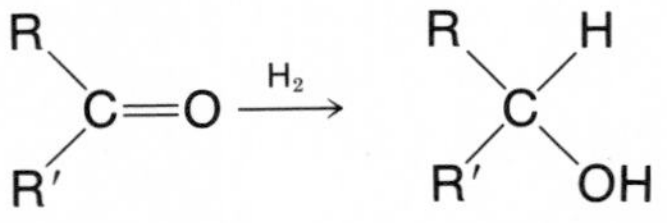

The addition of hydrogen to a carbonyl group is a common occurrence in biological systems, where it is accomplished with the aid of enzymes. An example is the addition of hydrogen to acetaldehyde to produce ethyl alcohol, which takes place in fermentation of sugar by yeast:

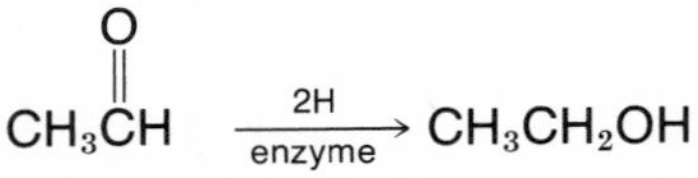

$$\underset{\text{Acetaldehyde}}{CH_3\overset{\overset{\textstyle O}{\|}}{C}H} \xrightarrow[\text{enzyme}]{\text{2H}} \underset{\text{Ethanol}}{CH_3CH_2OH}$$

REPLACEMENT OF α HYDROGEN

Aldol condensation **Aldol condensation** is the joining of two molecules of an aldehyde or ketone by the union of the α carbon of one molecule with the carbonyl carbon of the other to form a hydroxy aldehyde or ketone. The aldehyde or ketone must contain α hydrogens; the reaction is catalyzed by dilute acid or base and by enzymes:

Step 1:

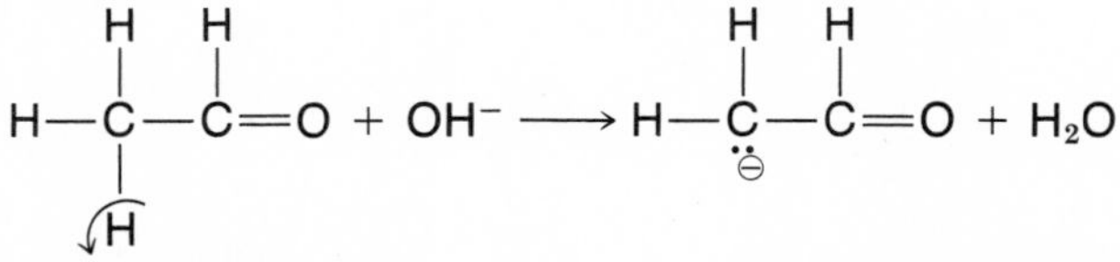

$$\underset{\text{Acetaldehyde}}{H-\overset{\overset{\textstyle H}{|}}{\underset{\underset{\textstyle H}{|}}{C}}-\overset{\overset{\textstyle H}{|}}{C}=O} + OH^- \longrightarrow \underset{\text{A carbanion}}{H-\overset{\overset{\textstyle H}{|}}{\underset{\underset{\textstyle \ominus}{..}}{C}}-\overset{\overset{\textstyle H}{|}}{C}=O} + H_2O$$

In step 1, the hydroxyl group abstracts an α hydrogen to produce a nucleophilic carbanion and water:

Step 2:

$$H-\overset{\overset{H}{|}}{\underset{\underset{H}{|}}{C}}-\overset{\overset{H}{|}}{\underset{\delta+}{C}}\overset{\delta-}{=}O + H-\overset{\overset{H}{|}}{\underset{\ominus}{C}}-\overset{\overset{H}{|}}{C}=O \longrightarrow H-\overset{\overset{H}{|}}{\underset{\underset{H}{|}}{C}}-\overset{\overset{H}{|}}{\underset{\underset{O^{\ominus}}{|}}{C}}-\overset{\overset{H}{|}}{\underset{\underset{H}{|}}{C}}-\overset{\overset{H}{|}}{C}=O$$

An alkoxide

In step 2 the carbanion becomes attached to the carbonyl carbon of the second aldehyde molecule to form an alkoxide, which in step 3 abstracts a hydrogen from water to produce the final product and regenerates the hydroxyl ion.

Step 3:

$$H-\overset{\overset{H}{|}}{\underset{\underset{H}{|}}{C}}-\overset{\overset{H}{|}}{\underset{\underset{O^{\ominus}}{|}}{C}}-\overset{\overset{H}{|}}{\underset{\underset{H}{|}}{C}}-\overset{\overset{H}{|}}{C}=O + H-O\overset{}{\underset{H}{\diagdown}} \longrightarrow H-\overset{\overset{H}{|}}{\underset{\underset{H}{|}}{C}}-\overset{\overset{H}{|}}{\underset{\underset{OH}{|}}{C}}-\overset{\overset{H}{|}}{\underset{\underset{H}{|}}{C}}-\overset{\overset{H}{|}}{C}=O + OH^-$$

β-Hydroxybutyraldehyde
(aldol, 3-hydroxybutanal)

Aldol is the condensation product of acetaldehyde from which the name of this reaction is derived. The analogous overall reaction for a ketone, with acetone as the example, is

$$2CH_3-\overset{\overset{O}{\|}}{C}-CH_3 \xrightarrow{OH^-} H_3C-\overset{\overset{CH_3}{|}}{\underset{\underset{OH}{|}}{C}}-CH_2-\overset{\overset{}{}}{\underset{\underset{O}{\|}}{C}}-CH_3$$

Acetone   4-Hydroxy-4-methyl-2-pentanone,
condensation product of acetone

We shall encounter this type of reaction again in discussing metabolism. Reactions of aldehydes and ketones are summarized in Figure 12.7.

REACTIONS OF ORGANIC ACIDS AND THEIR DERIVATIVES

Organic acids are compounds that contain the carboxyl group —COOH as a functional group. Organic acids are plentiful in nature and provide the pleasant acidic taste in many of the foods we eat. Organic acids are found in the metabolic processes, as components of lipids and phospholipids, as vitamins, and in the form of polyfunctional acids such as the amino acids, and many others. These acids are important not only in themselves but also as derivatives resulting from replacement or modification of the —OH group of the carboxyl group. The organic acids and their

However the simplest organic acid, HCOOH, or formic acid, is what puts the zing into the stings of bees, stinging nettles, and red ants; it was originally obtained from red ants.

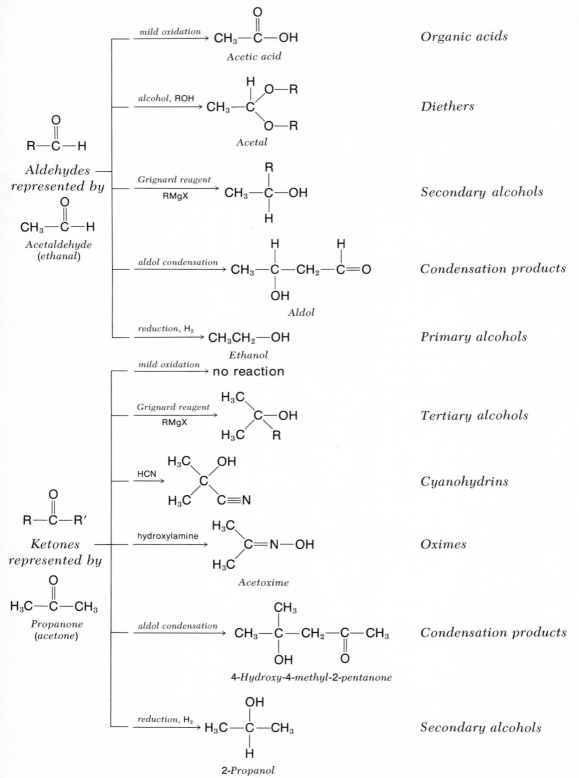

FIGURE 12.7 Reactions of aldehydes and ketones.

derivatives are useful to the organic chemist in synthetic work, but we also find them as components in the manufacture of soaps, plastics, fibers, pharmaceuticals, herbicides, and many other commodities of daily use.

The reactions of the organic acids may involve (1) the replacement of the hydrogen from the —OH group, (2) the replacement of the —OH group, and (3) the replacement of α hydrogens. All these reactions, summarized in Figure 12.8 (page 316), are affected by the presence of the carbonyl group:

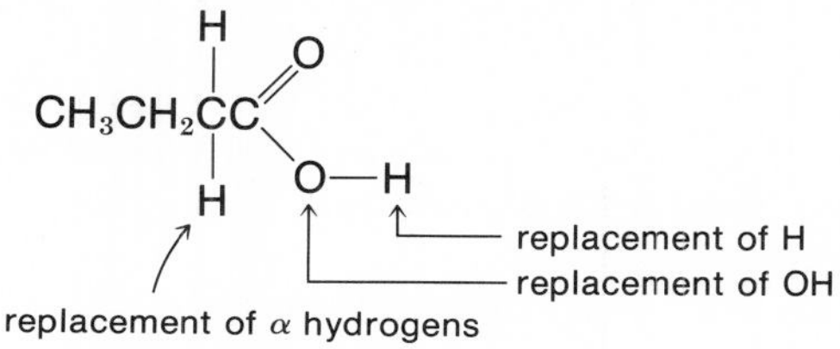

$$CH_3CH_2\underset{\underset{H}{|}}{\overset{\overset{H}{|}}{C}}C\overset{O}{\underset{O-H}{\diagup}}$$

—— replacement of H
—— replacement of OH

replacement of α hydrogens

REPLACEMENT OF THE CARBOXYL HYDROGEN; SALT FORMATION

The replacement of the hydrogen from the carboxyl group occurs in the reaction of an acid with a base such as NaOH, KOH, or NH_4OH, or with a basic salt such as a bicarbonate or a carbonate, to produce a salt and water:

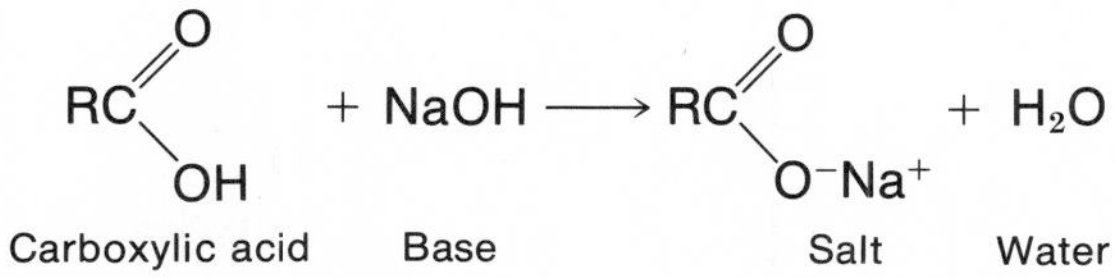

$$RC\overset{O}{\underset{OH}{\diagup}} + NaOH \longrightarrow RC\overset{O}{\underset{O^-Na^+}{\diagup}} + H_2O$$

Carboxylic acid   Base   Salt   Water

This is an acid-base neutralization which can be done quantitatively by titrating the acid with a standard base:

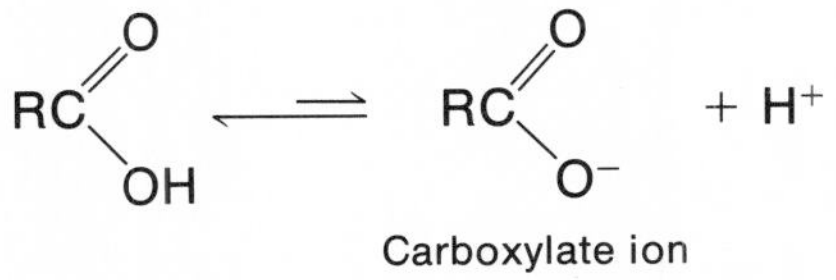

$$RC\overset{O}{\underset{OH}{\diagup}} \rightleftharpoons RC\overset{O}{\underset{O^-}{\diagup}} + H^+$$

Carboxylate ion

The stability of the **carboxylate ion** is enhanced by the electron-attracting effect of the carbonyl group. This promotes the separation of the hydrogen to form the carboxylate ion and H^+. The carboxylate ion can then associate with the positive ion of the base to form the salt, especially if the solution is concentrated. The organic acid can be liberated from the salt by the addition of

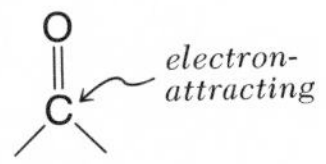

electron-attracting

a stronger acid such as HCl or H_2SO_4; this provides a way of separating organic acids from a reaction mixture:

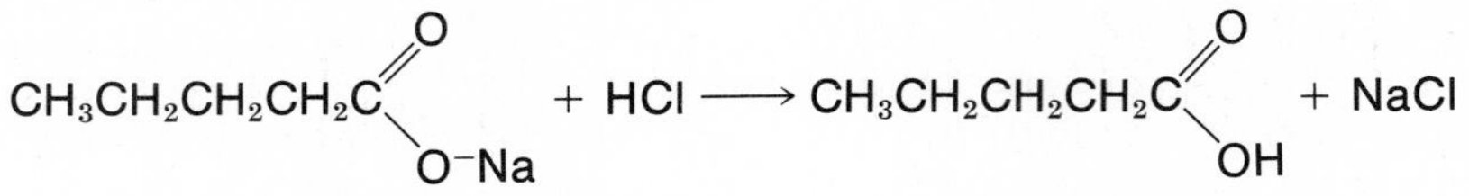

$$CH_3CH_2CH_2CH_2C\underset{O^-Na}{\overset{O}{\diagdown}} + HCl \longrightarrow CH_3CH_2CH_2CH_2C\underset{OH}{\overset{O}{\diagdown}} + NaCl$$

Sodium pentanoate   Pentanoic acid

Na⁺, K⁺, and NH₄⁺ salts are more soluble than the corresponding acid.

The salts of carboxylic acids have many uses. Sodium and potassium salts of long-chain fatty acids are soaps. Pharmaceutical preparations and drugs are often made up as the sodium or potassium salt; this serves to increase the solubility of the active acid component, rendering it more readily absorbed and more effective biologically.

REPLACEMENT OF THE HYDROXYL GROUP

The hydroxyl group of an organic acid can be replaced by several groups; the compounds so formed represent the bulk of organic acid derivatives. The general reaction is

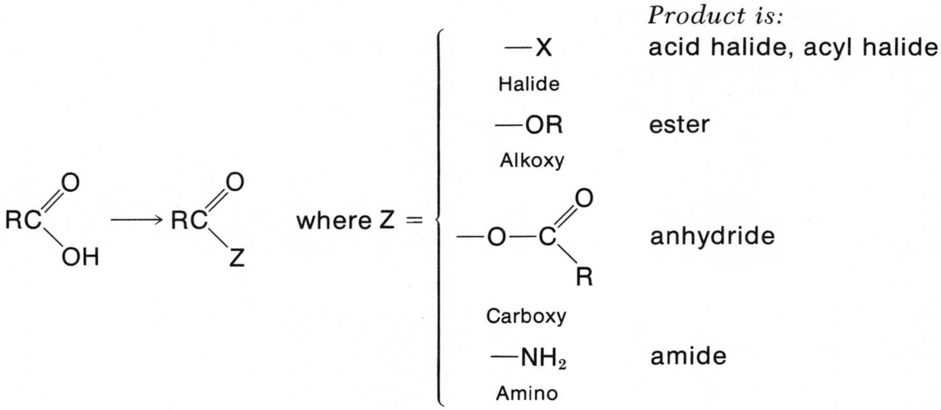

$$RC\underset{OH}{\overset{O}{\diagdown}} \longrightarrow RC\underset{Z}{\overset{O}{\diagdown}} \quad \text{where } Z =$$

Product is:

—X	acid halide, acyl halide
Halide	
—OR	ester
Alkoxy	
—O—C(=O)R	anhydride
Carboxy	
—NH₂	amide
Amino	

FORMATION OF ACYL HALIDES

The acyl halides are readily formed by the reaction of acids with phosphorus trichloride, PCl_3; phosphorus pentachloride, PCl_5; or thionyl chloride, $SOCl_2$. The reaction is a nucleophilic acyl substitution where RCO— is the acyl group on which the substitution is taking place. The reaction mechanism is similar to the one we saw before in nucleophilic substitution:

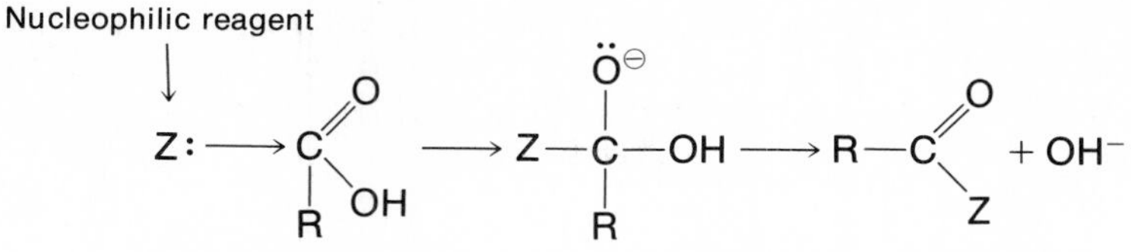

Nucleophilic reagent

Organic acid   Intermediate   Acyl halide

Since the acyl halides are more reactive than the corresponding acids, and since their reactions give high yields, they are used as intermediates in converting acids to other acid derivatives such as amides and esters:

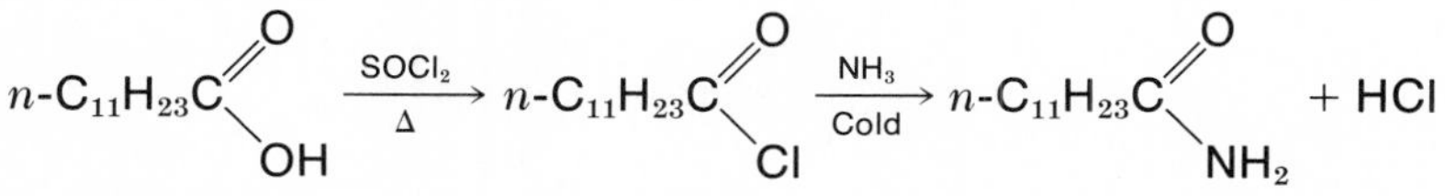

Lauric acid   Lauroyl chloride   Lauramide

The formation of amides from aromatic acids is difficult to achieve directly. Here again the conversion of the acid to the acyl chloride and subsequent reaction with the amine easily produces the amide in good yield and purity:

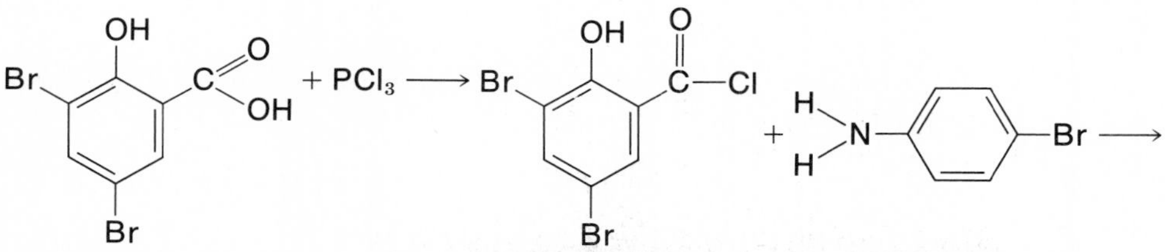

3,5-Dibromosalicylic   3,5-Dibromosalicyloyl   *p*-Bromoaniline,
acid   chloride   an aromatic amine

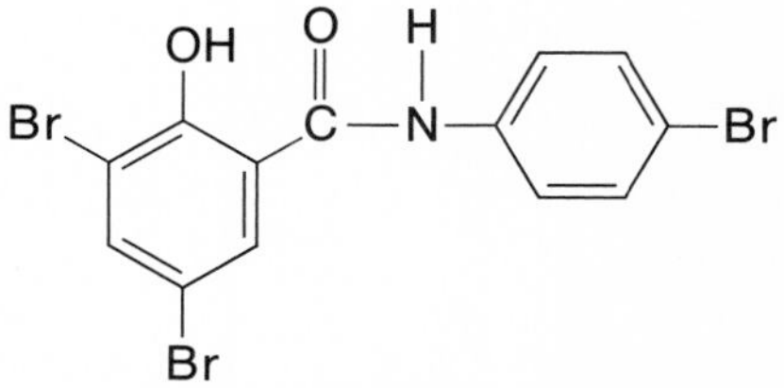

3,4′,5-Tribromosalicylanilide,
an excellent germicide

The acyl halides are also used for the preparation of esters difficult to prepare directly. An advantage of using two steps is that both the preparation of the acyl chloride and the subsequent amide or ester formation are essentially irreversible. On the other hand, the one-step reaction to form amide or ester is slow and reversible, and special pains must be taken to shift the equilibrium in the direction of the product in order to get a good yield.

FORMATION OF ESTERS

Structurally an ester may be derived from the corresponding acid by replacing the acidic hydrogen by an alkyl or aryl group. Many esters have fruity, flowery, or ethereal odors and are used as artificial scents and flavoring agents. They are found in paints, aerosols, cleaning agents, candies and ice cream, and many other products. Esters are present in biological systems as polysaccharides and in fats and oils as waxes and triglycerides; some alcohols are carried in the bloodstream as esters.

The esterification reaction requires a catalyst, usually H_2SO_4 or H_3PO_4, the H^+ ion, or proton, of the acid serving as the catalyst. In the simplest form the reaction can be written:

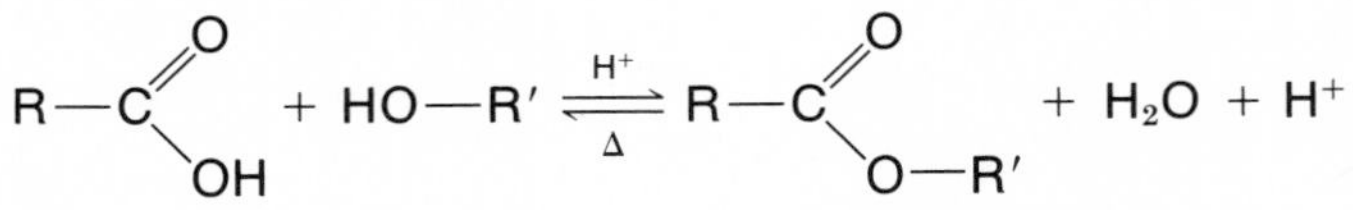

$$R-C\underset{OH}{\overset{O}{<}} + HO-R' \underset{\Delta}{\overset{H^+}{\rightleftharpoons}} R-C\underset{O-R'}{\overset{O}{<}} + H_2O + H^+$$

An acid   An alcohol   An ester   Water

The following reaction mechanism is generally accepted for primary and secondary alcohols:

Step 1:
$$R-C\underset{OH}{\overset{O}{<}} \overset{+H^+}{\rightleftharpoons} R-\overset{OH}{\underset{OH}{C^{\oplus}}}$$

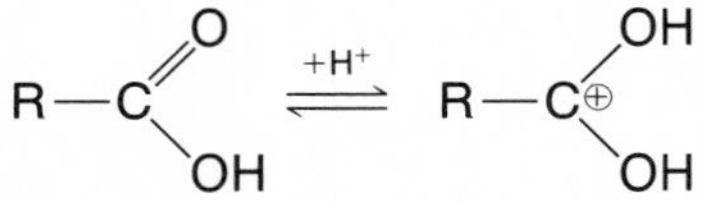

A carbonium ion

Step 1, the **protonation** of the carbonyl carbon of the acid, produces a carbonium ion, which reacts with the alcohol:

Step 2:
$$R-\overset{OH}{\underset{OH}{C^{\oplus}}} + \overset{O-R'}{\underset{}{\overset{|}{H}}} \rightleftharpoons R-\overset{OH}{\underset{OH\ H}{\overset{|}{C}}}-\overset{\oplus}{O}-R'$$

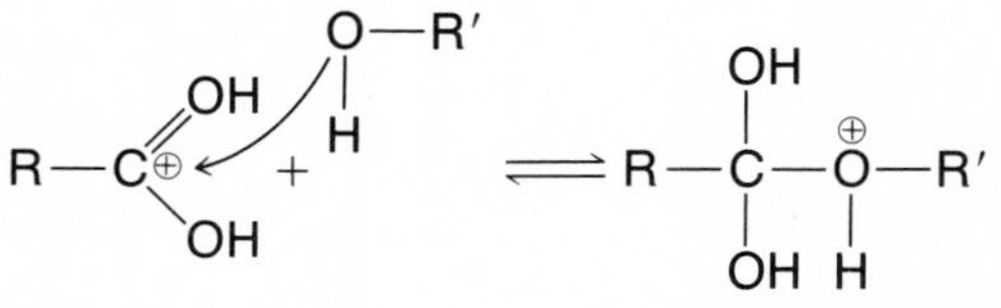

An oxonium ion

The positive ion formed in step 2, called an **oxonium ion,** in step 3 transfers the hydrogen from its oxonium oxygen to what was previously the carbonyl oxygen to form a new oxonium ion:

Step 3:

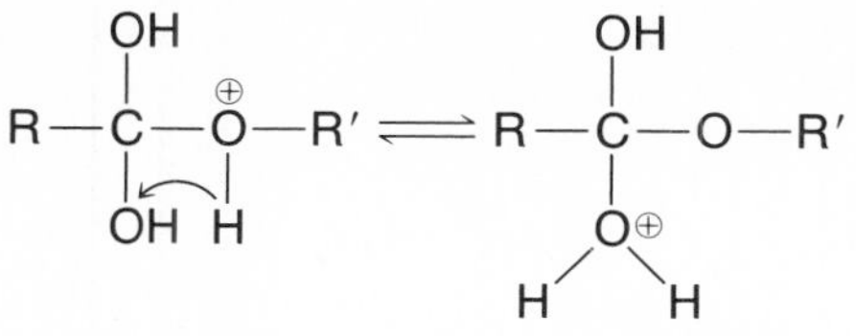

$$\begin{array}{c} OH \\ | \\ R-C-\overset{\oplus}{O}-R' \\ | \\ OH \ \ H \end{array} \rightleftharpoons \begin{array}{c} OH \\ | \\ R-C-O-R' \\ | \\ \overset{\oplus}{O} \\ H \ \ H \end{array}$$

An oxonium ion

This new oxonium ion then loses a molecule of water:

Step 4:

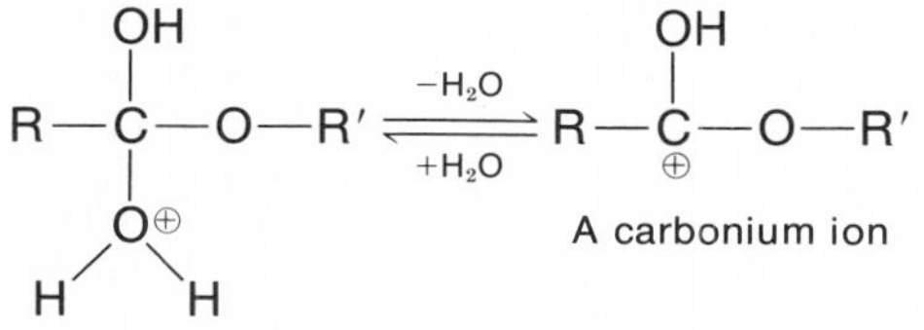

$$\begin{array}{c} OH \\ | \\ R-C-O-R' \\ | \\ \overset{\oplus}{O} \\ H \ \ H \end{array} \underset{+H_2O}{\overset{-H_2O}{\rightleftharpoons}} \begin{array}{c} OH \\ | \\ R-\overset{\oplus}{C}-O-R' \end{array}$$

A carbonium ion

The loss of a molecule of water produces a carbonium ion, which in step 5 pushes out a hydrogen ion:

Step 5:

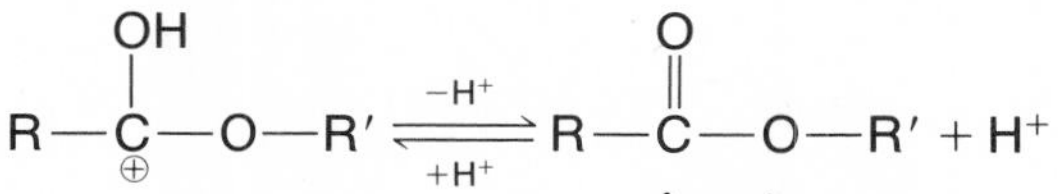

$$\begin{array}{c} OH \\ | \\ R-\overset{}{\underset{\oplus}{C}}-O-R' \end{array} \underset{+H^+}{\overset{-H^+}{\rightleftharpoons}} \begin{array}{c} O \\ \| \\ R-C-O-R' \end{array} + H^+$$

An ester

With the loss of H^+ the carbonyl group is regenerated, completing the formation of the ester as the final product. The hydrogen ion released is now free to participate in the next esterification. Note that all the reaction steps are *reversible*. The reverse of esterification is the addition of water to the ester, the hydrogen ion again acting as the catalyst. This reverse reaction, called **hydrolysis,** results in the formation of the parent acid and the parent alcohol.

Esters can also be prepared by the reaction of an acid anhydride with an alcohol. A case in point is the esterification used in the preparation of aspirin:

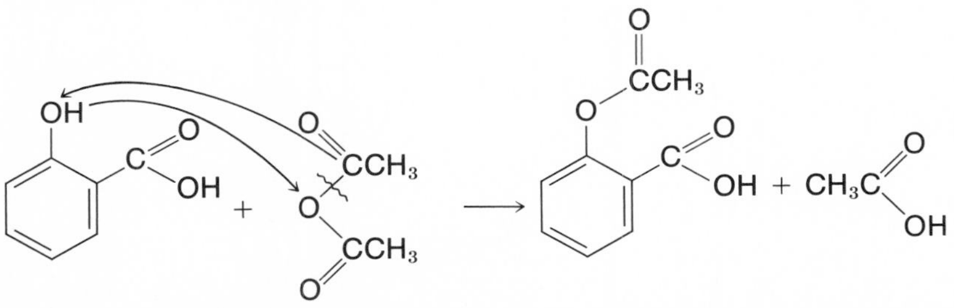

Salicylic acid   Acetic anhydride   Aspirin (acetylsalicylic acid)   Acetic acid

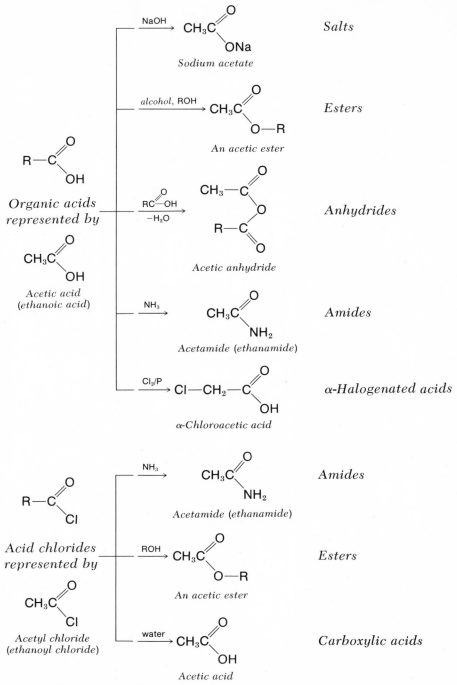

FIGURE 12.8 Reactions of organic acids.

PREPARATION OF AMIDES

An amide is structurally the product of replacing the hydroxyl group of a carboxylic acid with an amino or substituted amino group. The amides are important intermediates in manufacturing nylon, plastics, and pharmaceuticals, and the diamide **urea** is a valuable synthetic fertilizer. In biological systems, the proteins are polymeric amides formed by joining many amino acids by amide linkages, as will be seen in the discussion of proteins. Industrial preparations usually employ the ammonium salt of an acid to make the amide by **pyrolysis.**

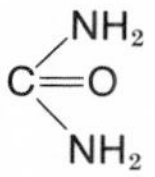

$$\begin{array}{c} NH_2 \\ | \\ C{=}O \\ | \\ NH_2 \end{array}$$

Urea

In the laboratory, amides are made from acyl halides or acid anhydrides by treating them with ammonia, a process called **ammonolysis.**

REPLACEMENT OF α HYDROGEN

The influence of the carbonyl group makes the hydrogen attached to the carbon adjacent to the carboxyl group easy to replace. Halogens such as Cl_2 or Br_2, in the presence of a little phosphorus as a catalyst, produce α-halogenated acids. The halogenated acids so produced are bifunctional compounds. They provide a convenient route to a number of other bifunctional acids.

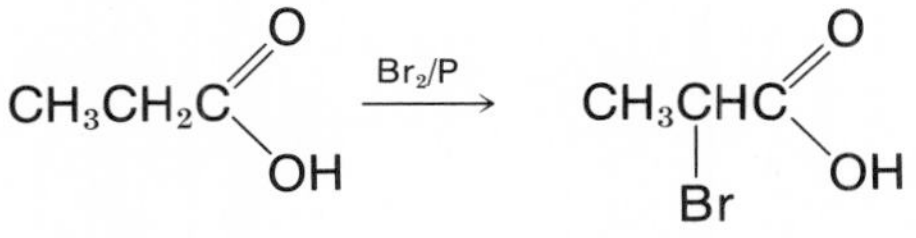

| Propanoic acid | | 2-Bromopropanoic acid, an α-halogenated acid |

The α-bromo acid will undergo most of the reactions discussed for alkyl halides, such as displacement and elimination, which can lead to many differently substituted carboxylic acids:

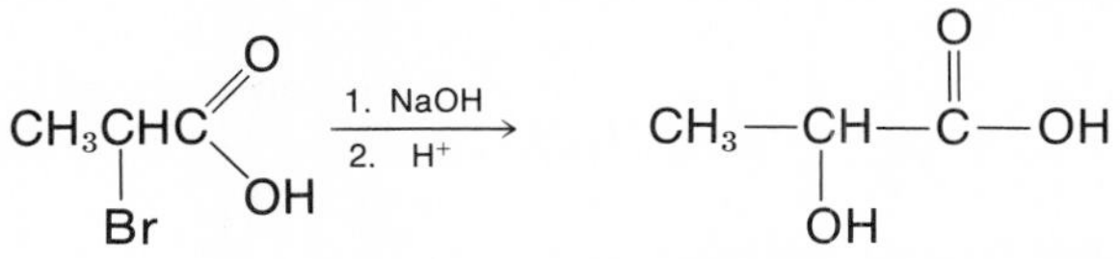

2-Hydroxypropanoic acid (lactic acid), a hydroxy acid

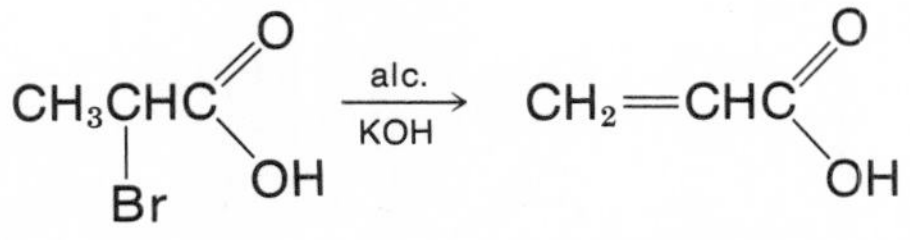

Propenoic acid, an α,β-unsaturated acid

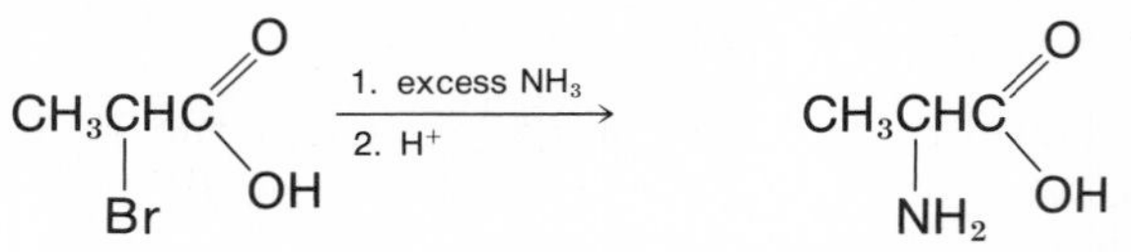

2-Aminopropanoic acid (alanine),
an amino acid

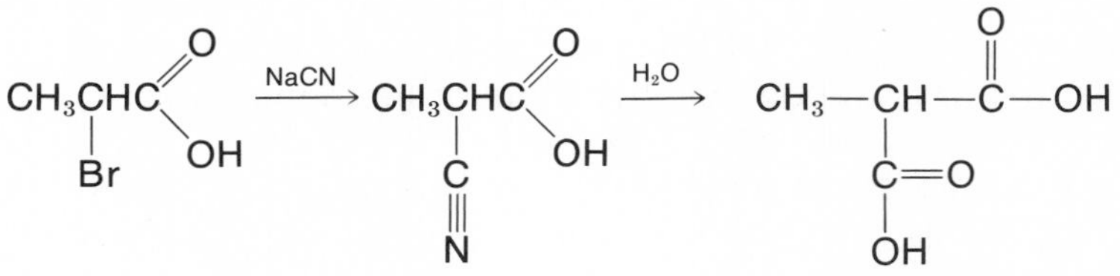

Methylmalonic acid
(2-methylpropanedioic acid)

In the dicarboxylic acid shown, malonic acid, the α hydrogens are especially reactive and the diethyl ester of the acid is successfully used in the synthesis of other organic acids and other intermediates. The anhydrides undergo reactions similar to those of the acid chlorides. The reactions of the halogenated acids were given on preceding pages.

REVIEW QUESTIONS

1 Where do reactions generally take place on an organic compound?
2 What is a functional group? Give examples.
3 Write the equation for the combustion of the alkane C_7H_{16}.
4 What are the possible products of the partial oxidation of alkenes?
5 What are petrochemicals?
6 Explain what is meant by a substitution reaction.
7 Describe the process of cracking. What are the products?
8 Write all the steps of the free-radical reaction of ethane with Cl_2. Identify all components in each step.
9 What is meant by homolytic cleavage and heterolytic cleavage of covalent bonds?
10 Discuss the bond strength between carbon and hydrogen of the different types of carbons. Relate this to predictions of possible products in free-radical reactions.
11 Make a list of the reactions of alkenes and give examples of each. Name reactants and products for each reaction shown.
12 How would you prepare 1,2-dichloropropane from propane with optimum yield?
13 Discuss the mechanism of halogenation of alkenes.

14 What are the products of the following reactions?

$$\underset{\overset{\displaystyle CH_3}{\displaystyle |}}{}$$

a $CH_3C\!\!=\!\!CHCH_3 \;\; + HCl \longrightarrow$

b $CH_3CH_2CH\!\!=\!\!CH_2 + HCl \longrightarrow$

15 Write the reaction mechanism for hydrohalogenation.

16 Explain Markovnikov's rule.

17 What is a carbonium ion?

18 Name the products of the following reactions:

a $CH_3CH_2CH_2CH\!\!=\!\!CH_2 \xrightarrow{H_2O/H^+}$

b $CH_3\overset{\overset{\displaystyle CH_3}{\displaystyle |}}{C}\!\!=\!\!CHCH_2CH_3 \xrightarrow{H_2O/H^+}$

19 Go to your library and look up in a chemical dictionary the different kinds of polymers. Make a list of the different polymers, their starting materials, and their uses.

20 How would you prepare 1,2-propanediol from propane?

21 What is the product of the reaction of 1-pentene with Cl_2 at 600°C?

22 Show the reaction mechanism of aromatic substitution. Use the chlorination of benzene as the example.

23 What is meant by ortho-para-directing and meta-directing groups? Give examples.

24 Name the products of the following reactions:

a

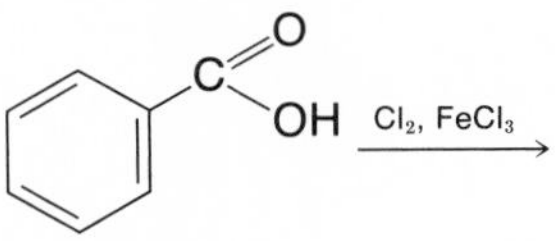

b

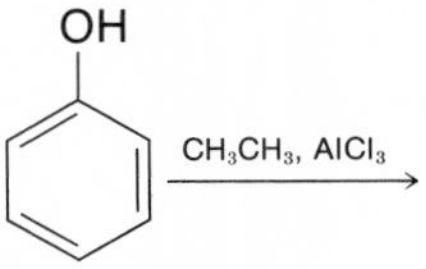

c

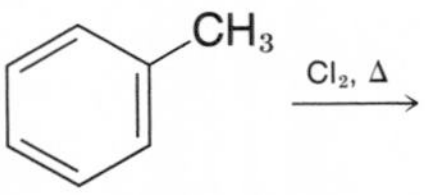

d

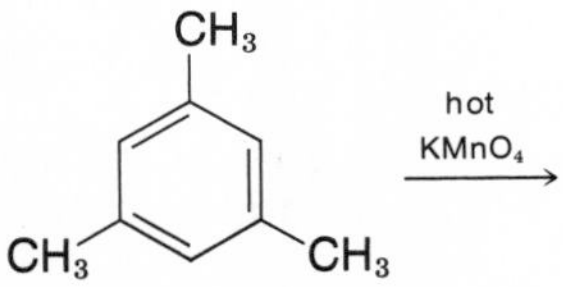

25 A number of halogenated hydrocarbons are mentioned in this chapter. Consult your library and the chemical dictionary and find more of the halogenated hydrocarbons, their uses, and their hazards.

26 Explain nucleophilic substitution.

27 Compare the S_N1 mechanism to the S_N2 mechanism.

28 Compare the E_1 mechanism with E_2 mechanism.

29 What is the transition state?

30 Name the products of the following reactions:

a
$$\underset{\underset{\displaystyle Br}{\displaystyle |}}{\overset{\overset{\displaystyle CH_3}{\displaystyle |}}{CH_3CH_2CCH_2CH_3}} \xrightarrow{\text{alc. KOH}}$$

b $CH_3CH_3CH_2{-}I \xrightarrow{\text{Mg, dry ether}}$

c Product from **b** $\xrightarrow[\text{2. } H_2O]{CO_2}$

d

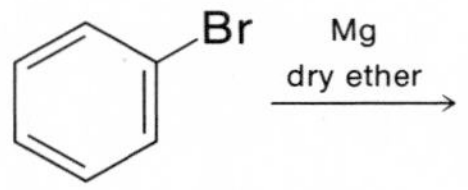

$\xrightarrow[\text{dry ether}]{\text{Mg}}$

e $CH_3CH_2Br + CH_3CH_2CH_2ONa \longrightarrow$

31 What is a tertiary alcohol?

32 What is a phenol?

33 Name the products of the following reactions:

a $CH_3CH_2CH_2OH \xrightarrow[160°C]{H_2SO_4}$

b $CH_3CH_2CH_2OH + Na \longrightarrow$

c
$$\underset{\underset{\displaystyle CH_3}{\displaystyle |}}{\overset{\overset{\displaystyle CH_3}{\displaystyle |}}{CH_3{-}C{-}OH}} + CH_3C\overset{\displaystyle O}{\underset{\displaystyle OH}{\Big\langle}} \xrightarrow[\text{heat}]{H^+}$$

d $CH_3CH_2CH_2OH + 2H_2SO_4 \longrightarrow$

e
$$\underset{\underset{\displaystyle CH_3}{\displaystyle |}}{\overset{\overset{\displaystyle CH_3}{\displaystyle |}}{CH_3{-}C{-}OH}} + O_2 \longrightarrow$$

34 Consult your chemical dictionary in your library and make a list of eight to ten alcohols, their uses, and some products in which they are found.

35 Start with 2-propanol; use any required reagents to synthesize: **a** 2-bromopropane, **b** 1,2-dibromopropane, **c** propane, **d** isopropylmagnesium bromide.

36 What reaction(s) can be used to make carbon chains longer?

37 Consult your chemical dictionary and make a list of 12 to 15 aldehydes and ketones, with the odor and uses for each.
38 What is the carbonyl group, and in what compounds is it contained?
39 What are the uses of formaldehyde?
40 What are the α hydrogens, and why are they significant in carbonyl
compounds?
41 Describe the mechanism of nucleophilic addition.
42 What is the oxidation product of **a** an aldehyde, **b** a ketone?
43 Name the products of the following reactions:

a $CH_3CH_2CH_2C\overset{O}{\underset{CH_3}{<}}$ + HCN $\longrightarrow$

b $CH_3\overset{O}{\overset{\|}{C}}CH_3$ + CH_3MgBr $\longrightarrow$

c $CH_3C\overset{O}{\underset{CH_3}{<}}$ + H_2NOH $\longrightarrow$

d $CH_3CH_2\overset{O}{\overset{\|}{C}}CH_3$ $\xrightarrow[\Delta \text{ pressure}]{H_2}$

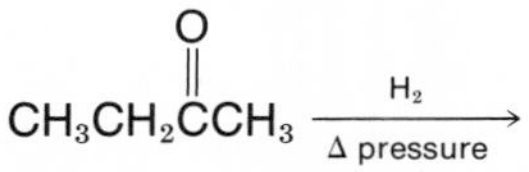

44 What is the aldol condensation, and what is its mechanism?
45 From your list of labels from food packages, containers, etc., make a
list of organic acids that are added to foods. Identify the reason for
the addition of the specific acid.
46 Name the products of the following reactions:

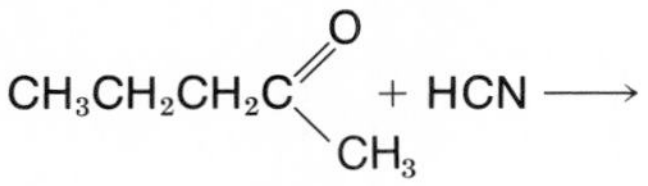

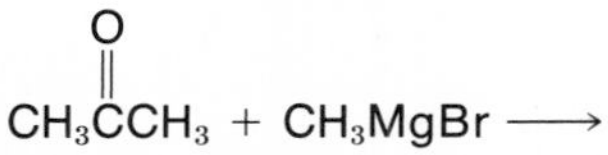

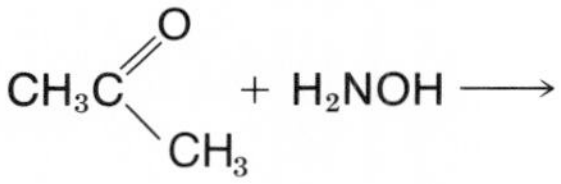

a $CH_3CH_2CH_2C\overset{O}{\underset{OH}{<}}$ + $NaHCO_3$ $\longrightarrow$

b $HC\overset{O}{\underset{OH}{<}}$ + (benzyl alcohol CH_2OH) $\longrightarrow$

c $CH_3CH_2CH_2CH_2C\overset{O}{\underset{OH}{<}}$ + PCl_3 $\longrightarrow$

d $CH_3CH_2CH_2C\overset{O}{\underset{OH}{<}}$ $\xrightarrow{Cl_2,\ p}$

e $CH_3C\overset{O}{\underset{Cl}{<}}$ + $CH_3CH_2CH_2NH_2$ $\longrightarrow$

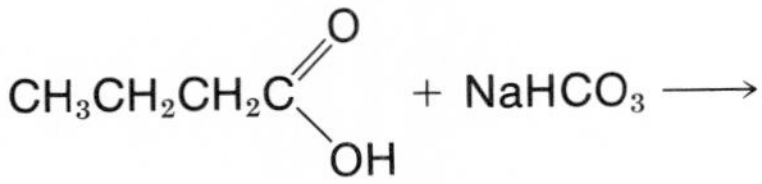

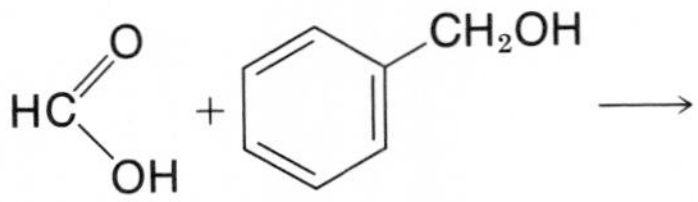

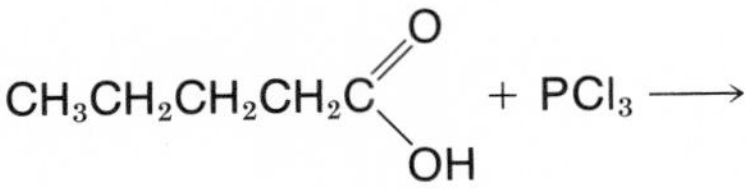

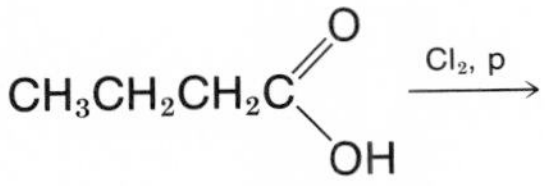

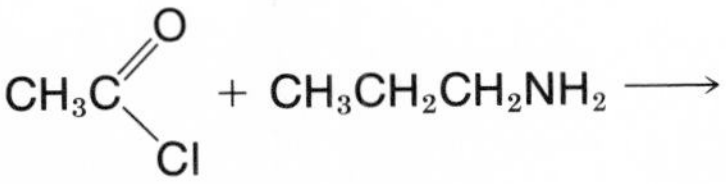

47 Write the reaction mechanism of the ester formation of an acid with a primary alcohol.

48 What is an oxonium ion?

Esters are also compounds with pleasant odors. Find some esters in the chemical dictionary. Write out the structure, the type of odor, and how the ester is used.

50 Prepare the following compounds from butanoic acid and any other reagent needed: **a** 2,3-dibromobutanoic acid, **b** butanoyl chloride, **c** butanoic anhydride, **d** 2-aminobutanoic acid, **e** 2-ethylpropanedioic acid.

THE CARBOHYDRATES
The Sweet and Sticky

13

Carbohydrates are the primary fuel for providing the energy the body needs, and they also do other jobs, as we shall see. These compounds are composed of carbon, hydrogen, and oxygen in a ratio of 1:2:1, that is, CH_2O. The name is derived from this relationship and means hydrated carbon, $C(H_2O)$. Carbohydrates are produced by what ecologists call **producer organisms,** those which provide the primary food stuff in the ecological system. These are **chlorophyll-containing** organisms and include the plants and certain microorganisms such as algae. These organisms produce carbohydrates by **photosynthesis,** the most important chemical process on this planet.

PHOTOSYNTHESIS

In this process, carbon, hydrogen, and oxygen are fixed by the use of solar energy into carbohydrates and other compounds, which can give up their stored energy to other organisms. The overall reaction of photosynthesis can be written

$$nCO_2 + nH_2O \xrightarrow[\text{chlorophyll}]{\text{sunlight, energy}} (CH_2O)_n + nO_2$$

$$6CO_2 + 6H_2O \longrightarrow C_6H_{12}O_6 + 6O_2$$

The * on the oxygen atom indicates that it is the heavy isotope of oxygen, ^{18}O.

For a long time, it was believed that the O_2 evolving in this process came from CO_2, but by using ^{18}O-labeled water it was found that the oxygen comes from the water molecule:

$$CO_2 + 2H_2O* \xrightarrow{h\nu} CH_2O + H_2O + O_2^*$$

The energy necessary for the reaction is obtained from light, the radiant energy from the sun. This light energy is trapped by the green chlorophyll in conjunction with other pigments such as carotenes and then "packaged" in the form of carbohydrates. In consumer organisms, such as animals and people, these carbohydrates are metabolized, that is, oxidized to carbon dioxide, water, and energy. Thus, the energy originally fixed by producer organisms has been used by consumer organisms in order to maintain themselves:

$$C_6H_{12}O_6 \quad + \quad 6O_2 \quad \longrightarrow \quad 6CO_2 \quad + 6H_2O + \text{energy}$$

Glucose, a carbohydrate	Oxygen	Carbon dioxide	Water

Two compounds are the parent structures for the simple carbohydrates, the trioses **glyceraldehyde** and **dihydroxyacetone.** The simple carbohydrates, or **monosaccharides,** contain from three to seven carbons. In the photosynthetic process the parent carbohydrates are connected to other groups such as phosphates, but for simplicity we shall show these compounds as unattached.

The first parent compound, glyceraldehyde, has the molecular formula $C_3H_6O_3$ or $(CH_2O)_3$, and the structural formula shown. In biological systems this compound appears as glyceraldehyde 3-phosphate, and we shall encounter it as such in glucose metabolism. Let us take a closer look at glyceraldehyde. As the name indicates, we are looking at an aldehyde with the aldehyde functional group —CHO at carbon atom 1. At carbon 2 we see a secondary hydroxyl group and at carbon 3 a primary hydroxyl group. The compound contains three carbons and is a monosaccharide. For these reasons we call it a **triose,** indicating by the ending *-ose* that it is a sugar and by the prefix *tri-* that the compound contains three carbons. The asterisk on carbon 2 indicates another important feature. This carbon has four different groups attached to it:

Glyceraldehyde

Glyceraldehyde 3-phosphate

Glyceraldehyde

1 —H
2 —OH
3 —CH₂OH
4 —COOH

OPTICAL ACTIVITY

A carbon atom with four different attachments is called an **asymmetric carbon atom.** Compounds containing asymmetric carbon atoms are dissymmetric molecules and exhibit optical activity. For each asymmetric carbon atom we can have a + and a − isomer. The + sign refers to a right, or clockwise, rotation of plane-polarized light, and the − sign to a left, or counterclockwise, rotation.

What is meant by optical activity? Light is a wave which vibrates with a given wavelength and a given amplitude. Light may also be described as a train of photons, extremely minute particles, which travel in a stream and vibrate with a given amplitude and wavelength. If we could slow down a photon and watch it as it travels directly toward us, we would see it vibrate in all planes, or 360° around its axis of motion (Figure 13.1). When such freely vibrating light passes through a tourmaline crystal, only light vibrating in a *single plane* emerges. This is called **plane-polarized light** (Figure 13.2). When a solution containing an optically active compound is placed in the path of plane-polarized light, the plane of polarization is turned either to the right (+) or to the left (−) and the compound is designated as **dextrorotatory** (right-turning) or **levorotatory** (left-turning). The optical effect can be determined and measured in a **polarimeter** (Figure 13.3).

Determining optical rotation is an important tool for identifying compounds and their concentration in solution. For each asym-

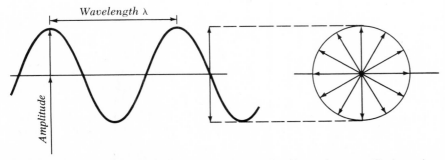

FIGURE 13.1 A photon vibrates in all planes and with a given amplitude and wavelength.

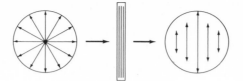

FIGURE 13.2 Nonpolarized light vibrating in all planes passes through a tourmaline crystal and emerges as polarized light vibrating in a single plane.

metric carbon in a compound we can have two optical isomers. The maximum number of optical isomers for a given compound, found from the number of asymmetric carbon atoms, is 2^n, where n is the number of asymmetric carbon atoms. For example, for a compound with three asymmetric carbons, $n = 3$ and the formula becomes

$$2^3 = 2 \times 2 \times 2 = 8 \text{ number of possible isomers}$$

The concept of optical activity and optical isomers is very important in relation to biological activity. Very often one isomer is biologically active while the other is not active or has very little activity.

Coming back to our triose glyceraldehyde, let us write the structures for the two isomers, where again the asterisk designates an optically active center, an asymmetric carbon atom:

L form, L structure   D form, D structure

Glyceraldehyde

The conventional way of writing the structure of glyceraldehyde is shown. By placing the —OH group on the asymmetric carbon to the *right* we have the structure of D-glyceraldehyde, and by placing the —OH group to the *left* we have the structure of L-glyceraldehyde. The designations D and L refer only to the position of *that* —OH group. To indicate the direction of the optical rotation we must add + or −. Since for our discussion the structural arrangements are the important considerations, we shall designate a structure as D or L, with the understanding that the optical rotation may be + or − in either case; where the optical rotation is important, it will be indicated. Otherwise, the student is referred to literature sources.

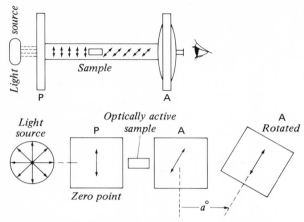

FIGURE 13.3   A polarimeter measures the direction and extent of rotation that an optically active material imposes on polarized light. P is the polarizing crystal and A the analyzer crystal.   As the plane-polarized light passes through the sample, it is rotated as shown.   The analyzer is then similarly turned $a°$ and so remains parallel to the rotated light.

BUILDING UP THE MONOSACCHARIDE STRUCTURES

ALDOSES

The two glyceraldehydes are the parent compounds of what is known as the D and L series of the **aldoses.**   The series (shown in Table 13.1) comprises the monosaccharide aldehydes with three to seven carbons.   Let us build up one of these series from the parent aldehyde.   Remember that the difference between each two successive sugars is a CH_2O unit, which can be represented as either HO—C—H or H—C—OH.   We use the D series because most natural sugars are D sugars.

H   O

C

1

H—C*—OH → D

2

H—C—OH

3

H

D-Glyceraldehyde

By inserting the group H—C—OH or HO—C—H between carbon 1 and 2, we now get two D-tetroses; as the structure shows, both are D sugars and have asymmetric carbon atoms 2 and 3:

TABLE 13.1   Aldoses

	Number of Carbon Atoms	Molecular Formula
Trioses	3	$C_3H_6O_3$
Tetroses	4	$C_4H_8O_4$
Pentoses	5	$C_5H_{10}O_5$
Hexoses	6	$C_6H_{12}O_6$
Heptoses	7	$C_7H_{14}O_7$

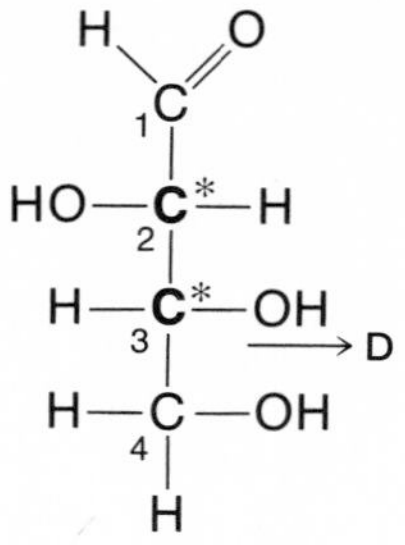

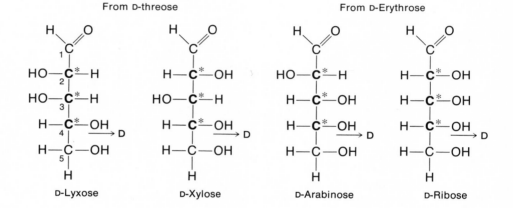

D-Threose

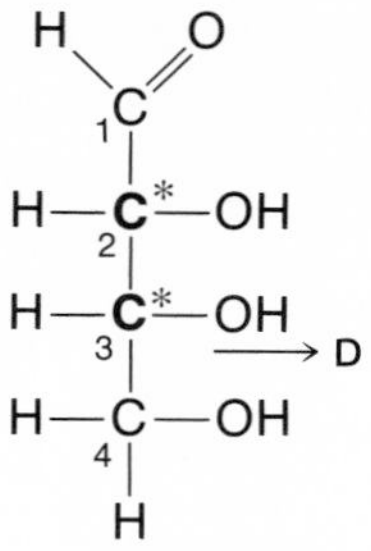

D-Erythrose

By inserting into these structures the group H—C—OH or HO—C—H between carbons 1 and 2 we get four D-pentoses:

From D-threose

From D-Erythrose

D-Lyxose

D-Xylose

D-Arabinose

D-Ribose

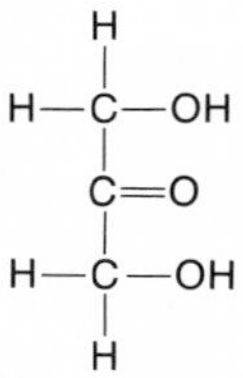

Note that these four pentoses are also aldoses, with three asymmetric carbons each. The asymmetric centers are at carbon atoms 2, 3, and 4. By insertion of a H—C—OH unit in the same fashion as before between carbons 1 and 2, we get eight aldohexoses with four asymmetric carbon atoms each (Figure 13.4). These are the eight isomers belonging to the D series, but there are also eight isomers for the L series, or a total of 16 isomers, as predicted for four asymmetric carbon atoms: $2^4 = 2 \times 2 \times 2 \times 2 = 16$.

Dihydroxyacetone

KETOSES

As we saw in Chapter 11, the aldehydes are isomeric with ketones. The molecular formula for glyceraldehyde, $C_3H_6O_3$, also applies to its ketone isomer, dihydroxyacetone. Dihydroxyacetone is a *keto* triose and is the parent compound for all keto sugars, the **ketoses.** This compound does *not* contain an asymmetric carbon atom. When the keto sugars, or keto monosac-

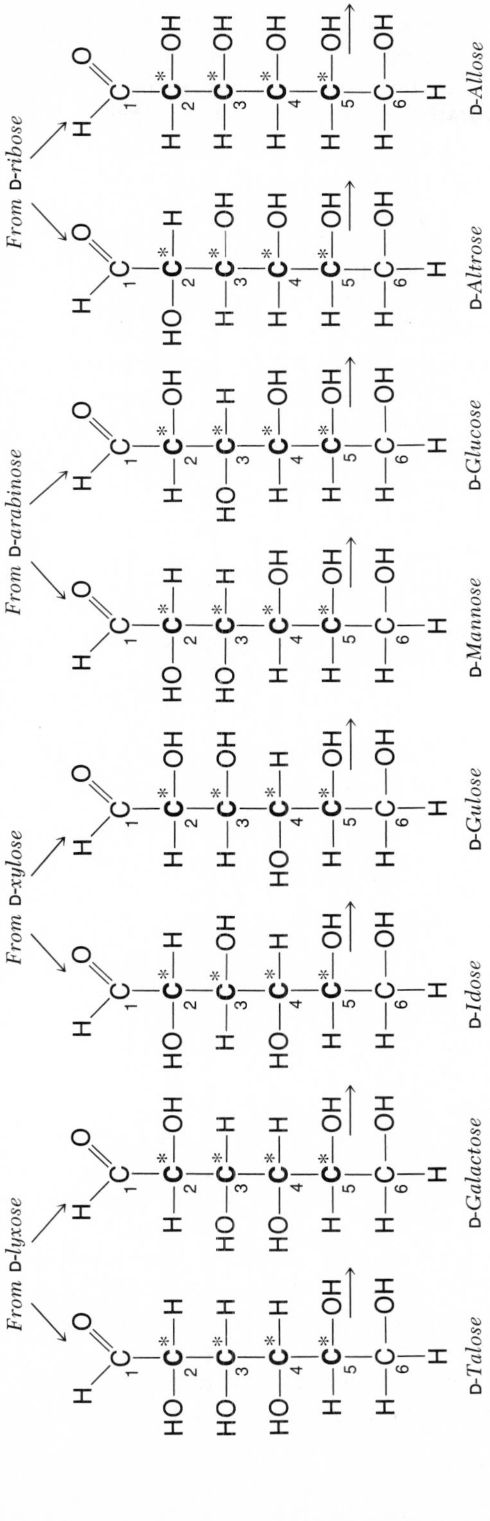

FIGURE 13.4 The eight isomers of the D series.

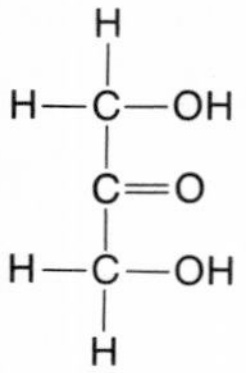

H
|
H—C—OH
|
C=O
|
H—C—OH
|
H

Dihydroxyacetone,
a triose

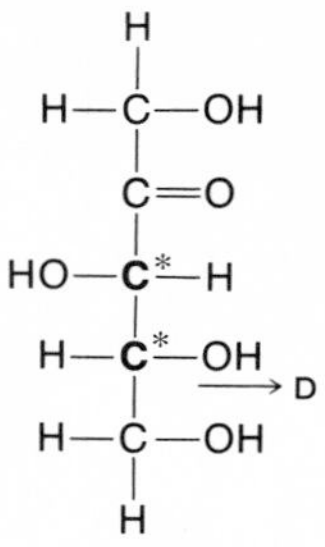

H
|
H—C—OH
|
C=O
|
HO—C*—H
|
H—C*—OH $\longrightarrow$ D
|
H—C—OH
|
H

D-*Xylulose,*
a pentose

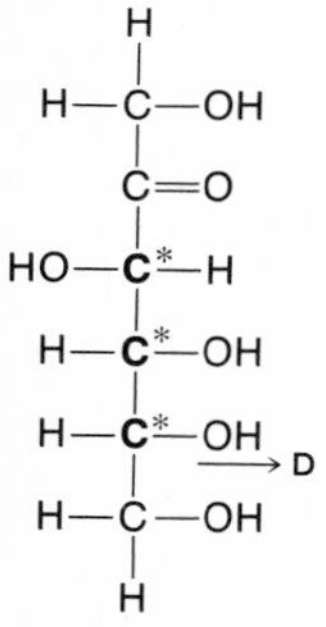

H
|
H—C—OH
|
C=O
|
HO—C*—H
|
H—C*—OH
|
H—C*—OH $\longrightarrow$ D
|
H—C—OH
|
H

D-*Fructose, a*
hexose

H
|
H—C—OH
|
C=O
|
HO—C*—H
|
H—C*—OH
|
H—C*—OH
|
H—C*—OH $\longrightarrow$ D
|
H—C—OH
|
H

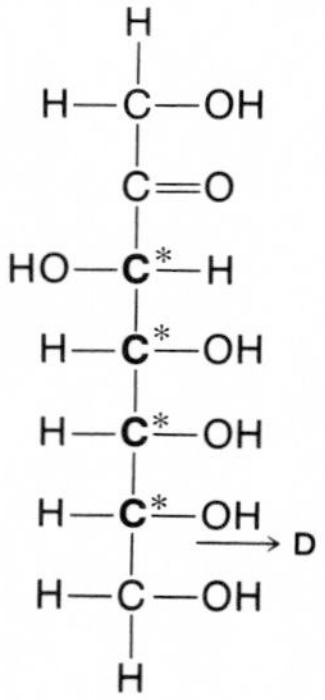

D-*Sedoheptulose,*
a heptose

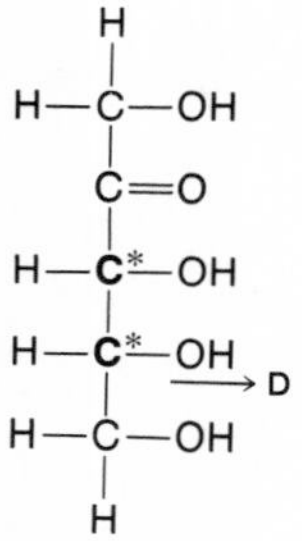

H
|
H—C—OH
|
C=O
|
H—C*—OH
|
H—C*—OH $\longrightarrow$ D
|
H—C—OH
|
H

D-*Ribulose,*
a pentose

charides, are built up, the four-carbon sugars have one asymmetric carbon atom, the five-carbon sugars have two, and the six-carbon sugars have three. This means that there are eight possible isomers for a ketohexose structure ($2^3 = 2 \times 2 \times 2 = 8$ isomers). Structural formulas of some important keto monosaccharides are shown in the margin.

FROM CHAIN TO RING AND VICE VERSA

The structures of the aldoses and the ketoses we have seen so far are open-chain forms, but in nature these compounds, especially the pentoses and the hexoses, exist as almost 100 percent cyclic structures. In the open form (the linear chain) the sugars are reactive aldehydes and ketones. In the cyclic form the aldehyde or ketone function is modified to form an oxygen bridge within the structure and yield a cyclic arrangement. The cyclic structures formed by aldehyde sugars are called **hemiacetals,** and those formed by ketones are called **hemiketals.**

You will remember that heterocyclic compounds are cyclic structures containing carbon and one or more *other* elements in the ring. Two such structures whose names are used to describe monosaccharide structures are

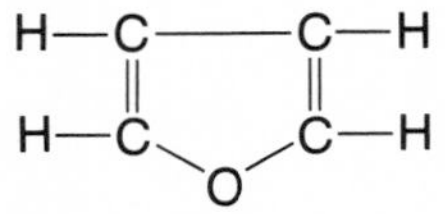

Furan, four carbons and one oxygen
in a five-membered ring

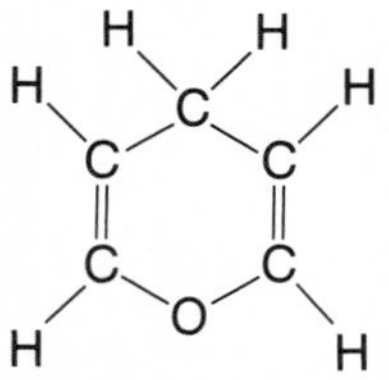

Pyran, five carbons and one oxygen
in a six-membered ring

Mnemonic devices: furan =
five-membered ring; if you
know one, you know the other.

When a sugar forms a five-membered ring, it is known as a **furanose;** when if it forms a six-membered ring, it is known as a **pyranose.** The ending -*ose* indicates a sugar, and *furan-* and *pyran-* name the ring size and type.

Let us examine the changes when a sugar cyclizes from an open aldehyde or ketone structure to a ring structure:

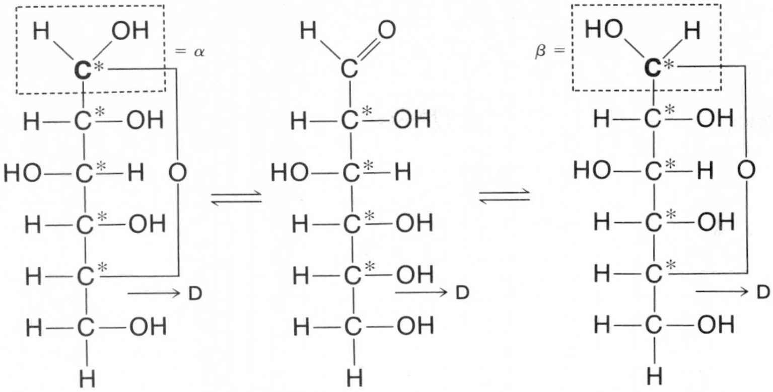

| α-D-Glucopyranose | Open aldehyde form of D-glucose | β-D-Glucopyranose |

By looking closely at the new structure, the ring structure, we can see that the aldehyde function has been lost. Instead on carbon atom 1 there is a hydroxyl group and an oxygen bridge to carbon 5. This represents an ether linkage R—O—R. The combination now present at carbon 1 we recognize (see Chapter 10) as a hemiacetal. We now have a cyclic structure made up of five carbons and one oxygen which is related to the pyran structure and therefore called a pyranose. Since the example chosen is glucose, the structure is known as a glucopyranose. The designation D signals the right-hand position of the —OH group on carbon 5. This is now represented by the oxygen bridge, which for a D structure is on the right-hand side. A new spatial difference is produced by the ring structure, the new forms being designated as α (alpha) and β (beta). The formation of the pyranose ring structure results in a new asymmetric center at carbon 1 with two possible orientations of the —OH group. If the new —OH group is on the same side as the oxygen bridge, we designate the structure as α-D-glucose. If the —OH group is on the opposite side of the oxygen bridge, it is β-D-glucose. This new —OH group is called the **glycosidic** —OH group. This —OH group is commonly involved in linkages with other sugars and other hydroxy compounds.

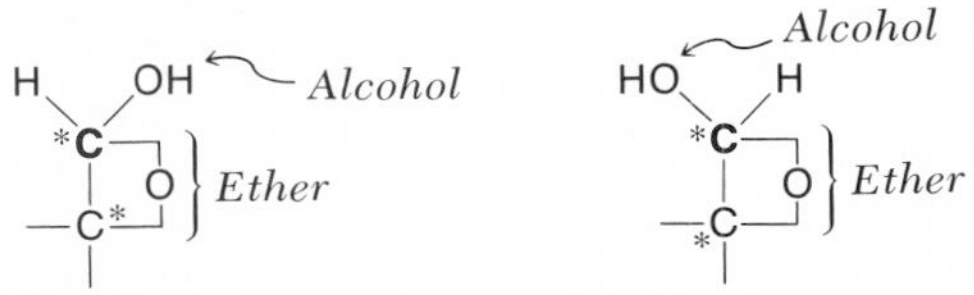

| α-D-Hemiacetal | β-D-Hemiacetal |

Using fructose as another example, we have the following structures:

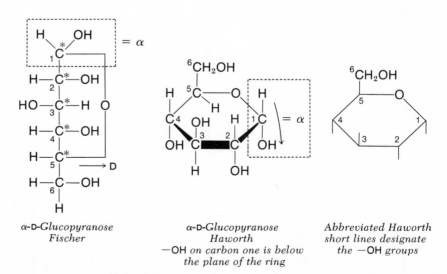

α-D-Fructofuranose   Open keto form of   β-D-Fructofuranose
                              D-fructose

Notice that carbon atom 2 of the keto form has again become an optically active center, an asymmetric carbon atom.

The ring structures shown in Figure 13.5 for glucose and fructose are known as **Fischer projection formulas.** A geometrically better presentation of the ring structures was devised by Sir Walter Haworth, an English chemist. The **Haworth projections** for D-glucose and D-fructose are also given in Figure 13.5 and compared with the Fischer structures.

α-D-*Glucopyranose*   α-D-*Glucopyranose*   *Abbreviated Haworth*
Fischer   *Haworth*   *short lines designate*
                             —OH *on carbon one is below*   *the* —OH *groups*
                             the plane of the ring

FIGURE 13.5 Haworth and Fischer structures for D-glucose and D-fructose.

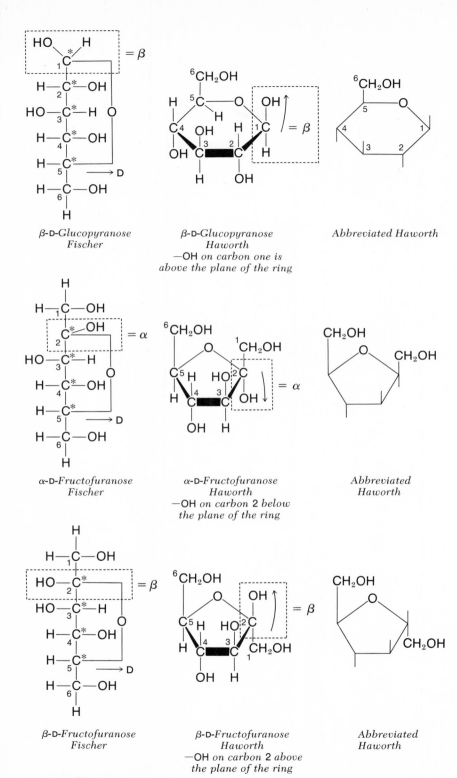

β-D-*Glucopyranose*
Fischer

β-D-*Glucopyranose*
Haworth
—OH *on carbon one is*
above the plane of the ring

Abbreviated Haworth

α-D-*Fructofuranose*
Fischer

α-D-*Fructofuranose*
Haworth
—OH *on carbon 2 below*
the plane of the ring

Abbreviated
Haworth

β-D-*Fructofuranose*
Fischer

β-D-*Fructofuranose*
Haworth
—OH *on carbon 2 above*
the plane of the ring

Abbreviated
Haworth

FIGURE 13.5 (continued)

IMPORTANT MONOSACCHARIDES

The trioses glyceraldehyde and dihydroxyacetone are found as phosphate esters in glucose metabolism. D-Erythrose is also involved as the phosphate ester in carbohydrate metabolism. That is, each of these exists in nature only as the corresponding phosphate.

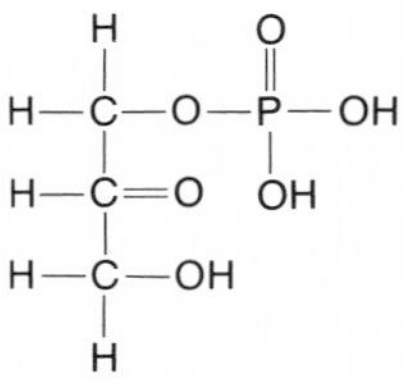

Dihydroxyacetone phosphate

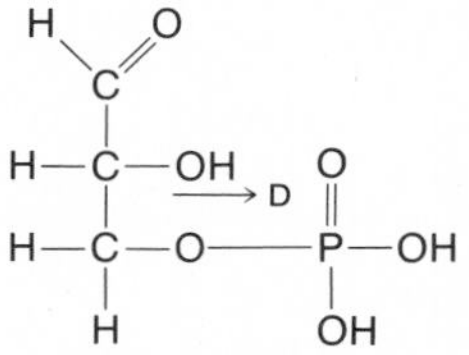

D-Glyceraldehyde 3-phosphate

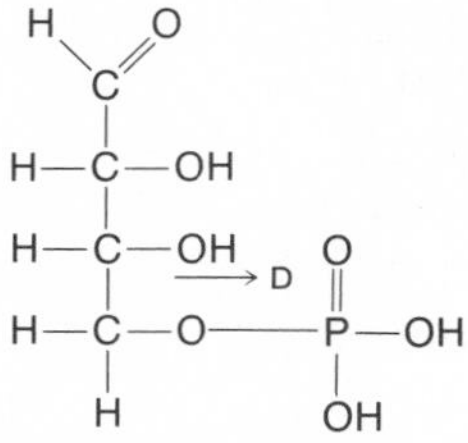

D-Erythrose 4-phosphate

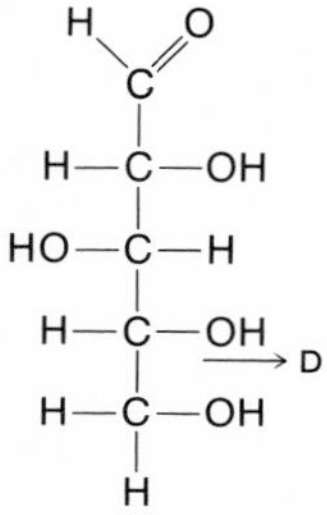

The pentoses xylose and arabinose are found in nature only as building blocks of polysaccharides. D-Xylose appears in plants as the polysaccharide xylan and can be obtained by hydrolysis from corncobs, straw, and wood. Arabinose is found in nature as an L-arabinose polymer in gums such as gum arabic; it can also be isolated from some hemicelluloses. The pentoses D-ribose and deoxyribose are part of the makeup of the genetic molecules ribonucleic acid (RNA), containing ribose, and deoxyribonucleic acid (DNA) containing 2-deoxyribose.

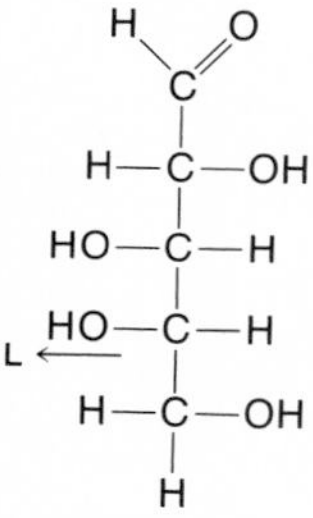

D-Glucose, also known as **dextrose** because it rotates polarized light to the right, is probably the most abundant and most

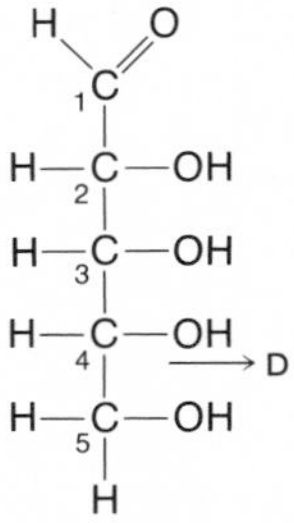

D-Xylose

L-Arabinose

D-Ribose, open form

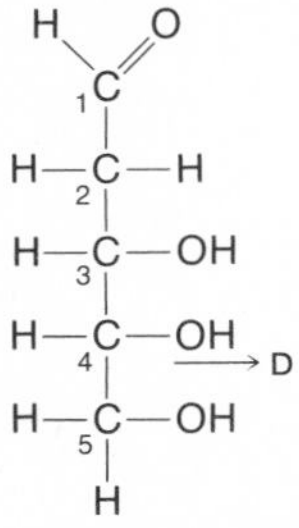

D-2-Deoxyribose, open form, oxygen missing from carbon 2

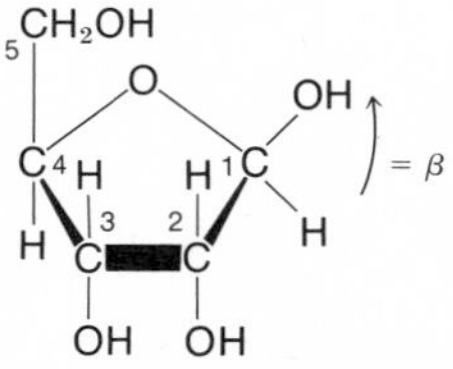

β-D-Ribofuranose, 1,4 oxygen bridge, β form, as found in RNA

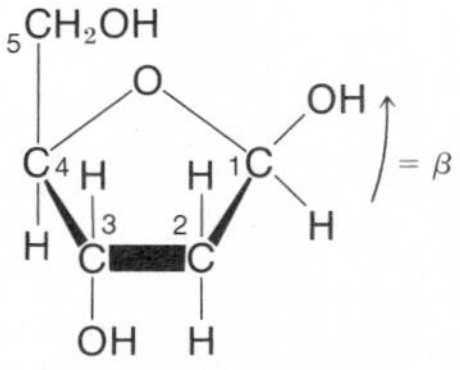

β-D-Deoxyribofuranose, 1,4 oxygen bridge, β form, as found in DNA important sugar. Since it is the sugar found in the bloodstream and carried to all parts of the body for energy production, it is called blood sugar. Glucose is also the principal sugar found in honey and many plant and fruit juices. The disaccharides and many of the polysaccharides are made up entirely or in part of the monosaccharide glucose. Important examples are starch and cellulose, where glucose is the sole building block of the molecule.

The monosaccharide D-**galactose** is a component of the milk sugar **lactose,** which it forms in conjunction with glucose. Galactose can be found associated with fats and proteins in the body. Here galactose combines, for instance, with lipids to form the galactocerebrosides, important constituents of nerve and brain tissue. Galactose can be obtained by hydrolysis from some polysaccharides such as agar-agar, a material isolated from seaweed. Galactose is also a constituent in pectins, mixed polysaccharides used in the food industry.

The aldohexose D-**mannose** can be obtained from the polysaccharide mannan by hydrolysis and also occurs in some vegetables and berries. It has been found associated with protein as a glycoprotein.

The keto sugar D-**fructose** is also called **levulose** because it rotates polarized light to the left. Fructose is the sweetest sugar, and it exists in the free form in honey and in many fruits. In combination with glucose, it forms the disaccharide sucrose, or table sugar. A pure polymer of fructose is inulin, a plant starch occurring in Jerusalem artichokes and the roots of dahlias. The phosphoricesters of fructose are important intermediates in the metabolism of glucose.

Some monosaccharide phosphates

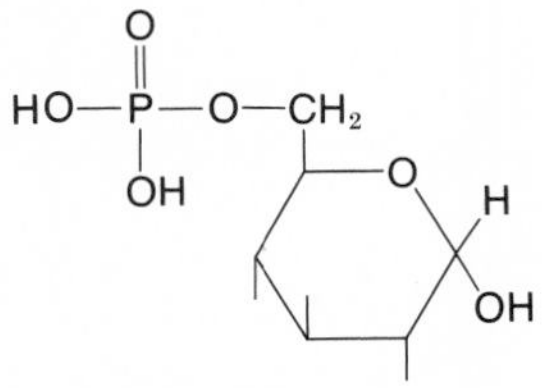

Glucose 6-*phosphate* 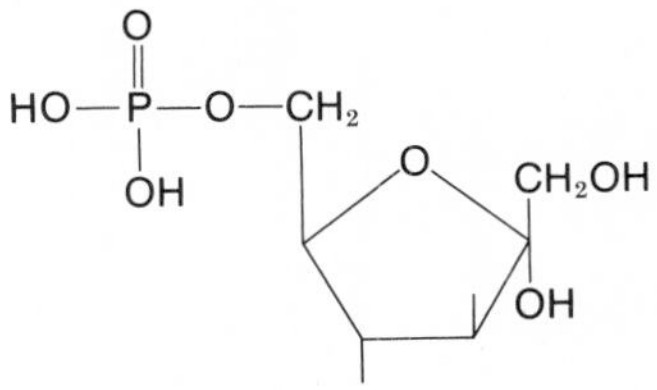   *Fructose* 6-*phosphate* 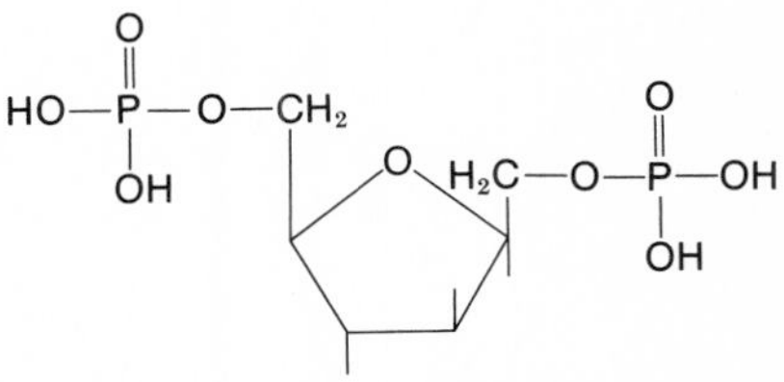   *Fructose* 1,6-*diphosphate*

DISACCHARIDES

The general formula for all the common **disaccharides** is $C_{12}H_{22}O_{11}$, which represents the combination of two units of hexose monosaccharides with the elimination of one molecule of

$$\begin{array}{r} C_6H_{12}O_6 \\ + \ C_6H_{12}O_6 \\ \hline C_{12}H_{24}O_{12} \\ - \quad H_2O \\ \hline C_{12}H_{22}O_{11} \end{array}$$

water. Two monosaccharides are **condensed** with the elimination of a molecule of water. The reverse process, by which a molecule of water is added to a molecule of a disaccharide and produces 2 moles of monosaccharides, is known as **hydrolysis.**

The important disaccharides and the monosaccharides from which they are formed are

$$\text{Glucose} + \text{fructose} \ \ = \text{sucrose} + H_2O$$
$$\text{Glucose} + \text{galactose} = \text{lactose} \ \ + H_2O$$
$$\text{Glucose} + \text{glucose} \ \ \ = \text{maltose} + H_2O$$

Sucrose and **lactose** are two disaccharides that appear in nature in the free form, whereas **maltose** is obtained by the hydrolysis of starch and does not appear free in nature. Sucrose is isolated from sugarcane and sugar beets and is the sugar used for cooking, baking, and sweetening beverages in the home. It is also a constituent of many fruits. Sucrose is formed from α-D-glucose in the pyranose form and β-D-fructose in the furanose form. Let us now take a look how the monosaccharides are connected to form the disaccharides.

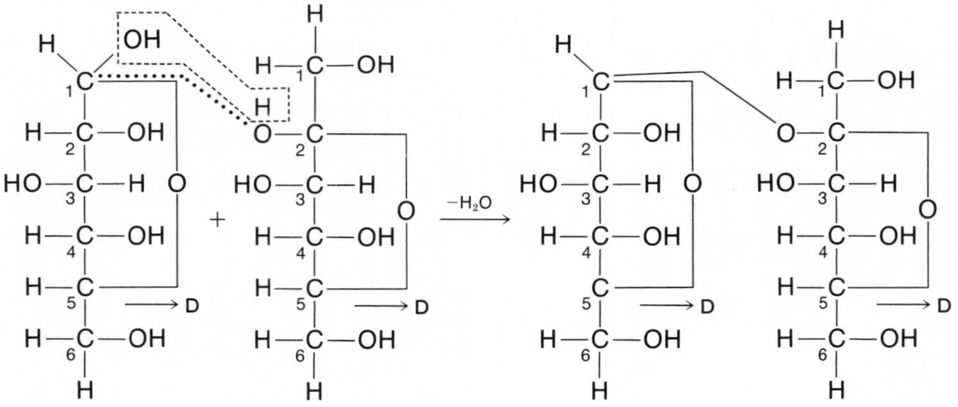

α-D-Glucose   β-D-Fructose   Sucrose

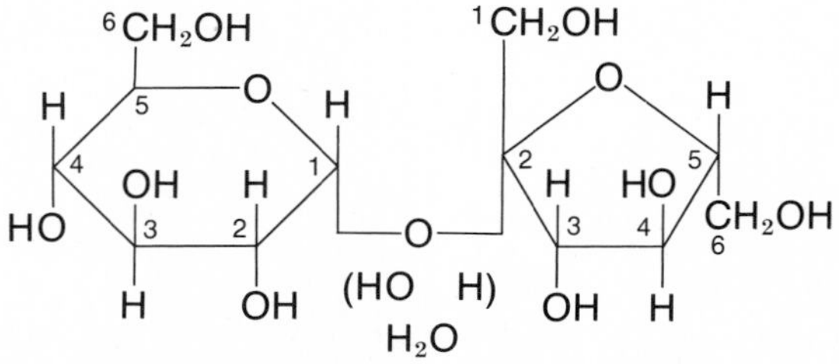

α-D-Glucopyranosyl-β-D-fructofuranoside,
sucrose; note 1,2 linkage

The junction between the two monosaccharides is by way of the oxygen bridge from carbon 1 of glucose to carbon 2 of fructose and the elimination of a molecule of H_2O. The result is that the two glycosidic —OH groups from the monosaccharides are eliminated, making this structure the only one possible for sucrose. This eliminates the possibility of rearrangement to the aldehyde or ketone form of the disaccharide and explains why sucrose is a **nonreducing sugar.** It simply does not have and cannot develop either of the groups that cause the reaction. This is also the reason why sucrose crystallizes more readily than monosaccharides and the other disaccharides. When sucrose is hydrolyzed, the result is a mixture of glucose and fructose, called **invert sugar.**

The formation of maltose may be written

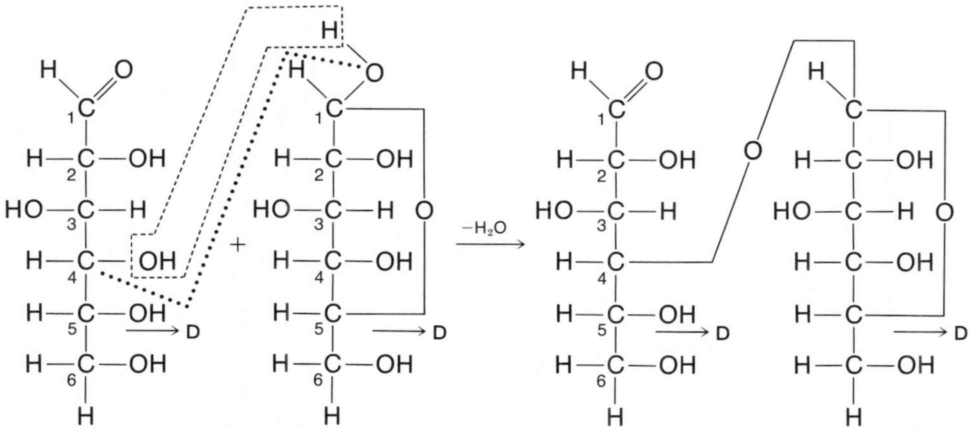

D-Glucose   α-D-Glucose   Maltose, open aldehyde form,
                                                              1,4 bridge

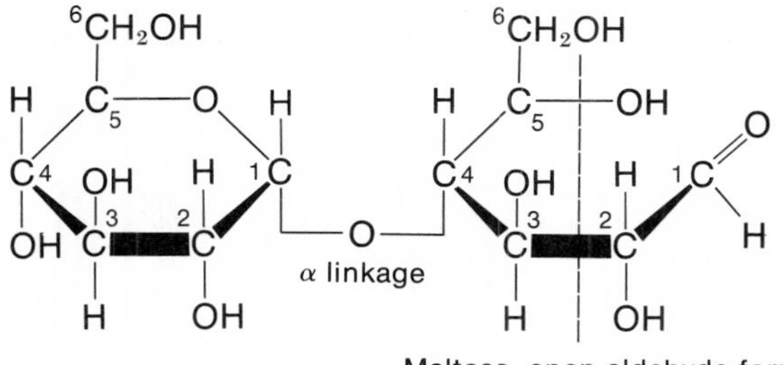

Maltose, open aldehyde form

The systematic name for maltose is 4-D-α-D-glucopyranoside; its two glucose units are joined by an oxygen bridge from carbon 1 to carbon 4 with the elimination of H_2O. This eliminates one glycosidic —OH but permits the other glucose unit to exist in either of three forms, the open aldehyde form, the α form, or the β

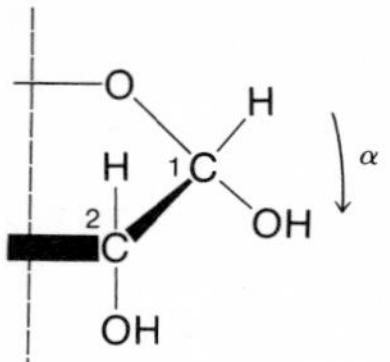

α-Maltose,
closed form

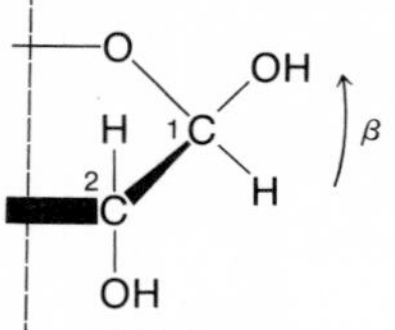

β-Maltose,
closed form

form, depending on whether the —OH group is above or below the plane of the ring structure. The term α **linkage** is a new designation for the linkage between the two glucose units. The reason for this is that the glucose in the pyranose form is in the α configuration when joining the other glucose unit. This conformation of the linkage is very important, since people can digest only polysaccharides that are made by hooking up glucose units by α linkages, as starch is. An important polysaccharide analogously made from glucose units connected by β linkages is **cellulose,** and, as you know, we cannot digest wood. The disaccharide formed by joining two glucose units by a β linkage is called **cellobiose:**

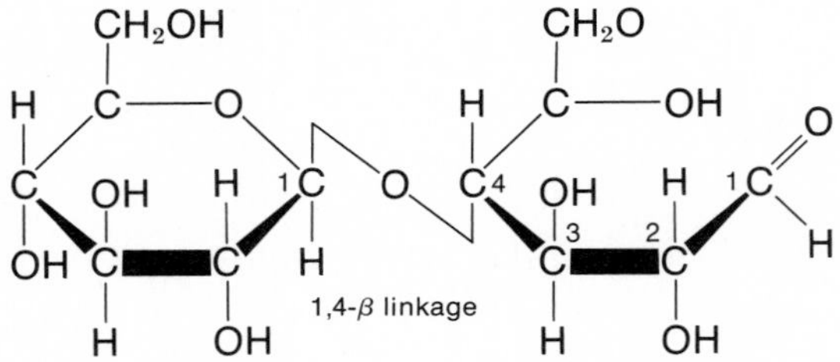

CH₂OH ... CH₂O

1,4-β linkage

β-D-Glucose   D-Glucose

4-D-Glucose-β-D-glucopyranoside, cellobiose

Thus open aldehyde forms of cellobiose can also exist as the α- or β-pyranose form.

As you can see, the 1,4 oxygen bridge again joins the two glucose units, but this time the pyranose is in the β form, giving a β linkage. If the first glucose unit is revolved 180° around a horizontal axis, the connection will be square but carbon 6 will be below the plane of the ring structure:

β linkage

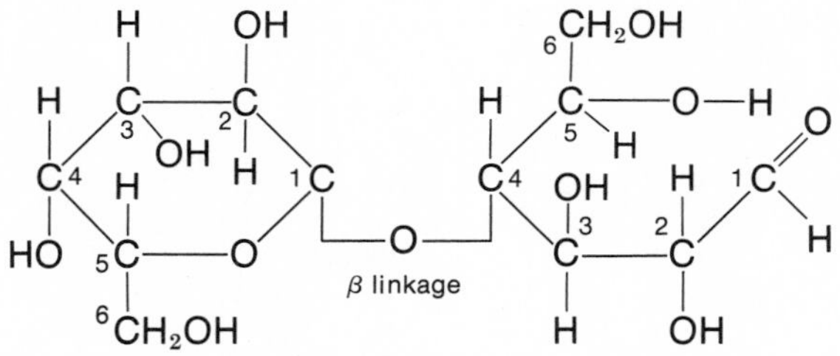

Cellobiose

Another disaccharide that is joined by a β linkage is lactose, or milk sugar. Consisting of the units glucose and galactose, this sugar is found in the milk of mammals in about 4 to 6 percent concentration:

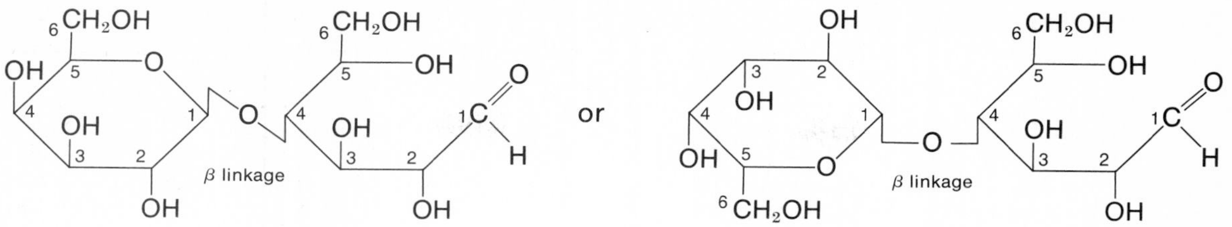

Lactose

Two equivalent projections of the open form of lactose are shown, with a β 1,4 linkage between the glucose and galactose units. Lactose can also exist in the α or β ring form. It is the only sugar not fermented by yeast, a fact used in bacteriology to identify specific organisms. Lactose in urine may indicate pregnancy.

CHEMICAL PROPERTIES OF THE MONO- AND DISACCHARIDES

The chemical properties of the mono- and disaccharides are basically the reactions of the functional groups, namely, the aldehyde and keto functions and the hydroxyl groups. Reactions of carbohydrates referred to by specific names are summarized in Table 13.2.

POLYSACCHARIDES

The **polysaccharides** are polymers of monosaccharides produced by condensation of various types of monosaccharides and the elimination of water:

$$n C_6H_{12}O_6 \longrightarrow (C_6H_{10}O_5)n \cdot H_2O + (n-1)H_2O$$

n monosaccharide molecules $-(n-1)$ water molecules $\longrightarrow$
a polysaccharide of n monosaccharide units

By the reverse process of **hydrolysis** the polysaccharides can be split into the monosaccharides from which they were made:

$$(C_6H_{10}O_5)_n \cdot H_2O + \text{water} \xrightarrow[\text{or acids}]{\text{enzymes}} n C_6H_{12}O_6$$

The polysaccharides can be classified into two major groups according to the product or products they yield upon hydrolysis. The simple polysaccharides, or homopolysaccharides, yield only one kind of monosaccharide, whereas the mixed polysaccharides, or heteropolysaccharides, yield more than one type.

TABLE 13.2 Some important carbohydrate tests and reactions

Reaction	Reagent	Product
Mohlish test, a general test for carbohydrates	Conc. H_2SO_4 and α-naphthol	H_2SO_4 dehydrates carbohydrate to a furfural, which forms a *purple-violet* product with the α-naphthol
Oxidation; for example, of glucose 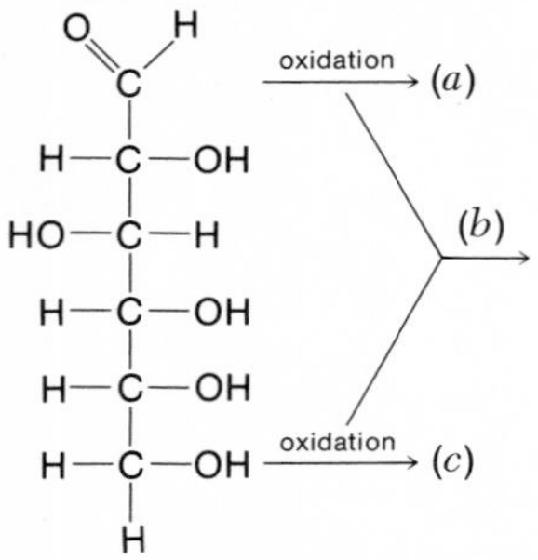	(a) $Br_2 + H_2O$ gives gluconic acid (b) O_2 + an enzyme gives glucuronic acid (c) HNO_3 gives saccharic acid	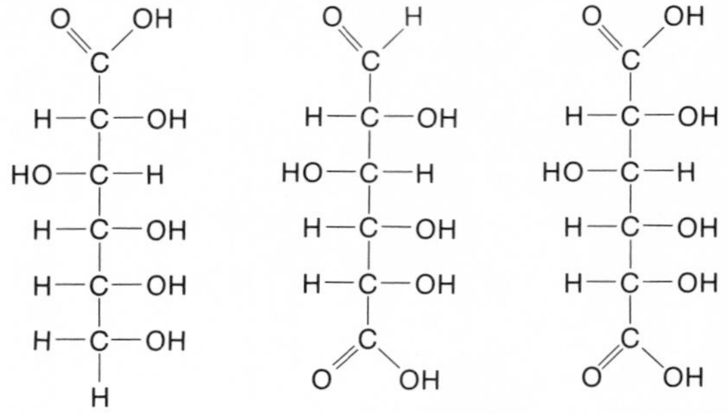 (a) Gluconic acid, an aldonic acid; (b) Glucuronic acid, a uronic acid; (c) Saccharic acid, a dicarboxylic acid
Benedict test for presence of a reducing sugar	Alkaline $CuSO_4$ solution; blue because of Cu^{2+} ion	Blue Cu^{2+} ion reduced by the reducing sugar to brick-red Cu^+ ion; reducing sugar broken down into smaller fragments
Barfoed test to distinguish between mono- and disaccharides	Acidic $CuSO_4$ solution, blue because of Cu^{2+} ion	Red precipitate of copper oxide settles out at bottom of test tube; it occurs more rapidly with monosaccharides than with disaccharides and is thus a way of distinguishing between them
Seliwanoff test to detect ketoses	Resorcinol in HCl	If a ketose is present, e.g., fructose, a wine-red color is obtained; the test distinguishes between ketoses and aldoses
Bial test to detect pentoses	Orcinol in HCl	Pentoses react with the reagent to give a mahogany-red color; hexoses do not
Condensation of mono- and disaccharides with phenylhydrazine	Phenylhydrazine	Reagent acts on the different sugars to produce distinctive crystalline solids; the differences in the microscopic appearance of the crystals is used to identify the different sugars
Reduction of monosaccharides	Electrolysis of acidified sugar solution	Each specific sugar is reduced to the corresponding polyhydroxy alcohol
Formation of methylglycosides of monosaccharides, e.g., glucose	Methanol + acid catalyst	Formation of acetals: 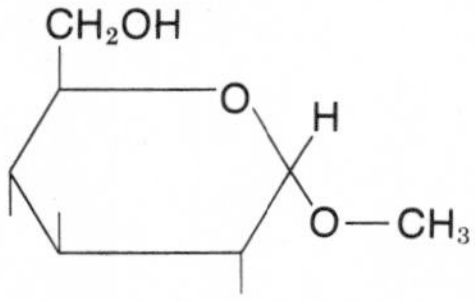 α-Methylglucoside, 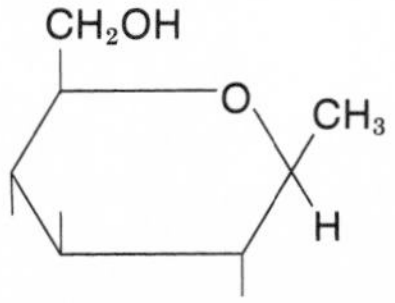 β-Methylglucoside

TABLE 13.2 (continued)

Reaction	Reagent	Product
Esters of mono-saccharides formed with phosphoric acid, e.g., glucose	Enzymes + phosphate	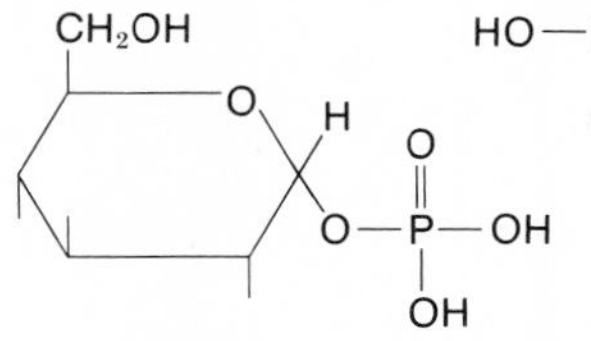 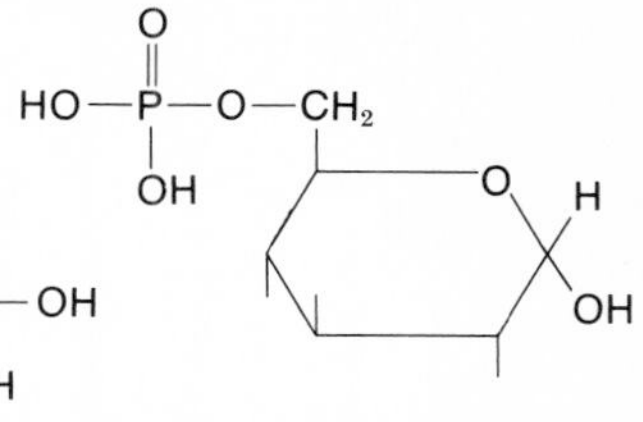

Glucose 1-phosphate   Glucose 6-phosphate

Polysaccharides are plentiful in nature and are used for many purposes by plants and animals. Many are structural components, like cellulose in plants and chitin in the exoskeleton of beetles and other insects. Others, like starch and glycogen, are used as energy-storage compounds in plants and animals.

SIMPLE OR HOMOPOLYSACCHARIDES

The most abundant simple polysaccharides in nature are those in which glucose is the building block. These are the different **plant starches, cellulose,** and the animal starch **glycogen.** These compounds are also referred to as **glucosans** since they are made from glucose alone. The general formula for the glucosans (and for the **fructosans,** made from fructose alone) is $(C_6H_{10}O_5)_n \cdot H_2O$.

Starch is the storage food obtained from cereal grains (corn and rice), from legumes (beans and peas), and from the tubers of potatoes. Two different types of molecules can be identified in starch. They appear in varying amounts in the starch granule depending on the source. The two molecules **amylose** and **amylopectin** do not differ in composition but in structure and molecular weight and size.

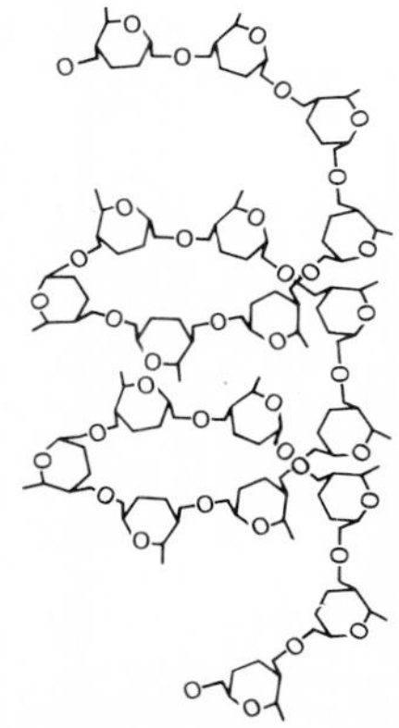

In amylose about 200 to 400 glucose units are hooked up into a straight chain like pearls on a string, where each pearl represents a glucose unit. It is believed that the molecule is arranged in a helix, which permits iodine to be entrapped when it is added to starch. The resulting blue-black color indicates starch $+ I_2$. If the blue-black amylose-iodine system is heated, the color fades and finally disappears completely because the amylose molecule swells on heating and as the entrapped iodine becomes free, the color disappears. The glycosidic linkage connecting the individual glucose units is a 1,4 linkage; the glucose units are α-D-glucose units, so that amylose is simply a sequence of glucose units connected by 1,4 α linkages:

Amylose helix

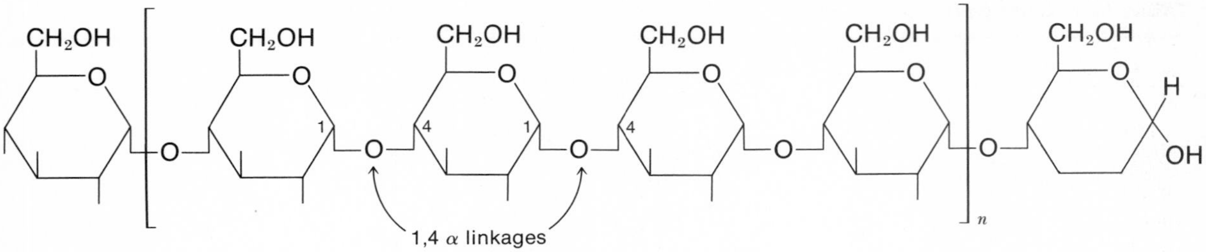

1,4 α linkages

Amylopectin is a *branched* molecule consisting of several hundred to several thousand glucose units which are connected by 1,4 α linkages to make a long chain. In addition, 1,6 α linkages occur at the branching points in the molecule, which are approximately 24 to 30 glucose units apart. The branches continue in linear sequence via the usual 1,4 α linkages:

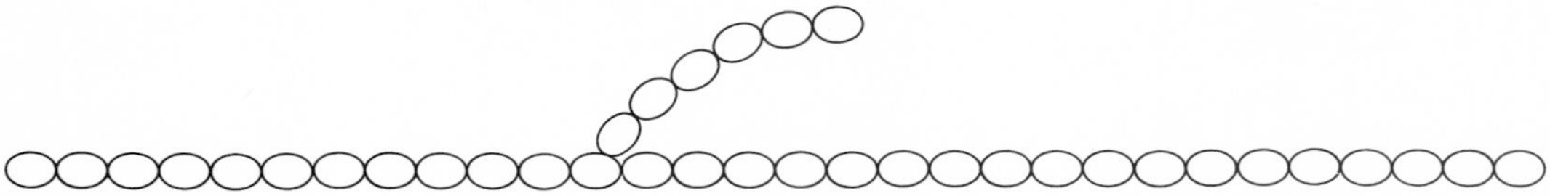

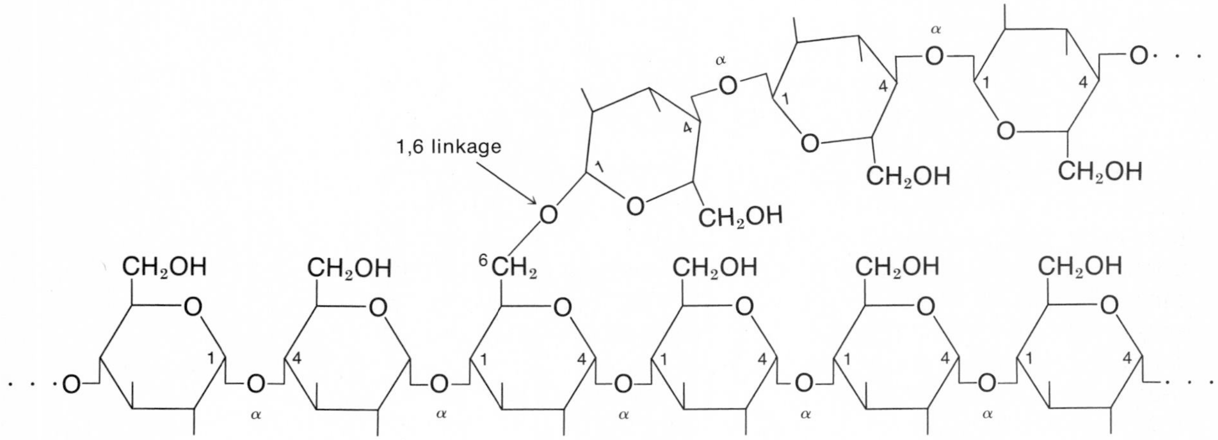

Pure amylopectin gives a purplish color with iodine rather than the blue-black of amylose.

Both starch components are insoluble in cold water but can be made into colloidal dispersions by heating the mixture. Amylose in such dispersions tends to aggregate because of its linear arrangement and the hydrogen bonds that can form between the amylose chains. In consequence, amylose will set into gels, which slowly **retrograde;** that is, the amylose precipitates. Because of their branched structure, amylopectins will not do this, and in the food industry such differences are important. The starches are used to thicken sauces, gravies, pies, ice cream, and salad dressings. The amylose and amylopectin content of starches used in different products affects their appearance, uniformity, and shelf life. For instance, the retrogradation of amylose is believed to be responsible for the staling of bread. To prevent such undesirable effects, starches are modified by chemical means. Introduction of side groups onto the polysaccharide chain, such as phosphates in the form of phosphate esters, prevents some of the hydrogen bonding and also *increases* solubility. Partial hydrolysis modifies starches to make them more suitable for various purposes.

Hydrogen bonds (. . .) between amylose chains

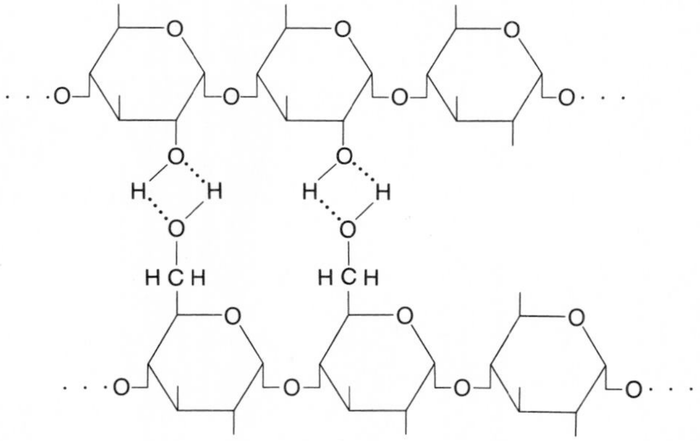

The hydrolysis of starch yields compounds that progressively contain fewer and fewer glucose units. By the use of enzymes, the reaction can be made to stop at maltose, the disaccharide compound with two glucose units. When acid catalysts are used, the end product is glucose. Starch and the intermediate products of its hydrolysis do not give a positive Benedict test but produce colors with iodine by which the hydrolysis progress can readily be followed (Table 13.3).

The **dextrins** are obtained by the incomplete hydrolysis of starch and are used as adhesives on postage stamps, envelopes, and labels because they form a sticky paste when moistened and are nonpoisonous. Bread, toasted food products, and prepared cereals contain dextrins produced by the heating process. A product of starch hydrolysis in the form of a mixture of lower-molecular-weight dextrins and primarily maltose and glucose is marketed as corn syrup.

Glycogen Glycogen, or animal starch, is the storage polysaccharide in animal tissue. Excess glucose in the bloodstream is

TABLE 13.3 Acid hydrolysis of starch

Compound†	Color with Iodine	Reaction with Benedict Reagent
Starch	Blue-black	Negative
Amylodextrin	Reddish-blue	Negative
Erythrodextrin	Reddish	Negative
Achroodextrins	No color (brown, the color of the iodine solution)	Negative
Maltose	No color (brown, the color of the iodine solution)	Positive
Glucose	No color (brown, the color of the iodine solution)	Positive

† Reading from top to bottom gives the sequence of hydrolysis.

used by the body to make glycogen to be stored. When glucose is needed again, glycogen is hydrolyzed to glucose. Glycogen is concentrated in liver and muscle tissue but occurs in small amounts in all tissue. Different animals have different concentrations of glycogen in their tissues. For instance, horse meat is much richer in glycogen than meat from other animals, and this fact is used to check adulteration of meat products with horse meat. Oysters are also rich in glycogen.

The glycogen molecule is formed by connecting α-D-glucose units by 1,4 and 1,6 α linkages, which makes it a branched molecule. The number of glucose units is less than in amylopectins, which it resembles. Glycogen is relatively highly branched, a branch appearing about every 12 to 18 glucose units. It is more soluble than the other starches, which accounts for its slight sweet taste.

Cellulose Cellulose is the structural polysaccharide found in plants. It can be compared in structural importance to proteins in animals. The amount of cellulose produced each year by plants puts it well in the lead as far as synthesis of organic compounds is concerned. Cellulose is a linear polymer of D-glucose joined by 1,4 β linkages; on the average it contains about 3000 glucose residues.

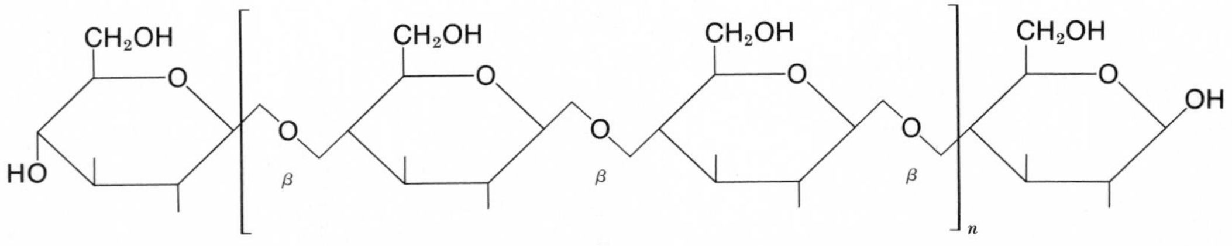

or

Structure of the cellulose molecule If we compare the structure of amylose and cellulose, we can see that their basic difference is in the linkage. A 1,4 α linkage is present in amylose and a 1,4 β linkage in cellulose. Many higher animals (including human beings) lack the enzymes to break the 1,4 β linkage, so that they cannot digest cellulose. The value of cellulose and other non-digestible polysaccharides in the diet is that they provide bulk and roughage, necessary for normal functioning of the digestive tract. Herbivorous animals like cows, deer, and others utilize such polysaccharides indirectly. Microorganisms in their intestines break down the 1,4 β linkage of cellulose, and the products of hydrolysis can then be digested. Some snails and termites also depend on microbes for enzymes that break down cellulose to provide food for both organisms.

The linear cellulose molecule is found in nature arranged in bundles or fibers. The uniformity of the molecules enables them to line up closely and pack side by side. This close proximity permits the formation of hydrogen bonds between the fibers, which stabilize and strengthen the total cellulose structure. Since the hydrogen bonding is not continuous, a certain amount of flexibility is maintained.

In cotton the cellulose content is about 90 percent, the highest in any natural material. In wood, composed of between 25 and 50 percent cellulose, the fibers are connected by a hemicellulose, which is another polysaccharide, or lignin, a polymeric coniferyl alcohol. Hemicelluloses are derived from mannose, galactose, and some pentoses and uronic acids. Cellulose will also combine with other organic substances, as in the fatty substances cutin, as **adipocellulose** in corky tissue, or with pectins as **pectocellulose,** which occurs in various fruits such as citrus fruits and apples.

Coniferyl alcohol

Cellulose in the natural state is quite insoluble. It is attacked by strong sulfuric acid and hydrolyzed to glucose. It is soluble in the Schweitzer reagent, an ammoniacal cupric hydroxide solution. A mixture of sodium hydroxide and carbon disulfide will also dissolve cellulose; the solution can be spun and coagulated into the synthetic fiber **rayon.** Since early times, man has used cellulose and the combinations of cellulose with other organic substances for his purposes including building materials, tools, paper, boats, and generating heat; as protection against the climate man has used wood, leaves, and straw. Fibers obtained from cotton, hemp, sisal, and flax are good in clothing, fabrics, and ropes. Modern technology modifies cellulose by chemical reactions, mainly esterification, oxidation, and partial hydrolysis, into whole new families of cellulose products. Cellulose esterified with acetic anhydride yields **cellulose acetate,** used to make fibers for clothing, carpets, and many other fabrics. Cementing layers of cellulose acetate between glass gives a shatterproof glass used in cars, planes, and store windows. **Cellulose nitrate,** a product of high esterification with nitric acid, is highly explosive and used in the manufacturing of smokeless gun powder. A modified cellulose is **carboxymethylcellulose** (CMC), which is more soluble in water and increases the viscosity of a solution. These properties make it useful as a thickening and emulsifying agent. It is often added to foods such as ice creams, toppings, and salad dressings, as well as to cosmetics and medicinal preparations, to which it gives uniformity and smoothness.

MIXED POLYSACCHARIDES

Many natural products contain polymers that include monosaccharide units but also other kinds of units that are not saccharides. The presence of these different components changes the physical and chemical properties of these polymers. The arrangement in the polymer chain becomes more irregular, which prevents close fitting between the molecules, and the solubility is increased. Colloidal dispersions are more readily formed, and water is imbibed into the polymer; this helps the retention of water in plants. The polymers have other physiological purposes in different organisms.

Common monosaccharide constituents are glucose and galactose and two pentoses arabinose and rhamnose. These mixed polysaccharides contain such sugar derivatives as glucuronic and galacturonic acids and N-acetylglucosamine and N-acetylgalactosamine. Some of the mixed polysaccharides are esterified and appear as the methyl esters of the sugar acids, whereas others are esterified with sulfuric acid. The appearance of the uronic acids and the sulfates gives such molecules an acidic property in solution. Moreover, and not surprisingly, these compounds are often found as salts, usually as calcium, magnesium, or potassium salts. The mixed polysaccharides can be grouped into gums, mucilages, hemicelluloses, and compound polysaccharides. The last group is made up from saccharides and other classes of organic substances.

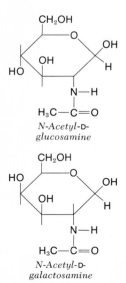

N-Acetyl-D-glucosamine

N-Acetyl-D-galactosamine

Gums   Gums are obtained from various plants which, when injured, exude a heavy thick sap, rich in gums. It slowly hardens, covering and protecting the injured portion of that plant, rather like the formation of a scab on a wound. Gums are collected and purified for many commercial uses. Early man knew how to use many such gums. Gum arabic contains arabinose as one of its prime constituents, and the pentose is named after its source. Other gums are gum ghatti and gum tragacanth. General constituents of the gums are pentoses, hexoses, and uronic acids. The gums have only limited solubility in water and form viscous colloidal dispersions. They are used widely in the food and cosmetic industry, for medical preparations, and by many other manufacturers. They provide adhesive and gelling properties and act as stabilizers and emulsifying agents in such products as dyes, inks, creams, lotions, beer, ice creams, and other desserts.

β-L-Arabinose

α-D-Galacturonic acid

Mucilages   Mucilages are found in many plants, where they are believed to help retain water. They are primarily derived from seeds of leguminous plants, such as guar bean and locust bean, and from seaweeds. Seaweed mucilages are the well known agar-agar, the algins, and carrageenin. The mucilages are polysaccharides formed from galactose and mannose units. The seaweed mucilages contain varying amounts of sulfate esters. The mucilages produce slippery colloidal sols in hot water, which set into gels on cooling. Agar-agar is used extensively in bacteriology for the preparation of culture media. Mucilages are used as stabilizers and gelling agents in the food, pharmaceutical, and cosmetic industries. Mucilages are found in such diversified products as bread, health foods, laxatives, photographic emulsions, paints, and fluids used in drilling of oil wells.

D-Xylopyranose

Hemicelluloses   Hemicelluloses are not celluloses, as the name might imply, but are polymers of xylose, mannose, or galactose. They appear as structural components with cellulose in the cell wall of plants. A nonfibrous polymer of this type composed of 80 to 150 units of the monosaccharide fills the intermolecular spaces between cellulose fibers in the cell wall and acts like a cement. The content of hemicellulose in wood is around 15 percent and in corncobs, cornstalks, and straw up to 40 percent.

Another kind of polysaccharide associated with cellulose is **pectin,** isolated principally from fruit and berries. These polymers are made up primarily of galacturonic acid units. The acid function is partially esterified, mostly as the methyl ester. Some sources for pectins are green apples, quinces, and citrus fruits. They are soluble in water and form colloidal dispersions. Under acidic conditions and the presence of sugar, they form

Galacturonic acid

Methyl ester of galacturonic acid

stable gels. This property is used in manufacturing jams and jellies.

Several types of high-molecular-weight polysaccharides which are connected with proteins are known collectively as the **mucopolysaccharides.** They appear in the cell wall of bacteria and the connective tissues of animals, as well as in body fluids and special tissues. As structural units, they are supporting and binding proteins in cell walls and connective tissue. Their affinity for water permits them to bind water in the interstitial spaces and provide elasticity and lubrication in ligaments and tendons. In some body fluids, such as the vitreous humor in the eye, synovial fluid, and fluid of the umbilical cord, they provide lubrication, viscosity, and control of electrolytes. Specialized ones act as anticoagulants in the blood. Connected to the blood proteins they make for differences in blood types A, B, and O, and for the Rh factor. They influence the immune responses in connection with the globulins in the blood. Research is going on today to elucidate the physiological role of the mucopolysaccharides.

The building blocks of the mucopolysaccharides are primarily hexoses, sugar acids, and amino sugars. Several types of mucopolysaccharides contain the sulfate ester group. Some names and functions of mucopolysaccharides follow. **Hyaluronic acid** is widely distributed in the body of animals and occurs in cell walls, connective tissue, and viscous body fluids. In the cell walls, it acts as a cementing substance for proteins and protection against bacterial invasion. **Keratin sulfate** and **chondroitin sulfates** appear in cartilage, the cornea of the eye, and the skin. They lubricate and connect these specialized tissues and help import toughness and flexibility. **Heparin,** the natural anticoagulant of the blood, is concentrated in the liver and the arterial walls. The building block has a 1,4 α linkage:

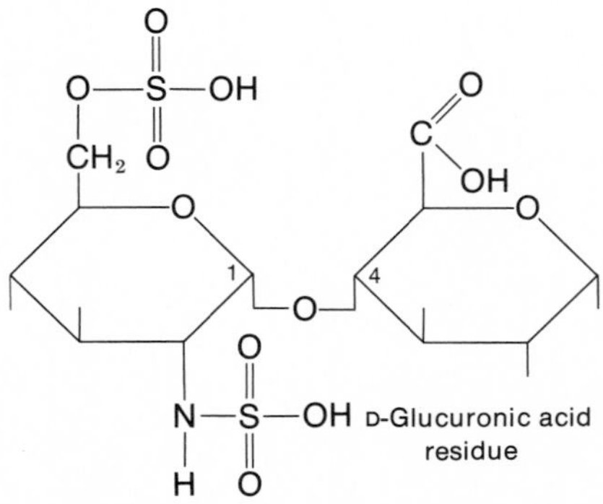

Sulfate ester of D-glucosamine *N*-sulfate

Building block of heparin

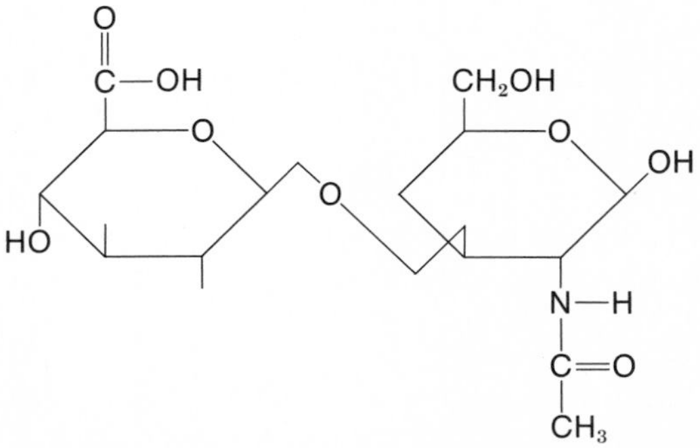

D-Glucuronic acid   *N*-Acetyl-D-glucosamine

Building block of hyaluronic acid

REVIEW QUESTIONS

1 Collect labels from food packages and containers and identify all carbohydrates. List them as **a** monosaccharides, **b** disaccharides, **c** simple polysaccharides, and **d** mixed polysaccharides.

2 What purpose and function do the carbohydrates from Question 1 have in the food preparations?

3 Describe the process of photosynthesis and its ramifications.

4 Write the structures of **a** a tetrose, **b** a pentose, **c** a hexose. Label all carbons and identify the attachments to the individual carbons. Determine which carbons are asymmetric carbons. Identify all functional groups.

5 If an organic compound has seven asymmetric carbons, how many possible isomers has it?

6 Write all the isomers of L-glyceraldehyde for L-tetroses, L-pentoses, and L-hexoses.

7 Look closely at the structure given, identify all carbons by number, and answer the following questions:
  a Is this an aldose or ketose?
  b Is it a furanose or pyranose?
  c Is the configuration α or β?
  d How many asymmetric carbon atoms are there?
  e Give the position of primary hydroxyl groups (—OH groups).
  f Give the position of secondary —OH groups.
  g Give the position of the glycosidic —OH group.

8 Point out the differences between ribose and deoxyribose.

9 Write glucose as an α-pyranose and β-furanose structure.

10 Write the general formulas for **a** all hexoses, **b** all disaccharides, **c** polysaccharides (starch).

11 Explain why the sugar sucrose crystallizes readily.

12 Write the Haworth structures for **a** sucrose, **b** α-maltose, **c** β-cellobiose.

13 Make a list of all the tests indicated for carbohydrates. Use the list to answer the following: What test(s) can you use to identify **a** fructose, **b** sucrose, **c** starch, and **d** dextrin?

14 Write the formulas for the different kinds of sugar acids.

15 Write the structure of the methyl ester of glucuronic acid.

16 What is a glycosidic linkage, and which part of a sugar molecule is involved in it?

17 By examples differentiate between homopolysaccharides and heteropolysaccharides.

18 Use structures to show the difference between starch and cellulose.

19 What is the effect of temperature on the starch I_2 complex?

20 Use structures to differentiate between **a** amylose, **b** amylopectin, and **c** glycogen.

21 What is meant by retrograde starch?

22 What are the hydrolysis products of starch by enzymatic action?

23 Which one of the homopolysaccharides is slightly sweet?

24 What is the function of glycogen in the body?

25 Discuss the benefits of nondigestible carbohydrates in our diet.

26 Discuss the function of cellulose in plants.

27 Make a list of products and materials you know that contain cellulose or cellulose derivatives.

28 What are gums? What is **a** their function in plants and **b** their uses?

29 What are hemicelluloses, and how do we use them?

30 What are the functions of mucopolysaccharides, and where do we find them?

LIPIDS
The Greasy and Tasty

14

The next major group of compounds found in the living tissues of plants and animals is the **lipids** and lipid-related compounds. These substances have the common physical characteristic of being insoluble in water. They are soluble in nonpolar and weakly polar solvents, including benzene, petroleum ether, halogenated hydrocarbons (such as chloroform and carbon tetrachloride), and ethers. Simple lipids contain carbon, hydrogen, and oxygen only; more complex lipids may contain phosphorus and nitrogen as well. The major structural units found in fats, waxes, phospholipids, other related lipids are the aliphatic **fatty acids,** which contain carbon chains, generally long. The resulting high concentration of carbon and hydrogen with relatively little oxygen accounts for the hydrophobic, or water-disliking, characteristics of lipids and lipid-related substances. We can also apply the old adage that like dissolves like to explain the fact that lipids, with their high carbon and hydrogen content, dissolve in hydrocarbon solvents. Structurally, lipids exist as single molecules (rather than as polymers, as do many of the carbohydrates). In water they aggregate to form droplets or films because they prefer each other's company to that of water; hanging

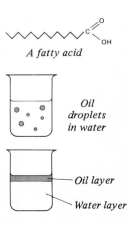

A fatty acid

Oil droplets in water

Oil layer

Water layer

together is a way of protecting each other. The importance of the lipids will be seen in their uses and occurrence — as foods, as components of various cellular structures, as lubricants, as thermal insulators, and in conjunction with other substances as specialized biological agents. Lipids can be classified according to their composition, structure, physiological functions, and physical properties:

1 Simple lipids, composed of alcohols and fatty acids connected by ester linkages
2 Compound lipids, composed of alcohols and fatty acids connected by ester linkages and further linked with phosphorus- and nitrogen-containing organic molecules
3 Conjugated lipids, composed of lipids joined to proteins and carbohydrates
4 Lipid-related compounds, which although not true lipids have the same solubility characteristics as lipids and are therefore generally discussed along with them

SIMPLE LIPIDS

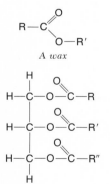

A wax

A triglyceride; a simple lipid

Composed of alcohols and fatty acids connected by an ester linkage, simple lipids have two subgroups: **waxes,** esters of a long-chain alcohol and a fatty acid, and **triglycerides,** where the specific alcohol is glycerol, forming a triester with three fatty acids.

WAXES

These are simple (though high-molecular-weight) esters with general formula RCOOR′; each is composed of a long-chain monohydric alcohol esterified with a long-chain fatty acid. They are relatively inert chemically and insoluble in water. Their hydrolysis requires strong alkali solutions, such as an alcoholic KOH solution, probably because of their low solubility. This alkaline hydrolysis is generally termed a **saponification** since one of the products of the reaction is the salt of the liberated fatty acid, a soap; the second hydrolysis product is the long-chain alcohol.

Waxes are biologically useful. They are found in plants, primarily in leaves and fruits, where they act as protection against water loss and invasion by microorganisms. In picking and handling apples, pears, and citrus fruit, great care is taken not to destroy the waxy surface. Recent evidence indicates that many simple forms of marine life use waxes as a fuel for producing energy. In animals and microorganisms, waxes are usually oily secretions which serve as a water barrier. Aquatic animals like ducks and geese would drown without this protection. These oily waxes also keep hair and feathers flexible. There is a wax in the blood whose alkyl portion is an alcohol called **cholesterol.** These cholesteryl esters may deposit in abnormally high concentrations in arteries and thereby cause arterial sclerosis.

Many waxes are isolated and used commercially for the preparation of creams, candles, polishes, and coating paper and cardboard. **Carnauba wax,** obtained from the carnauba palm in Brazil, is one of the hardest plant waxes. It is used primarily in car waxes, where it provides long-lasting protection against moisture and climate. **Beeswax,** of which the bees construct their honeycomb, finds uses in candles, lipsticks, floor waxes, artificial fruit, and anatomical specimens. **Lanolin,** a wax isolated from wool, is used in ointments, soaps, and many other cosmetic preparations. **Sperm oil,** a liquid wax from the head of the sperm whale, is one of the best natural lubricants and is used to lubricate watches and precision instruments.

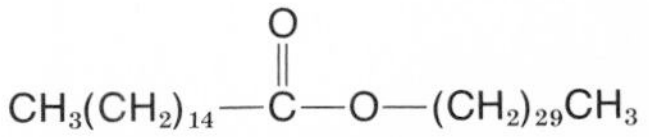

$$CH_3(CH_2)_{14}-\overset{\overset{\displaystyle O}{\|}}{C}-O-(CH_2)_{29}CH_3$$

Myricyl palmitate, beeswax

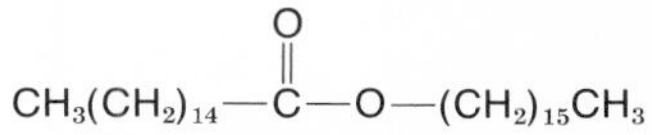

$$CH_3(CH_2)_{14}-\overset{\overset{\displaystyle O}{\|}}{C}-O-(CH_2)_{15}CH_3$$

Cetyl palmitate, spermaceti

TRIGLYCERIDES

The triglycerides, the true fats and oils, are composed of fatty acids esterified with glycerol. Their general formula can be represented as shown in the margin. R, R', and R'', the alkyl chains of the fatty acids, may be alike or different. α, β, α' indicate the position of the several alkyl chains in the fat molecule. The molecule can be considered as being formed by the esterification of the trihydroxy alcohol glycerol (glycerin) by three fatty acids with the loss of three molecules of water:

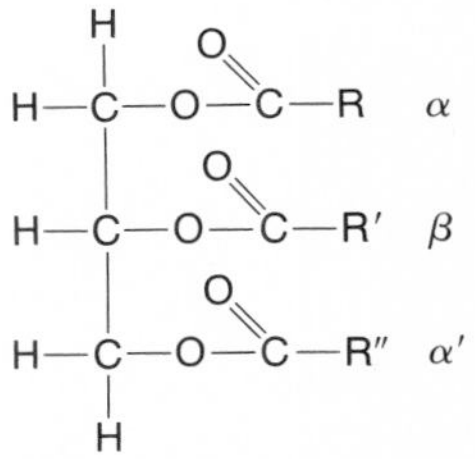

General formula for triglycerides

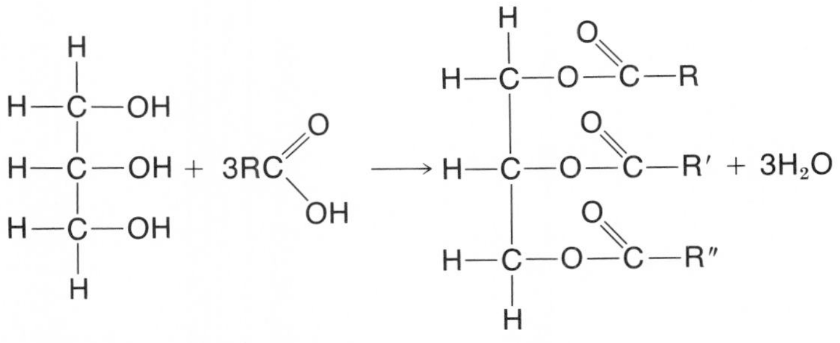

Glycerin 3 Fatty acids Triglyceride

The reverse reaction, hydrolysis or saponification, can also be carried out; alkaline conditions are preferable, but it also takes place in acid solution:

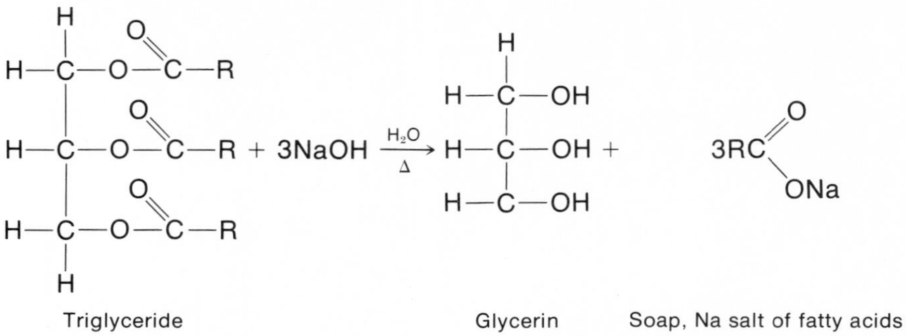

Triglyceride   Glycerin Soap, Na salt of fatty acids

The trihydroxy alcohol glycerine, or glycerol, is an oily, clear, and viscous sweet liquid readily soluble in water. As we know from our discussion of organic compounds, fatty acids yield hydrogen ions and give an acidic reaction in solution. When glycerin and the fatty acids are combined to form the triglyceride, the solubility of glycerin and the acid nature of the fatty acids are lost and the result is a water-insoluble neutral compound: a triglyceride or neutral fat.

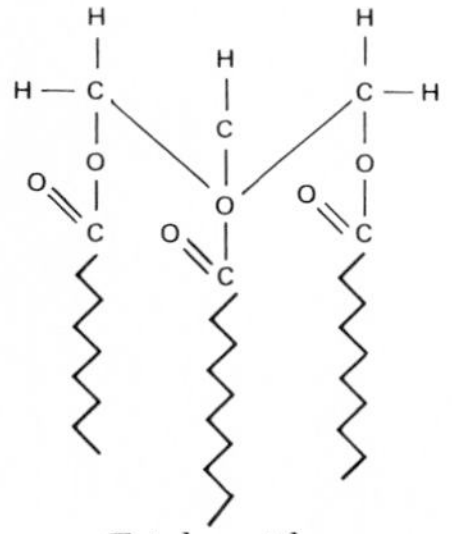

Fatty acids Since the physical and chemical characteristics of the fat molecule are primarily determined by the alkyl chains of the fatty acids in the molecule, we should take a look at the fatty acids found in the natural triglycerides. Their principal characteristic is the presence of only *one carboxylic acid* group,

Triglyceride

TABLE 14.1 Saturated fatty acids

Number of Carbon Atoms	Formula	IUPAC Name	Common Name	Melting Point, °C
4 †	$CH_3(CH_2)_2COOH$	Butanoic acid	Butyric acid	−4.7
6 ‡	$CH_3(CH_2)_4COOH$	Hexanoic acid	n-Caproic acid	−1.5
8 §	$CH_3(CH_2)_6COOH$	Octanoic acid	n-Caprylic acid	16.5
10	$CH_3(CH_2)_8COOH$	Decanoic acid	n-Capric acid	31.3
12	$CH_3(CH_2)_{10}COOH$	Dodecanoic acid	Lauric acid	43.6
14	$CH_3(CH_2)_{12}COOH$	Tetradecanoic acid	Myristic acid	58.0
16	$CH_3(CH_2)_{14}COOH$	Hexadecanoic acid	Palmitic acid	62.9
18	$CH_3(CH_2)_{16}COOH$	Octadecanoic acid	Stearic acid	69.9
20	$CH_3(CH_2)_{18}COOH$	Eicosanoic acid	Arachidic acid	75.2
22	$CH_3(CH_2)_{20}COOH$	Docosanoic acid	Behenic acid	80.2
24	$CH_3(CH_2)_{22}COOH$	Tetracosanoic acid	Lignoceric acid	84.2

† Soluble in water.
‡ Slightly soluble in water.
§ All other acids insoluble in water.

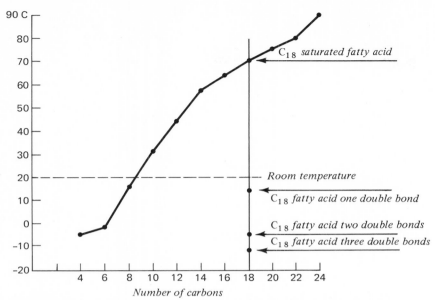

FIGURE 14.1 The curve shows the increase of melting point with increasing number of carbon atoms of *single*-bonded fatty acids. For comparison, the melting points of three 18-carbon fatty acids with one, two, and three double bonds, respectively, are also shown.

—COOH, and a long carbon chain, which may consist of 4 to 24 carbons or even more. The number of carbons in each chain is always *even*, such as, 4, 6, 8, 10, 12, 14, 16, 18, and the carbons are generally arranged in *straight chains*, with a few exceptions. Sometimes, a cyclic structure occurs but rarely the kind of branching we know so well from alkane structures. Finally the fatty acids can be *unsaturated*, that is, have one, two, three, or even four double bonds along the carbon chain. This may result in cis and trans isomerism, which is physiologically important. In some triglycerides a fatty acid has an—OH group attached to the carbon chain and is optically active.

The names and formulas of common fatty acids and their melting points are listed in Tables 14.1 and 14.2. Figure 14.1 gives a graphical representation of melting point versus carbon content of the saturated fatty acids. Of special interest are the unsaturated fatty acids since they provide differences in structure and reactivity of the alkyl chain. A saturated carbon chain can be visualized as

TABLE 14.2 Unsaturated, hydroxy, and cyclic fatty acids

Number of Carbon Atoms	Formula and Name	Melting Point, °C
	Unsaturated Acids	
18	$CH_3(CH_2)_7CH{=}CH(CH_2)_7COOH$ *cis*-9-Octadecenoic acid, oleic acid	14.0
	$CH_3(CH_2)_7CH{=}CH(CH_2)_7COOH$† *trans*-9-Octadecenoic acid, elaidic acid	44.0
	$CH_3(CH_2)_4CH{=}CHCH_2CH{=}CH_2(CH_2)_7COOH$ *cis,cis*-9,12-Octadecadienoic acid, linoleic acid	−5.0
	$CH_3CH_2CH{=}CHCH_2CH{=}CHCH_2CH{=}CH(CH_2)_7COOH$ *cis,cis,cis*-9,12,15-Octadeca-trienoic acid linolenic acid	−11.0
20	$CH_3(CH_2)_4CH{=}CHCH_2CH{=}CHCH_2CH{=}CHCH_2CH{=}CHCH_2)_7COOH$ *cis,cis,cis,cis*-5,9,11,14-Eicosatetraenoic acid arachidonic acid	−49.5
	Hydroxy	
18	$CH_3(CH_2)_5CH(OH)CH_2CH{=}CH(CH_2)_7COOH$ *cis*-12-Hydroxy-9-octadecenoic acid, ricinoleic acid	5.5
24	$CH_3(CH_2)_{21}CH(OH)COOH$ 2-Hydroxytetracosanoic acid, cerebronic acid	
	Cyclic Acids	
16	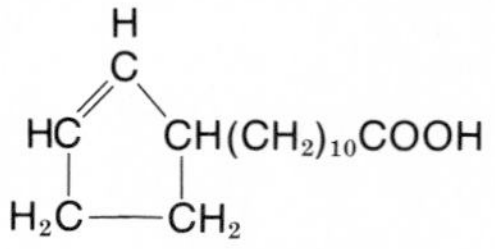 Hydnocarpic acid	
18	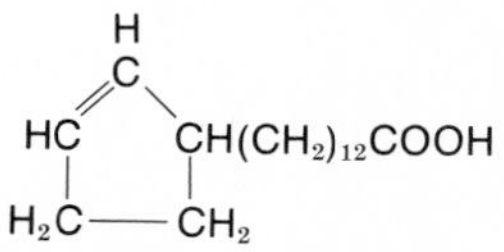 Chaulmoogric acid	68.5

† Does not occur naturally.

whereas the a double-bonded one can be

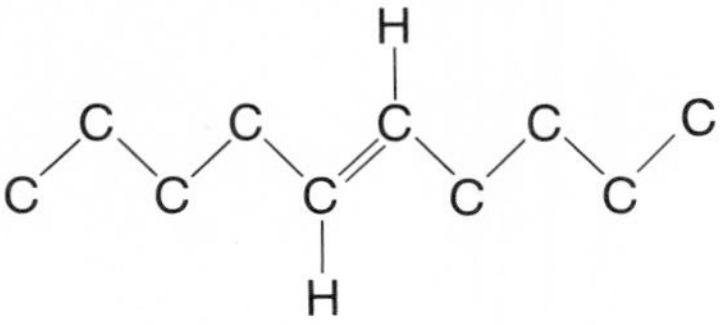

Trans   or   *Cis*

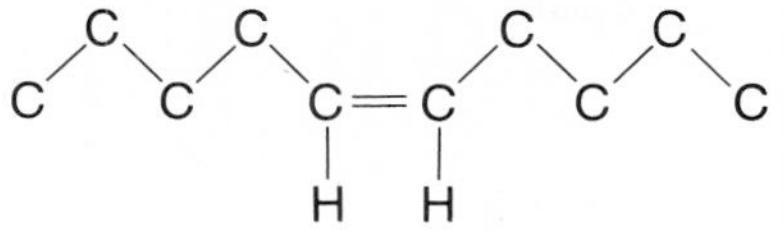

When two double bonds are in the chain, it can be *cis, cis:*

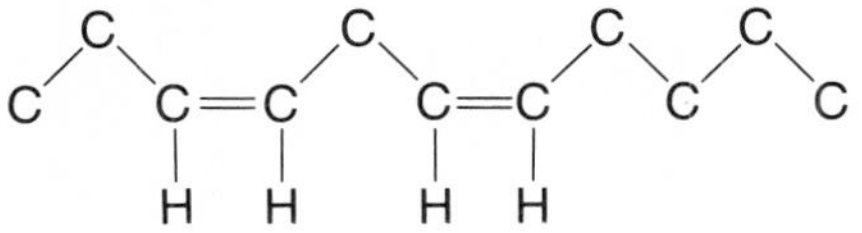

cis-cis

In most of the naturally occurring fatty acids of lipids, the unsaturated chains are in the less stable cis configuration. When two or more double bonds are present, they are separated from each other by 2 or more methylene groups, $-CH_2-$, which makes it a nonconjugated double-bond system. In some instances, the double bonds in plant fats have a conjugated (alternating) arrangement, which increases the reactivity of the system.

$$C-C=C-C-C=C$$
Nonconjugated double bonds

$$C-C=C-C=C-C$$
Conjugated double bonds

The presence of the double bonds leads to addition reactions, such as halogenation, hydrogenation, polymerization, and oxidation, typical of the double bond, as discussed in Chapter 12. Unsaturated fatty acids that contain more than one double bond cannot be synthesized by the human body and must be supplied by food. For this reason, they are called **essential fatty acids.**

Essential Fatty Acids
Linoleic
Linolenic
Arichidonic

Solids and liquid fats As Figure 14.1 shows, the melting points of fatty acids vary greatly with chain length and unsaturation. Since the neutral fats reflect this, we have both *solid fats* and *liquid fats* (usually called **oils**), where 20°C (room temperature) is the dividing line. Usually, animal fats are solids, whereas plant fats are liquids at room temperature. Animal fats are richer in saturated long-chain fatty acids while plant fats are richer in unsaturated fatty acids. Some fats such as coconut oil and especially butter are also rich in short-chain fatty acids, favoring the liquid state. These two factors affect the physical existence of a fat as a liquid or a soft solid; in butter they combine to make it a soft solid.

The commercially important animal fats are butterfat, beef tallow, and lard. Butterfat obtained from milk is a fat with an unusually high content of short-chain fatty acids. This content ranges from 10 to 14 percent, which gives butter its softness at room temperature. Butter is used exclusively for human consumption. The other animal fats are used also primarily as human food, either directly or by incorporation into margarines and cooking and frying fats. Margarines today are manufactured from many different fats and oils modified for the purpose. Incorporation of vitamins, butter flavors, and fats containing essential fatty acids makes margarines tough competition for butter.

Plant oils are obtained from the seeds or fruits of many different plants. They are primarily storage fats and have been harvested since early civilizations for use as food, fuel for lamps, protective coatings, medications, etc. We can classify these oils into three groups according to their content of unsaturated fatty acids:

1 Nondrying oils, rich in oleic acid
2 Semidrying oils, rich in oleic and linoleic acids
3 Drying oils, rich in linoleic and linolenic acids

In the first group we find such oils as coconut oil, olive oil, peanut oil, and palm oil. Coconut oil and palm oil are major raw materials for manufacturing soaps and detergents, whereas olive oil and peanut oil are familiar edible oils.

In the second group, corn oil, cottonseed oil, soybean oil, and many others are mainly used in manufacturing margarine, frying oils, and mayonnaise.

Under the heading of drying oils, we find linseed oil, tung oil, hempseed oil, poppy-seed oil, and others whose main use is in paints and other protective and decorative coatings. Fatty acids with two and three double bonds per fatty acid chain in the drying oils will cause them to "dry" under atmospheric conditions. In the presence of light, the double bonds are activated and will combine with oxygen and other fatty acid chains. The overall result is a polymerization of the oil, producing a tough and protective surface. The "drying" action is not a release of water but the production of a hard film that is no longer liquid. You can observe this reaction when you apply an oil-based paint to a surface. It will take some time to dry, but when it does, the polymerized oil finish is no longer liquid but hard, waterproof, and durable.

CHEMICAL REACTIONS OF TRIGLYCERIDES

The chemical reactions of the triglycerides are basically the reactions of the ester linkage, especially hydrolysis, and the reactions of the double bonds. Reactions of the triglyceride molecule can also result in rancidity; let us see how this happens and what is done to reduce it.

Basically, rancidity is the spoilage of triglycerides by chemical reactions, and two types can be recognized, hydrolytic and oxidative rancidity.

Hydrolytic rancidity Hydrolytic rancidity most often occurs in fats with short fatty acid chains, like butter and coconut oil. This is due to the formation of short-chain volatile free fatty acids which have unpleasant tastes and odors. The reaction is caused by enzymes present in the fat or by enzymes of microorganisms growing on its surface. The enzymes are lipases, hydrolytic enzymes capable of splitting the fat into glycerin and free fatty acids.

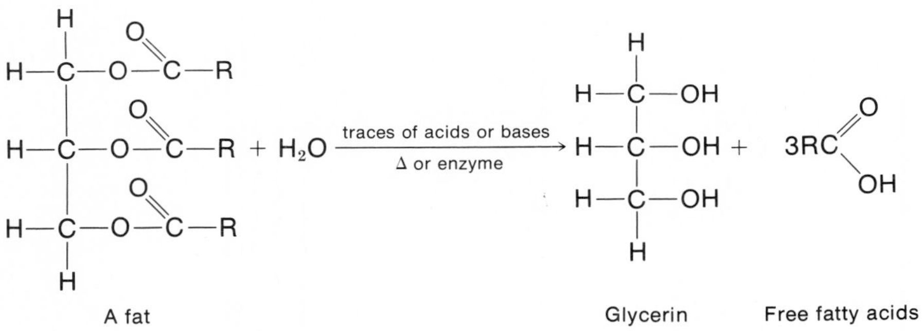

A fat   Glycerin   Free fatty acids

Prolonged heating of a fat will also promote hydrolysis, especially in the presence of moisture. The glycerine produced by hydrolysis can undergo a further reaction at high temperatures; it is dehydrated to produce the unsaturated aldehyde acrolein, or propenal, which has a very obnoxious and irritating odor, which you may have detected from fat used too long for deep frying.

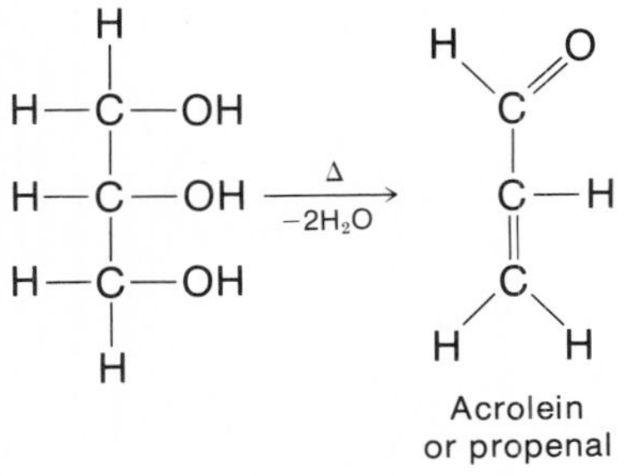

Acrolein
or propenal

Hydrolytic rancidity can be prevented by pasteurizing the fat and keeping it under sterile conditions. Refrigeration retards enzymatic reactions; salt or other antimicrobial components added to the fat also help prevent this type of rancidity. Naturally, the exclusion of moisture will do the same thing.

Synthetic Antioxidants

Butylated hydroxyanisole (BHA)

Propyl gallate

Nordihydroguiaretic acid (NDGA)

Dilauryl thiodipropionate

Butylated hydroxytoluene (BHT)

Oxidative rancidity Oxidative rancidity involves the reaction of oxygen, O_2, at or near the double bonds in the fatty acid chains of an unsaturated fat. Whereas hydrolytic rancidity may occur in any fat, *oxidative* rancidity can develop only in *unsaturated* fats, so that plant oils, richer in unsaturated fatty acids, are more subject to this type of damage than animal fats. The reaction produces oxygenated short-chain aldehydes, ketones, and acids that often have unpleasant odors and tastes. To prevent oxidative rancidity, natural and synthetic antioxidants are employed in the manufacture of foods, oils, cosmetics, and medications (such as salves and lotions). The antioxidant has a high affinity for oxygen, thereby undergoing oxidation in place of the unsaturated parts of the lipid molecule.

Many herbs and spices, such as clove, sage, allspice, and oregano, have antioxidant properties in food preparations. Since the antioxidant property in natural compounds is often destroyed in processing, synthetic antioxidants have been developed. They are primarily complex phenols which readily lose hydrogens to become quinones. Some amines and sulfur-containing amino acids also show this activity and are in use. The above compounds are commonly used as additives not only in foods, for food packaging, but as antioxidants in gasoline, lubricants, plastics, and rubber products. The amount of such an additive in food is controlled by strict regulation of the Federal Food and Drug Administration (FDA).

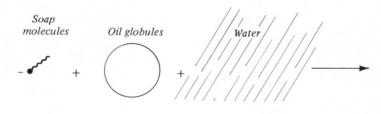

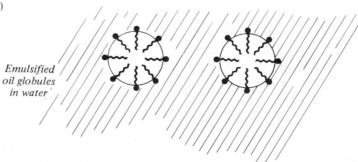

FIGURE 14.2 In the emulsifying action of soap the polar end of the soap molecule faces the water, and the nonpolar end reaches for the nonpolar lipid.

Mixing two antioxidants often gives higher antioxidant activity than using them alone. This synergistic effect boosting the activity permits lower concentrations and more safety in a product. This effect also occurs when phosphates or citrates are mixed with antioxidants. Ascorbic acid, vitamin C, is an antioxidant in its own right and an excellent synergistic agent.

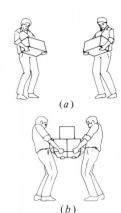

(a)

(b)

Soapmaking and soap action The oldest reaction performed on fats is the preparation of soap. The Romans made soap, and in medieval Europe soap was a big trade item. The early history of the Colonies and the United States often refers to soap manufacture. The process by which soap is made has not changed except that modern technology is more efficient than home production. A fat is boiled in the presence of water and a sodium or potassium alkali to split the triglyceride into free glycerin and the salts of the fatty acids, the soap:

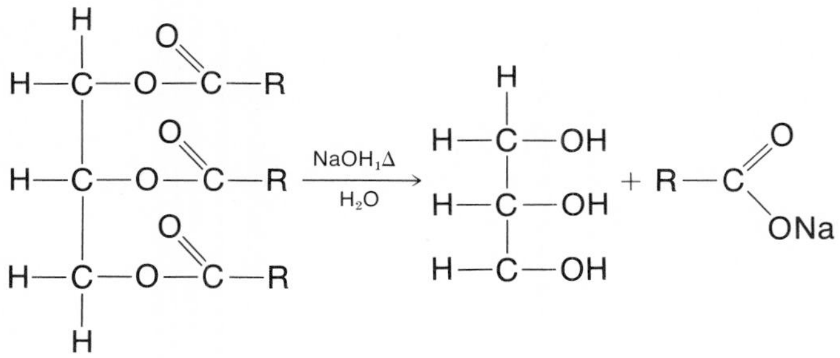

$$\underset{\text{H}}{\overset{\text{H}}{\mid}}$$

Sodium soaps are harder than potassium soaps, and the physical characteristics of the soap formed are influenced by the origin of the fat. The action of soap basically is that of an emulsifying agent (Figure 14.2). After an oily substance is removed by mechanical action from skin or fabric, it is emulsified by the soap molecules in the water and thereby suspended in the water and prevented from being redeposited. Ultimately, the emulsified particle is carried away by the water.

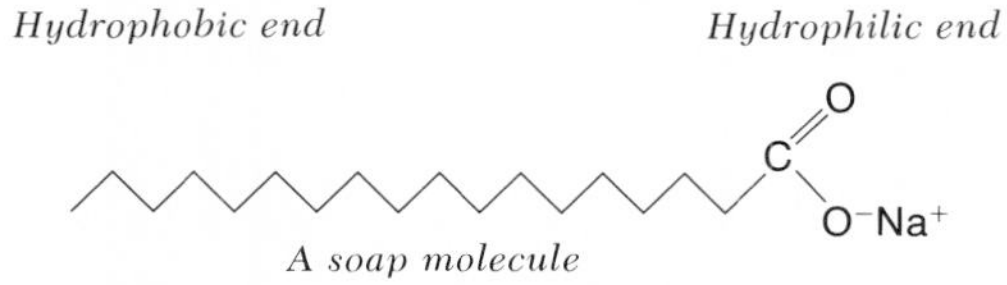

Hydrophobic end   *Hydrophilic end*

A soap molecule

The hydrophobic end of the soap molecule attaches to the surface of the lipid, and the hydrophilic end of the soap molecule reaches out to the water, thereby dispersing the lipid globules in the water so that they are easily rinsed off. Two important factors in this process should be mentioned: (1) if too much soap is used, it may not all dissolve, which represents a waste, and (2) in high concentration soap molecules, like detergents, aggregate to form **micelles.** These bunches of many soap molecules will not be available for emulsifying action and represent a loss of soap effectiveness. In washing operations it is better to use a smaller amount several times than large amounts once.

A further problem with soaps is that many mineral ions react with the fatty acid ions to form insoluble soaps. The positive ions found in water, such as Mg^{2+}, Ca^{2+}, and Fe^{2+}, do just that; the result is not only loss of emulsifying action but formation of an undesirable and insoluble scum. If you take a bath under such conditions, you may come out dirtier than when you went in. Modern technology today produces synthetic emulsifiers called **detergents** as replacements for soaps. They act like soaps, but because they do not contain the carboxylate ion, $-COO^-$, they do not react with the ions present in hard water. The result is improved emulsification and no scum formation in hard water. This has been a blessing in places where the water is hard. Detergents are made today from such raw materials as plant fats and petroleum derivatives. They are much more versatile than soaps, which they replace in many instances. The poor biodegradability of some detergents containing branched carbon chains, however, has raised problems in sewage-disposal plants.

Hydrogenation In another reaction performed on the triglyceride molecules, called hydrogenation, hydrogen gas, H_2, is added to the double bonds of a fat molecule. Elevated pressure and temperature and a catalyst are required. Hydrogenation modifies the physical properties of plant fats by making them more saturated and thus "harder"; it also improves their resistance to autooxidation. These hydrogenated plant fats are used for the manufacture of margarines and shortenings.

Another modification is the reaction of triglycerides with glycerin in the presence of some alkali to produce a mixture of mono- and diglycerides, which are used as emusifying agents and dispersing agents in foods, oils, cosmetics, and other products.

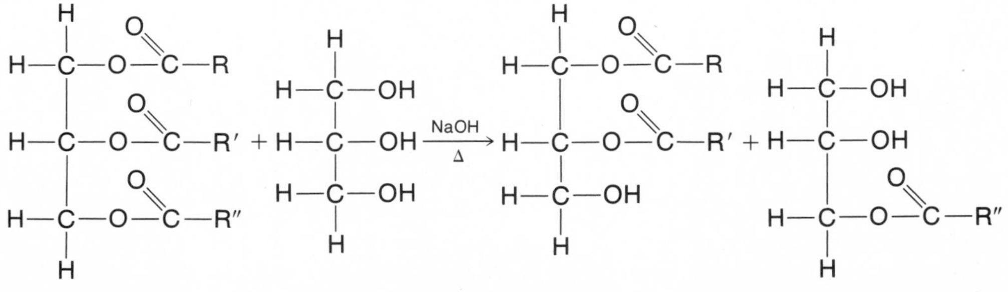

Triglyceride   Glycerin   Diglyceride   Monoglyceride

COMPOUND LIPIDS

Compound lipids are of physiological, rather than nutritional or industrial, importance. The most common are the **phospholipids,** so called because they contain phosphorus in the form of a phosphate ester connected to the glycerol moiety of the molecule.

In a second group of phospholipids replacement of the alcohol, glycerol, by the compound sphingosine gives rise to the sphingolipids. Table 14.3 lists the various phospholipids and their hydrolysis products. Because of their composition, phospholipids are amphipatic. Compounds containing hydrophobic as well as polar or hydrophilic groups will exhibit this characteristic. The phospholipids are physiologically important compounds and are distributed throughout living matter. They act as emulsifying agents in cellular fluids, enter into the regulation of the transfer of ions across cell walls, provide insulation and a barrier for nervous tissue, and perform other specialized functions.

Phosphate ester of a diglyceride α-phosphatidic acid; complete hydrolysis of this molecule yields glycerol, two fatty acids, and phosphoric acid

LECITHINS

The nitrogen-containing organic base in lecithins is choline, $HOCH_2CH_2N(CH_3)_3^+$. The nitrogen in the lecithin molecule is a quaternary nitrogen and thus has a positive charge. At the same

polar end, hydrophilic

nonpolar, hydrophobic end

α-Lecithin or α-phosphatidylcholine (If the choline-phosphate group is attached to the second carbon atom, we have β-phosphatidylcholine.) The fatty acid in the β position is usually unsaturated. The molecule can have two polar groups on the hydrophilic end, + and −.

TABLE 14.3 Compound lipids

Name	Hydrolysis Products	Occurrence
Phosphatidyl choline, a lecithin	Glycerin, fatty acids, phosphoric acid, choline	Higher plants and animals
Phosphatidyl serine, a cephalin	Glycerin, fatty acids, phosphoric acid, serine	Higher plants, brain tissue
Phosphatidyl ethanolamine, a cephalin	Glycerin, fatty acids, phosphoric acid, ethanolamine	Higher plants, brain tissue
Phosphatidyl serine or choline or ethanolamine plasmalogens	Fatty aldehydes, fatty acids, glycerin, phosphoric acid and serine, choline, or ethanolamine	Membranes of muscle and nerve cells
Phosphoinositides, phosphatidylinositol	Fatty acids, glycerin, phosphoric acid, inositol	Brain tissue, cellular membranes
Sphingolipids	Sphingosine, fatty acids, phosphoric acid, choline	Membranes of plants and animals, brain and nerve tissue

time, the phosphate can provide a negative charge, so that there are two polar groups at one end of the molecule. Because of its excellent emulsifying properties, lecithin is produced commercially from soybeans and other sources for use as an emulsifying agent, a wetting agent, and an antioxidant in chocolates, candies, margarine, baked goods, and animal feeds. Lecithins are widely distributed in plants and animals and are believed to help many physiological functions. They are found in large amounts in nervous tissue, together with other phospholipids and related compounds. An enzyme contained in snake venom, called lecithinase A, can hydrolyze the α-lecithins to lysolecithins:

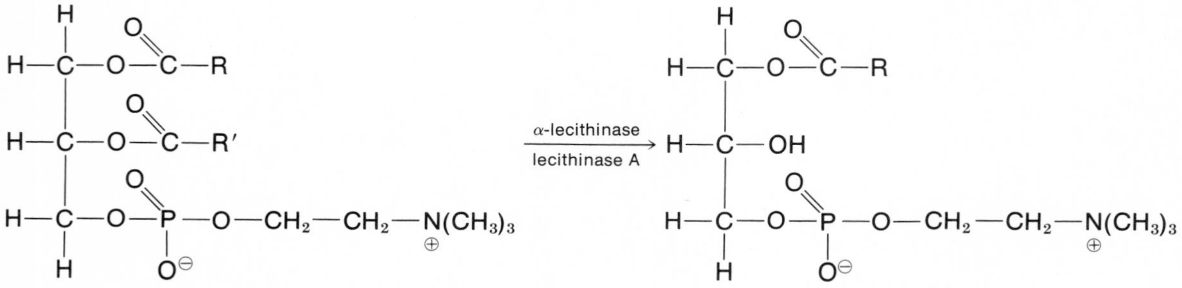

$$\xrightarrow[\text{lecithinase A}]{\alpha\text{-lecithinase}}$$

α-Lecithin

Lysolecithin

Lysolecithins have hemolytic activity; that is, they can destroy red blood cells. This property of snake venom explains why snakebite can be so dangerous.

CONJUGATED LIPIDS

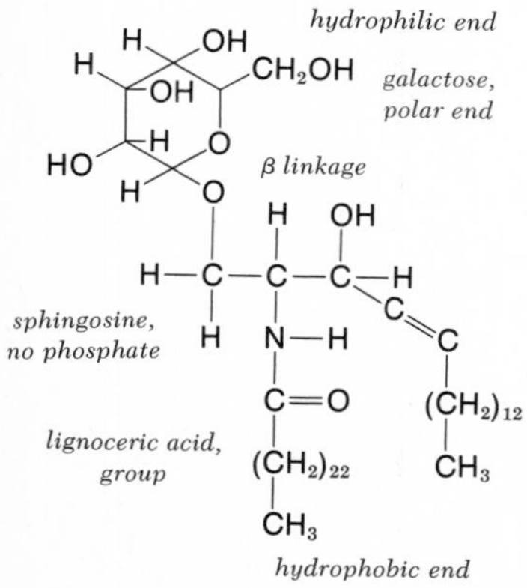

Under this heading we find a number of compounds that are made by joining lipids with either proteins or carbohydrates. Glycolipids are those containing lipid and sugar moieties. The sugars are usually either glucose or galactose or some of their derivatives. Examples of glycolipids are the cerebrosides and gangliosides, in which sphingolipids are joined with the sugars indicated. The gangliosides are similar to the cerebrosides but usually contain more sugar residues, about three to five units of a mixture of galactose and glucose units. These compounds are found in the gray matter in the brain and spinal cord as well as in many other organs. The glycolipids do not contain a phosphate residue.

The combinations of lipids with proteins known as the lipoproteins are under extensive study today. With the availability of new and better techniques, they can be isolated and their physiological importance demonstrated. They are found in cell walls as

A glycolipid, a cerebroside, a cerebrogalactoside

hydrophilic end

galactose, polar end

β linkage

sphingosine, no phosphate

lignoceric acid, group

hydrophobic end

well as the nucleus and the mitochondria. Lipoproteins have been found in egg yolk and milk, and the transport of lipids in the bloodstream is effected by their complexing with proteins. A lipoprotein is also involved in the clotting of blood.

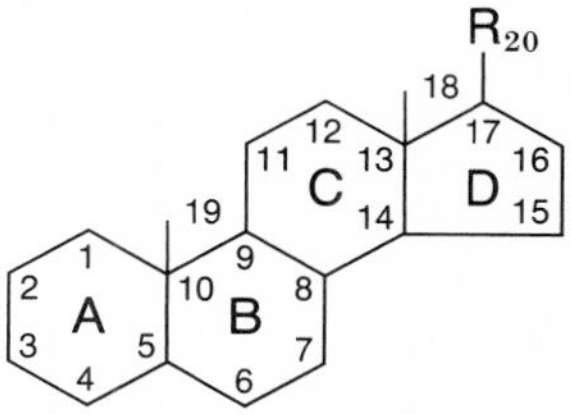

LIPID-RELATED SUBSTANCES

Compounds that fall in the category of lipid-related substances are the **steroids, essential oils, carotenoid plant pigments,** and **fat-soluble vitamins.**

STEROIDS

These compounds are characterized by having a central common structure, or nucleus, which is the cyclopentanoperhydro-phenanthrene structure represented in Figure 14.3. It consists of the three fused cyclohexane rings (A, B, and C) in the phenanthrene-like structure and the fused cyclopentane ring D. *Perhydro-* means that all unsaturation of the phenanthrene aromatic system has been removed by adding hydrogen. The numbering starts from the top of ring A in a counterclockwise direction around rings A and B, continues clockwise around C, and then finishes counterclockwise around ring D. There are usually methyl groups on carbons 10 and 13 and a longer carbon chain attached at carbon 17. The methyl-group carbons are then numbered 18 and 19, and any carbon chain at 17 will be numbered starting with 20. The attachments on carbon 17 vary in chain length and structure and are one of the main points of difference between steroids. The structure shown in the margin represents a **sterid** molecule, containing only C and H. The sterids are the parents of the steroids, biologically active compounds which also contain oxygen. The steroids can be arranged according to biological function into the following groups:

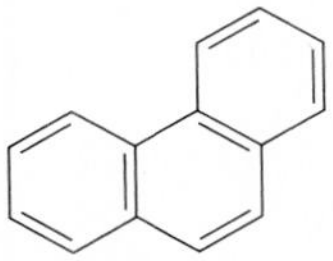

1 General steroids
2 Bile acids
3 Vitamins
4 Cardiac-active compounds
5 Adrenal hormones
6 Sex hormones

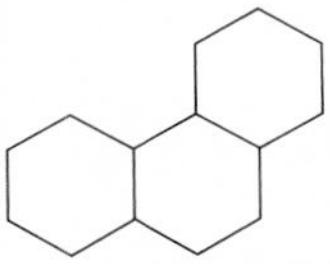

In looking at these compounds we should realize the similarities as well as the small differences in structure and attachments which make for the big differences in their biological activity. Fur-

Steroid nucleus

FIGURE 14.3 The common backbone structure of all steroids.

Phenanthrene

Perhydrophenanthrene

Sterid:
  C + H *only*

Steroid:
  Oxygenated sterid or
    C + H + O

thermore, we should know that very often even minute quantities of these compounds have great effects upon the organism. The stereochemistry of these molecules influences their properties. You will see that they contain sites for cis-trans isomerism and also quite a number of asymmetric carbon atoms, which make possible a considerable number of optical isomers. Fortunately, only one or two isomers are biologically active, that is, have the specific conformation required for the specific biological activity.

General steroids The general steroids appear in plants and animals alike but with differences in their side-chain structure (Figure 14.4). They are sterols, that is, steroid alcohols. The general steroids are considered the precursors for the other steroids and are found in greater abundance. Cholesterol is present in all animal tissue but is concentrated in bile fluid and nervous tissue. Gallstones consist mainly of cholesterol that has crystallized out from the bile fluid. Cholesterol is a chief contributor to arterial sclerosis. When the concentration of cholesterol in the blood is high, it can be deposited on the arterial wall and cause constriction of the artery and reduction of blood flow. The condition is an abnormal one, but the causes of the high levels of cholesterol are still not clear. Cholesterol is produced by the body besides being ingested in eggs, butter, and red meat. The main method for reducing blood cholesterol level is dietary restriction of cholesterol-rich foods. Some research has recently established the role of cholesterol in the transmission of nerve impulses, wherein it acts as a coenzyme. Under conditions of

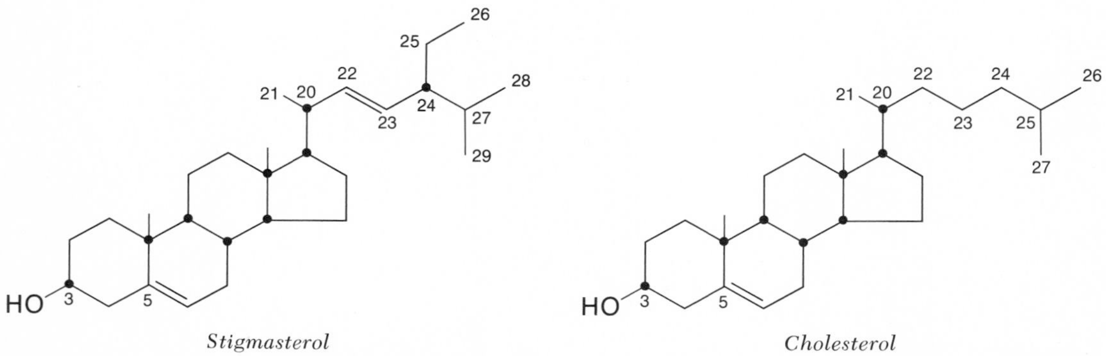

Stigmasterol   *Cholesterol*

FIGURE 14.4 Stigmasterol has an —OH group at carbon 3 and a double bond at carbons 5 and 22. The side chain has 10 carbons. The compound has 9 asymmetric carbons, which give 512 possible isomers. Cholesterol has a —OH group at carbon 3 and a double bond at carbon 5. The side chain has 8 carbons. The compound has 8 asymmetric carbons, which makes possible 256 different isomers.

nervous tension, the body produces more cholesterol because it needs more, thereby elevating the cholesterol levels. Apparently under such conditions, the body produces more cholesterol than needed, which in turn can cause other problems. Sensible living and proper nutrition can help regulate the production and intake of cholesterol.

Bile acids The bile acids are amides of steroid acids such as cholic acid, deoxycholic acid, and chenodeoxycholic acid. Note that cholic acid has —OH groups on carbons 3, 7, and 12 and a five-carbon side chain on carbon 17 with a carboxyl group at the end. The carboxyl groups provide the acid function for the amide linkage to glycine or taurine to form the complete bile acid:

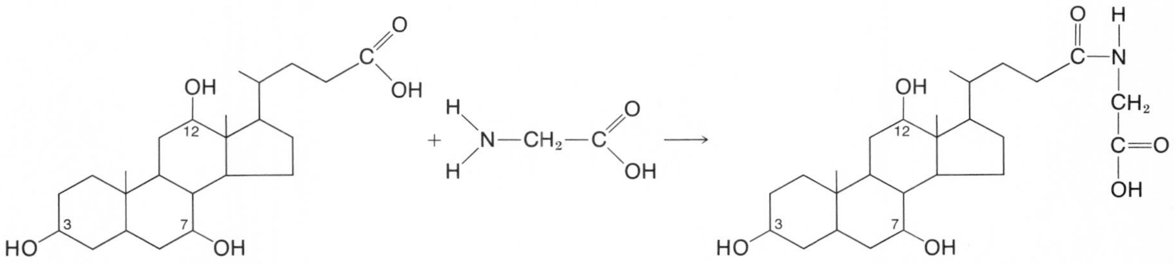

Cholic acid   Glycine   Glycocholic acid, complete bile acid

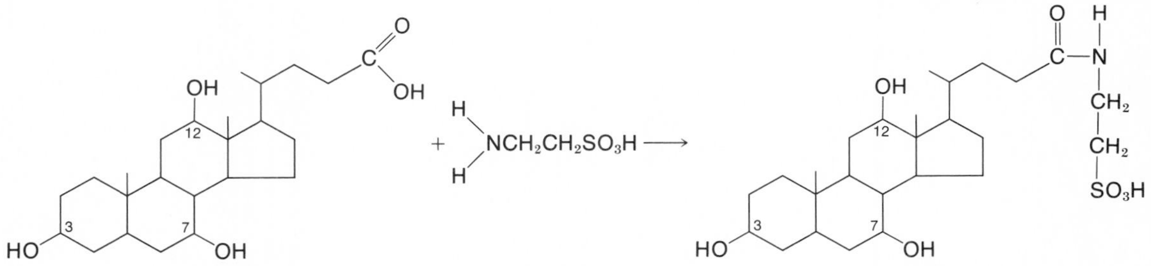

Cholic acid   Taurine   Taurocholic acid, complete bile acid

The bile acids act as emulsifiers for fats in the intestinal tract and so aid in the digestion and absorption of lipids. They activate the lipases and are essential for the uptake of the fat-soluble vitamins from the intestines into the body.

Vitamins The D vitamins have the steroid structure and can be obtained from plant or animal sources as precursors. The action of ultraviolet light changes the precursors to the active vitamins:

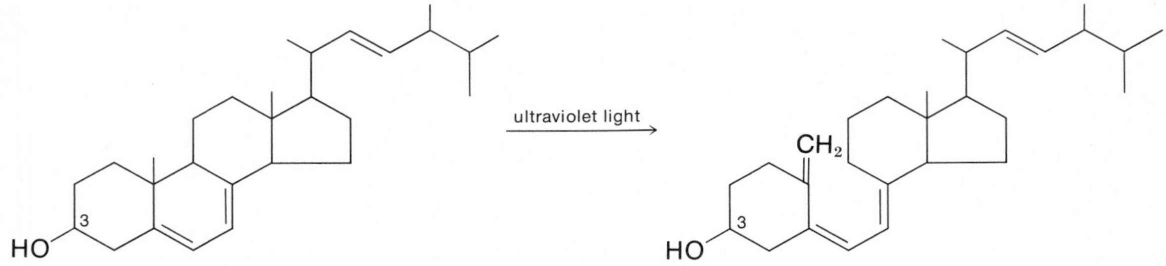

Ergosterol,
plant source

ultraviolet light

Vitamin D$_2$,
ergocalciferol

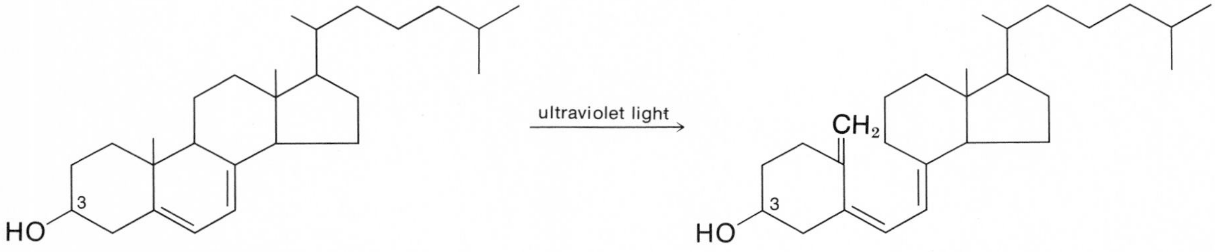

7-Dehydrocholesterol,
animal source

ultraviolet light

Vitamin D$_3$

The irradiation with ultraviolet light produces a split between carbons 9 and 10 which gives the molecule its vitamin activity. The vitamin aids in the absorption of calcium and regulates the metabolism of bone tissue.

Cardiac-active compounds: genins These compounds have been isolated from plants such as foxglove, *Digitalis purpurea*, and from toads. They are composed of a steroid nucleus with a sugar unit attached at the −OH group on carbon 3 to complete the molecule. The steroid is called the aglycon or genin part of the molecule, and the whole molecule can be classified as a glycoside. Such steroids resemble the cholic acids in structure except that the attachment on carbon 17 is cyclized. From three to five saccharide units (glucose, galactose, rhamnose, and others) are attached. These compounds in small quantities regulate the heartbeat and are invaluable for the treatment of certain heart conditions. Excess quantities cause death, and the dosage must be carefully controlled.

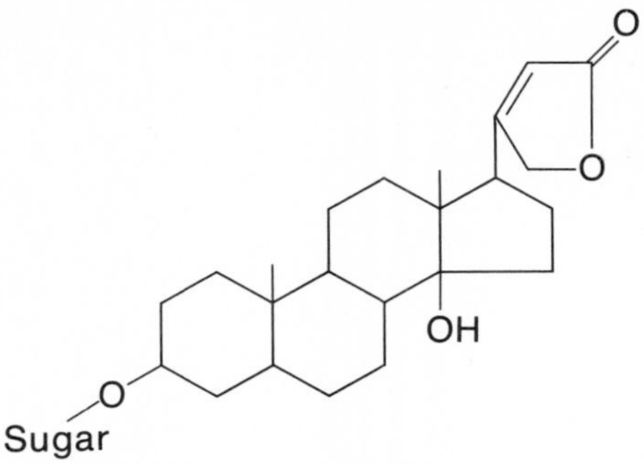

Digitoxigenin, the active component
in digitalis extract

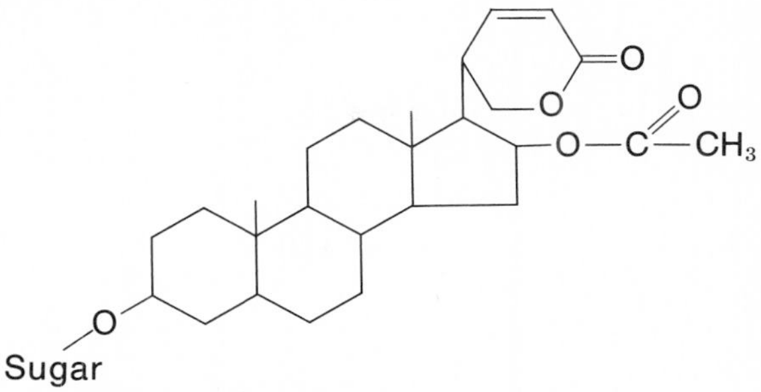

Bufotalin, a similar active compound,
obtained from toads

Adrenal hormones The adrenal glands are producers of vital hormones for the body. The secretion of the adrenal cortex, called **cortin,** is a mixture of steroid hormones. One group influences the glucose metabolism and is known as **glucocorticoids.** Another group controls the mineral and water balance in the body and is known as **mineralocorticoids.** A third group of steroids found in cortin are sex hormones, which are also produced by the testes or ovaries.

Glucocorticoids:

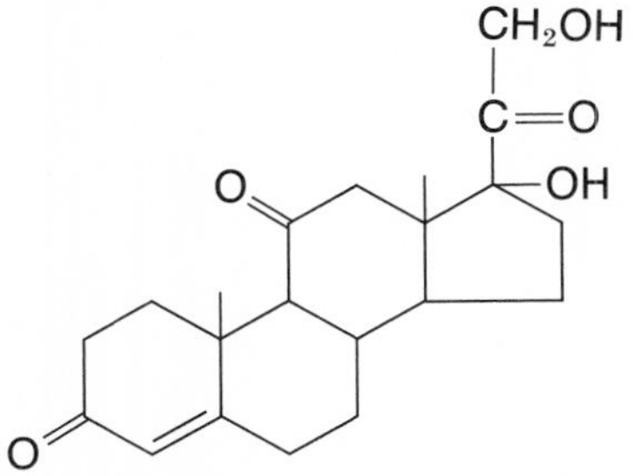

Cortisone, keto functions on
carbons 3 and 11 on nucleus
and on carbon 20 of side
chain; double bond at carbon 4;
hydroxyl functions at carbons 17 and 21

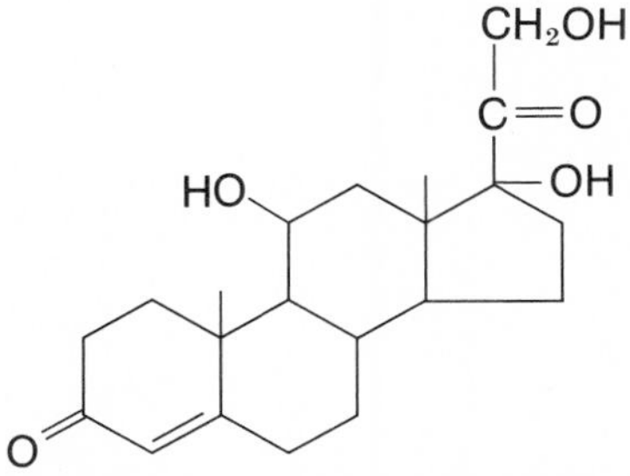

Hydrocortisone,—OH on carbon 11 instead
of keto group; more effective than cortisone

The two compounds shown here as well as similar synthetic ones are used today to treat inflammatory diseases, such as rheumatic arthritis, and allergies and skin inflammations.

Mineralocorticoids:

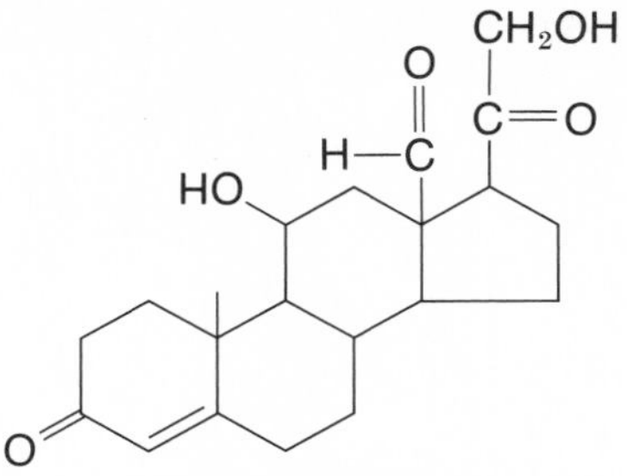

Aldosterone, ketosteroid with
aldehyde function on carbon
13 instead of methyl group

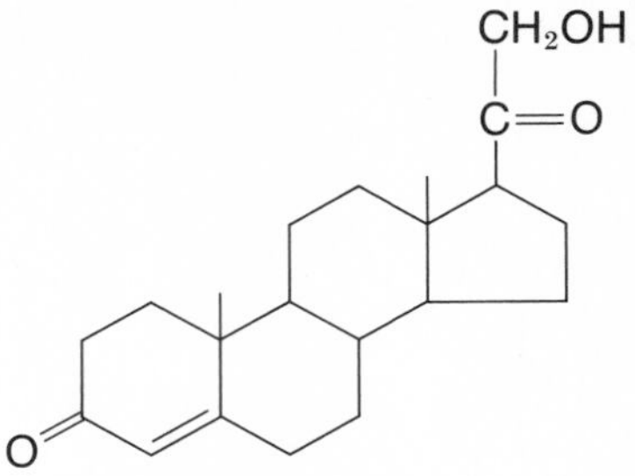

Deoxycorticosterone

The two compounds shown here control the Na$^+$ and K$^+$ balance
in the body. Aldosterone is much the more powerful of the two
by a factor of 100. By the control of the mineral balance these
hormones also affect the water balance in the body.

Sex hormones The sex hormones are also steroids and can be
divided into the male sex hormones, or androgens, the female sex
hormones, or estrogens, and the progestagens, which regulate
the menstrual cycle and pregnancy.

Male sex hormones:

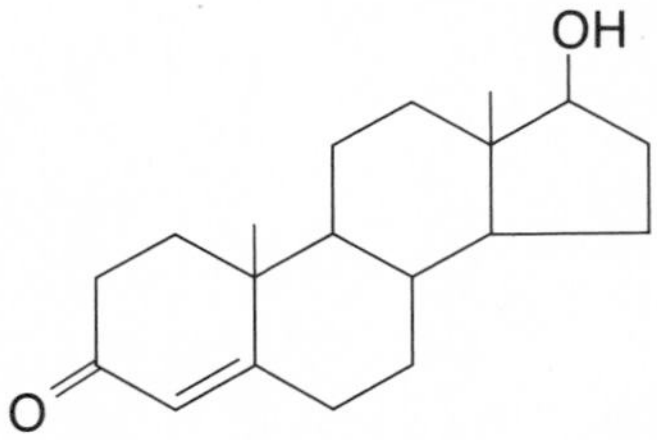

Testosterone (most potent male sex hormone),
keto function on carbon 3 and
double bond at carbon 4;
—OH at carbon 17

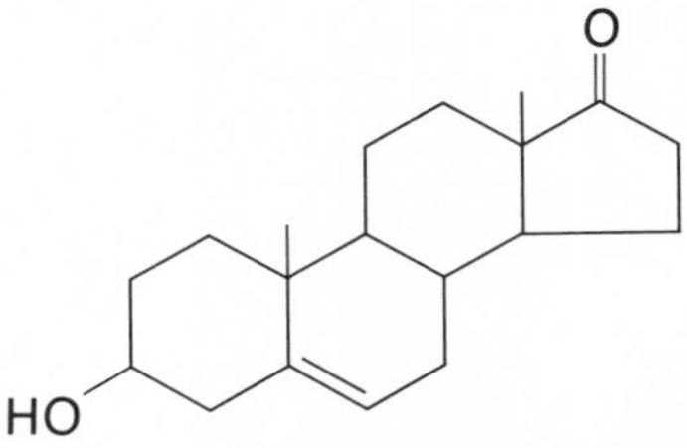

Dehydroandosterone (lesser male sex hormone),
—OH at carbon 3 and ketone function
at carbon 17; double bond from
carbon 5 to carbon 6

The hormones shown here control the production of sperm and
secondary sex characteristics such as hair, voice, and muscle
development but also affect metabolic functions.

Female sex hormones:

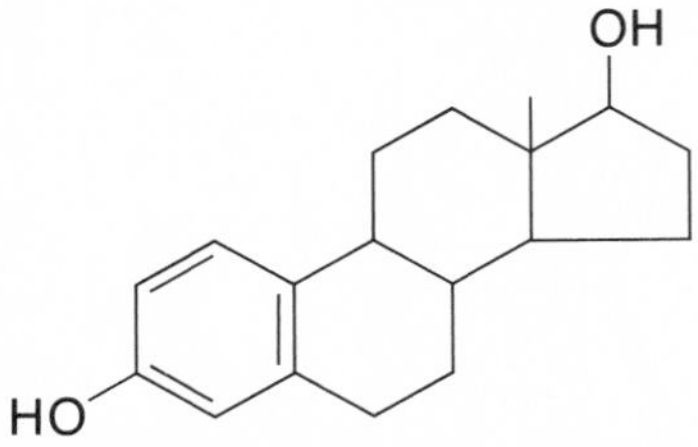

Estradiol, ring A in all female sex hormones, is
aromatic; the —OH group on
carbon 3 is therefore phenolic; hydroxyl
function on carbon 17 and no methyl
group at carbon 10

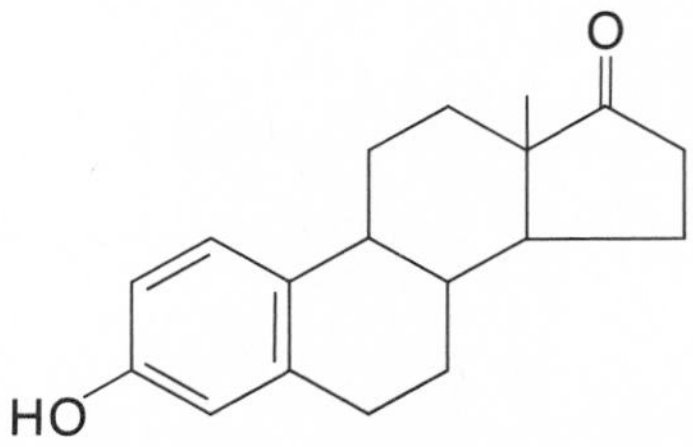

Estrone, like estradiol, but with a keto function
on carbon 17

The female sex hormones control the development of eggs and of secondary sex characteristics such as mammary glands and uterus.

Progestagen:

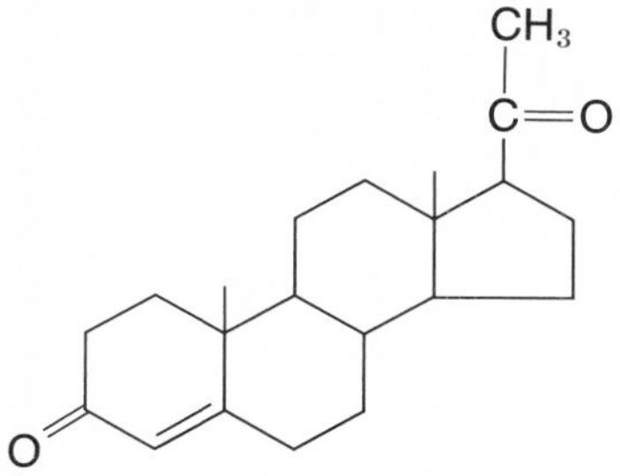

Progesterone keto function at carbons 3 and 20,
double bond at carbon 4
(note the similarity to some of
the hormones produced by the
adrenal cortex)

This compound is believed to be the primary gestogen controlling the menstrual cycle although it operates in conjunction with the other female sex hormones. During pregnancy it is needed for maintenance of the fetus. It stimulates the development of the mammary glands and inhibits ovulation. In connection with the sex hormones, it should be mentioned that many plant steroids have been modified by chemical means to have as much or even more potency than the naturally occurring compound. Such compounds are marketed in a variety of birth control pills which either promote ovulation or inhibit fertilization by affecting the menstrual cycle in such a way that no pregnancy occurs. Since all these compounds are steroids and can have side effects even in small quantities, medical supervision is always advisable when such drugs are taken.

   Active research is carried out on steroids both natural and synthetic by drug companies to elucidate their function and structure and find new and better drugs in the fight against diseases.

FAT-SOLUBLE VITAMINS

In the discussion of steroids we encountered the provitamin D and vitamin D structures. The other fat-soluble vitamins are A, E, and K. A group of fat-soluble substances called the prostaglandins and believed to be essential to the body are named vitamin F by some researchers. The precursors of these compounds are the essential fatty acids, and the relation between them is analogous to that between provitamin and vitamin.

Vitamin A   The precursors of vitamin A are the α, β, and δ carotenes. They are converted in the body to vitamin A by oxidative

---

cleavage as indicated. Vitamin A deficiency causes night-blindness due to the incorporation of the vitamin into the pigment rhodopsin in the eye. This pigment is responsible for vision at reduced light. Further indications of vitamin A deficiency are weight loss and low resistance to infections. Good sources of the carotenes are leafy vegetables, carrots, and apricots. Good sources of vitamin A are butter, cheese made from whole milk, liver, codliver oil, and eggs.

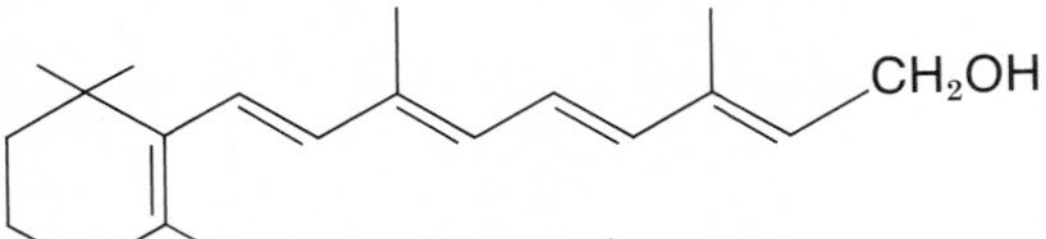

Vitamin A

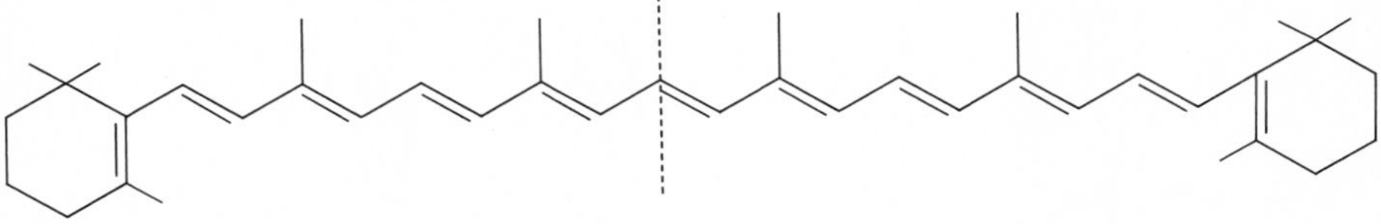

Provitamin A
β-Carotene

Vitamin E Several different forms of vitamin E are known today, as expressed in differences in its molecular structure. The cyclic nucleus of the molecule has methyl groups at several positions, as can be seen from the structure of α-tocopherol, the most potent vitamin E:

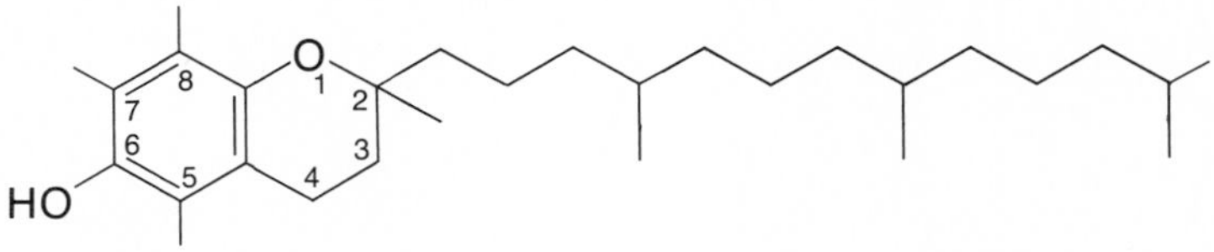

Vitamin E, α-tocopherol;
side chain consists of isoprene units;
methyl groups are on carbons 2, 5, 7, and 8

$\alpha = alpha$
$\beta = beta$
$\gamma = gamma$
$\delta = delta$
$\epsilon = epsilon$
$\zeta = zeta$
$\eta = eta$

Less potent tocopherols have fewer methyl groups on the cyclic part of the molecule and are known as β-, γ-, δ-, ϵ-, ζ-, and η-tocopherols (see margin for names of these Greek letters). The tocopherols, which have been isolated from wheat germ and other plant sources, are not identified with any deficiency disease in man, but their lack produces sterility in other animals and causes dystrophy of the muscles. As already indicated, the tocopherols are excellent natural antioxidants.

Vitamin K This vitamin contains an aromatic nucleus and an isoprene-type side chain. Differences in that side chain account for the existence of different K vitamins. The aromatic part of the molecule is a naphthoquinone structure,† which cannot be synthesized by mammals.

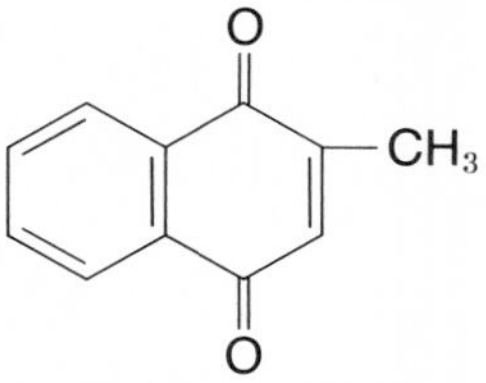

2-Methyl-1,4-naphthoquinone

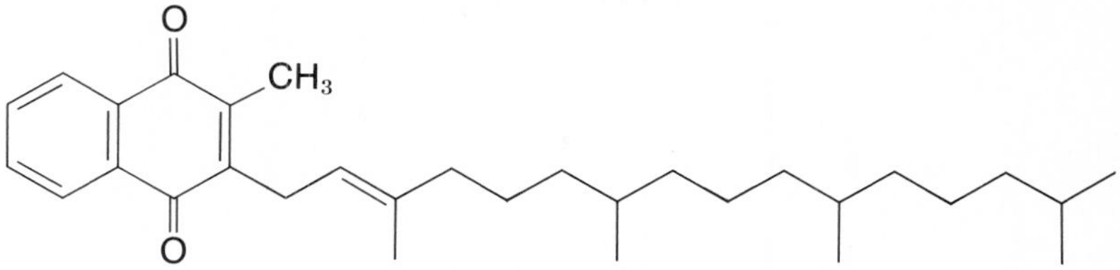

Vitamin K₁

2-Methyl-1-1,4-naphthoquinone has vitamin K activity because the body can add the isoprene chain to form the complete vitamin structure. Vitamin K activity is expressed in the control of blood clotting. A vitamin K deficiency produces a tendency to bleed, and the time for the formation of a blood clot is greatly increased. Good sources of vitamin K are green leafy plants such as spinach and alfalfa. Most plant oils are also rich in the vitamin. The intestinal mucosa produce this vitamin.

Prostaglandins Considerable interest in the prostaglandins has been generated since it became apparent in the 1960s that they are essential to body functions. They are distributed in many tissues and organs in the body such as muscle tissue, the genital glands, brain, thymus gland, pancreas, and kidney. They are effective in concentrations as low as 1 pg (10^{-12} g) per kilogram of body weight. The prostaglandins affect the functioning of smooth muscles, lipid metabolism, and blood pressure; they also influence such hormones as norepinephrine and vasopressin. Clinical studies indicate that they can induce human abortion. One interesting observation is that aspirin affects the prostaglandins and their activity, which is the first time that a specific site of action of aspirin has been determined. With further knowledge of the role and the potency of the prostaglandins, there is hope that they can be used as drugs against inflammatory diseases, arterial sclerosis, and high blood pressure.

The precursors of the prostaglandins are the highly unsaturated fatty acids with 18 or 20 carbon atoms such as homo-γ-linolenic acid (three double bonds), arachidonic acid (four double bonds), and eicosapentaenoic acid (five double bonds).

The prostaglandins are formed by cyclization of the fatty acid chain in the middle to form a five-membered ring, a cyclopentane structure, with two side chains. A further modification is oxida-

1 pg = 1 picogram
= 10^{-12} g

† An aromatic ketone.

TABLE 14.4

Precursor	*Prostaglandin*
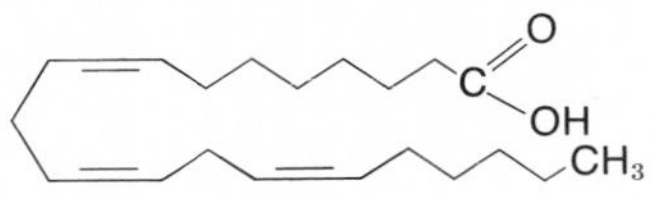 Homo-γ-linolenic acid, three double bonds	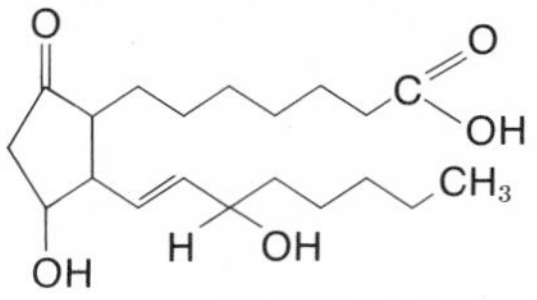 PGE₁
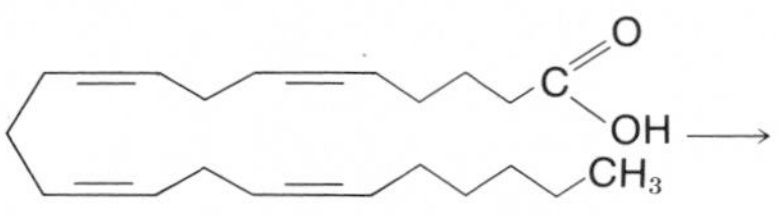 Arachidonic acid, four double bonds	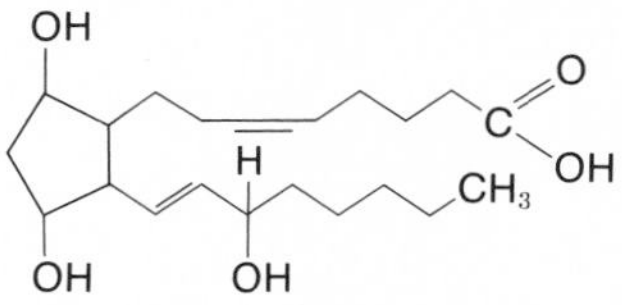 PGF₂ α
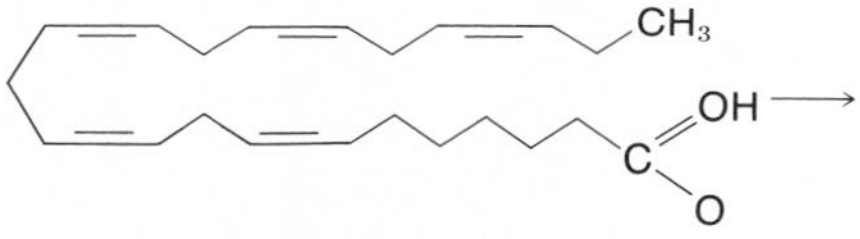 Eicosapentaenoic acid, five double bonds	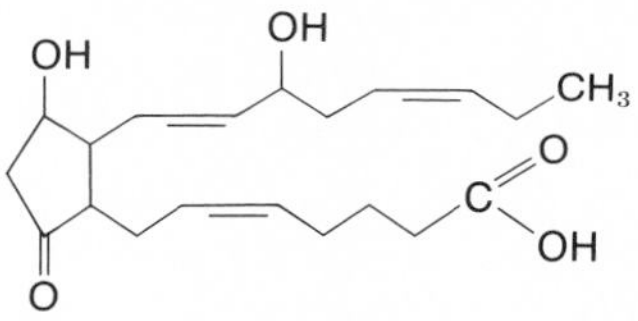 PGE₃

tion on the cyclopentane ring and at various positions on the side chains. Two types of prostaglandins are identified; the PGE series, with an alcohol and a ketone function on the cyclopentane ring, and the PGF series, where two hydroxyl groups are on the cyclopentane ring. Differences within the two series are variations in the positions of other hydroxyl groups and the locations of double bonds in the side chains. Some representative precursors and prostaglandins are given in Table 14.4.

REVIEW QUESTIONS

1 Collect labels from food packages and containers and list all lipid substances named. Group them according to simple lipids, compound lipids, conjugated lipids, and lipid-related substances. Identify the function and purpose of each compound in the food product.
2 What is a wax, and what is its function in biological systems?
3 From other sources such as the dictionary or encyclopedia find out as much as you can about the use of waxes.
4 Write the general formula of a molecule of a triglyceride and identify all components and their positions as well as the functional groups.

5 What is saponification, and what are the end products of that reaction when NaOH is used?

6 Discuss the characteristics of the fatty acids found in natural fats and oils.

7 Why are the fats and oils also called neutral fats?

8 What is the importance of essential fatty acids, and what are they?

9 Why are solid fats solid at room temperature and liquid fats or oils liquid at room temperature?

10 What are the functions of fats in **a** animals and **b** plants?

11 What is margarine, and what is its composition?

12 What is a drying oil, and what are its uses?

13 What is acrolein, and how does it appear in fats?

14 From your collected food labels make a list of compounds that act as antioxidants (sometimes indicated as oxygen interceptors). Identify them as **a** natural and **b** synthetic.

15 Describe the action of soap.

16 Compare soaps with detergents from the point of view of composition and their action in hard water.

17 Collect labels from boxes or bottles of cleansing products. Identify their components as **a** soaps, **b** detergents, **c** others.

18 What is the process of hydrogenation used on fats and oils, and what is its purpose?

19 What are mono- and diglycerides, and what uses do we have for them?

20 What compounds can be isolated if compound lipids are hydrolyzed?

21 What are the differences between lecithin and lysolecithin?

22 From memory draw the steroid structure and number all carbons.

23 Write structures of some of the different classes of steroids and compare their differences.

24 What is the role of bile acids in the body?

25 Which one of the steroids is a glycoside? Give its structure.

26 What are the medical uses of cortisone?

27 What is a mineralocorticoid, and what is its function?

28 Raid your mother's kitchen and smell the different spices you have in your house. Try to remember them by their odor and describe that odor.

29 Write the structure of α-tocopherol.

30 What is controlled by vitamin K?

31 Show how prostaglandins are formed from polyunsaturated fatty acids (essential fatty acids).

32 Name some of the effects of prostaglandins on the body.

15

THE PROTEINS
And Above All

One important distinction between plants and animals is their cell wall. An essential component of animal cell walls is **protein,** whereas plant cell walls are made up of polysaccharides, such as cellulose. The difference in cell-wall material is the difference between the hard and relatively rigid structure of plants and the softer, more pliable substance of animals. It does make a difference: have you ever tried to hug an oak tree?

The polysaccharides are polymers of monosaccharides, whereas the proteins are polymers of amino acids. Inside both plant and animal cells are a great variety of proteins, indispensable in function, and both plants and animals secrete protein materials into their body fluids necessary for the total organism. Life depends on the proteins, as expressed by the name for these compounds, from the Greek word *proteios*, meaning of the first rank.

Analysis of proteins shows that they contain carbon, hydrogen, oxygen, nitrogen, sulfur, and traces of phosphorus, iron, zinc, copper, manganese, and iodine (see Table 15.1).

THE AMINO ACIDS

When proteins are broken down, the building blocks of which they are made can be isolated. These building blocks are the **amino acids,** and the disassembly of a protein can be accomplished by hydrolysis with HCl or NaOH, or a **proteolytic** enzyme acting as the catalyst. Thus,

$$\text{Protein} \xrightarrow[\text{HCl or NaOH}]{\text{H}_2\text{O, } \Delta} \text{amino acids}$$

$$\text{Protein} \xrightarrow[\text{proteolytic enzyme}]{\text{H}_2\text{O, 40°C}} \text{amino acids}$$

In addition to the elements C, H, and O, which make up the carbohydrates and the lipids, the amino acids introduce the nitrogen atom to the proteins; some also contain sulfur. About 20 different amino acids are common to all living organisms, and a great number of other amino acids are specific to some proteins or specific to some given species of animal, plants, or microorganism. Some amino acids also occur free in various cells and tissues, and since they are not associated with proteins, they are known as nonprotein amino acids.

The general structure for the common amino acids is shown in the margin. The right-hand side of the form sketched is common to all the amino acids, so that the difference between them is in the R group. We classify these amino acids according to differences in the R groups as:

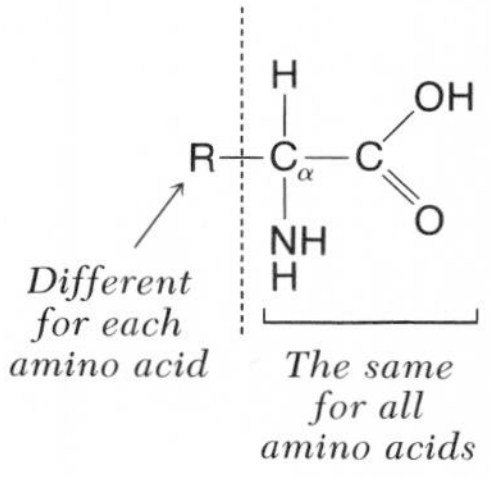

1 Aliphatic monoamino monocarboxylic acids
2 Aromatic
3 Heterocyclic
4 Sulfur-containing
5 Monoamino dicarboxylic acids
6 Diamino monocarboxylic acids

In a newer system of classifying the amino acids they are classified according to whether they have:

1 Nonpolar R groups and are therefore hydrophobic in character and less soluble in water
2 Polar R groups and are therefore less hydrophobic and more soluble in water
3 Negatively charged R group, an acid group at the physiological pH $\approx$ 7
4 Positively charged R group, a basic group at the physiological pH $\approx$ 7

TABLE 15.1 Average elementary composition of proteins

Element	%
C	50–55
H	6–8
O	20–24
N	15–18
S	0.0–3.5
P	0.0–1.5
Fe	0.0–trace
Zn	0.0–trace
Cu	0.0–trace
Mn	0.0–trace
I	0.0–trace

Different for each amino acid   *The same for all amino acids*

TABLE 15.2 Some plant and animal amino acids

Structure		Derived Name	Common Name†	Abbreviation
R Group	Portion Common to All Amino Acids			

Aliphatic Monoamino Monocarboxylic Acids

Structure	Derived Name	Common Name†	Abbreviation
H—C(H)(NH$_2$)—C(=O)OH	α-Aminoacetic acid	Glycine	Gly
H—C(H)(H)—C(NH$_2$)—C(=O)OH	α-Aminopropionic acid	L-Alanine	Ala
H—C(H)(OH)—C(NH$_2$)—C(=O)OH	α-Amino-β-hydroxypropionic acid	L-Serine	Ser
H—C(H)(H)—C(CH$_3$)—C(NH$_2$)—C(=O)OH	α-Amino-β-methylbutyric acid	L-Valine	Val
H—C(H)(H)—C(OH)—C(NH$_2$)—C(=O)OH	α-Amino-β-hydroxybutyric acid	L-Threonine	Thr
H—C(H)(H)—C(CH$_3$)—C(H)—C(NH$_2$)—C(=O)OH	α-Amino-γ-methylvaleric acid	L-Leucine	Leu
H—C(H)(H)—C(H)(H)—C(CH$_3$)—C(NH$_2$)—C(=O)OH	α-Amino-β-methylvaleric acid	L-Isoleucine	Ile

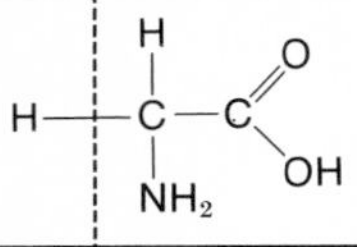

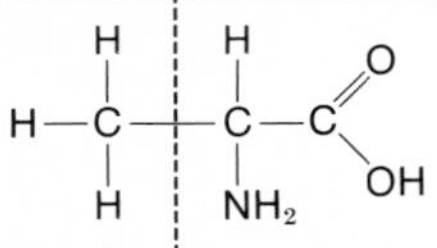

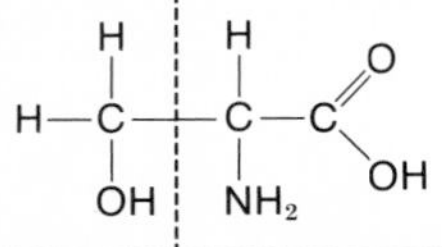

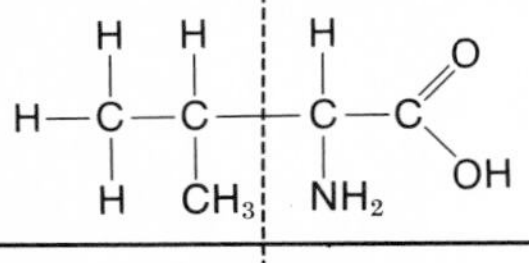

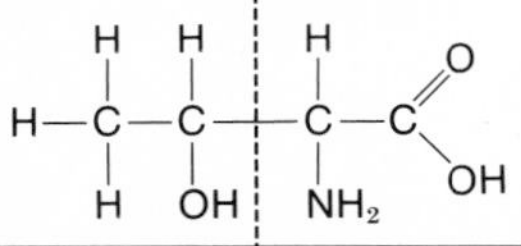

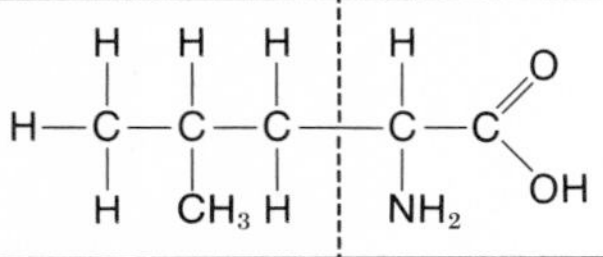

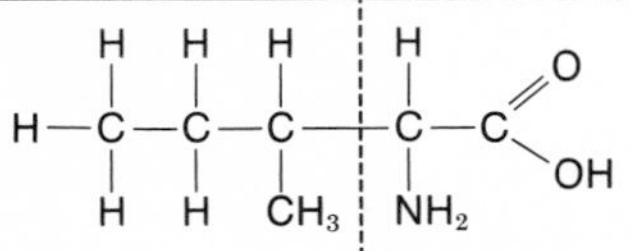

Aromatic Monoamino Monocarboxylic Acids

Structure	Derived Name	Common Name†	Abbreviation
C$_6$H$_5$—C(H)(H)—C(NH$_2$)—C(=O)OH	α-Amino-β-phenylpropionic acid	L-Phenylalanine	Phe
HO—C$_6$H$_4$—C(H)(H)—C(NH$_2$)—C(=O)OH	α-Amino-β-(p-hydroxyphenyl)-propionic acid	L-Tyrosine	Tyr

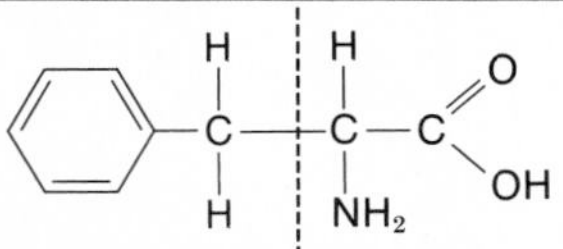

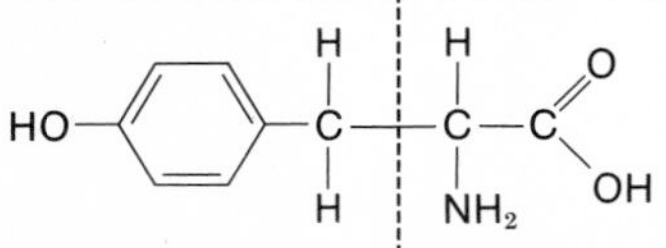

TABLE 15.2 *(continued)*

Structure		Derived Name	Common Name†	Abbreviation
R *Group*	*Portion Common to All Amino Acids*			

Heterocyclic Monoamino Monocarboxylic Acids

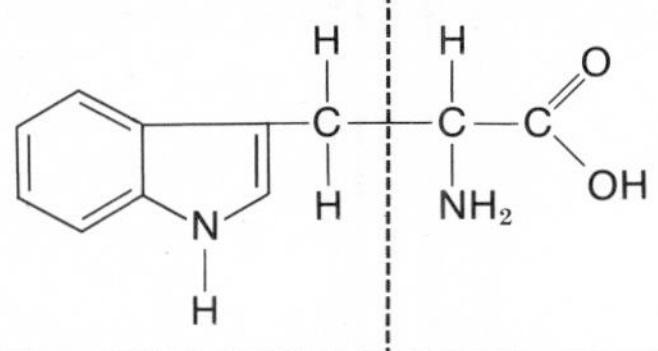	α-Amino-β-indole propionic acid	L-Tryptophan	Trp
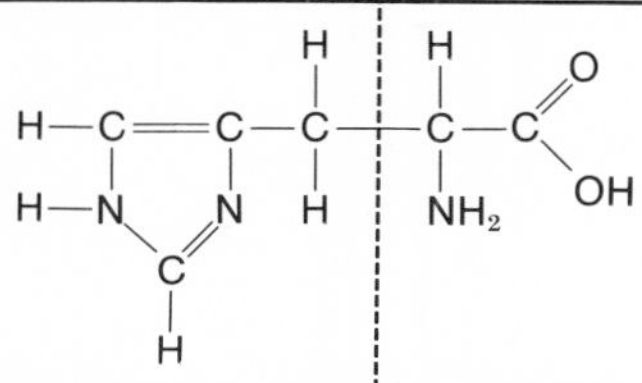	α-Amino-β-imidazole propionic acid	L-Histidine	His
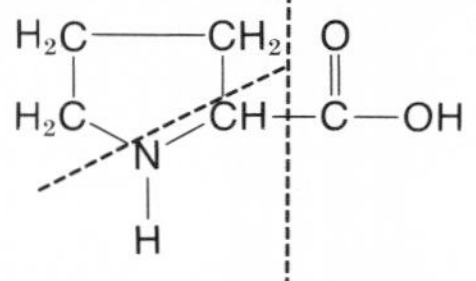	α-Pyrrolidinecarboxylic acid	L-Proline	Pro
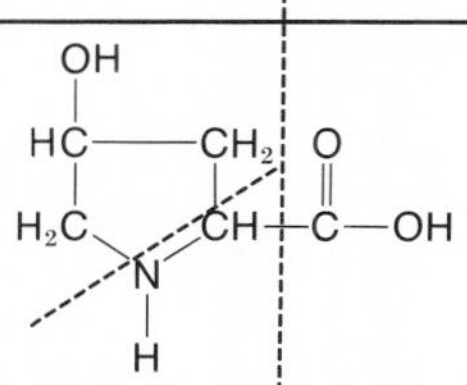	γ-Hydroxy-α-pyrrolidine-carboxylic acid	L-Hydroxyproline‡	Hyp

Sulfur-containing Monoamino Monocarboxylic Acids

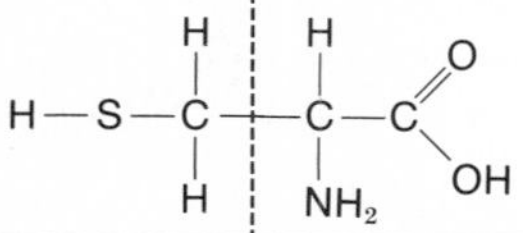	α-Amino-β-mercapto-propionic acid	L-Cysteine	Cys
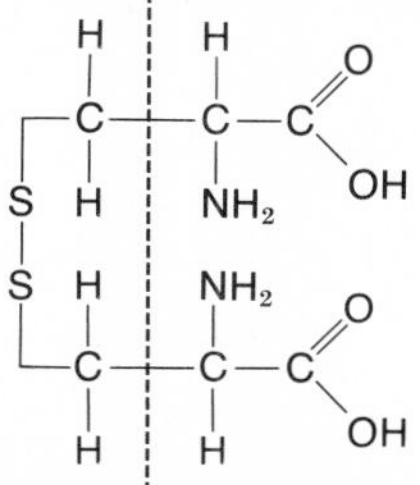	Di(α-amino)-β-mercapto-propionic acid	L-Cystine	Cys
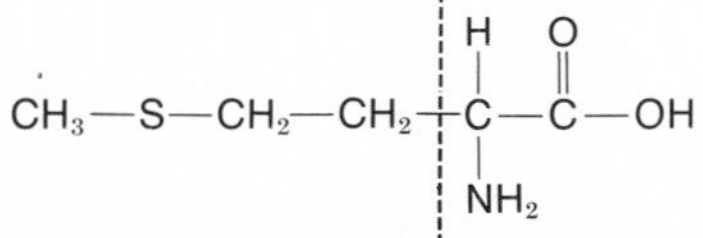	α-Amino-γ-methylbutyric acid	L-Methionine	Met

TABLE 15.2 (continued)

Structure		Derived Name	Common Name†	Abbreviation
R Group	Portion Common to All Amino Acids			
Monoamino Dicarboxylic Acids				
$HO{-}\overset{O}{\overset{\|}{C}}{-}\overset{H}{\underset{H}{\overset{\|}{C}}}{-}$	$\overset{H}{\underset{NH_2}{\overset{\|}{C}}}{-}C\overset{O}{\underset{OH}{}}$	Aminosuccinic acid	L-Aspartic acid	Asp
$\underset{H}{\overset{H}{N}}{-}\overset{O}{\overset{\|}{C}}{-}\overset{H}{\overset{\|}{C}}{-}$	$\overset{H}{\underset{NH_2}{\overset{\|}{C}}}{-}C\overset{O}{\underset{OH}{}}$	Amide of aspartic acid	L-Asparagine	Asn
$HO{-}\overset{O}{\overset{\|}{C}}{-}\overset{H}{\underset{H}{\overset{\|}{C}}}{-}\overset{H}{\underset{H}{\overset{\|}{C}}}{-}$	$\overset{H}{\underset{NH_2}{\overset{\|}{C}}}{-}C\overset{O}{\underset{OH}{}}$	α-Aminoglutaric acid	L-Glutamic acid	Glu
$\underset{H}{\overset{H}{N}}{-}\overset{O}{\overset{\|}{C}}{-}\overset{H}{\underset{H}{\overset{\|}{C}}}{-}\overset{H}{\underset{H}{\overset{\|}{C}}}{-}$	$\overset{H}{\underset{NH_2}{\overset{\|}{C}}}{-}C\overset{O}{\underset{OH}{}}$	Amide of glutamic acid	L-Glutamine	Gln
Diamino Monocarboxylic Acids				
$N{=}\overset{}{C}{-}\underset{NH}{}{N}{-}\overset{H}{\underset{H}{C}}{-}\overset{H}{\underset{H}{C}}{-}\overset{H}{\underset{H}{C}}{-}$	$\overset{H}{\underset{NH_2}{\overset{\|}{C}}}{-}C\overset{O}{\underset{OH}{}}$	α-Amino-δ-guanidinovaleric acid	L-Arginine	Arg
$\underset{H}{\overset{H}{N}}{-}\overset{H}{\underset{H}{C}}{-}\overset{H}{\underset{H}{C}}{-}\overset{H}{\underset{H}{C}}{-}\overset{H}{\underset{H}{C}}{-}$	$\overset{H}{\underset{NH_2}{\overset{\|}{C}}}{-}C\overset{O}{\underset{OH}{}}$	α,ε-Diaminocaproic acid	L-Lysine	Lys
$\underset{H}{\overset{H}{N}}{-}\overset{H}{\underset{OH}{C}}{-}\overset{H}{\underset{H}{C}}{-}\overset{H}{\underset{H}{C}}{-}\overset{H}{\underset{H}{C}}{-}$	$\overset{H}{\underset{NH_2}{\overset{\|}{C}}}{-}C\overset{O}{\underset{OH}{}}$	α,ε-Diamino-δ-hydroxycaproic acid	L-Hydroxylysine‡	Hyl

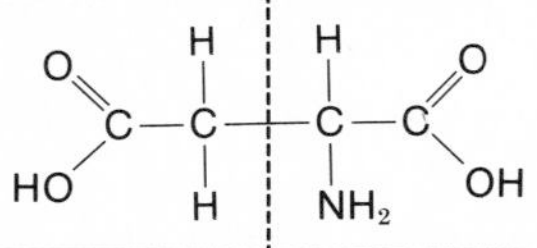

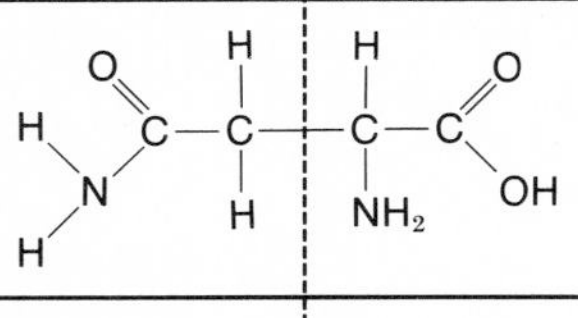

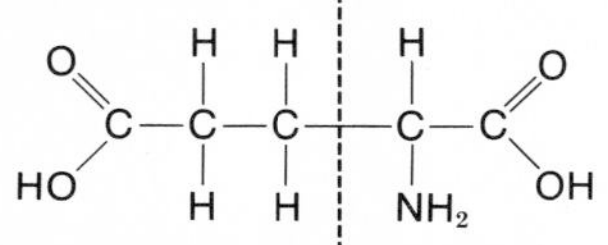

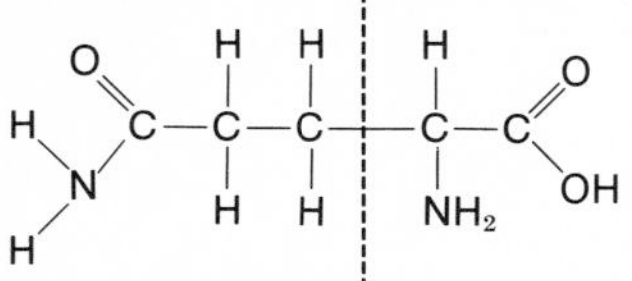

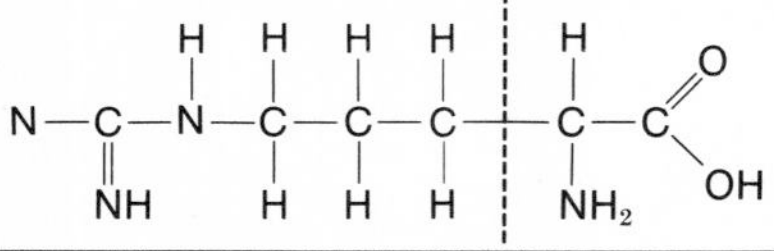

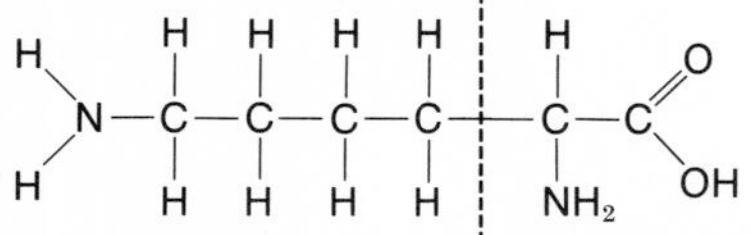

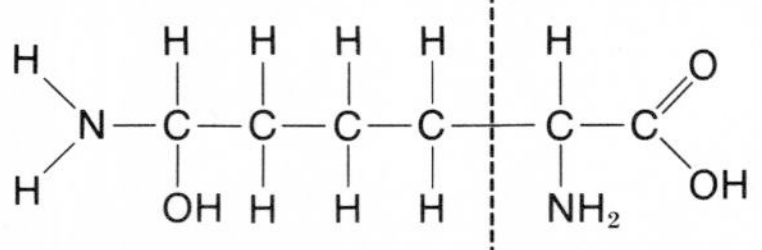

† Essential amino acids are underlined.
‡ Hydroxy derivatives primarily found in connective tissue such as collagen.

In Table 15.2, showing common amino acids and some of their derivatives as they appear in proteins, both systems are used. All the common amino acids are α-amino acids, and since the α carbon is asymmetric, they exhibit optical activity. The one exception is the amino acid glycine, whose R group is hydrogen, so that the carbon is not asymmetric. The naturally occurring amino acids are of the L configuration, although D-amino acids occur in some microorganisms.

The organic nomenclature system of the IUPAC uses numbers to label carbon atoms, but in the derived system Greek letters are used. The naming of the natural amino acids follows the derived system. Figure 15.1 shows the relationship between the two systems.

The amino acids are also known by their common names, which are abbreviated as shown in Table 15.2.

The convention adopted for the designation of the spatial configuration of the optically active carbon atom of glyceraldehyde was explained in Chapter 13. A similar idea is applied to the optically active amino acids (Figure 15.2). Remember that D and L refer only to the spatial configuration, whereas + and − designate the direction of the optical rotation. This means that an L configuration may be either a + or − and therefore rotate light to the right or left.

All the commonly occurring amino acids are necessary for good health. Of these, 10 are referred to as **essential amino acids,** not because they are any more essential than the others but because apparently the body does not produce them in sufficient amounts, so that they must be supplied by the diet. If they are absent from our food, nutritional deficiencies may result.

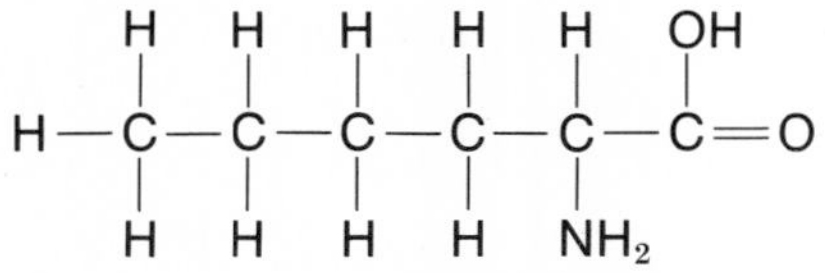

2-Aminohexanoic acid

IUPAC

6	5	4	3	2	1	*The carboxyl group is number 1.*
ϵ	δ	γ	β	α		

α-Aminocaproic acid

Derived

In the derived system the carbon adjacent to the carboxyl group is α.

FIGURE 15.1 The IUPAC and derived systems for naming amino acids.

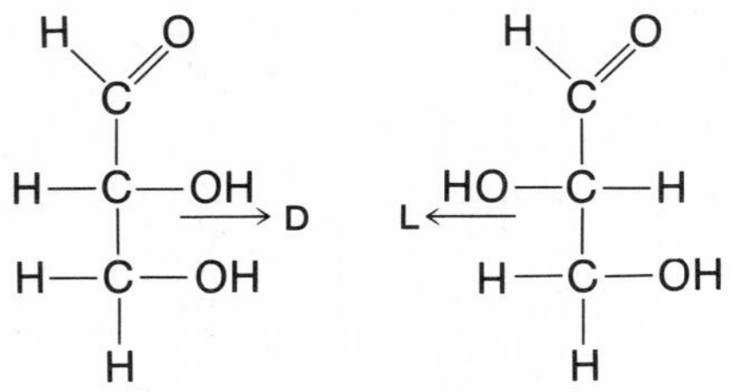

D-*Glyceraldehyde* L-*Glyceraldehyde*

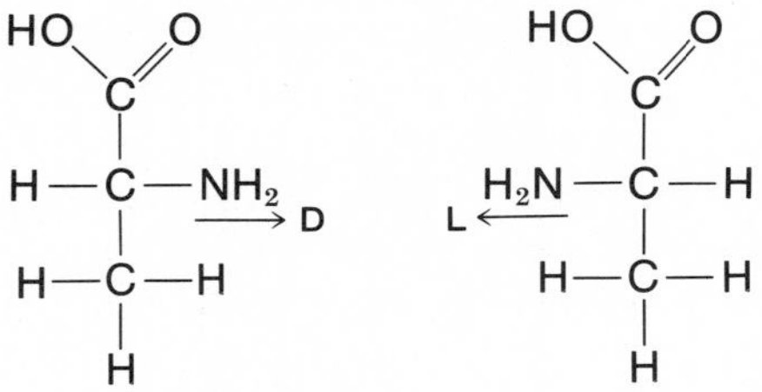

D-α-*Aminopropionic acid,* L-α-*Aminopropionic acid,*
D-*alanine* L-*alanine*

FIGURE 15.2 The optically active forms of glyceraldehyde and aminopropionic acid.

M *methionine*
I *isoleucine*
L *lysine*
L *leucine*
P *phenylalanine*
A *arginine*
T *threonine*
H *histidine*

T *tryptophan*
V *valine*

The mnemonic device in the margin will help you remember them.

As mentioned earlier, some special amino acids are found in specific proteins, and others are found free in cells and body fluids as nonprotein amino acids (Table 15.3).

PHYSICAL AND CHEMICAL PROPERTIES

Amino acids are polar molecules and more or less soluble in water, forming true solutions. Their melting points are high for

TABLE 15.3 Amino acids and their functions

Amino Acid	Function
Hydroxyproline	In connective tissue; collagen
Hydroxylysine	In connective tissue; collagen
Desmosine	In connective tissue; elastin
Phosphoserine	In casein in milk
Thyroxine	Thyroid-gland hormone, contains iodine
Citrulline	Intermediate in the urea cycle
Ornithine	Intermediate in the urea cycle
Homocysteine	Intermediate in amino acid metabolism
Canavanine	Plant amino acid, poisonous
β-Cyanoalanine	Plant amino acid, poisonous

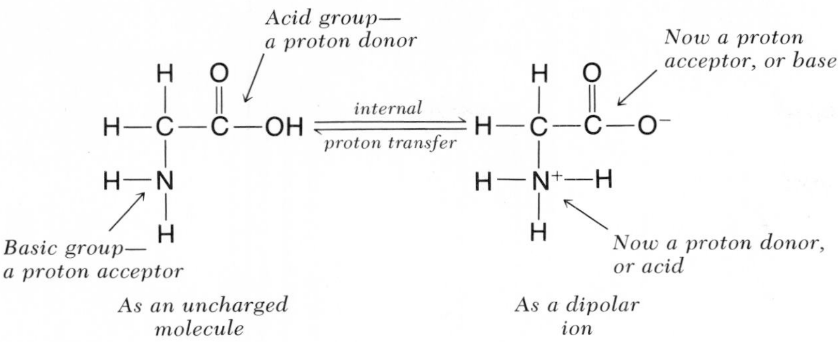

FIGURE 15.3 The amphoteric nature of glycine.

organic compounds, usually above 200°C. In aqueous solutions they migrate under the influence of a direct electric current.

Acid-base properties Let us take a closer look at the amino acids. Their interesting physical and chemical behavior is related to:

1 The amino and acid groups of the molecule
2 The structure of the R group

The amino acids have a dual nature, since there is a carboxylic acid group, —COOH, an acid function, and an amino group, —NH$_2$, a basic function. With both acidic and basic features, it is **amphoteric** (Figure 15.3). Note that the net charge on the **dipolar ion** (also called a **zwitterion**) is zero, although it contains both a positive and a negative part. An amino acid molecule generally exists in the dipolar form in aqueous solution, but this depends on the pH. As a base, it neutralizes an acid:

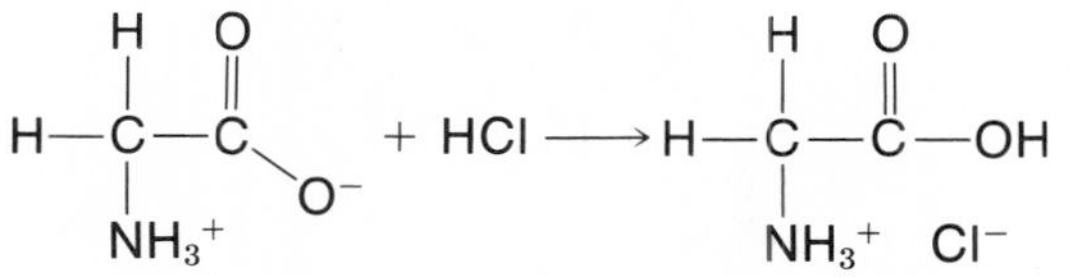

Glycine hydrochloride

As an acid, it neutralizes a base:

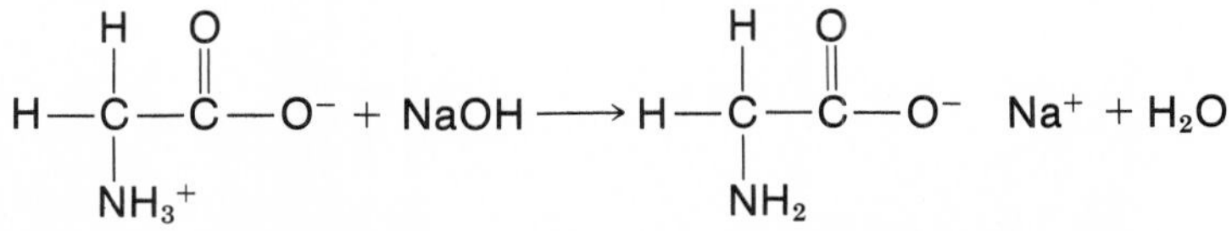

Sodium glycinate

At a certain pH amino acids exist almost entirely in the dipolar form. At this pH, which is different for each amino acid, it has as many positive charges as negative ones and is therefore **isoelectric.** Since the charges are equal and opposite, the net charge is zero; the pH at which this occurs is called the **isoelectric point.**

PROTEINS

Proteins are macromolecules with molecular weights ranging from several thousands to several millions. They are built up by successive condensation of amino acids, the amino acids being joined together by peptide bonds, and contain the elements C, H, O, N, S, and often P. Because of their large molecular size, they form colloidal dispersions instead of true solutions. They are kept in the colloidal state by the repulsive forces between their similar surface charges, which prevent aggregation and precipitation of the proteins. Colloidal dispersions of proteins are not

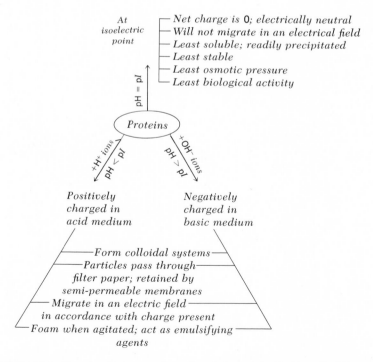

FIGURE 15.4 The physical properties of proteins related to the pH of the medium in which they are acting.

affected by gravitational forces but can be sedimented out of dispersion by ultracentrifugation, a powerful isolation and differentiation technique for the proteins.   Proteins are not removed from a true solution by filtration but can be retained on ultrafilters such as millipore filters, and they cannot pass through semipermeable membranes.   As discussed in Chapter 6, dialysis is a process whereby proteins can be separated from solutions; semipermeable membranes *retain* the protein but permit the passage through the membrane of dissolved particles.   Otherwise essential components of the blood, such as the plasma proteins, would constantly be leaking out.   Colloidal dispersions show the Tyndall effect, and when agitated, they foam.   Both phenomena are simple physical tests indicating the presence of proteins.   Under an ultramicroscope they show the typical brownian movement of colloidal particles.   In an acidic medium the proteins are positively charged, whereas in an alkaline medium they are negatively charged (Figure 15.4).   Like an amino acid, each protein has its own isoelectric point, where there is no net charge on the molecule.   At this point, they are most vulnerable and are readily precipitated.   Furthermore, at this point they produce the least osmotic pressure, have the least biological activity, and do not move under the influence of an electric field.

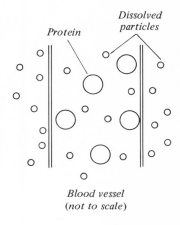

Protein   *Dissolved particles*

Blood vessel (not to scale)

Brownian movement is the random, zigzag motion of colloidal particles due to the continuous motion of the water molecules around them.

THE STRUCTURE OF PROTEINS

The organization of a protein can be divided into four successively higher structural levels, each characterized by specific features of bonding, atomic interaction, shape, and three-dimensional arrangement.   Since the proteins achieve their three-dimensional configurations spontaneously, we know that these arrangements are stable.   The specific organization of a protein is critical for the proper biological activity of that protein.   A loss of organized structure can mean a breakdown in the biological function of the protein.

THE PRIMARY STRUCTURE

The **primary structure** of a protein is the specific sequence of the amino acids in the protein.   The bonds that hold that sequence together are peptide bonds.   A polypeptide looks something like Figure 15.5.   It is important to specify each amino acid in the sequence of the primary structure; because protein molecules are so large, the abbreviated names for the amino acids are used:

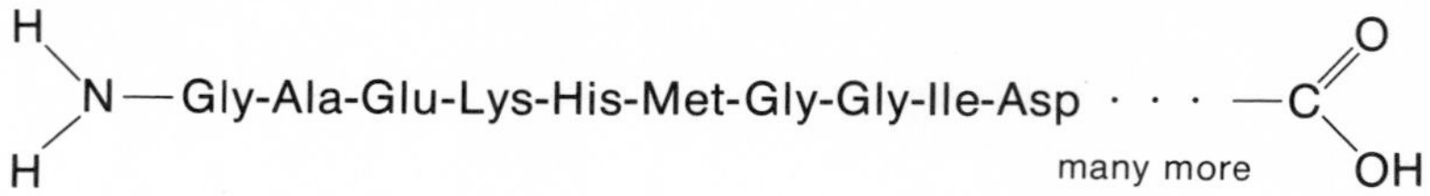

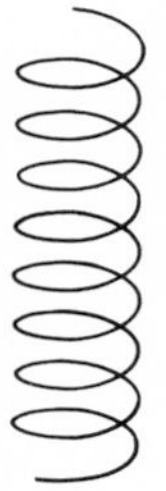

$$\begin{array}{c} H \\ | \\ N \\ | \\ H \end{array} \text{—Gly-Ala-Glu-Lys-His-Met-Gly-Gly-Ile-Asp} \cdots \text{—C} \begin{array}{c} O \\ \\ OH \end{array}$$

many more

Alpha-helix structure

intra = within

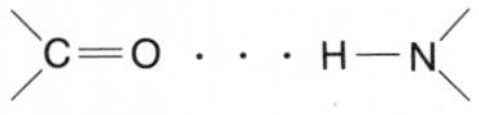

$\searrow C = O \cdots H - N \nearrow$

Carbonyl   Imino
group   group

THE SECONDARY STRUCTURE

The primary structure results in a proper amino acid sequence but not a biologically active protein. The secondary structure of a polypeptide chain is its further organization through hydrogen bonding. In general, this can be done in two ways. The first is for the chain to arrange itself in the shape of a spring, called an **alpha helix,** in which hydrogen bonds are established between successive turns of the helix; this *intra*molecular hydrogen bonding is formed between the carbonyl group and the imino group at intervals of three amino acid residues along the helix. The hydrogen bonds are the helix stiffeners (Figure 15.6).

The second way the polypeptide chain can achieve a stable structure is by hydrogen bonding between *different* polypeptide

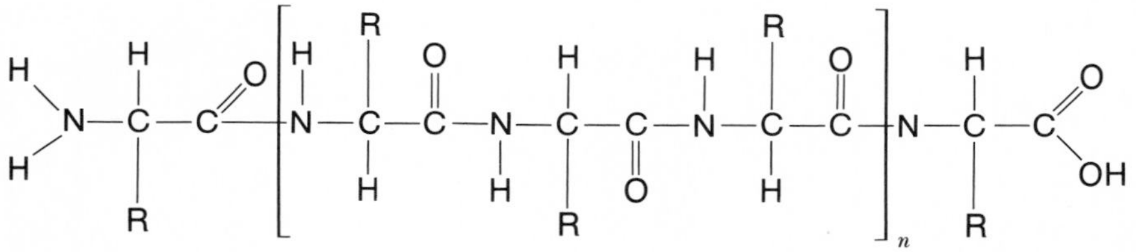

NH₂ or N—terminal
end of chain

—COOH or C—terminal
end of chain

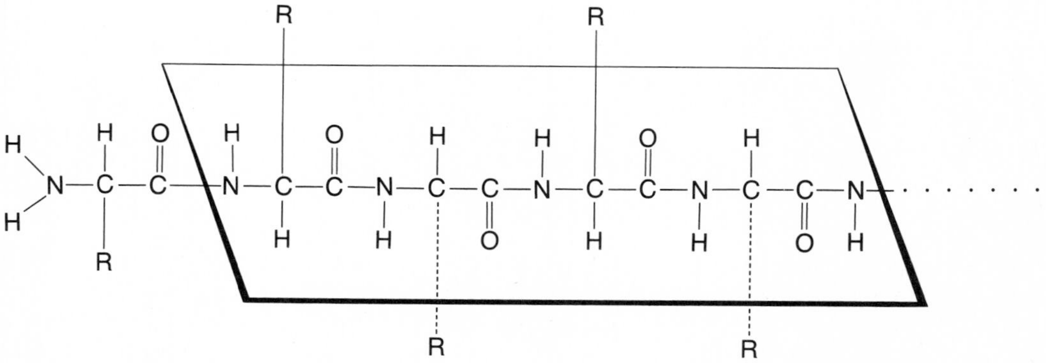

FIGURE 15.5 The backbone of a protein is its primary structure. The peptide bonds are located in a plane, with the R groups above and below the plane, and transrelative to each other.

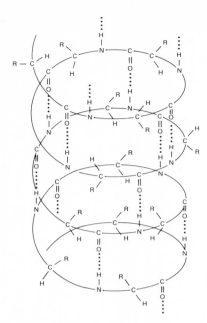

FIGURE 15.6 The alpha helix. Hydrogen bonds stabilize the alpha-helix structure, while the R groups protrude out of the helix.

chains, that is, by *inter*molecular hydrogen bonding. The polypeptide chains thereby stabilize each other and take on a conformation known as a pleated-sheet arrangement (Figure 15.7). The alpha-helix arrangement appears in all types of proteins, whereas the pleated-sheet arrangement is found in the fibrous or scleroproteins such as silk, hair, myosin of muscle, and elastin.

inter = between

THE TERTIARY STRUCTURE

The **tertiary structure,** which involves the bending and folding of the polypeptide chain, is found in conjunction with coiling in globular proteins. The picture becomes complicated, but basi-

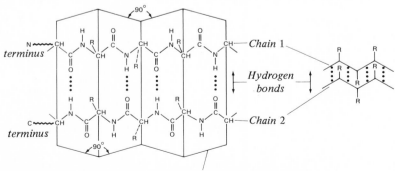

FIGURE 15.7 The pleated-sheet arrangement of the polypeptide chain.

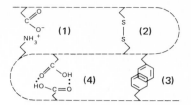

FIGURE 15.8 The tertiary structure of a polypeptide chain showing the four types of bonds that maintain the proper conformation: a, salt bridges, or ionic bonds, b, disulfide linkages, c, hydrophobic interaction, d, hydrogen bonds.

cally it represents a three-dimensional arrangement of the polypeptide chain (Figure 15.8). To ensure the proper conformation of such a structure several types of bonds, involving primarily the R groups of the amino acid residues, are of importance here:

1 Salt bridges, or ionic bonds
2 Disulfide linkages
3 Hydrophobic interactions
4 Hydrogen bonds

All these bonds (Figure 15.8) occur wthin the polypeptide chain, and the proper sequence of the amino acids is of utmost importance to ensure the conformation of the protein molecule essential for biological activity.

Salt bridges Salt bridges can be formed within a polypeptide chain between the R groups of lysine and arginine, each with a basic ($—NH_2$) group at the end, and the R groups of aspartic acid and glutamic acid, each with an acidic ($—COOH$) group at the end of the side chain (Figure 15.9).

Disulfide linkages This type of linkage is formed by the oxidation of two cystine residues in the polypeptide chain to produce cystine with a sulfur linkage (Figure 15.10).

Hydrogen bonds Hydrogen bonds have already been mentioned as the stabilizing influence in the secondary structure of a pro-

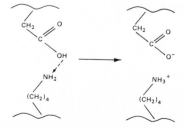

FIGURE 15.9 The formation of salt bridges.

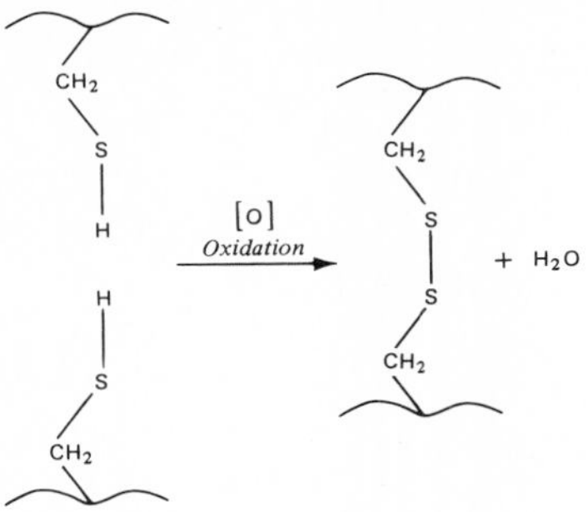

FIGURE 15.10 The formation of a disulfide linkage.

tein involving the carbonyl group and the imino group of the peptide bond. The polar R groups also provide possibilities for hydrogen-bond formation, as for the residues of histidine, tyrosine, serine, aspartic and glutamic acid, and others (Figure 15.11).

Hydrophobic interactions Hydrophobic interactions occur because of the polar nature of water, the general solvent in biological systems. A number of amino acids such as phenylalanine, leucine, isoleucine, and valine have nonpolar side groups. One of the characteristics of these R groups is that they are hydrophobic; they are nonpolar structures in a polar medium. These

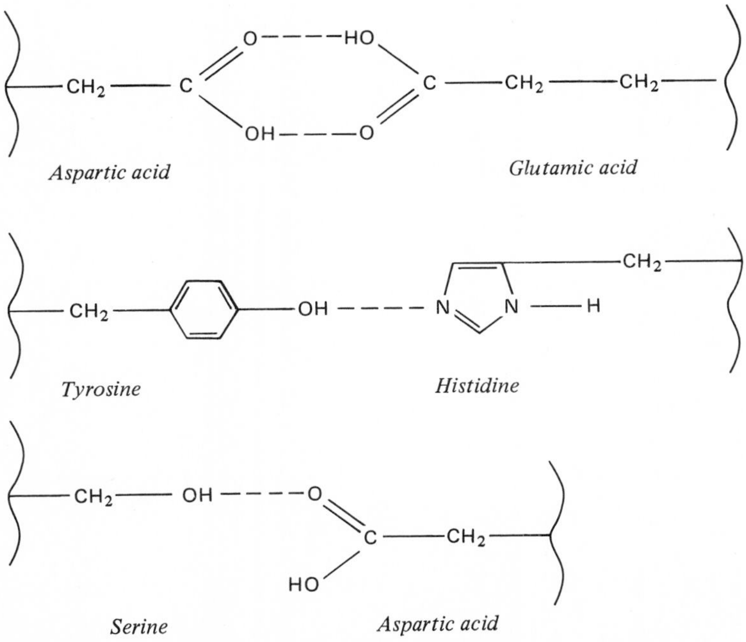

Aspartic acid   *Glutamic acid*

Tyrosine   *Histidine*

Serine   *Aspartic acid*

FIGURE 15.11 Hydrogen-bond formation by R groups.

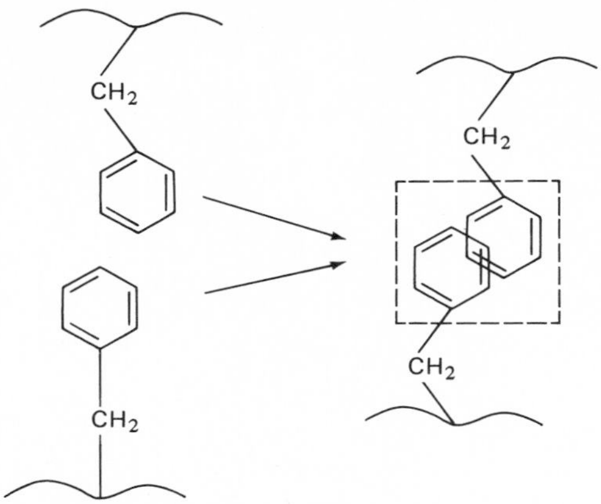

FIGURE 15.12 Hydrophobic interaction.

groups neither want the solvent water nor are wanted by it; when they come close enough together, they overlap and are held in this position by the action of water somewhat like two slices of salami held by bread in a sandwich (Figure 15.12) or two friends sticking together in a crowd of strangers.

QUATERNARY STRUCTURE OF PROTEINS

This structure represents two or more polypeptide chains joined into a larger unit. It may be made from identical units, in which case it is a **homogeneous** quaternary structure, or it may be made up of different units, in which case it is a **heterogeneous** quaternary structure. These multiple units are usually held together by electrostatic forces and by hydrophobic interaction between the different units. The quaternary structure is of importance for the biological activity of the protein. In certain instances only the complete quaternary structure is active; for example, the enzyme phosphorylase contains four subunits which must be held together for the enzyme to function. In other cases only the individual subunits are biologically active, whereas the quaternary structure is inactive. The change from active to inactive by forming and unforming the quaternary structure is a fast way of turning an enzyme on or off.

CLASSIFICATION OF PROTEINS

There are a great many types of proteins varying widely in properties, functions, and composition, so that no single classification fits them all. For this reason, several classifications are used, each contributing added insight into the nature of proteins.

From the standpoint of composition there are:

1 Simple proteins, which on hydrolysis yield only amino acids or their derivatives.
2 Conjugated proteins, which on hydrolysis yield amino acids plus another organic or inorganic compound. Examples are phosphoproteins, which yield amino acids and phosphates, and glycoproteins, which yield amino acids and carbohydrates.

Proteins can be further classified by function into such groups as enzymes, hormones, antibodies, and structural proteins. For separating proteins in a mixture their different solubilities in different solvents are utilized.

Protein *shape* is a simple and useful method of classification. In general terms a protein is either globular or linear. Although not exactly spherical, globular proteins have much less surface area than fibrous proteins. They are found in enzymes, blood, and inside and outside the cell. They are essential to the biological functioning of the organism. On the other hand, the fibrous, or linear, proteins are largely involved in the *structure* of the organism, being found in cell walls, skin, hair, ligaments, and tendons.

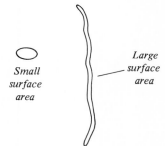

Small surface area

Large surface area

DENATURING OF PROTEINS

Denaturing of proteins is the disturbance or destruction of the native protein configuration with loss of biological activity. In many cases, this reaction is irreversible; in other cases, frequently in the presence of enzymes, the protein molecule will re-form its native structure. Table 15.4 lists denaturing agents and their effects on the protein molecule. These agents break hydrogen bonds, hydrophobic bonds, disulfide linkages, and salt bridges, resulting in the unfolding and opening of the polypeptide chain, which exposes the active groups contained in the molecule such as —SH, —OH, and —COOH. It also reduces the solubility of the molecule at the isoelectric point and thereby enhances precipitation. The denaturing agent does not break the primary structure of the protein, as the peptide bonds are not ruptured. Denaturation is the normal process that begins the digestion of a protein food, and denatured proteins are more readily digested because they are open and unfolded and more vulnerable to attack by enzymes. No change in the nutritive value of the protein occurs by denaturing. In the food industry the knowledge of denaturing is valuable because it has effects on flavors and texture of foods containing proteins.

TABLE 15.4   Denaturing agents and their effect on proteins

Agent	Effects on Protein Molecule
Heat	Breaking hydrogen and hydrophobic bonds; unfolding chain; coagulation
Strong acids and alkalis	Breaking salt bridges, unfolding polypeptide chain
Hydrophilic organic solvents: alcohol, acetone, and others	Breaking hydrogen bonds; removal of bound water from protein molecule; coagulation
Urea, detergents	Breaking hydrogen bonds; unfolding chain
Trichloroacetic, tannic, phosphotungstic, and phosphomolybdic acid	Breaking salt bridges; precipitation of protein
Ultraviolet radiation, X-rays	Adding energy to protein molecule, causing bonds to split
Vigorous shaking and stirring	Unfolding of the polypeptide chain; formation of foam, greater insolubility of protein

PHYSICAL AND CHEMICAL PROPERTIES OF PROTEINS

These properties are often used to isolate and identify the protein and to determine its composition from the point of view of amino acid content and the presence of prosthetic groups in the molecule. They are also used to determine the sequence of the amino acids in the protein.

PHYSICAL PROPERTIES

As indicated previously, proteins are macromolecules and colloidal in size. The physical properties of colloids are used in protein studies for the determination of molecular weight, separation, and qualitative and quantitative observations. The following properties are useful in determining molecular weight:

1 Sedimentation by ultracentrifugation
2 Light scattering, use of Tyndall effect
3 Osmotic pressure
4 Viscosity
5 Diffusion rate

The most useful method for the determination of the molecular weight is sedimentation. By placing a protein dispersion in an ultracentrifuge and subjecting it to 100,000 times the earth's gravitational force g (obtained by spinning the centrifuge rotor at 60,000 rpm) the protein is made to settle out. The protein is denser than water and is forced to the bottom of the tube. The rate of sedimentation can be recorded by optical methods, and the molecular weight can be determined from the sedimentation constant. If several proteins with different molecular weights are present, a sedimentation constant for each can be obtained as well as separation of the proteins.

All the above methods are used together to get the most accurate results. The greatest error is caused by the shape of the protein molecules, which greatly affects the motion of the protein in any medium.

CHEMICAL PROPERTIES

In addition to a general test to indicate whether a material is a protein or not, there are specific tests to confirm the presence or absence of various R groups. These color reactions, such as the Millon and Hopkins-Cole tests, serve to identify the amino acid components of the native protein. The general test for the presence of a protein is the biuret reaction. In a strongly alkaline medium proteins will give a purple color with a dilute copper sulfate solution. The color is produced by the complexing of copper(II) ion with peptide linkages. The color reaction can also be used to obtain quantitative results by measuring its intensity by means of an electrophotometer, since intensity varies with the amount of protein present (Figure 15.13).

The determination of total protein is often of great importance in the food industry, in nutritional studies, and related areas. A standard method for this determination is known as the **Kjeldahl test;** several modifications of this test are used, but essentially it is based on the fact that "digestion" of an organic sample with concentrated sulfuric acid, which is a strong oxidizing agent, will

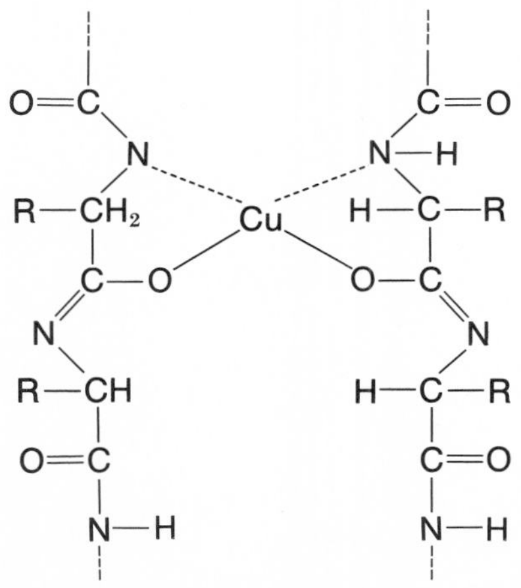

FIGURE 15.13 The copper-peptide color complex formed as the product of the biuret reaction.

oxidize the carbon and hydrogen in the sample to carbon dioxide and water but leave the nitrogen as ammonia. Being basic, the ammonia reacts with the sulfuric acid to form ammonium sulfate, $(NH_4)_2SO_4$, while SO_2 is produced by the reduction of the acid:

$$\text{Organic compound} + H_2SO_4 \xrightarrow{\text{boil}} CO_2 + H_2O + (NH_4)_2SO_4 + SO_2$$

When digestion of the sample is complete, the acidic solution is made alkaline, thereby liberating ammonia, which is distilled into a standard acid. Since the liberated ammonia will neutralize the acid, the amount of ammonia obtained from the sample is measured by a titration of the remaining unneutralized acid against a standard base. The percentage of nitrogen in the sample is then given by

% N

$$= \frac{\text{ml acid} \times N \text{ of acid} - \text{ml of base} \times N \text{ of base} \times 0.014 \times 100}{\text{sample weight}}$$

$$N \approx 16\% = \frac{16}{100} = \frac{1}{6.25}$$

Hence, % $protein$ = 6.25 (%N)

The average nitrogen content in a protein is about 16 percent. By dividing 16 into 100 we get the factor 6.25. By multiplying the percentage of nitrogen obtained in the Kjeldahl determination by the factor 6.25 we find the percentage of protein in the sample. Since in natural products nitrogen compounds other than proteins are also found, the Kjeldahl test will give the total nitrogen content, including the nonprotein nitrogen. For this reason, the term **crude protein** is used for the values obtained by the Kjeldahl test.

REVIEW QUESTIONS

1 What elements are contained in amino acids?
2 Give the general structure of the α-amino acids.
3 Make a list of the different types of amino acids and write the structure of several amino acids for each type.
4 Differentiate between D-amino acids and L-amino acids.
5 What is a β-amino acid?
6 From memory write as many essential amino acids as you can.
7 List some of the nonprotein amino acids found in the body and their functions.
8 What is meant by essential amino acids?
9 Explain the amphoteric character of the amino acids.
10 What is the isoelectric point of an amino acid?
11 Take three different amino acids and connect them by the peptide linkage. Name the product so obtained.
12 What elements are found in proteins?

13 Discuss some of the properties of proteins.

14 What is the definition of the primary structure of a protein?

15 What type of bond is found in the secondary structure of a protein?

16 What is the alpha-helical structure of a protein?

17 Use structures to show the bonds found in the tertiary structure of proteins.

18 What can you say about the shape of different protein molecules?

19 Make a drawing of a parallel-pleated-sheet structure of a protein.

20 Describe the different functions that proteins perform in the body.

21 Proteins are easily denatured. What is meant by denaturing, and how can it be done?

22 Different physical methods are used to determine the molecular weight of proteins. What are they?

23 What is the general test for proteins, and what is the reaction?

24 The following data were obtained in the test for protein in food sample:

$$\text{Sample weight} = 0.75 \text{ g}$$
$$\text{Amount } 0.12 \, N \text{ acid used} = 50 \text{ ml}$$
$$\text{Amount } 0.14 \, N \text{ base used} = 28 \text{ ml}$$

What are the percentages of nitrogen and protein in the sample?

16

ENZYMES
The Invaluable Busybodies

ENZYMES AS BIOLOGICAL CATALYSTS

The chemical reactions we have discussed have been either exothermic or endothermic. In either case, an initial investment of energy is generally required to get the reaction going, the energy of activation. For organic reactions especially, the energy of activation can be a decisive factor. If the energy barrier is too high, even a spontaneous reaction may occur much too slowly to be useful, since too few of the reacting molecules are able to get over the barrier and react. Every reaction involved in the chemistry of living matter is associated with an organic catalyst, called an **enzyme,** which speeds up the reaction so that it can contribute to the overall chemical process (Figure 16.1). The action of enzymes is therefore essential in keeping the traffic moving along the chemical pathways that characterize living organisms.

Being catalysts, enzymes behave like catalysts:

1 The enzyme speeds up the reaction, nothing else.
2 The enzyme facilitates the reaction: It helps it happen.

3 The enzyme does not alter the amount or the kind of products formed.
4 The enzyme does not affect the amount of energy released by the reaction (if exothermic) or the amount of energy absorbed (if endothermic).
5 Since the enzyme is itself not used up in the reaction, the amount of enzyme needed can be very small and its effects very big.

As the biological catalysts that facilitate biological reactions inside and outside the living cells of plants and animals, enzymes are unique in that no chemist has been able to duplicate or even closely imitate enzymatic action under the conditions under which enzymes work. Let us look at a simple, everyday example. The hydrolysis of a protein by chemical means requires boiling the protein with strong HCl or strong NaOH solutions for 48 to 72 h. The body does more with less drastic methods. Our digestive juices have enzymes that break down ingested proteins in a relatively short time, 2 to 6 h, and they do this at body temperature. Without these enzymes we would presumably have to follow up a delicious meal with a stiff drink of hydrochloric acid and then sit on a hot stove. What a revolting thought, and thank you enzymes! How enzymes manage to do so much so well is a major subject of chemical study. Even if many things are still unclear, much has been learned of the structure of enzymes—how they work and the factors that influence their performance.

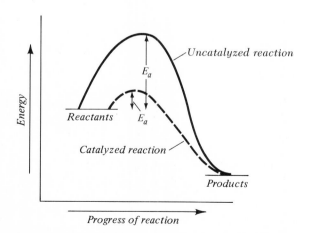

FIGURE 16.1 Catalysis lowers the energy of activation of a reaction.

COMPOSITION OF ENZYMES

The major feature of enzyme composition is that they are protein materials, in whole or for the most part. Another central feature of enzymes is that a relatively small change in their composition can turn an enzyme on or off, can make it biologically *active* or *inactive*. In its active form the enzyme is complete, or whole, and is called a **holoenzyme;** its composition and structure fit it for its job. In its inactive form, part of the enzyme is missing, or it needs some change in structure or composition. In general, the conversion of an enzyme from its inactive to its active form is accomplished by an **activator,** and on this basis there are three major types of enzymes.

Type 1 The entire enzyme is a globular protein, and in its inactive form is called a **zymogen** or a **proenzyme.** The role of the activator is to modify the structure of the protein, converting it to the active holoenzyme form:

$$\text{Zymogen} + \text{activator} = \text{holoenzyme}$$
$$\text{(Inactive)} \qquad\qquad\qquad \text{(Active)}$$

Type 2 The enzyme consists of a globular protein, called the **apoenzyme,** plus an organic part called a **coenzyme,** which is a **prosthetic group.** If the coenzyme is removed from the protein portion, the remaining apoenzyme is not active. If the coenzyme is restored to the protein part, enzymatic activity is also restored. It is interesting and important that many coenzymes are vitamins or nucleotides.

$$\text{Apoenzyme} + \text{coenzyme} = \text{holoenzyme}$$
$$\text{(Inactive)} \qquad \text{(Activator)} \qquad \text{(Active)}$$

Type 3 As before, the apoenzyme is a globular protein, but the activator is a **metal ion,** rather than an organic structure. Most of the trace elements found in the body have been identified as such metal-ion activators. Among the many ions that act as activators are K^+, Ca^{2+}, Cu^{2+}, Fe^{2+}, Mg^{2+}, Co^{2+}, Mn^{2+}, Cl^-, and I.

$$\text{Apoenzyme} + \text{metal-ion activator} = \text{holoenzyme}$$
$$\text{(Inactive)} \qquad\qquad\qquad\qquad \text{(Active)}$$

Since the holoenzymes here contain a metal atom, they are also called **metalloenzymes.**

The enzymes that facilitate the digestion of foods in the stomach and the intestinal tract belong to the first, all-protein type of

TABLE 16.1 Some all-protein enzymes and activators

Inactive Zymogen	Activator	Active Enzyme
Pepsinogen	H$^+$ or pepsin	Pepsin
Trypsinogen	Enterokinase	Trypsin
Prothrombin	Prothrombinase + Ca^{2+}	Thrombin
Fibrinogen	Thrombin	Fibrin

enzyme. The activators of these digestive enzymes include a special group of organic materials called **kinases.** The digestive enzymes are secreted in the body as inactive zymogens, so that there is no unwanted and damaging digestive action between the time and place the enzymes are produced and the time and place where they must do their work. Only at the proper digestive site does the activator convert the zymogen to the corresponding active enzyme (Table 16.1).

ENZYME CLASSIFICATION

Enzymes can be classified in several different ways:

1 According to the compound, or **substrate** on which they are acting
2 According to the reaction they catalyze
3 According to the source from which they are obtained and the location within the source where they work

ACCORDING TO SUBSTRATE

The enzymes are named by adding the ending *-ase* to the name of the substrate, meaning the compound on which the enzyme acts.

Very general:
 A protease, acting on proteins A lipase, acting on lipids
 A carbohydrase, acting on carbohydrates
Very specific:
 Sucrase, acting on sucrose
 Urease, acting on urea

ACCORDING TO REACTION BEING CATALYZED

In 1962 the International Union of Biochemistry (IUB) recommended a classification of enzymes into six groups according to the reactions they catalyze.

Oxidoreductases These enzymes are associated with oxidation-reduction reactions; for example, oxidases and dehydrogenases, which remove or add hydrogens.

Transferases These enzymes transfer chemical groups from one molecule to another. In this process, the chemical group being transferred is not in a free state. Examples are **transaminases,**

which transfer —NH$_2$ groups (amino groups); one-carbon transferases, which transfer one-carbon fragments in the form of a CH$_3$— methyl group or a —CHO, formyl group; and phosphate-transfer enzymes, which transfer phosphate groups.

Hydrolases The enzymes that catalyze hydrolysis reactions in which bonds are broken by the introduction of water are called hydrolases. The enzymes in this group participate mostly in the breakdown of foods and include carbohydrases, lipases, and proteases.

Lyases These enzymes catalyze the addition or removal of a chemical group by other means than hydrolysis or oxidation-reduction. They differ from transferases in that the chemical group is liberated or taken up in the free state, which is not the case with transferases. Examples of lyases are decarboxylases, which remove a carboxyl group and thus liberate CO$_2$; dehydratases, which remove water; and hydratases, which add water as in changing citric acid into isocitric acid in the Krebs cycle.

Isomerases Enzymes that catalyze the intramolecular rearrangement of molecules to form an isomer of the original molecule without gain or loss of electrons are called isomerases. Examples of isomerases are retinene isomerase, which changes *all-trans*-retinene to 11-*cis*-retinene (a cis-trans isomerase), and triosephosphate isomerase, which changes dihydroxyacetone phosphate to glyceraldehyde 3-phosphate.

Ligases Also called synthetases, these enzymes join two molecules with simultaneous breaking of a pyrophosphate bond. They are needed for the synthesis of the body's own proteins, lipids, and other compounds.

Finally, some enzymes are still known by the names given to them by early investigators in the nineteenth century. They are the specific enzymes associated with digestion and fermentation which were the subject of early enzyme investigation. Table 16.2 gives the names, sources, and reactions of some of these enzymes.

ACCORDING TO SOURCE

It is often advantageous when working with enzymes to know their origin, whether animal or plant, and even from which part of the animal or plant they were obtained. This provides information about the optimum conditions for use of the enzymes. The source is then expressed in the name, and we see such names as

TABLE 16.2 Digestive enzymes

Enzyme	Source	Substrate	Reaction	Product(s)
Ptyalin	Saliva	Amylose	Hydrolysis	Dextrins, maltose
Diastase of malt	Barley seeds	Starch	Hydrolysis	Dextrins, maltose
Amylopsin	Pancreas	Starch	Hydrolysis	Dextrins, maltose
Steapsin	Pancreas	Lipids	Hydrolysis	Glycerin, fatty acids
Pepsin	Stomach	Proteins	Hydrolysis	Proteoses, peptones
Trypsin	Pancreas	Proteins	Hydrolysis	Proteoses, peptones, amino acids
Chymotrypsin	Pancreas	Proteins	Hydrolysis	Proteoses, peptones, amino acids
Rennin	Stomach	Casein	Coagulation	Paracasein

salivary amylase, an amylase found in saliva; pancreatic lipase, a lipase found in the pancreas; or diastase of malt, a carbohydrase from the sprouts of barley seeds. We shall see that each enzyme has its own set of reaction conditions which influence the activity of the enzyme.

ENZYME SPECIFICITY

Every cell has many different enzymes, each of which can do some things and not others—can act on some substrates and not on others. In other words, enzymes are *specific* in their behavior. This is of great importance when we realize how many different chemical and biological reactions go on simultaneously in living cells and the whole organism. If every enzyme could catalyze every reaction, there would be chaos. What is seen on closer study is a well-organized and controlled action of the individual enzymes. Each enzyme is specific to the job it can and must perform for the well-being of the organism. Nevertheless, there are different types of specificities according to the reaction catalyzed and the structure of the substrate. Hydrolases such as the proteases, lipases, and carbohydrases catalyze only the hydrolysis of their respective substrate and no other reaction. Transferases transfer only the specific groups which form their catalytic objective, and so on. This type of specificity we can call **reaction specificity.**

There are several kinds of **substrate specificity.** A **general substrate specificity** is exhibited by some enzymes, like the hydrolases. A lipase will work on lipids but not on carbohydrates or protein; the proteases and carbohydrases similarly affect only their respective substrates. But then the lipases (also called esterases) are rather nonspecific in their action because they will break (hydrolyze) almost any ester linkage associated with the lipids.

Lipases (esterases)

Low general specificity of enzymes

If we now turn our attention to other hydrolases, such as the proteases and carbohydrases, we find a higher degree of specificity, expressed as stereospecificity and group specificity.

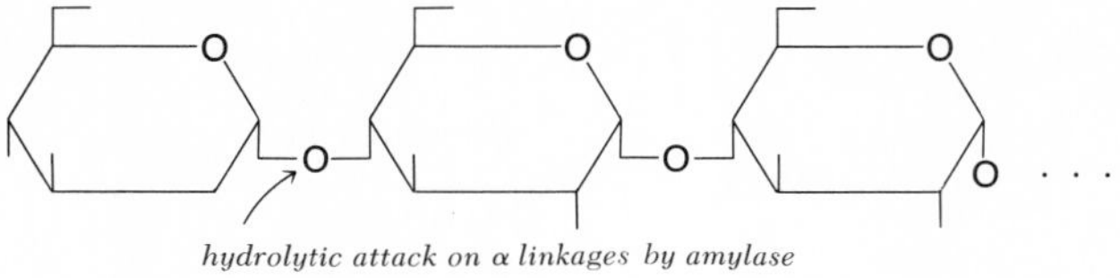

hydrolytic attack on α linkages by amylase

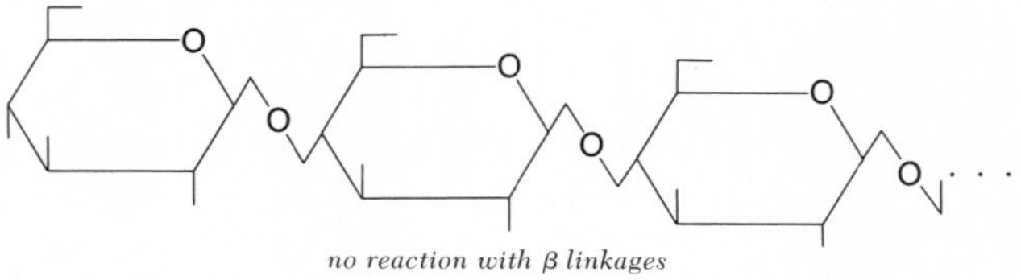

no reaction with β linkages

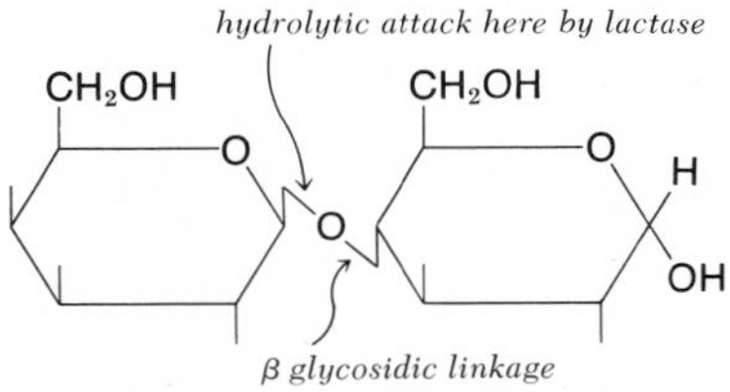

hydrolytic attack here by lactase

CH₂OH   CH₂OH
                              H
                              OH

β glycosidic linkage
Lactose

Stereospecificity is exhibited by the amylases that hydrolyze starch. They hydrolyze only α glycosidic linkages, found in starch and dextrins and not the β glycosidic linkages found in cellulose. The proteases provide a good example of **group specificity.** Chymotrypsin hydrolyzes a polypeptide chain at a point where one of the amino acids involved in the peptide within the chain contains an aromatic R group, as in phenylalanine and tyrosine:

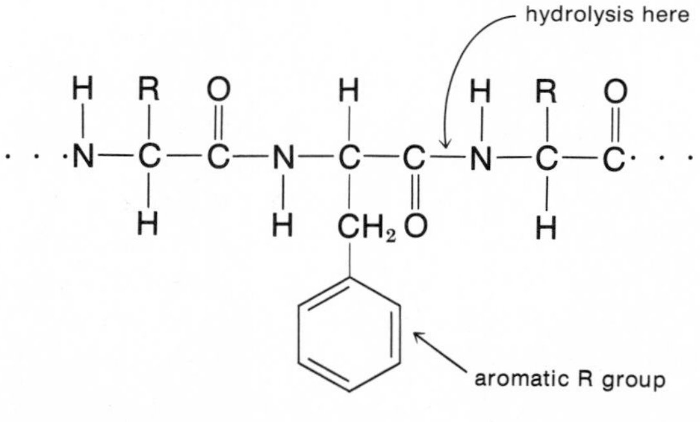

hydrolysis here

H R O   H   H R O
| | || | | | ||
· · ·N—C—C—N—C—C—N—C—C· · ·
| | | | |
H H   H CH₂ O   H

aromatic R group

Trypsin does the same thing at a point in the chain where a peptide bond involves an amino acid with a free —NH₂ group in the

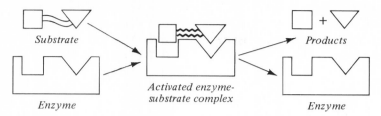

Mechanism of enzymatic reaction

Substrate

Enzyme

Activated enzyme-
substrate complex

Products

Enzyme

FIGURE 16.2 The mechanism of enzymatic reaction.

side chain, as in lysine and arginine. Furthermore, the amino acids must be L-amino acids and not D-amino acids.

In the highest degree of specificity, called **absolute specificity,** one enzyme catalyzes a single reaction of one substrate. Examples are sucrase, which works only on sucrose to hydrolyze it to glucose and fructose; lactase, which catalyzes only hydrolysis of lactose to yield glucose and galactose; and urease, which hydrolyzes urea to carbon dioxide and ammonia.

The overall reaction between an enzyme and the substrate on which it acts can be expressed as in the following equation and in Figure 16.2.

$$E + S \rightleftharpoons ES \rightleftharpoons E + P$$

| Enzyme | Substrate | Enzyme-substrate activated complex | Enzyme | Product(s) |

The enzyme and substrate meet to form an enzyme-substrate activated complex. This complex is the transition state for the reaction, resulting in the formation of the product or products and the liberation of the complete and intact enzyme. The key word is the activated complex; it expresses the lowering of the activation energy and the speeding up of the reaction.

Most enzymatic reactions (but not all) are reversible, as indicated by the two arrows. We can also see that there are two distinctive parts of the overall reaction: (1) combining the enzyme with the substrate to form the enzyme-substrate activated complex and (2) the actual reaction of the complex to yield the product or products and to restore the enzyme to its original condition.

In this sequence, the enzyme and the substrate must have the right geometries, or fit, and are believed to have a mutual attraction in the form of opposite charges on the surface of the enzyme and substrate or mutual attraction of certain groups such as aromatic rings. Also, hydrogen bonding can be involved, as well as

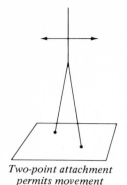

Two-point attachment permits movement

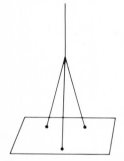

Three-point attachment does not permit movement

charges from the metal-ion activators and their complexing capabilities, as found in the metalloenzymes. The enzyme and the substrate must attain a close fit, which requires at least a three-point attachment between the enzyme and the substrate. This close fit is necessary to result in the enzyme-substrate activated complex. You can try this yourself by putting two fingertips on a flat surface. Although you press down, a back-and-forth motion is still possible. If you now add the tip of your thumb, you have a three-point stable attachment to the surface.

The part of the enzyme involved in forming the activated complex is called the **active site.** After formation of the complex, the reaction can begin, which, as in any chemical reaction, involves bond breaking and bond formation. This action disturbs the combination of enzyme and substrate, which actually repel each other, causing the enzyme to "peel" off the substrate. The products formed also go their own way. This repulsion may have various causes, such as reversal of polarity or loss of attraction due to the bond breaking and bond formation. Also, the geometry of the substrate changes in the reaction process, which enhances the repulsion process and prevents restoration of a close fit. The entire sequence of enzyme-substrate union followed by enzyme-product separation is known as the **lock-and-key** theory, where the enzyme represents the key and the substrate represents the lock. This is a good analogy since for each substrate (the lock) we must have the fitting key (the appropriate enzyme, Figure 16.3).

ROLE OF ENZYME ACTIVATORS

Kinases Usually protein, the kinases induce in the inactive proenzyme a change in the conformation which exposes the active site of the enzyme so it can now work on the substrate. Once the enzyme is activated, it can activate its own proenzyme to the active enzyme. It is like turning a switch; electricity flows and work can be done.

Coenzymes Usually these combine with the enzyme to complete the formation of the active site on the enzyme. Furthermore they often carry away part of the product or bring components to be added to the reaction site.

Metal activators These combine with the enzyme at the active site to make the proper three-point attachment of the enzyme to the substrate possible. Without this close fit the reaction sequence could not be carried to conclusion. The metal ac-

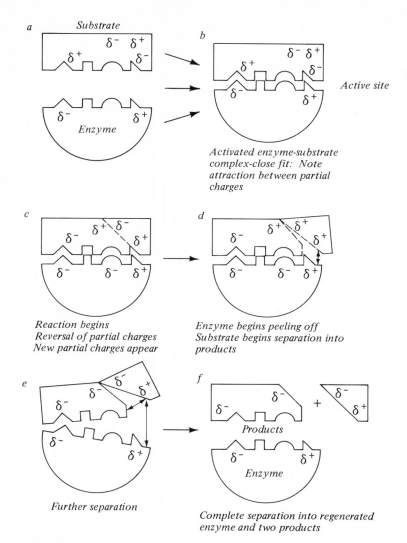

a Substrate

δ^- δ^+

δ^+ δ^-

δ^- δ^+

Enzyme

b

δ^- δ^+

δ^+ δ^-

δ^-

δ^+

Active site

Activated enzyme-substrate complex-close fit: Note attraction between partial charges

c

δ^+ δ^-

δ^- δ^+

δ^- δ^- δ^+

Reaction begins
Reversal of partial charges
New partial charges appear

d

δ^+ δ^+

δ^- δ^+

δ^- δ^- δ^+

Enzyme begins peeling off
Substrate begins separation into products

e

δ^-

δ^- δ^+

δ^-

δ^-

δ^+

Further separation

f

δ^-

δ^-

$+$

δ^-

δ^+

Products

δ^-

δ^+

Enzyme

Complete separation into regenerated enzyme and two products

FIGURE 16.3 Schematic representation of the lock-and-key theory of enzyme/substrate interaction.

tivator stays with the enzyme. Coenzymes and metal activators can be removed from the enzyme by dialysis or precipitation.

An amazing feat is performed by enzymes when they catalyze reactions, as we have seen above. Equally amazing is the speed with which these reactions are carried out. This is expressed in the **turnover number,** which gives the number of molecules of substrate one molecule of enzyme has converted in 1 min. Some of the turnover numbers determined in the laboratory show that enzymes surpass all other known catalysts in their effectiveness. Table 16.3 gives the turnover number of several enzymes.

TABLE 16.3
Turnover number of some enzymes

Enzyme	Turnover Number, Molecular Reactions per Minute
Carbonic anhydrase	36,000,000
Catalase	5,000,000
β-Amylase	1,100,000
Sucrase	40,000
Carboxylase	800

FACTORS INFLUENCING ENZYMATIC REACTIONS

Since enzymes are basically proteins, they are affected by the many physical and chemical agents that act on proteins (Table 15.4). The effects are often on the shape and stereoarrangement, as well as on the charges on the protein molecule. These factors affect the rate of enzymatic reactions greatly. More and more use of this important knowledge is made in working with enzymes, as we shall see. The most common factors affecting enzyme action are:

1 pH (hydrogen-ion concentration)
2 Temperature
3 Substrate concentration
4 Enzyme concentration
5 Concentration of coenzymes and metal-ion activators
6 Inhibitors
  a Competitive
  b Noncompetitive

EFFECT OF pH

Enzymes are very sensitive to pH changes, and the rate of reaction drops rapidly when the hydrogen-ion concentration is raised or lowered by 1 or 2 pH units on either side of the optimum pH value. Figure 16.4 shows this relationship.

EFFECT OF TEMPERATURE

As a general rule, a rise of 10°C in a chemical reaction will about double the reaction rate. This is also true of enzyme-catalyzed reactions but only until they reach an optimum temperature, after

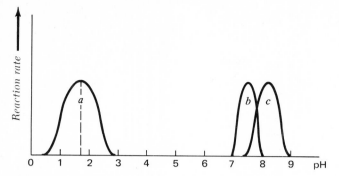

FIGURE 16.4 Effect of pH on reaction rate. *a* Optimum conditions for enzymes working in the stomach; *b* optimum conditions for enzymes in the blood; *c* optimum conditions for enzymes working in the small intestine.

which the reaction rate declines. If the temperature continues to rise, the enzyme will be destroyed. The breakdown of an enzyme with heat is used in the pasteurization process, which is a heat treatment sufficient to destroy natural enzymes and bacteria in milk and other foods. Blanching is the same thing. On the other hand, if we lower the temperature, the reaction rate also drops and can finally reach zero, but the enzyme is not destroyed, only rendered inactive.

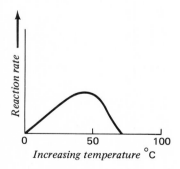

EFFECT OF SUBSTRATE CONCENTRATION

The substrate concentration affects the rate of the enzymatic reactions so that for a given enzyme concentration the rate increases as the substrate concentration is increased. This will go on until a maximum rate is achieved, beyond which further increase of substrate concentration will not affect the rate.

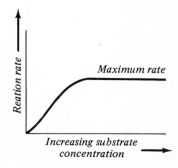

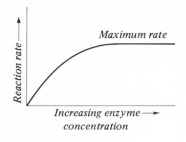

EFFECT OF ENZYME CONCENTRATION

Increasing enzyme concentration affects the reaction rate for a given substrate concentration in the same manner. Given a fixed amount of substrate, addition of more enzymes will result in reaction-rate increases until saturation of enzyme-substrate complexes is reached and with it the maximum reaction rate. No further increase in enzyme concentration will have any effect.

An important control mechanism is related to two enzyme-substrate systems. If for some reason an accumulation of end products occurs which is potentially dangerous to the biological system, the enzyme can be turned off. This is a safety mechanism like a governor on an engine or the fuse in an electric circuit which prevents overloads. Called an **end-product feedback mechanism,** it can slow down or even stop the enzymatic reaction. It acts in the following way. When a high concentration of end product develops, the end-product molecules themselves become attached to the enzyme at some point and by this attachment cause a deformation of the active site, so that no further enzyme-substrate complex can be formed and no further reaction

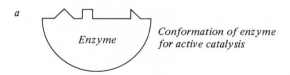

a

Conformation of enzyme for active catalysis

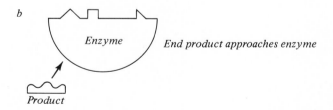

b

End product approaches enzyme

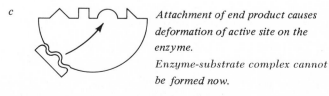

c

Attachment of end product causes deformation of active site on the enzyme.
Enzyme-substrate complex cannot be formed now.

FIGURE 16.5 Allosteric feedback mechanism.

can take place. When the end-product concentration is lowered, the end-product molecules leave the enzyme and further reactions can be catalyzed by the enzymes. This is known as the **allosteric feedback mechanism.** Allosteric means other steric, which refers to the alteration of the spatial conformation of the active site. A case in point is alcoholic fermentation. This process is limited to an alcohol end-product concentration of 12 to 14 percent, depending on the strain of yeast used. When this concentration of ethanol is reached, the alcohol apparently combines with the enzymes in the yeast to inhibit further production of ethanol. Removal of alcohol results in the continuation of the fermentation process and the production of alcohol. Figure 16.5 illustrates the allosteric mechanism schematically.

EFFECT OF METAL-ION ACTIVATORS AND COENZYMES

As seen in Figure 16.6, enzymes function poorly, if at all, without activators. Their role in the enzymatic reaction was described earlier, and more will be said later about coenzymes. Metal-ion activators and coenzymes are quite mobile and can pass through the semipermeable membranes of the cells. Since enzymes require activators to do their job effectively, the cells must have an adequate supply of metal ions and vitamins, many of which are incorporated into the enzyme as coenzymes.

EFFECT OF INHIBITORS

Competitive inhibition The inhibitory effect of end products of enzymatic reaction in the biological systems discussed before is regulatory and acts as a safety device for the organism. This type

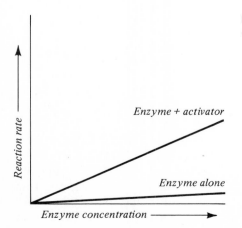

FIGURE 16.6 Effect of activators on the rate of enzymatic reactions.

Reaction rate

Enzyme + activator

Enzyme alone

Enzyme concentration

of restraint is classified as competitive inhibition, which in the more general case is the reversible combination of biological antagonists, or antimetabolites, with an enzyme in competition with the substrate. In this way, the amount of enzyme complexing with the substrate is reduced and the reaction rate diminished. Excessive amounts of such inhibitors can stop the reaction completely, although this can be reversed by adding more substrate. A good illustration of such a beneficial competitive inhibition can be seen in the action of sulfa drugs. One of the vitamins needed in many organisms is folic acid. Folic acid is available to the human body in sufficient amounts in a regular diet, but the body cannot manufacture it. Many microorganisms, including some pathogenic bacteria, make their own folic acid from pteryl alcohol, p-aminobenzoic acid (PABA), and glutamic acid (Figure 16.7). The sulfa drugs interfere with this folic acid synthesis in bacteria, which need it for growth. Sulfa drugs consist of the aromatic compound sulfanilamide modified by attachment of an R group. The structure of a sulfa drug is shown in Figure 16.7 and in size and shape it resembles the structure of PABA. In the presence of sulfa drugs, the formation of folic acid is competitively inhibited because the bacteria use some of the sulfa drug instead of the PABA. The result is no folic acid and fewer bacteria. If a patient has a bacterial invasion, administered sulfa drug will check the excessive growth of the pathogenic organism in the way described and the body defenses can than contain and destroy the invading bacteria. Needless to say, the sulfa drug in the concentration administered does not harm the patient.

Penicillin is another antimetabolite which suppresses bacterial growth by interfering with the normal cell-wall formation of the bacteria. The cell walls are now more readily attacked and broken down by the leucocytes, the white blood cells in the body. This insight into the action of chemotherapeutic agents has greatly stimulated drug research for other antimetabolites. Nevertheless, success in this area is slow, especially since a drug of this kind must be selective, acting against the invader and not the host.

Noncompetitive inhibition Noncompetitive inhibitors combine irreversibly with enzymes, thereby preventing any further use of the enzyme. The emphasis here is on *ir*reversible. If the inhibitor concentration equals the enzyme concentration, the enzymatic reaction is completely stopped. Most metabolic poisons are noncompetitive inhibitors. When heavy-metal ions, such as those of silver, mercury, lead, arsenic, and thallium, are taken into the body, they irreversibly block vital enzymatic reactions and

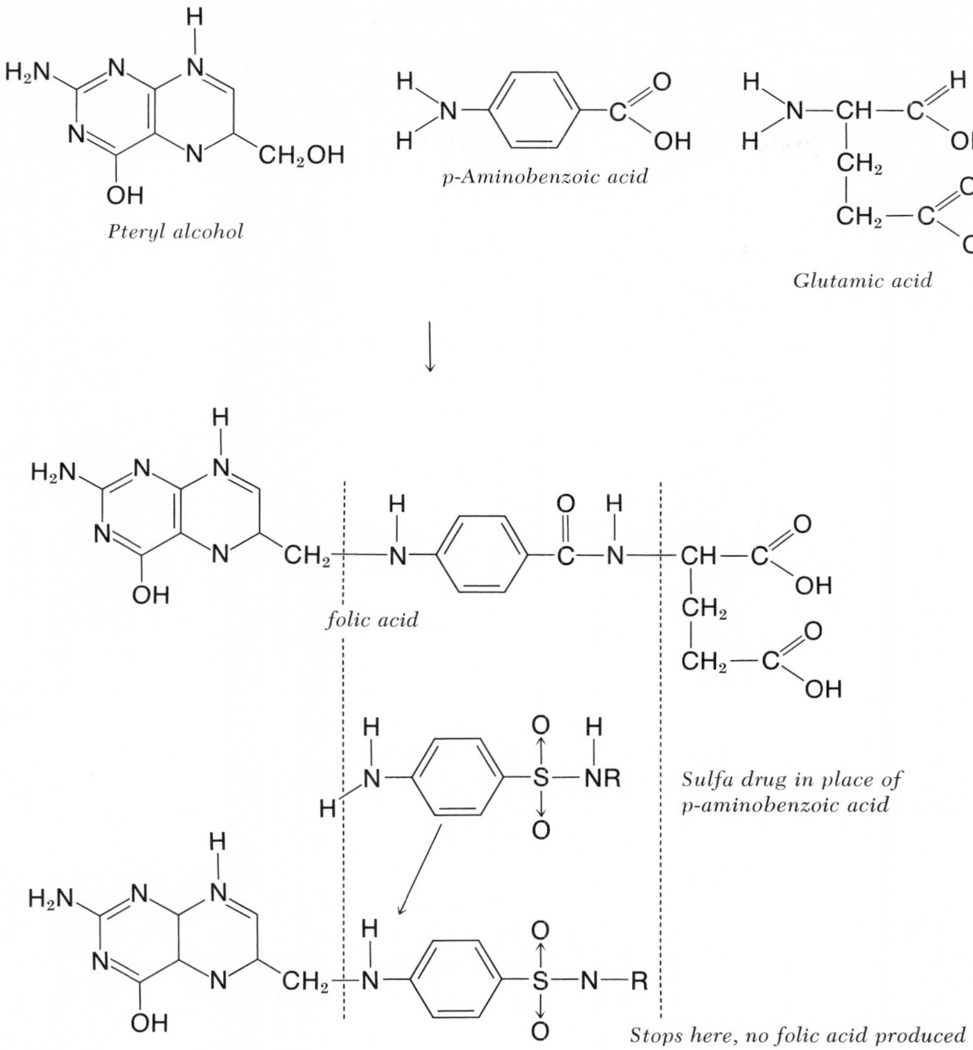

FIGURE 16.7 Folic acid synthesis and its inhibition by a sulfa drug.

cause death. Methanol cannot be metabolized by the body; it or its derivative, formaldehyde, noncompetitively inhibits the enzymes involved in vision and causes blindness. In higher concentrations it also affects brain functions and causes death. In carbon monoxide, CO, poisoning the gas cuts into the oxygen-carrying capacity of hemoglobin. Carbon monoxide combines with hemoglobin more readily than oxygen. The more CO absorbed and the more hemoglobin occupied, the less oxygen can be supplied to the body, which will succumb to oxygen starvation. Modern inventions such as nerve gases and related insecticides (mostly organic phosphates) interfere with the functioning of the nervous system, noncompetitively inhibiting the enzyme acetylcholinesterase, causing paralysis and death.

COENZYMES

Many enzymes require coenzymes, so that their composition and how they contribute to enzyme action is an important consideration. The functions of the coenzymes are illustrated in Figure 16.8, showing how the coenzyme completes the enzyme, which becomes the holoenzyme. Note that the coenzyme and the apoenzyme must have the right fit. The holoenzyme can now engage in the formation of enzyme-substrate activated complex. Observe also that proper fitting is involved between all the component parts of the activated complex. The second part of the catalysis can now proceed with the actual reaction. After this is completed and the enzyme has peeled away, we are left with three components: namely, the apoenzyme again, one kind of product, and a second kind of product still associated with coenzyme. The coenzyme then releases its part of the product and re-

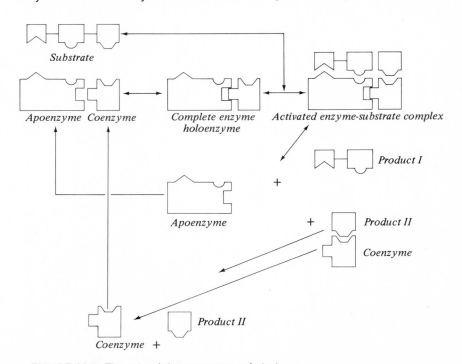

FIGURE 16.8 The role of the coenzyme substrate.

turns to another substrate molecule to help again in the same fashion. Figure 16.8 shows the breakup of a molecule in an enzymatic reaction with the aid of the coenzyme. In synthesis reactions this process goes the other way, in the reverse of the sequence shown. The function of the coenzyme is a dual one. It first participates in the completion of the enzyme and then acts as a carrier of a part of the product. A competitive inhibition can occur here also if the coenzyme in its role as a carrier of product becomes saturated; the enzymatic action is inhibited until the congestion is relieved, which is another safety mechanism in biological systems.

THE USE OF ENZYMES

Until about 100 years ago, enzymatic reactions were used only unconsciously with the aid of microorganisms. We know from early history that the use of fermented food and beverages for human consumption is very old. The fermentation process used in various forms not only provided new tastes but also added stability and made storage of these products possible. Milk was made into cheese, while cucumbers, cabbage, and many other vegetables could be preserved. Food that was plentiful in the summer became available throughout the winter. Beer, wine, and other fermented beverages not only gave a kick because of the alcoholic content but also provided nourishment in the form of carbohydrates and proteins contained in them.

As the knowledge of enzymes expands, it is used in more ways. Enzymes used in industry are obtained from animals, plants, and microorganisms. Some proteolytic enzymes, such as papain from papaya and bromelin from pineapple, are used as meat tenderizers, in cheese making, and in the chillproofing of beer. Diastase of malt, obtained from germinating barley and wheat, is an amylase used to prepare corn syrups, to modify cereal starches in precooked cereals, and in brewing. Lipases, which come chiefly from animal sources, are used to enhance the flavor of cheeses, margarine, and chocolate products. The best sources for industrial enzymes are microorganisms. Strains can be selected which are rich in a specific enzyme, and the organism is then grown for its production. The enzymes are isolated and harvested, thanks to our increased knowledge of them. Industry prefers the use of enzymes wherever possible since they are better catalysts than other materials. The enzymes produced from microorganisms are of excellent quality, and it is easy to produce them in large quantities. A process recently developed for using enzymes holds great promise for the future: Enzymes can be chemically attached to a rigid surface and still retain their activity. The substrate can be passed over the enzyme. At one end you add the substrate and move it over the enzyme in a suitable solvent and at a suitable rate; the product comes out at the other end.

Enzymes are added to household detergents to improve their cleaning power. These enzymes are proteases and amylases stable to alkaline conditions in washing machines. They partially digest food and other organic stains, so that they can be removed by the detergent.

In recent years, the use of coal tar dyes on citrus fruit was prohibited. Today, oranges are stored for a short while after picking, and some ethylene gas is added to the storage compartments. The ethylene triggers an enzymatic reaction in the skin of the oranges which destroys all the chlorophyll, and the oranges take on a uniform orange-red color. This process is referred to as artificial ripening. It works on bananas too.

Many steroids and antibiotics produced today are the work of enzymes. A living microorganism is used to modify plant steroids so that they become potent drugs. Antibiotics such as penicillin are harvested from microorganisms that produce them. Certain diseases and heart injuries can be diagnosed by monitoring enzymes in the blood or other body fluids. A proteolytic enzyme is sometimes used to remove cataracts from the cornea of the eye. Digestive enzymes can be given to patients with gastrointestinal problems when digestive enzymes fail to act properly.

REVIEW QUESTIONS

1 Define enzyme and catalyst.
2 What is an enzyme composed of?
3 What are zymogens and apoenzymes?
4 What is a kinase, and what is its role in relation to enzymes?
5 What are metal activators, and what are their roles in relation to enzymes?
6 Enzymes can be classified by different criteria. Make a list of the different classifications and point out their individual usefulness.
7 Identify some hydrolases shown in various biochemistry texts. **a** Where do they work? **b** What are the products of the reactions they help?
8 Use examples to show the different kinds of enzyme specificity.
9 Make a list of the different types of enzymes. Indicate where they are found and what substrate they work on. Indicate further the product(s) of the enzyme-catalyzed reaction.
10 Make a diagram of the enzyme mechanism as we visualize it today.
11 What is the activated complex?

12 What is the lock-and-key theory?
13 Why must the enzyme and the substrate attain a three-point attachment?
14 What is meant by the active site of an enzyme?
15 What is the effect of temperature on enzyme-catalyzed reactions?
16 What other factors influence enzymatic reactions?
17 What is competitive inhibition? Give examples.
18 What is noncompetitive inhibition? Give examples.
19 What is the allosteric feedback mechanism?
20 Make a list of coenzymes and indicate which ones are vitamins.
21 How can you separate an enzyme from a coenzyme when they are in solution?
22 How could you demonstrate the effect of a metal activator on enzymatic reactions?
23 How are enzymes used by man?
24 What is papain, where does it come from, and what is its use?
25 How does a meat tenderizer work?
26 What do you know about fermentation? Give examples.

DIGESTION AND ABSORPTION
Making Little Ones Out of Big Ones

The three classes of foodstuffs discussed earlier, namely, carbohydrates, lipids, and proteins, are the materials used by the body for the production of energy, and for the structure and function of the organism. However, the gross particle sizes, as well as the molecular sizes, of the foods we ingest are much too large to be used directly for these purposes. In addition, foods are generally mixtures of the three categories of nutrients and need to be sorted out accordingly in order to do their jobs properly. The process whereby the massive molecules of the foods we eat are broken down into their smaller components is called **digestion.** The reduction of large nutrient molecules to their smaller and varied chemical units allows them to be absorbed into the bloodstream for distribution to the various organs and cells of the body as needed. The process of digestion is, therefore, an essential activity of the body, and we will now consider how this process is accomplished.

Although the digestion of food begins with its mechanical breakup in the mouth, it occurs mostly in a series of chemical baths that hydrolyze the macromolecules to their smaller units. It is essential to keep in mind that these hydrolytic reactions, which begin in the mouth and continue in the stomach, the small intestine, and the large intestine, are all catalyzed by specific enzymes. Furthermore, they take place in steps, each one leading to the next. As a matter of fact, digestion can begin even before the food is eaten. Food preparation, including cooking, fermentation, aging, and marinating, result in some degree of predigestion. The cooking of carbohydrates under acidic conditions (most food preparations are slightly acidic) will first open up starch granules to release amyloses and amylopectins, and then begin the hydrolysis of these molecules to form dextrins as the first hydrolytic products. In addition, disaccharides are also hydrolyzed to monosaccharides. The cooking of meat softens it by changing the tough collagen fibers into gelatin. The tenderizing of meat is also achieved by meat tenderizers and by aging, where enzymes hydrolyze some of the structural proteins. Pickling is another process for tenderizing tough meat. Many proteins are denatured and coagulated in the cooking process and become easier to digest, the cooking of eggs being a case in point. Upon storage, many fruits and vegetables become easier to eat, and taste better owing to enzymatic processes that soften their tissues.

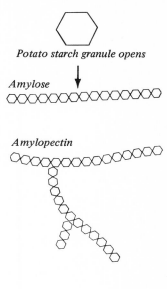

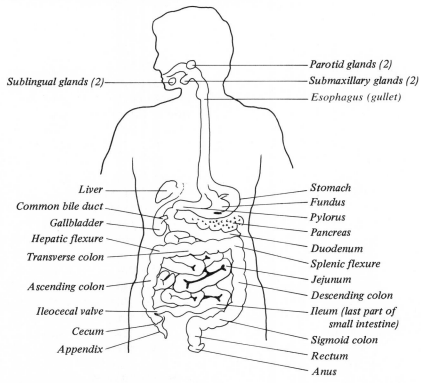

FIGURE 17.1 The digestive system.

It should also be noted that cooking destroys bacteria and other microorganisms that can cause poisoning, especially in high-protein foods. There are other ways whereby food preparation affects digestion, including the addition of spices, the production of flavors, and the physical appearance of the food. All of these have definite physiological effects on digestion by stimulating the flow of digestive juices into the body's alimentary tract. Has your mouth ever watered at the sight, aroma, or taste of various foods? The flow of saliva in the mouth is stimulated by these sensations and, know it or not, so is the flow of gastric fluid in the stomach, as well. The body is getting ready to accept and digest the foods. The reverse is true if the food is poor in appearance, taste, and aroma. In well-run hospitals, special attention is given not only to the proper nutritive value of the food served patients, but also to the factors just mentioned.

The use of beverages in moderation, such as wine, cocktails, and apéritifs, may serve to help digestion and appetite where such stimulation is needed.

DIGESTION IN THE MOUTH

Several distinctly different processes take place when food enters the mouth. The first is the mechanical action of the teeth whereby chewing breaks up the food into smaller and smaller pieces. By this process, the surface area of the food components is increased tremendously, thereby making the digestive enzymes more effective by giving them a larger area to attack. Poorly chewed food will stay longer in the stomach than well-chewed food. The second process is mastication, which is the action of **saliva** combined with the action of the teeth to make the food into a pastelike substance for easier swallowing. Saliva contains a glycoprotein called **mucin** which makes the food slippery for easy swallowing. The third process is a digestive action on carbohydrates contained in the food. The enzyme ptyalin in saliva (see Table 18-1, which lists the digestive enzymes) is an amylase which works on polysaccharide substrates such as starch and dextrins. The enzymatic action is on the inside of the polysaccharide chain, where it cleaves α-1,4 linkages by hydrolysis. The products of the action are **dextrins, erythrodextrins, achroodextrins,** and maltose (Fig. 17.2). The chloride ion is the activator for this enzyme, which acts over a pH range of 4.0 to 9.0 with maximum action at a pH of about 6.5. Below pH 4.0 the enzyme is completely inactivated. You can detect the hydrolysis reaction by chewing a piece of coarse bread, such as rye bread, for a while; you will find that it becomes slightly sweet. The sweetness is that of the maltose resulting from the hydrolysis.

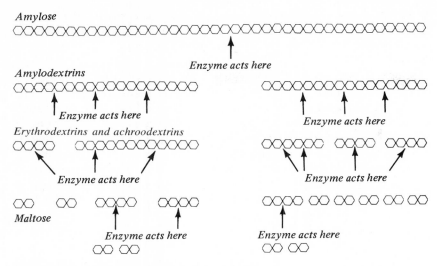

FIGURE 17.2 The action of pytalin on amylose.

A total of about 1500 ml of saliva is secreted daily by three different pairs of glands into the mouth: the sublingual pair located under the tongue, the submaxillary pair located under the jaw, and the parotid pair located under the ear. Salivary amylase is principally found in the secretion of the parotid glands whereas the other two are rich in mucin. Saliva consists of about 99.5% water; the rest is amylose, mucin, and inorganic components such as chlorides, bicarbonates, sulfates, and phosphates. The average pH of saliva is about 6.6 but it may be anywhere between a pH 5.2 and 7.2. The flow of saliva is stimulated by the sight and aroma of food as well as the taste sensation of spices, salts, and acids. It is further stimulated by the mechanical action of chewing. It is also of interest that the body secretes into saliva compounds taken into the body, either orally or by injection, and these can be detected in the saliva shortly afterward. When alcohol is consumed it will show up in saliva, which can then be used to determine intoxication.

DIGESTION IN THE STOMACH

The masticated food passes from the mouth through the esophagus into the upper, large portion of the stomach called the fundus. Here the food will be stored from 1 to 5 h depending on the composition of the food and its digestibility by the **gastric juices.** The lower, smaller portion of the stomach is called the pylorus, and is essentially a tube connecting the stomach with the intestines. The stomach is a muscular pouch which is collapsed when empty and expands to accept and store food. Three different secretions can be identified in the gastric fluid, each coming from different cells in the stomach wall. The parietal cells produce and secrete a strongly acidic solution containing HCl of a normality of 0.15 and a pH of 0.87. The chief cells secrete a mixture of inactive enzymes, the **zymogens.** These inactive enzymes are primarily **pepsinogen, prorennin,** and a small amount of **gastric lipase.** The combined secretions of the parietal and the chief cells are collectively known as the gastric juice, which has a pH between 1.2 and 2.2. A third secretion is produced by mucous cells in the stomach wall called stomach mucus. This secretion is rich in **mucin,** a **glycoprotein,** and has an alkaline reaction. The mucus bathes and protects the stomach wall against the secreted HCl and the digestive action of active pepsin, because mucin is not digested by pepsin. The mucus secretion itself has no digestive action. The flow of gastric secretions is stimulated by the sight and smell of food, and by the arrival of food in the stomach. This serves to release a hormone called gastrin, which, in turn, stimulates the production of gastric juice.

THE PRODUCTION AND ACTION OF GASTRIC HCl

The parietal cells that generate the HCl secreted in the gastric juice are surrounded by blood vessels. H^+ and Cl^- ions pass through the membranes that separate the blood vessels from the cells where the ions accumulate, and are then secreted as a strong HCl solution. This transfer of H^+ ions from the blood to parietal cells causes a slight rise in the blood pH despite the buffers present. This is so especially after a meal, and results in the so-called alkalinity wave which results in a feeling of comfort and general euphoria. The formation of HCl can be generally considered as taking place as follows:

$$NaCl + H_2O \longrightarrow HCl + NaOH$$

As NaOH forms it is neutralized by buffers such as $NaHCO_3$ or NaH_2PO_4

$$NaOH + NaH_2PO_4 \longrightarrow Na_2HPO_4 + H_2O$$

The hydrochloric acid secreted in gastric juice has very little digestive action, but has many other essential functions, including:

1 Denaturing and breaking the inter- and intramolecular bonding in proteins. This results in the opening up of protein molecules for better enzyme attack.

2 Activating pepsinogen to the active enzyme pepsin

$$\text{Pepsinogen} + \text{HCl} \longrightarrow \text{pepsin}$$

3 Providing suitable pH condition for pepsin, for which the optimum pH is 1 to 2.5.

4 Possibly causing hydrolysis of disaccharides such as sucrose, maltose, and lactose to the respective monosaccharides.

5 Exerting a germicidal effect on bacteria and other microbial organisms entering the stomach.

6 Solubilizing iron compounds so that iron can be absorbed into the body.

7 Stimulating the secretion of the hormone secretin by the intestines, which, in turn, promotes the flow of intestinal juices.

Sucrose + $H_2O \rightleftharpoons$
glucose + *fructose*
Maltose + $H_2O \rightleftharpoons$
glucose + *glucose*
Lactose + $H_2O \rightleftharpoons$
glucose + *galactose*

THE ACTION OF ENZYMES IN THE GASTRIC JUICE

The enzymes found in gastric juice are primarily the **protease pepsin,** the **coagulase rennin,** and small amounts of a **gastric lipase.** The gastric lipase hydrolyzes fats that reach the stomach in an emulsified form only. More will be said about the action of lipases in connection with digestion in the intestine. The enzyme rennin, which is secreted as prorennin, is activated by H^+ ions to rennin and acts on the protein casein as follows:

$$\text{Casein} \xrightarrow[+ H_2O]{\text{rennin}} \text{paracasein} + \text{rennin}$$

then, (Insoluble) (Soluble)

$$\text{Paracasein} + \text{Ca}^{2+} \longrightarrow \text{calcium paracaseinate}$$
(Insoluble)

As an enzyme, the rennin catalyzes the reaction, but is not used up.

The calcium paracaseinate is further altered by pepsin. Rennin is the principal enzyme present in the stomach of infants. As one ages the amount of rennin produced diminishes, and in adults it disappears completely.

The most important enzyme in the gastric juice is the activated enzyme pepsin. It is a protease that hydrolyzes proteins to **proteoses** and **peptones.** The enzyme is an **endopeptidase,** which means it works on peptide linkages *within* the protein molecule (Fig. 17.3). The enzyme preferentially attacks peptide linkages that contain the amino acids phenylalanine and tyrosine. This results in the formation of protein fragments called proteoses and peptones, but not many free amino acids. The enzyme is especially active on proteins that are rich in sulfur-containing amino acids. Not all proteins are attacked by pepsin. Under the strongly acidic conditions in the stomach pepsin curdles milk, as rennin does, and in the presence of Ca^{2+} ions insoluble paracasein is

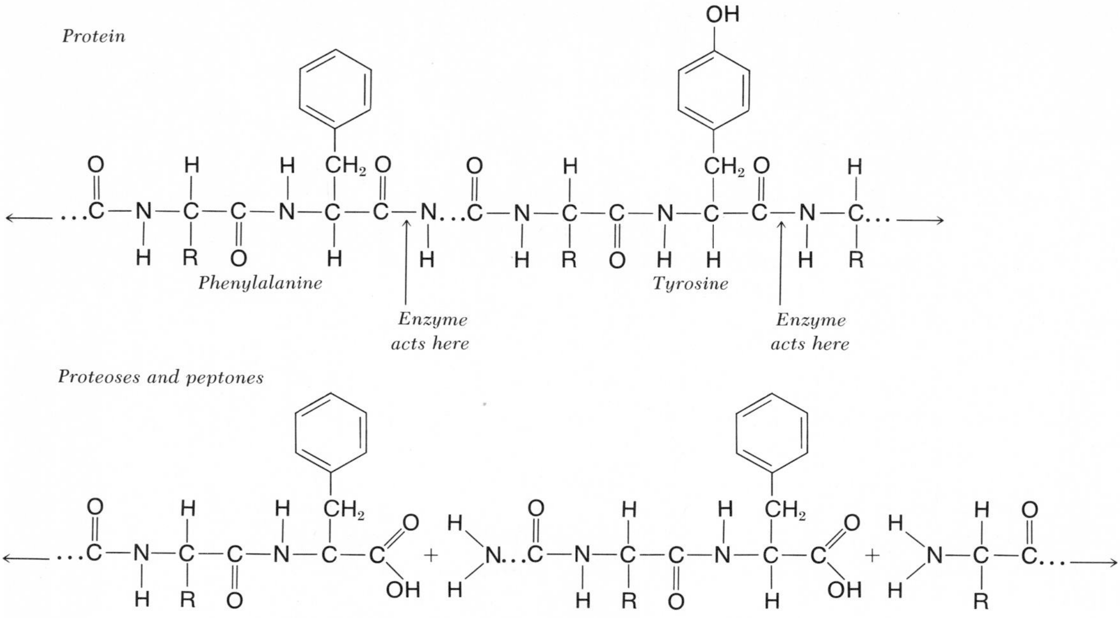

Protein

Phenylalanine

Enzyme acts here

Tyrosine

Enzyme acts here

Proteoses and peptones

FIGURE 17.3 The action of pepsin on a protein.

formed. This makes casein digestible by pepsin without the action of rennin.

So far, we have followed the digestion of polysaccharides to smaller polysaccharides and maltose in the mouth, and the digestion of proteins to proteoses and peptones in the stomach. Since the saliva and the salivary amylase is carried into the stomach along with the food, the digestion of carbohydrates continues in the stomach until the α-amylase is inactivated by the acidity of the gastric juice, which can take from 10 to 20 min. About 2 to 3 l of gastric juice is secreted over a 24-h period; it is 99.4 percent water, and 0.6 percent enzymes and inorganic components such as HCl, NaCl, and phosphates. The addition of the gastric juice gives the food a liquid consistency in which the smaller hydrolysis products of proteins can dissolve more readily. This liquefied mixture is referred to as the **acid chyme.** Contraction of the stomach walls causes a peristaltic wave which forces the acid chyme through the pylorus into the intestines. The peristaltic wave also mixes the food in the stomach, which aids in the digestive process.

DIGESTION IN THE INTESTINE

From the pylorus of the stomach the food passes into the duodenum, the jejunum, and then the ileum of the small intestine. This is followed by the cecum, colon, and the rectum of the large intestine, which complete the alimentary tract. The main digestion and absorption of food nutrients take place in the small intestine, with its conveniently large surface area due to the tiny protrusions from the intestinal wall called villi. It is through this large working area that the end products of digestion pass into the bloodstream and on to the cells.

Three digestive fluids enter the small intestine: the **pancreatic juice,** the **bile fluid,** and the **intestinal juice.** The first two enter the digestive tract at the beginning of the duodenum. The intestinal juice is produced by small glands located in the wall of the small intestine, and is secreted directly into the intestine. Most of it is produced and secreted in the duodenum, with smaller amounts from the jejunum and ileum. All three fluids are alkaline, and neutralize the acid chyme to the optimum pH of 7.2 to 8.4 for the action of the digestive enzymes in the small intestine. The stimulus for the production and secretion of the digestive fluids arises in different ways. Intestinal juice production is stimulated by the mechanical action of food moving along the intestinal walls. This also releases a hormone from the intestinal mucosa called **enterocrinin,** which stimulates the mucosal glands to produce intestinal juice. The hormone **secretin** is released from the mucosal wall by stimulation with HCl and causes the production of pancreatic juice, the production of bile in the liver, and the production of intestinal juice. Yet another hormone released is **cholecystokinin,** which causes contraction of the gall bladder so that bile fluid is forced into the duodenum. The hormone **pancreozymin** stimulates the flow of pancreatic juice. Secretions of all these are stimulated by the nervous system, primarily through the vagal nerve.

Villi

THE ACTION OF PANCREATIC JUICE

As the name implies, this digestive fluid is produced by the pancreas. It has a pH of 7.5 to 8.2; between 500 and 800 ml are produced daily. The juice is 98.7 percent water and 1.3 percent solids; its organic components are the most powerful enzymes in the digestive system. They are proteases, amylases, lipases, and nucleases. The inorganic components are carbonates, bicarbonates, and alkaline phosphates, which produce the high pH in the pancreatic juice that acts to neutralize the acid chyme.

CO_3^{2-}, HCO_3^-, *and* PO_4^{3-} *are all proton acceptors, or bases.*

Proteases The proteases in the pancreatic juice are **trypsinogen, chymotrypsinogen,** and **carboxypeptidase.** The first two require activation as follows:

Trypsinogen + enterokinase $\longrightarrow$ trypsin
(From the   (Active)
intestinal mucosa)

Chymotrypsinogen + trypsin $\longrightarrow$ chymotrypsin
(Inactive)   (Active)

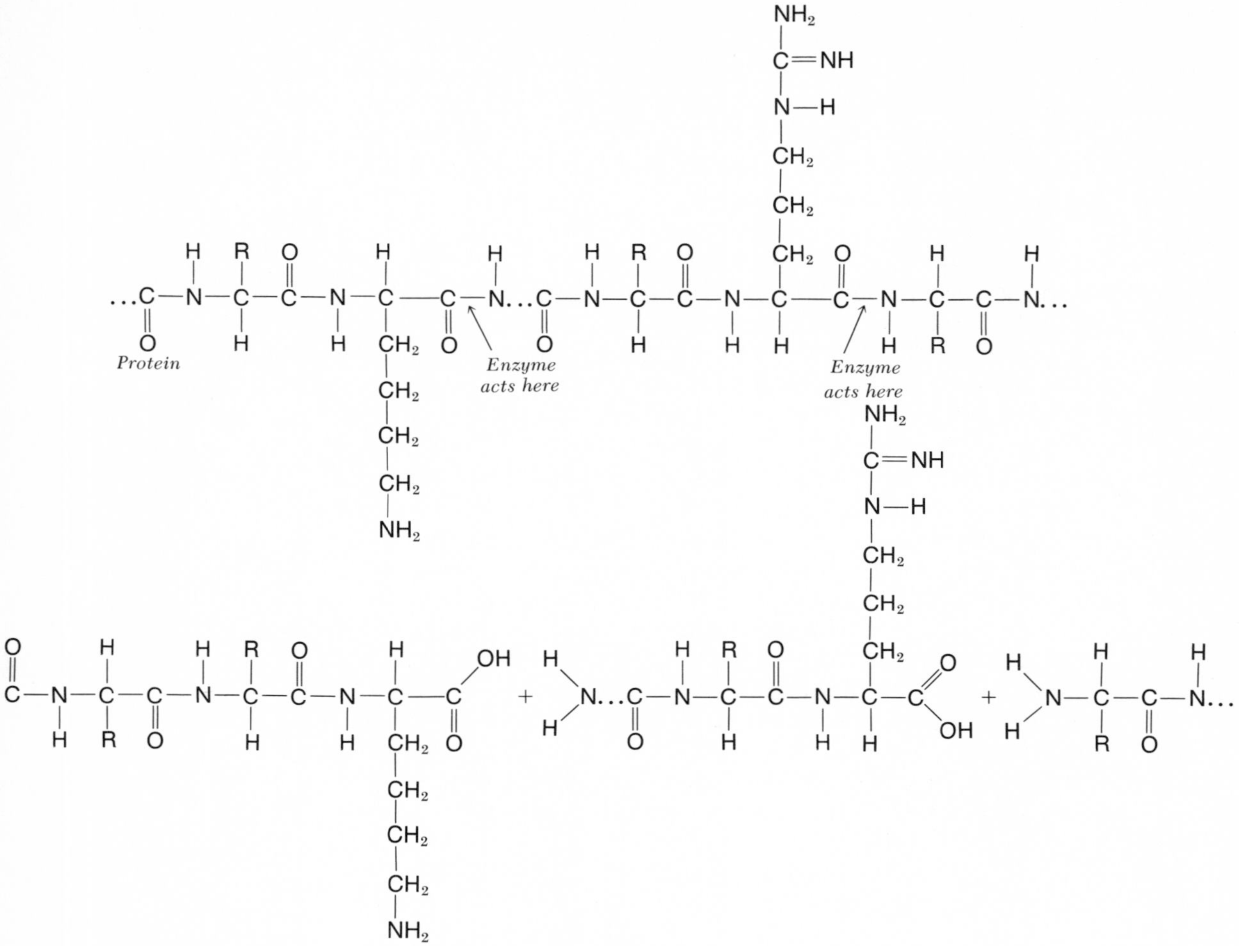

FIGURE 17.4 The action of trypsin on proteins, proteoses, and peptones.

Trypsin is an **endopeptidase,** attacking all proteins and protein fragments. It preferentially hydrolyzes peptide bonds within a polypeptide chain containing the amino acids argenine and lysine in their makeup (Fig. 17.4). The products are smaller protein fragments and some free amino acids. **Chymotrypsin** is also an endopeptidase, but it preferentially hydrolyzes polypeptide chains where peptide bonds are formed with tryptophan, tyrosine, and phenylalanine (Fig. 17.5). The result is again the production of smaller polypeptides and some free amino acids. Chymotrypsin also effectively curdles milk, as did rennin and pepsin.

The third protease found in the pancreatic juice is a **carboxypeptidase.** It is released in its active form and hydrolyzes peptide

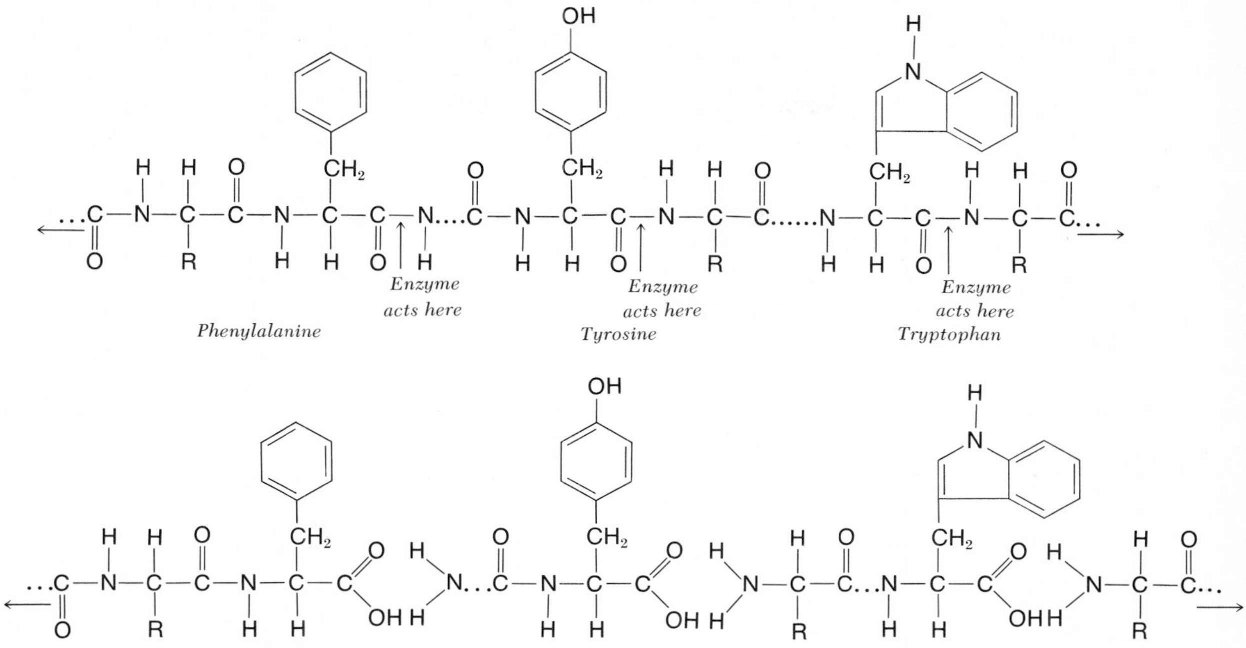

FIGURE 17.5 The action of chymotrypsin producing proteins, proteoses, and peptones.

bonds starting from the free end of a polypeptide chain having a free carboxyl group (Fig. 17.6). It also is called an **exopeptidase,** as it works from the outside or the end of the polypeptide chain, attacking any peptide linkage as long as there is a free carboxyl group. The result is the production of free amino acids as the polypeptide chain continues to shorten.

Pancreatic amylase Pancreatic amylase, or **amylopsin,** is the most powerful amylase in the digestive system. Its action is identical with, but faster than, the salivary amylase ptyalin (Fig. 17.7). It will also attack the 1,6 linkages in branched polysaccharides.

Pancreatic lipase The pancreatic lipase, or **steapsin,** is an esterase, and is secreated in a mildly active form. However, in the presence of bile salts, Ca^{2+} ions, and other compounds in the intestine that have emulsifying action, its enzymatic activity is greatly enhanced and it hydrolyzes ester linkages in fats and oils, the triglycerides (Fig. 17.8). However, before this reaction can take place to any significant degree, the fats must be emulsified; this is done by bile acids, as will be discussed later. The result of the hydrolysis is the production of mono- and diglycerides, free fatty

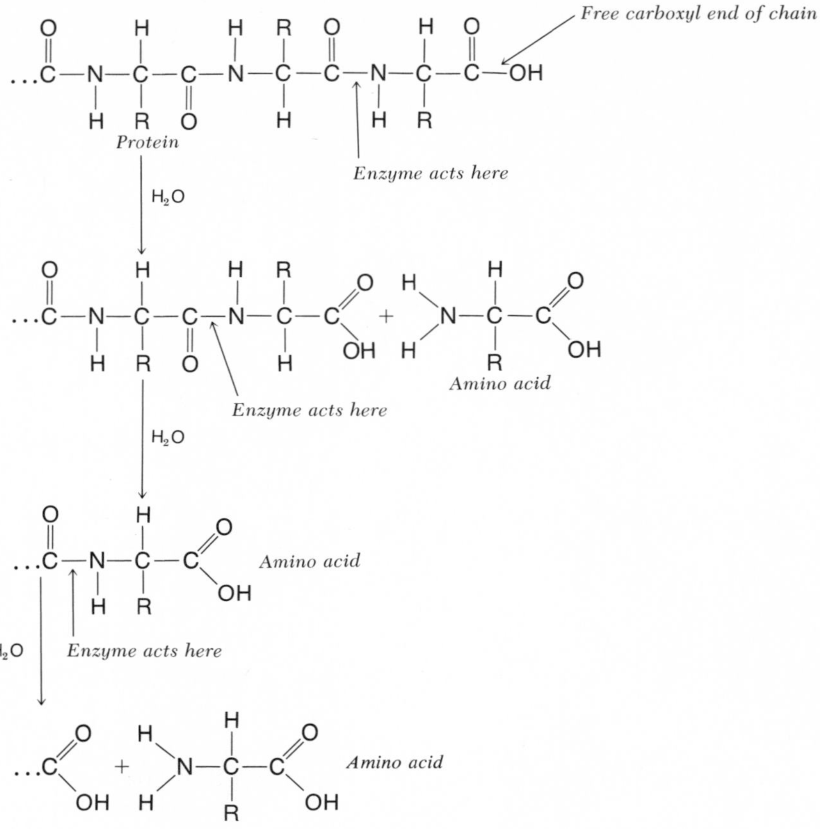

Free carboxyl end of chain

Protein

Enzyme acts here

H_2O

Enzyme acts here

Amino acid

H_2O

Amino acid

H_2O *Enzyme acts here*

Amino acid

FIGURE 17.6 The action of carboxypeptidase on proteins, proteoses, and peptones. Individual amino acids are removed in sequence and the chain continues to diminish in length.

acids, and glycerin. The enzyme acts best on lipids with long-chain fatty acids and unsaturated fatty acids in their makeup, and triglycerides are hydrolyzed faster than diglycerides. It is believed today that the action of the enzyme stops at the stage where monoglycerides are formed. The alkaline conditions in the small intestine are best for the action of this lipase.

Phospholipases and nucleases Other esterases in the pancreatic juice are **phospholipase A** and **phospholipase B,** which remove the two fatty acid residues from lecithins and cephalins by hydrolysis. The products are glycerophosphoryl-choline, glycero-phosphorylethanolamine, and glycerophosphorylserine (Fig. 17.9). The presence of bile acids for the hydrolytic reaction is also essential here. Two different **nucleases** in the pancreatic juice

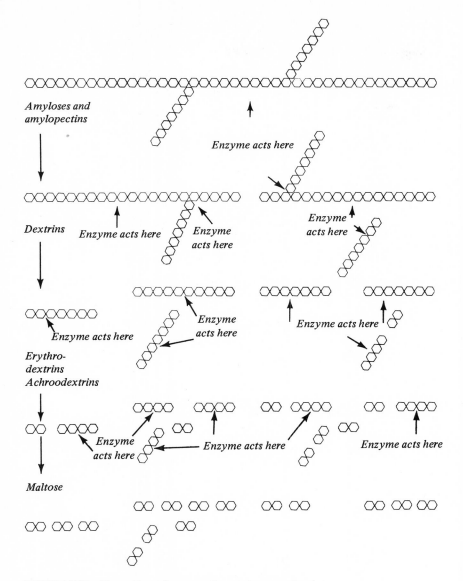

FIGURE 17.7 The action of pancreatic amylase on starch.

hydrolyze the nucleic acids, ribonucleic acid (RNA) and deoxyribonucleic acid (DNA), to their respective nucleotides (Fig. 17.10).

THE ACTION OF THE INTESTINAL JUICE

The important intestinal juice is produced by several million microscopic glands located in the mucosa of the wall of the small intes-

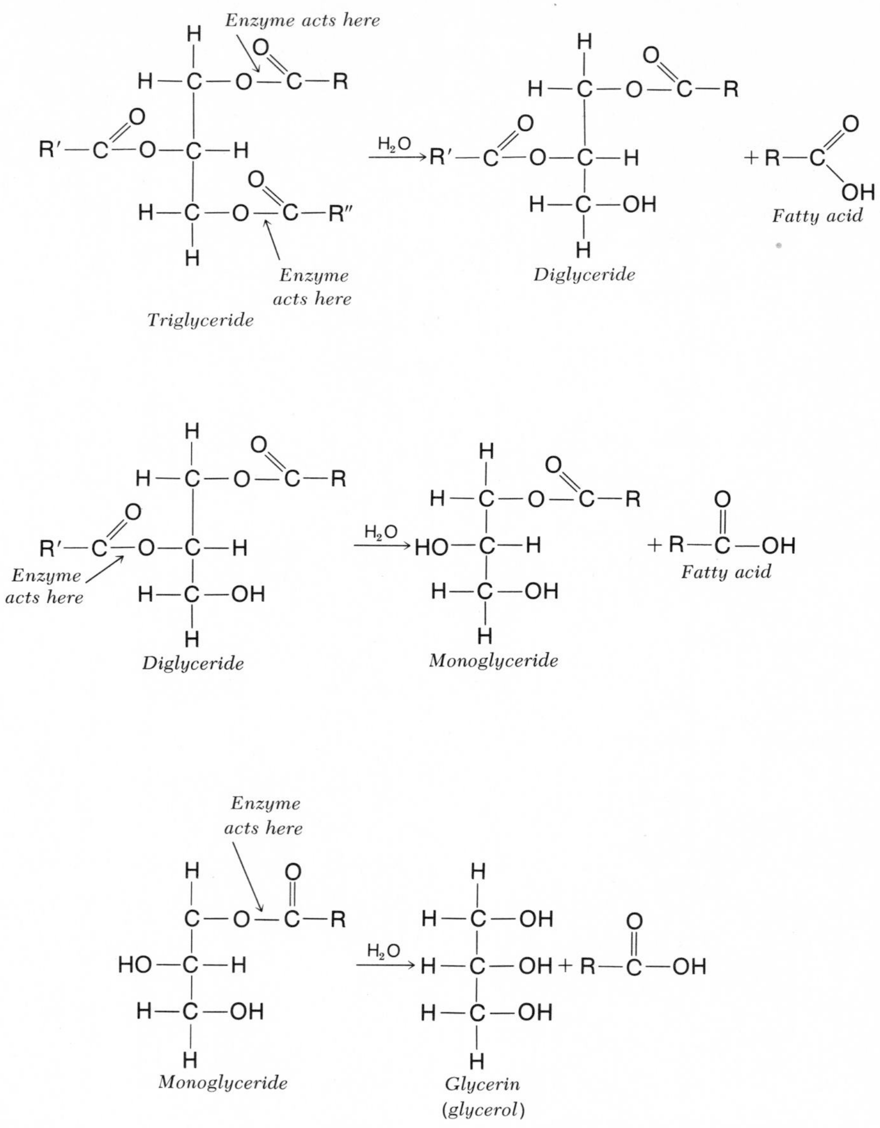

FIGURE 17.8 The action of pancreatic lipase. It is believed the reaction stops with the formation of monoglycerides.

tine. Between 5 and 10 l of this juice is produced in 24 h, with a pH between 7.0 and 8.0. The enzymes contained in the intestinal juice finalize the digestion of the foods into components that can be absorbed by the body. The enzymes are peptidases, disaccharidases, nucleosidase, phosphatases, and enterokinase. Enterokinase is the activator for the proenzyme trypsinogen, as indicated earlier.

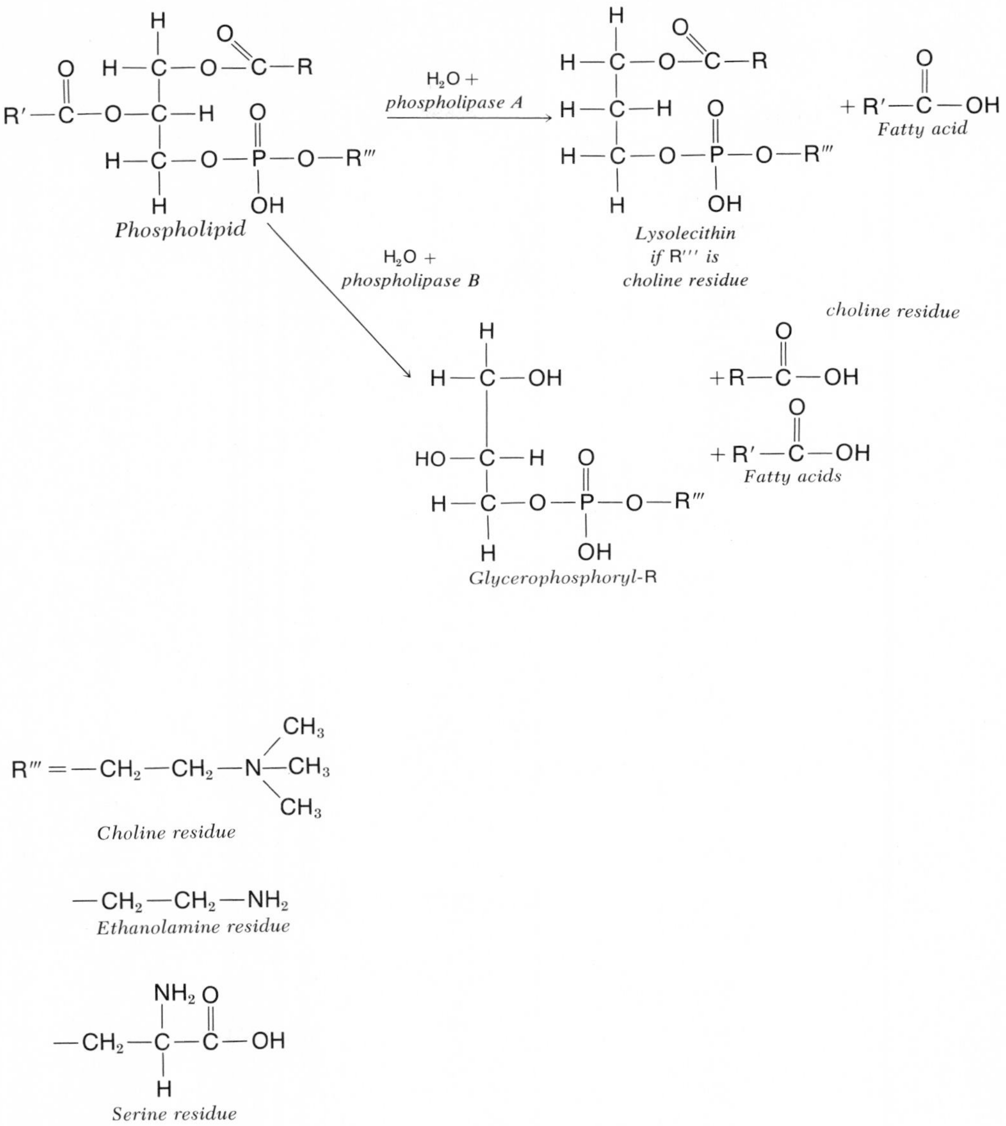

H_2O +
phospholipase A

Phospholipid

*Lysolecithin
if R''' is
choline residue*

$+ R' - C - OH$
Fatty acid

H_2O +
phospholipase B

choline residue

$+ R - C - OH$

$+ R' - C - OH$
Fatty acids

Glycerophosphoryl-R

$$R''' = -CH_2 - CH_2 - N \begin{matrix} CH_3 \\ CH_3 \\ CH_3 \end{matrix}$$
Choline residue

$$-CH_2 - CH_2 - NH_2$$
Ethanolamine residue

$$-CH_2 - \underset{H}{\overset{NH_2}{C}} - \overset{O}{C} - OH$$
Serine residue

FIGURE 17.9 The action of phospholipase A and phospholipase B.

Peptidases These enzymes complete the work of the proteolytic enzymes of the pancreas. They also hydrolyze peptide linkages, but act only on small polypeptides, not on complete proteins. The two kinds of peptidases found here are an **aminopeptidase** and a **dipeptidase.** The aminopeptidase is an exoenzyme which hydrolyzes peptide bonds at the free amino acid end of a polypeptide and chops off one amino acid unit after the other until the dipep-

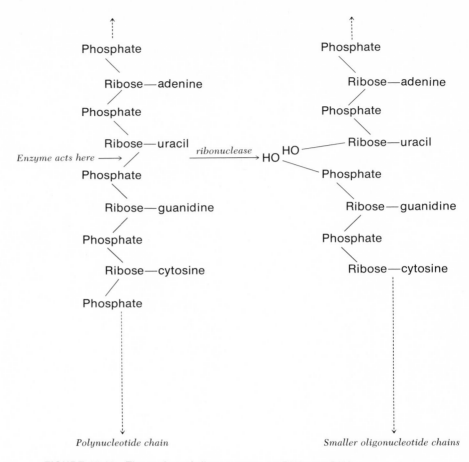

FIGURE 17.10 The action of ribonuclease on RNA and DNA.

tide stage is reached (Fig. 17.11). The main products are free amino acids and dipeptides. The final breakdown to free amino acids is completed by the dipeptidases, which hydrolyze dipeptides to two free amino acids, which can now pass through the intestinal wall (Fig. 17.12).

Disaccharidases The three enzymes that complete the breakdown of carbohydrates into monosaccharides for absorption into the body are sucrase, maltase, and lactase. Sucrase, or invertase, splits sucrose into glucose and fructose, maltase splits maltose into glucose and glucose, and lactase splits lactose into glucose and galactose. The first two are present at all times, but lactase is present in adequate amounts only in infants and children. In adults, lactase concentration and its activity gradually diminishes.

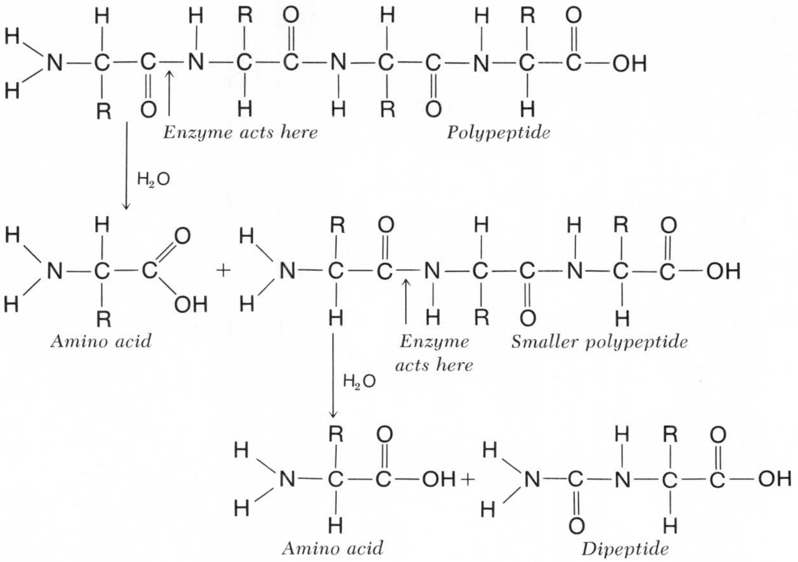

FIGURE 17.11 The action of aminopeptidase on polypeptides.

A process has been recently proposed whereby lactose in milk is hydrolyzed without affecting other components in the milk. This would greatly increase the nutritive and digestive value of milk for adults who have insufficient lactase. The glucose, fructose, and galactose resulting from the action of the disaccharidases are all absorbed by the body. The nucleosidases and phosphatases in the intestinal juice complete the breakdown of nucleic acids started by the nucleases of the pancreatic juice (Fig. 17.13).

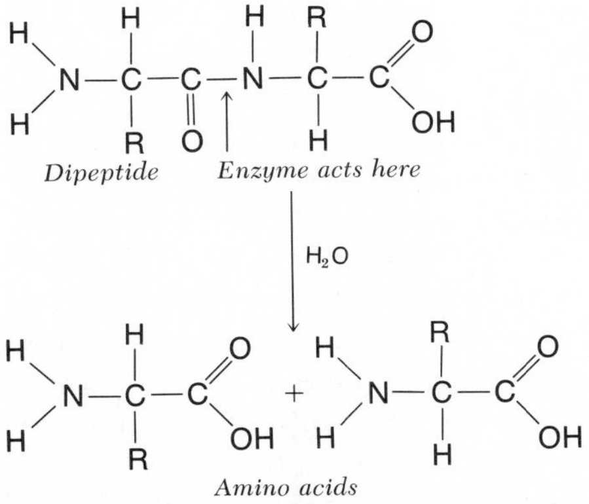

FIGURE 17.12 The action of dipeptidase on dipeptides.

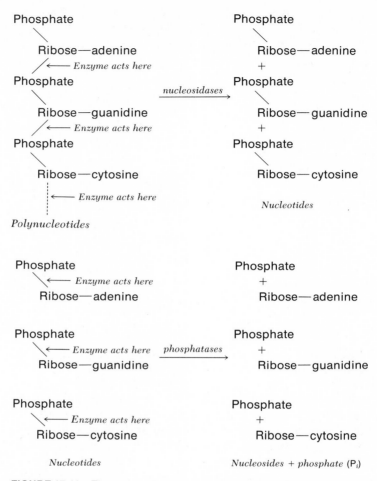

FIGURE 17.13 The action of nucleosidases and phosphatases on oligonucleotides and nucleotides.

THE COMPONENTS AND ACTION OF BILE FLUID

The third fluid entering into the duodenum of the small intestine is the bile fluid. It performs no digestive action itself, but contributes greatly to digestion and absorption in the small intestine. Up to 1 l of bile is produced by the liver in 24 h on a continuous basis. When bile is not needed it is stored in the gall bladder where it is somewhat concentrated by reabsorption of water. The pH of the bile fluid ranges from 7.0 to 8.4 and it contains organic compounds such as bile pigments, bile acids (salts), and cholesterol. The inorganic components are primarily bicarbonates, chlorides, sodium, and potassium, which account for the alkaline character of the bile. Bile has a bitter taste; the bile pigments are responsible for its greenish color. It is both a secretion and an excretion at the same time. Some components, such as the bile pigments and other toxic substances, are excreted by the liver and carried by the bile fluid to the alimentary tract, where they are excreted with the feces. Other components, such as the bile acids (salts), are secretions which aid in digestion and are readsorbed and returned to the liver.

Bile pigments   The bile pigments are the waste products of the breakdown of hemoglobin excreted by the liver. When blood cells die, the heme from the hemoglobin is degraded by the removal of the central iron atom (Fig. 17.14). The iron is stored by the body

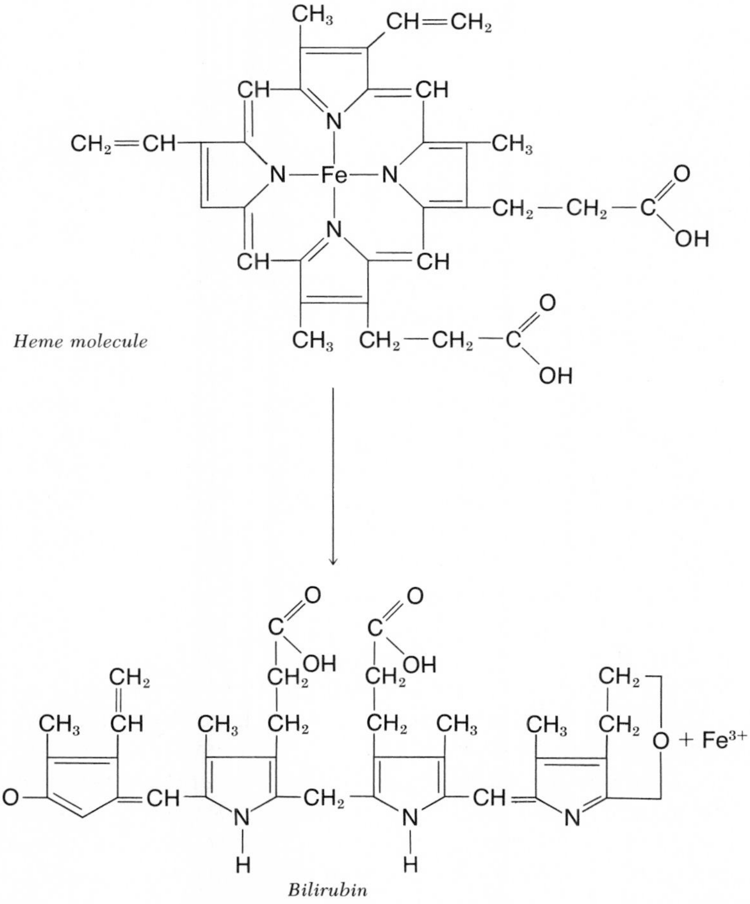

FIGURE 17.14   The removal of the iron atom from the heme molecule and the formation of bilirubin.

for further use, but the rest of the heme molecule is opened and becomes biliverdin, which is excreted. By oxidation and reduction other bile pigments can be formed.

$$\text{Biliverdin} \xrightarrow{\text{oxidation}} \text{bilirubin}$$
$$\text{(Green)} \qquad\qquad \text{(Yellow)}$$

$$\text{Bilirubin} \xrightarrow{\text{reduced}} \text{stercobilin}$$
$$\text{(Yellow)} \qquad\qquad \text{(Brown)}$$

Stercobilin is the pigment found in the feces that produces the typical color. The bile pigments do not perform any special function but abnormal formation of these pigments is used for the diagnosis of diseases of the liver and of the blood. You can follow the formation of these pigments when you have a bruise. It starts with a blue-black coloration produced by blood clotting. As the clotted blood is removed, the coloration changes to green, yellow, and brown, which are the colors of the bile pigments formed as the hemoglobin breaks down.

Bile acids (salts) The bile acids previously discussed under lipids appear primarily as the sodium salts of taurocholic acid and glycocholic acid. To a limited extent potassium salts of the acids are present. These salts are important for the pH of the bile, and help keep cholesterol in solution. They also aid in the neutralization of the acid chyme entering the small intestine. The bile acids perform several important functions.

When the acid chyme enters the small intestine, the carbohydrates and proteins are essentially emulsified and solubilized. However, this is not so for the fats, which appear as large fat globules in the watery chyme. (Remember that fats are hydrophobic and don't like water.) The bile acids act as powerful emulsifying agents and break up the large fat globules into many smaller ones, thereby providing a large surface area for the action of the enzyme **steapsin.** Simultaneously, the bile acids activate the steapsin to full activity. After the fats have been hydrolyzed by the steapsin the products formed—mono- and diglycerides and free fatty acids—are kept in suspension by the bile acids. The bile acids combine with these products and this is believed essential in the absorption of lipids and lipid-related substances. After absorption, the bile salts separate from the lipids and are returned to the liver (Fig. 17.15). Bile salts also aid in solubilizing the fat-soluble vitamins (vitamins A, D, E, K, and the carotenes) in the digestive tract. These vitamins can then be more readily absorbed.

Cholesterol Cholesterol, also discussed under lipids, is found in larger amounts in the bile fluid than in any other fluid in the body.

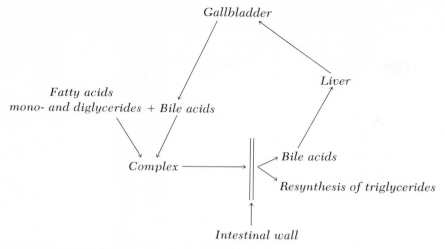

FIGURE 17.15 The circulation of the bile acids.

Its presence here is not very well understood, and again can be considered an excretion from the liver. No function for cholesterol has been identified in the digestive process, and it is excreted with the feces. Precipitation of cholesterol, bile pigments, and calcium carbonate produces gallstones. Gallstones contain up to 98 percent cholesterol, but the reason for their formation is not clear.

THE ABSORPTION OF FOOD COMPONENTS

No absorption of food components takes place in the mouth. The stomach can absorb some water, inorganic ions, monosaccharides, and free amino acids when present, as well as ethanol and other small organic substances. Highly water-soluble medications can also be absorbed through the stomach wall. All other absorption takes place in the small intestine through the villi. The end products of protein digestion, the free amino acids, are directly absorbed by the villi and passed into the bloodstream, where they join the amino acid pool of the body. The end products of the breakdown of the carbohydrates, the monosaccharides glucose, fructose, and galactose, are also absorbed by the villi. To some extent fructose and galactose are converted to glucose in the villi, so that mostly glucose and only little fructose and galactose enter the bloodstream. The absorption is believed to go by two routes. One is by direct diffusion of the monosaccharides from the intestine into the villi and then on to the bloodstream. In a second process, which is not yet fully understood, the glucose molecule is carried by some chemical agent across the barrier into the bloodstream. This is called the active transport mechanism of glucose diffusion. The transport of lipid fractions into the villi, as previously indicated, involves the combining of mono- and diglycerides and free fatty acids, with the bile acids. Once inside the villi, the bile acids become separated from the lipid fractions and the lipid fractions are recombined into triglycerides and then pased on to the bloodstream. Triglyceride globules that are smaller than 0.5 μm in size when emulsified are called **chylomicrons** and can pass directly from the intestines through the villi into the bloodstream. Phospholipids are handled in a similar manner. Water-soluble vitamins and inorganic ions are absorbed readily through the villi. The fat-soluble vitamins need the solubilizing aid of the bile acids for absorption.

REACTIONS IN THE LARGE INTESTINAL TRACT

The peristaltic action of the intestines moves the contents of the small intestine into the large intestine. At this point, the whole mixture is in semifluid state and consists of undigested foods, nondigestible food components such as cellulose and other fibrous material, the remains of the digestive juices, and a large amount of water. Very little digestion and hardly any absorption take place here except the reabsorption of water into the body. This reabsorption converts the contents of the large intestine into a semidry mass called the feces. The bulkiness of the feces is due to its nondigestible materials and this bulkiness helps the peristaltic action to remove the feces on a regular basis. If this is not done, the health of the individual may be affected. This is especially true in developed countries where food is more and more refined and contains less bulk material. Because of its smaller bulk, the feces is retained longer in the large intestine, and bacterial and biochemical changes may have more time to occur, with results that the body may not be able to handle. Chemical reactions in the large intestine are caused primarily by bacteria, which are present in large numbers. We can divide these reactions into **fermentation, putrefaction,** and **detoxification.**

Fermentation is due to certain microorganisms that ferment undigested carbohydrates into fragments which the microorganisms use for food and energy. The by-products of fermentation are organic acids such as acetic, butyric, lactic, and oxalic acids and gases such as carbon dioxide and methane. The by-products are excreted into the large intestine by the bacteria. The action of bacteria on proteins breaks them down into free amino acids, and certain amino acids are then further degraded by decarboxylation, deamination, oxidation, and other degradation reactions to some

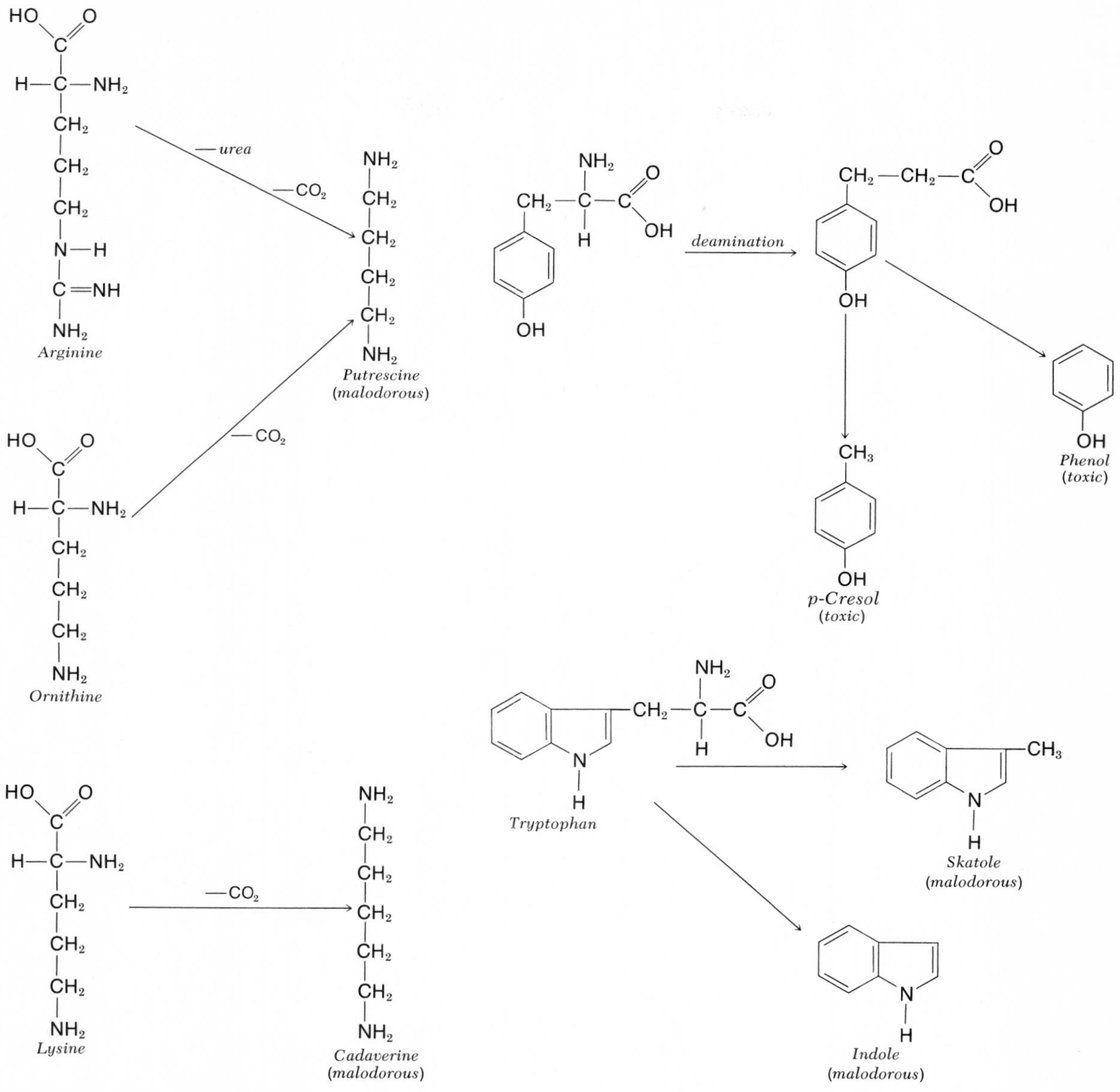

FIGURE 17.16 Intestinal putrefaction.

unpleasant and often toxic substances. These reactions are collectively known as putrefaction; a few examples will illustrate them (Fig. 17.16) *Cadaverine* and *putrescine* have the odor of decaying meat and are especially obnoxious. *Skatole* and *indole* also possess bad odors and are responsible for the odor of feces. *Cresols* and *phenols* are also toxic substances.

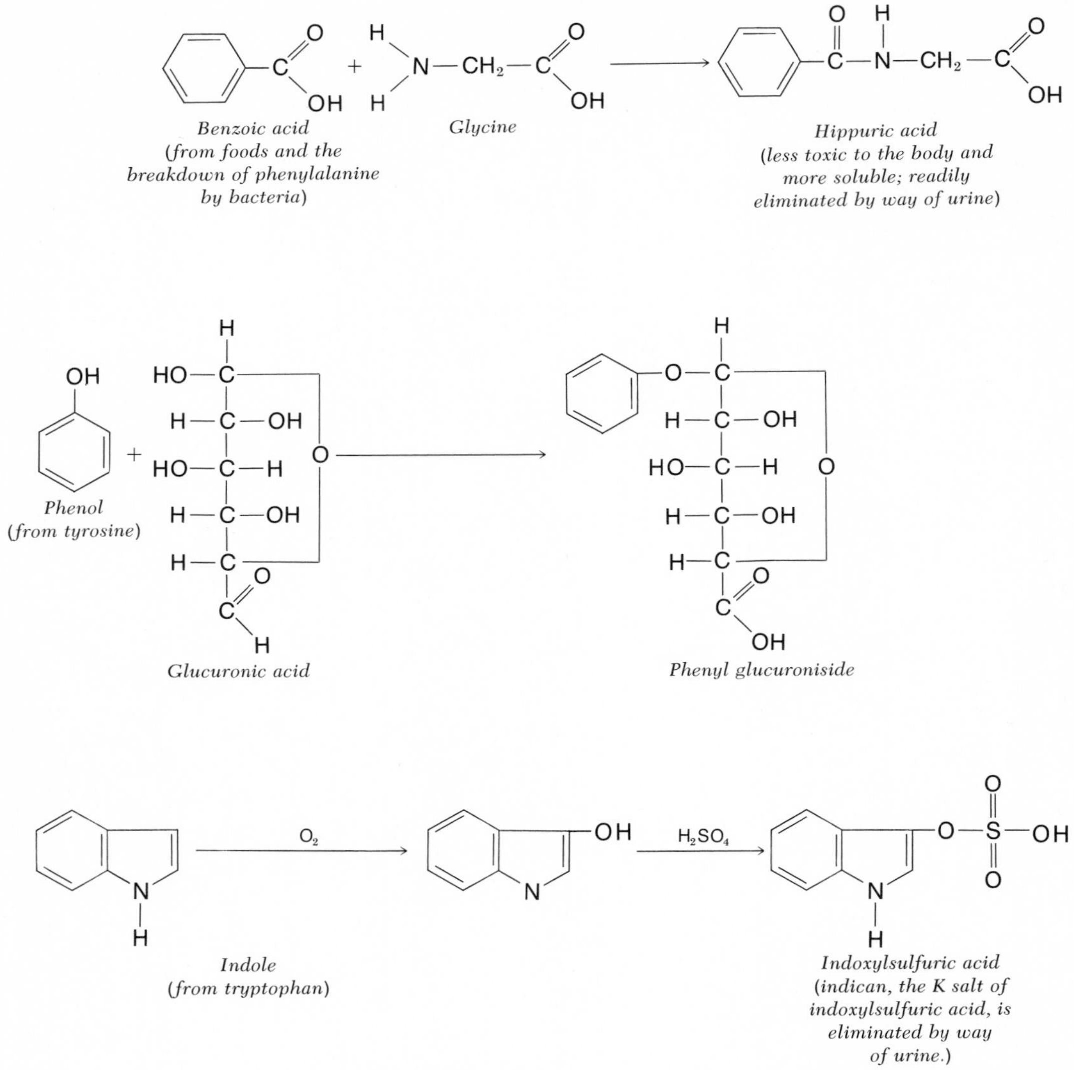

FIGURE 17.17 Detoxification reactions in the large intestine. H_2SO_4 in the detoxification of indole is obtained from phosphoadenosine-phosphosulfate.

Fortunately, several things happen to combat the accumulation of these substances. First, many of the bacteria die from their own poisons, which keeps the pollution caused by the bacteria from reaching a dangerous level. Second, detoxification reactions take place which render these substances harmless (Fig. 17.17).

REVIEW QUESTIONS

1 Why is digestion necessary?

2 The basic process in digestion is the hydrolytic breakdown of the food components. What are the end products of the hydrolysis of **a** proteins; **b** carbohydrates; **c** lipids; **d** maltose; **e** nucleic acids?

3 Identify the enzymes used in the hydrolysis of the food components listed in Question 2.

4 Do you know of any predigestion of natural products before they are eaten? List such natural products and indicate the breakdown that has taken place.

5 Describe the effect of cooking upon various kinds of food.

6a What external factors stimulate the digestive processes in humans? **b** Internal factors?

7 How does the acidity of the gastric juice help in digestion?

8 What type of digestion and what other action upon food takes place in **a** the mouth; **b** the stomach; **c** the small intestinal tract?

9 Classify all the enzymes used in the digestive processes listed in Question 8.

10 List all the secretions entering the stomach and the small intestinal tract. Identify what they are and where they come from.

11 List the functions of the secretions from your answer to Question 10.

12 Why are the digestive enzymes secreted as proenzymes?

13 Give examples of how different proenzymes are activated.

14 What is an endopeptidase?

15 What is an exopeptidase?

16 Discuss the components of bile fluid and their functions.

17 Where and how do food components enter the body?

18 What chemical reactions take place in the large intestinal tract?

19 Describe some detoxification reactions.

18

METABOLISM
Burning and Building — Gently

A chemical reaction conducted in the laboratory in some suitable vessel will sooner or later reach equilibrium. Unless the system is disturbed, there will be no net change in the amount of reactants or products or in its energy, which has come to a minimum. This is a **closed system,** for which equilibrium means the end of change. Obviously, true equilibrium cannot be the condition of living organisms, for whom change is a constant feature. Although the *tendency* toward equilibrium drives forward the chemical reactions of living matter and is behind such biologically necessary phenomena as diffusion and osmosis, equilibrium itself is never attained. Living organisms are **open systems,** constantly receiving energy and building materials from the environment in the form of nutrients, transforming and rearranging them as needed, and discarding what cannot be used. The sum total of the chemical events involved is **metabolism,** our concern in this chapter.

The foods we ingest are, in the main, the huge molecules of carbohydrates, proteins, and lipids discussed in previous chapters. Before these nutrients can pass into the bloodstream

TABLE 18.1 Substrates, products, coenzymes or activators, and pH for digestive enzymes

Enzyme	Substrate	Product	Coenzyme or Activator	pH
Ptyalin, α-amylase salivary amylase	Amylose	Dextrins, maltose	Cl^- ions	4.0–9.0
Pepsin, gastric protease	Proteins	Proteoses, peptones	H^+ ions	1.0–2.5
Rennin coagulase	Casein	Paracasein	H^+ ions, Ca^{2+} ions	1.0–2.5
Steapsin, gastric lipase	Lipids	Glycerin, fatty acids		1.2–3.0
Trypsin	Proteins	Polypeptides, proteoses, peptones	Enterokinase	7.2–8.5
Chymotrypsin	Proteins	Polypeptides, proteoses, peptones	Trypsin	7.2–8.5
Carboxypeptidase	Proteins, proteoses, peptones	Amino acids		7.2–8.5
Pancreatic amylase, Amylopsin	Amylose, amylopectin	Dextrins, maltose		7.0–8.2
Pancreatic lipase	Lipids	Glycerin, fatty acids, mono- and diglycerides		7.0–8.2
Phospholipase	Phospholipids	Fatty acids, glycerylphosphoryl choline, glycerylphosphoryl ethanolamine, glyceryl-phosphoryl serine		7.0–8.2
Nuclease	RNA, DNA	Nucleotides		7.0–8.2
Aminopeptidase	Polypeptides	Dipeptides, amino acids		7.0–8.2
Dipeptidase	Dipeptides	Amino acids		7.0–8.2
Lactase	Lactose	Glucose, galactose		7.0–8.2
Sucrase	Sucrose	Glucose, fructose		7.0–8.2
Maltase	Maltose	Glucose		7.0–8.2
Nucleosidase	Oligonucleotides	Mononucleotides		7.0–8.2
Phosphatase	Nucleotides	Phosphate, nucleosides		7.0–8.2

for transport to the cells, where metabolism takes place, they must be reduced to smaller molecules. The breakdown of very large molecules to smaller component molecules is called **digestion.** Chemically, the process of digestion is a sequence of hydrolyses, each hydrolysis being catalyzed by a specific enzyme. Table 18.1 lists these enzymes together with the substrates on which they act and the conditions required for their action.

CATABOLISM AND ANABOLISM

The digestion and absorption of nutrients are really the first, preparatory steps of metabolism. The enzymes involved in breaking down ingested foods into their simpler and smaller components are **extracellular,** doing their work outside the cells that produced them. When the products of digestion are absorbed through the intestinal wall and transported by the bloodstream to the cells, they enter into the process of metabolism and become **metabolites.** Metabolism takes place in the cells, and the many enzymes needed are **intracellular,** that is, produced and used within the cell. All living matter needs energy to carry on its activities, and all living matter has a structure. Metabolism, therefore, has two aspects, one that leads to the production of energy and another that leads to the synthesis of the chemical materials the organism needs for its structure and function. The first is called **catabolism,** and the second is called **anabolism.** The nutrients carried to the cells are available both for energy and for structure, but they move where they are most needed and best used. The primary building units for anabolism are the amino acids, with smaller amounts of fatty acids and monosaccharides also used. The needs of the body for structural proteins, blood proteins, enzymes, cell replacement, and the many chemical materials involved in maintaining and running the organism are provided by anabolism.

Catabolism produces energy.
Anabolism uses energy.

NH$_2$
|
C$=$O
|
NH$_2$
Urea

Under normal conditions, the bulk of the digestion products of carbohydrates and lipids (and to a lesser extent of the proteins) enter into **catabolic pathways,** which are specific sequences of chemical changes in the course of which energy is *extracted* from the nutrients by converting them into low-energy end products, namely, CO_2, H_2O, and urea. Since monosaccharides, fatty acids, and amino acids are chemically different, the metabolic path each type follows at the beginning is different, but, as we shall see, they all lead into a *final* and *common* path, the **Krebs cycle,** which is the mainstream of the catabolic process. In the course of the Krebs cycle, the metabolites are broken down into CO_2 and H atoms; the latter are then carried by coenzymes through a series of oxidation-reduction reactions called the **electron-transport system,** which ends with the formation of water by the union of hydrogen and atmospheric oxygen. For this reason, this sequence, which works in conjunction with the Krebs cycle, is also called the **respiratory chain.** It is along this chain that the bulk of the energy from the original nutrients is delivered for use by the body. The Krebs cycle also contributes to anabolism by providing intermediate chemical products that are used for building up final products. Since energy is the maker and mover of things, catabolism will be given major consideration in our discussion of metabolism.

Figure 18.1 is a sort of aerial view of what has been said so far. The representation is much simplified, and only the main roads are sketched in; as we consider each in turn, and in more detail, we should not lose sight of the essential unity of the entire metabolic process.

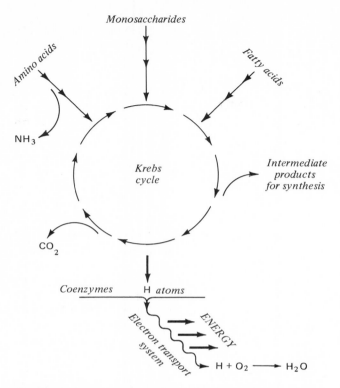

FIGURE 18.1 Simplified representation of catabolism in cells. The multiple arrows →→→ signify the preliminary chemical pathways taken by the monosaccharides, fatty acids, and amino acids.

HIGH-ENERGY COMPOUNDS

Catabolism releases energy to the body in two forms, thermal energy, or heat, and chemical energy. The heat serves to maintain body temperature but passes out of the body and into the atmosphere, becoming unavailable for doing work. On the other hand, chemical energy is fixed in a molecular structure that stays in the cell and can be called upon as needed. The call for energy never stops; even when we are at rest, substantial amounts of energy are required to keep all the vital functions going. This is referred to as **basal metabolism.** When we exert ourselves physically, the call for energy increases accordingly.

If our food intake is more than the body needs and can use, we store this excess in the form of fat, which acts as a nutrient reserve. Since cells cannot store chemical energy against all

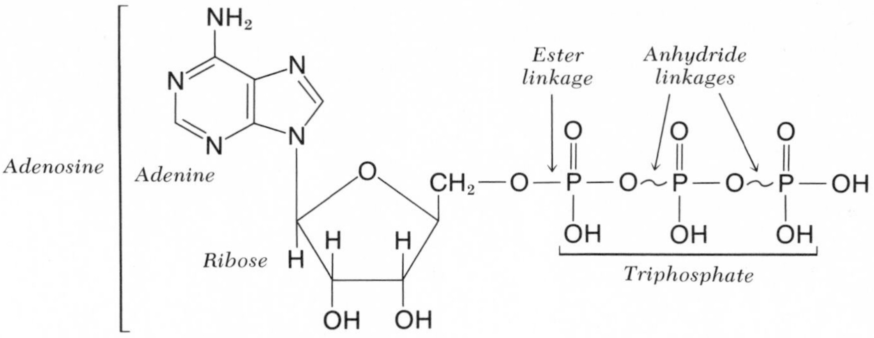

FIGURE 18.2 Adenosine triphosphate (ATP).

possible needs, like money in a bank, a mechanism is needed that responds quickly when additional chemical energy is required. The energy itself comes from burning the food nutrients via the catabolic pathways of Figure 18.1 but is delivered, in the main, in the form of a high-energy compound called **adenosine triphosphate** (ATP). Like all high-energy compounds, ATP is an unstable structure with a high potential energy; when it breaks up by hydrolysis to smaller and more stable parts, energy is given off. The structures of ATP and its component parts are shown in Figure 18.2; linkages where breakup can occur by hydrolysis are marked. Note that the two anhydride linkages are represented by squiggles, which indicate a bond where hydrolysis occurs with a substantial release of free energy. The same symbol will be used in other high-energy compounds. The bond is not in itself the source of the energy released but merely the point at which the compound breaks up into smaller parts of lesser free energy. This energy difference is the energy released.

ATP can undergo three successive hydrolyses, each time losing a phosphate group to water and releasing energy. If we abbreviate the structure but show the phosphates, the first hydrolysis can be represented as

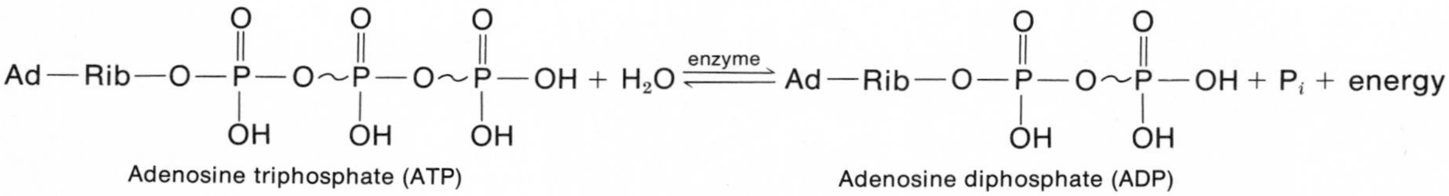

Adenosine triphosphate (ATP)   Adenosine diphosphate (ADP)

$$\tag{18.1}$$

The hydrolysis of ATP therefore results in the formation of **adenosine diphosphate** (ADP) plus inorganic phosphate (P_i), and energy. The other two hydrolyses take place analogously, and all can be summarized as follows:

$$ATP + H_2O = ADP + phosphate + 7.3 \ kcal/mole \qquad (18.2a)$$

$$ADP + H_2O = AMP + phosphate + 7.3 \ kcal/mole \qquad (18.2b)$$

$$AMP + H_2O = adenosine + phosphate + 3.4 \ kcal/mole \qquad (18.2c)$$

AMP stands for **adenosine monophosphate.** It can be seen that the energy released by the hydrolysis of its ester linkage is less than the energy obtained by the hydrolysis of the anhydride linkages of ATP and ADP. The significant feature of these reactions, apart from the energy released, is that they are reversible. When ATP is hydrolyzed, releasing energy for some energy-using process, it can be regenerated from ADP by the addition of inorganic phosphate and energy. In Figure 18.1 the energy needed for recycling ADP back to ATP is seen leaving the electron-transport system at three points. These are the major payoff points, where the energy provided by catabolism is delivered and stored as ATP. For the most part, energy transactions in living matter are made through the ATP-ADP cycle (Figure 18.3). ADP receives energy and becomes ATP; ATP gives up energy and becomes ADP. In this give and take, a phosphate group is lost by ATP and gained by ADP. It is via this mechanism that the energy released by catabolism or the radiant energy received by plants is transferred to the energy-using activities of living matter.

The type of diagram shown in Figure 18.3 is often used to represent **coupled reactions,** in which one reaction sets another in motion, somewhat like a series of meshed gears. In the diagram the receipt of energy by ADP is coupled to the release of energy by ATP. As we shall see later, the electron-transfer system is a prime example of a sequence of coupled reactions, and many of the chemical reactions of metabolism *require* coupling with an

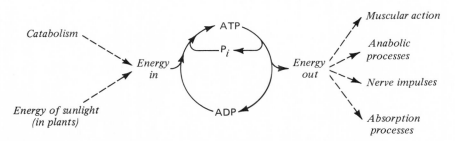

FIGURE 18.3 The ATP-ADP cycle acts as the intermediary between energy-producing processes in living matter and energy-using processes. (P_i = inorganic phosphorus.)

energy-releasing cycle to make them go. Keeping in mind that these reactions take place in the cells, we can appreciate the advantage of coupled reactions:

1 They are economical in materials and space; to make each ATP molecule from fresh adenine, ribose, and phosphate would require impossibly large reserves of these materials.
2 If ADP were *not* recycled back to ATP, the cell would very soon be jammed up with accumulated ADP molecules and phosphate.
3 In a sequence of coupled reactions those which are energy-producing drive those which are energy-consuming, thus keeping the entire process going.

OXIDATION-REDUCTION REACTIONS IN METABOLISM

As the form in which chemical energy is stored, ATP plays a central role in metabolism. However, neither ATP nor any of the other high-energy compounds is the source of that energy, only its *carrier*. It was mentioned before that the bulk of the energy derived from the foods we ingest is drawn off along the respiratory chain, or electron-transport system. This is a chain of oxidation-reduction reactions, that is, of electron transfer from compound to compound. For the most part, this chain is fed from the Krebs cycle.

As the metabolites are carried through the Krebs cycle, they yield a pair of hydrogen atoms at several steps (not really free hydrogen atoms but readily transferrable ones). Since the loss of a pair of hydrogens, each with its electron, is an oxidation, if a metabolite is generally represented as MH_2, the reaction is

$$MH_2 \longrightarrow M + 2H \qquad (18.3)$$

Reduced metabolite → Oxidized metabolite

But an oxidation implies a reduction, a compound that will take the hydrogen atoms and their electrons. This function is served by a number of coenzymes, of which **nicotinamide adenine dinucleotide** (NAD^+) is the most common (Figure 18.4). We can now write the general oxidation-reduction reaction between a reduced metabolite and NAD^+:

$$MH_2 + NAD^+ \rightleftharpoons M + NADH + H^+ \qquad (18.4)$$

Reduced metabolite | Oxidized form | Oxidized metabolite | Reduced form

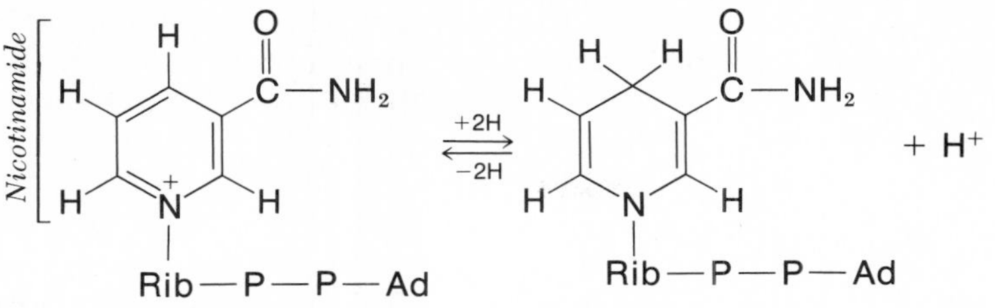

NAD$^+$ (*oxidized form of the coenzyme*) NADH + H$^+$ (*reduced form of the coenzyme*)

FIGURE 18.4 NAD$^+$ as an oxidizing agent. The moieties that make up NAD$^+$ are ribose, adenine, two phosphates, and nicotinamide. It is the nicotinamide part, derived from the vitamin niacin, that is the site for the oxidation-reduction reaction. In taking two hydrogen atoms (and their two electrons) from a metabolite, the oxidized form of the coenzyme becomes the reduced form. The positive charge on the NAD$^+$ structure is at the nitrogen atom, since a nitrogen atom with four bonds carries a plus charge, whether it is in NAD$^+$ or NH_4^+.

The stepwise release of energy through the chemical reactions of catabolism is a central and necessary feature of living organisms. The most common source of energy in animal life is glucose, and the simplest way to obtain the energy is to burn the glucose in air. The reaction is

$$C_6H_{12}O_6 + 6O_2 \longrightarrow 6CO_2 + 6H_2O + 673 \text{ kcal/mole}$$

A reaction of this sort, with its rapid release of all the energy at one time, may be suitable for running a steam engine but would never do for living organisms. Not only would the rapid release of so much thermal energy raise body temperature abnormally, but the energy would escape before it could be used. What is needed is *chemical* energy in the form of energy-rich ATP molecules whose hydrolysis releases energy *when* and *where* it is called for. The stepwise energy release of the catabolic process makes this possible.

We now turn our attention to the three preliminary pathways that lead digested nutrients into the Krebs cycle, beginning with the conversion of glucose to pyruvic acid. At first sight these reactions may seem confusing, but if you keep in mind the ultimate outcome and the purpose they serve, you will find them both simple and ingenious.

THE EMBDEN-MAYERHOF PATHWAY: ANAEROBIC GLYCOLYSIS

The sequence of chemical events that breaks down a glucose molecule into two pyruvic acid molecules is called the Embden-Mayerhof pathway or **glycolysis.** These reactions are **anaerobic**

Glyceric acid 1,3-diphosphate

Glucose 6-phosphate

ADP

ATP

Glucose

Pyruvic acid,
a keto acid

(do not require oxygen), which suggests that they were evolved in that early period of living matter when there was little if any oxygen in the atmosphere. The fact that this anaerobic pathway is common to all organisms that use glucose for energy further supports the idea that glycolysis was the original energy-producing system. As conditions and organisms changed, higher forms of life later supplemented glycolysis with the more efficient, oxygen-using Krebs cycle. However, some simple microorganisms still use glycolysis as their only route for producing energy, extending it somewhat to convert the pyruvic acid into smaller molecules such as acetic acid. With yeast, the process (called **fermentation**), ends by converting pyruvic acid into carbon dioxide and ethanol. Figure 18.5, a map of the Embden-Mayerhof pathway, can be best read by dividing it up into three parts.

PART ONE

The pathway can be considered to begin at glucose or glycogen, the storage form of glucose. Both lead to glucose 6-phosphate, which is the metabolically active form of glucose, after which part one continues on through steps 2 to 5 and the formation of two molecules of glyceric acid 1,3-diphosphate. As Figure 18.5 shows, this is a turning point. The reactions of part one are energy-consuming; from then on they are mainly energy-releasing.

PART TWO

Taken together, steps 6 to 9 are downhill on the energy scale, and the energy is released both as thermal energy and as chemical energy in the form of ATP. Part two ends with the formation of two molecules of pyruvic acid from the original glucose molecule. Retracing the sequence glucose $\longrightarrow$ 2 pyruvic acid and counting the ATPs used up and those produced by the reactions shows that there is a net gain by the organism of two ATP molecules of stored energy.

PART THREE

The glycolysis map shows pyruvic acid standing at a crossroads; the particular direction it takes depends on the type of organism involved and the availability of oxygen. If the organism is yeast, pyruvic acid moves down step 10, which actually consists of two parts, and becomes converted into ethanol, in which case the entire pathway is called a fermentation. Although ethanol is the most important fermentation product, there are others, including

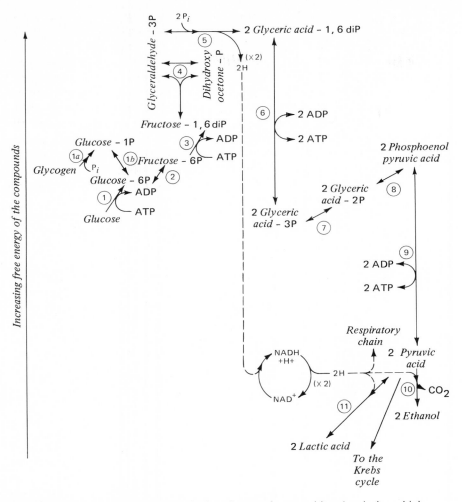

FIGURE 18.5 The Embden-Mayerhof pathway of anaerobic glycolysis, which shows the sequence of reactions as well as the relative free energy of the product formed in each case. If the reaction moves a product up the free-energy scale, an input of energy is needed; if the reaction moves a product down, a release of energy occurs. Although many of the reactions are reversible, the entire sequence is driven from left to right by the net release in free energy that results from the breakdown of glucose.

acetic acid, lactic acid, and citric acid, each involving specific microorganisms.

On the other hand, if the organism is aerobic and depends mainly on atmospheric oxygen for its energy-releasing reactions, pyruvic acid goes on toward the Krebs cycle. However, the Krebs cycle is tied to the respiratory chain and the availability of oxygen, so that when there are prolonged or intense calls for energy, as in heavy labor or athletics, the rate of oxygen supply may not keep up with demand. In such cases the pyruvic acid molecules seek an additional way out by being reduced to lactic acid in step 11. The result is an accumulation of lactic acid in muscle tissue, which can cause fatigue, pain, and stiffness. Step 11 is reversible. When a race is over, the runners still keep breathing heavily for a time, providing the additional oxygen needed to reverse step 11, changing lactic acid back to pyruvic acid and then down to the Krebs cycle. The oxygen needed to bring the lactic acid produced during intense physical effort back toward the oxidative Krebs cycle is sometimes called the **oxygen debt.**

Steps 10 and 11 share an important feature in common: each needs a pair of hydrogen atoms. Both in yeast cells and in our body cells, these hydrogens come from step 5 of glycolysis, and, as Figure 18.5 shows, they are delivered via a coupled reaction with NAD^+ to produce ethanol in one case and lactic acid in the other. However, when there is no need for the body to produce lactic acid or when the oxygen debt has been paid off and the reversal of step 11 returns the two hydrogens, they enter the electron-transport system as indicated schematically in Figure 18.6.

The individual glycolysis reactions will now be outlined, with the necessary enzymes and activators given for each reaction.

Step 1: Glucose phosphorylated and activated Glycolysis in the cell begins with the formation of glucose 6-phosphate, obtained either from glycogen via steps 1*a* and 1*b* or from glucose:

Glucose   Glucose 6-phosphate

Hexokinase also catalyzes the phosphorylation of other monosaccharides such as fructose and mannose.

Of the 7.3 kcal of free energy obtained from the hydrolysis of ATP to ADP, about 3.3 kcal goes to form the glucose 6-phosphate and the remaining 4 kcal is lost as heat, so that the reaction is essentially irreversible. Irreversible reactions of metabolic pathways are often called **pacemaker reactions** since they control the rate of travel down the pathway. If a pacemaker reaction is blocked or slowed down, the reactions behind it and ahead of it will be affected accordingly.

Step 2: Glucose 6-phosphate isomerized to fructose 6-phosphate This reaction prepares the molecule for the next step

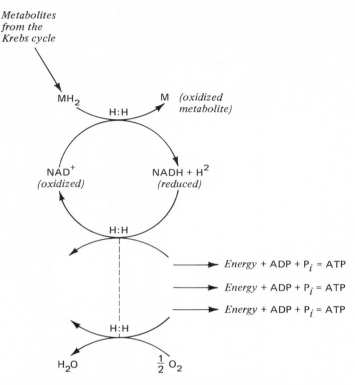

FIGURE 18.6 The metabolic oxidation of hydrogen atoms to water. This is an abbreviated version of the electron-transport system with only the first and last steps shown. So long as the metabolites from the Krebs cycle make hydrogen atoms available, the sequence of coupled reactions keeps turning out water and energy. Each enzyme, beginning with NAD^+, receives and then passes on hydrogen atoms and so is alternately reduced and oxidized. In the final step the hydrogens combine with oxygen to produce the water of metabolism.

by making the —OH group more available through the pyranose structure of fructose:

Glucose 6-phosphate   ⇌ (Mg^{2+}, phosphoglucoisomerase)   Fructose 6-phosphate

Since only a rearrangement is involved, the energy change is small and the reaction is readily reversible.

Step 3: Fructose 6-phosphate phosphorylated to fructose 1,6-diphosphate The addition of a second phosphate group requires an investment of about 3.6 kcal of free energy, and, as in step 1, the hydrolysis of ATP to ADP provides both the energy and the phosphate group needed:

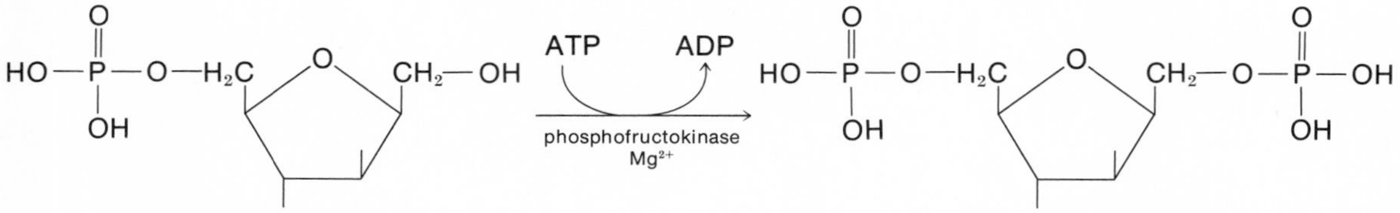

Fructose 6-phosphate   Fructose 1,6-diphosphate

In the liver, step three can be reversed by energy and a special enzyme, enabling the body to synthesize glucose.

Since, as in step 1, the free energy obtained from the hydrolysis of ATP exceeds the energy needed to form the phosphate ester, there is a net release of heat and the reaction is irreversible and is a pacemaker. The heat energy that passes out spontaneously cannot do work for the organism, but it does help to maintain body temperature. The six carbon atoms of the original hexose are still present in the diester formed by this reaction, but now the structure is ready to break up into two parts.

The condensation of the aldehydes gives a structure with an aldehyde and an alcohol group; hence the term aldolase. The same enzyme catalyzes the reverse of condensation.

Step 4: Fructose 1,6-diphosphate split into two three-carbon molecules The reactions of catabolism generally involve degrading metabolites to smaller units, and this is the first such instance. As shown below, the fructose 1,6-diphosphate splits into two three-carbon units. For clarity, the fructose 1,6-diphosphate is shown in its open keto form:

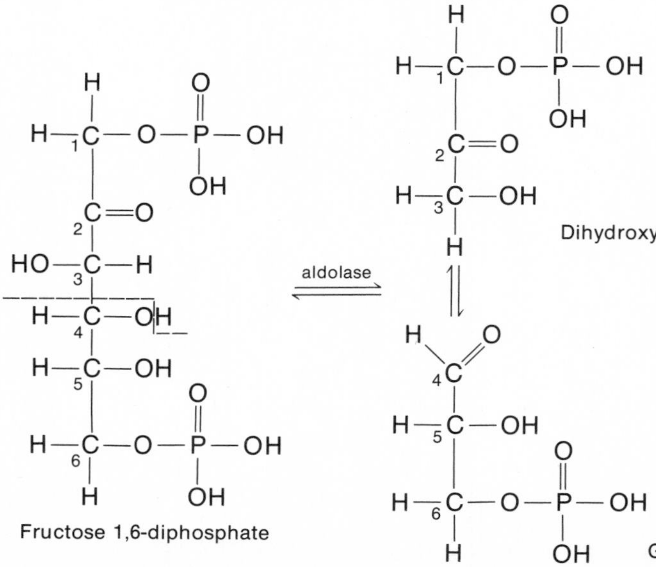

Fructose 1,6-diphosphate

Dihydroxyacetone phosphate

Glyceraldehyde 3-phosphate

Although dihydroxyacetone phosphate and glyceraldehyde 3-phosphate are both formed, only the second compound continues down the glycolysis pathway. However, the dihydroxy-acetone phosphate is not wasted because it is isomerized to glyceraldehyde-3-phosphate:

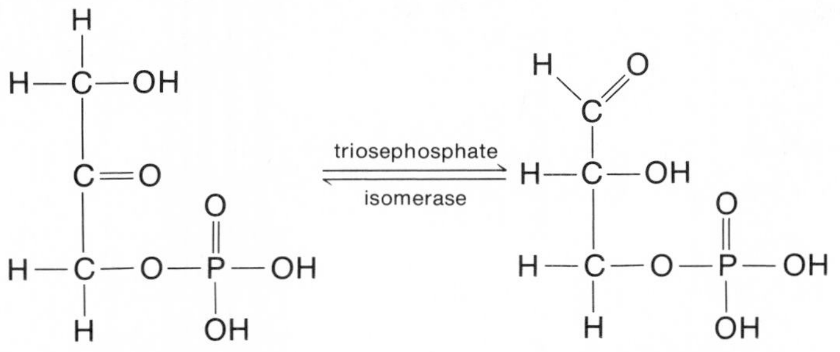

Dihydroxyacetone phosphate

Glyceraldehyde 3-phosphate

The reactions of step 4 are significant for several reasons:

1 The degradation of one molecule of fructose 1,6-diphosphate results in two molecules of glyceraldehyde 3-phosphate that continue down the pathway.
2 At this point the glycerol obtained from the hydrolysis of lipid triglycerides enters glycolysis and catabolism.
3 This step also provides the starting material for making the glycerol used in synthesizing the body's own lipids.

Like other reactions we shall meet later, this reaction is therefore a crossing point between catabolic and anabolic pathways, a junction at which tendencies for feedback and feed-in can be expressed.

Step 5: Glyceraldehyde 3-phosphate oxidized to glyceric acid 1,3-diphosphate Besides being oxidized, the glyceraldehyde is also phosphorylated, with the net result that a high-energy compound is produced. This oxidation is anaerobic and is effected through the loss of a pair of hydrogens, that is, two H^+ ions and two electrons. Remember that oxidation is electron loss. The additional phosphate group that is taken on comes from inorganic phosphate, shown as phosphoric acid in the reaction:

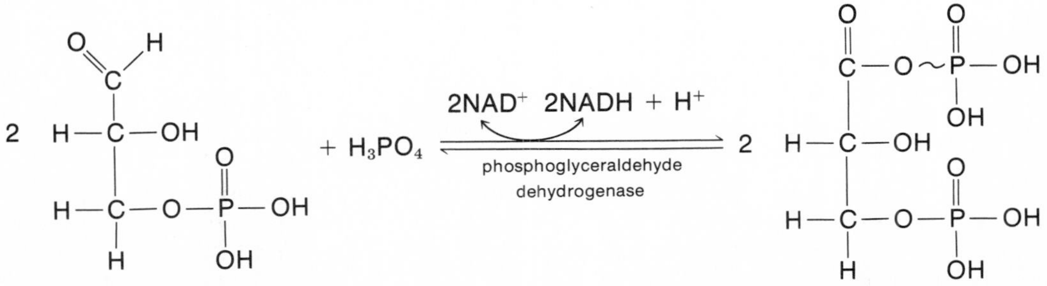

Glyceraldehyde 3-phosphate

Glyceric acid 1,3-diphosphate

It should be noted that the loss of hydrogens that constitutes the oxidation here depends on their acceptance by NAD^+, which is thereby reduced to $NADH + H^+$.

Step 6: Glyceric acid 1,3-diphosphate converted into glyceric acid 3-phosphate Chemical energy is stored as ATP, which is formed by the transfer of a phosphate group from the diphosphate to ADP:

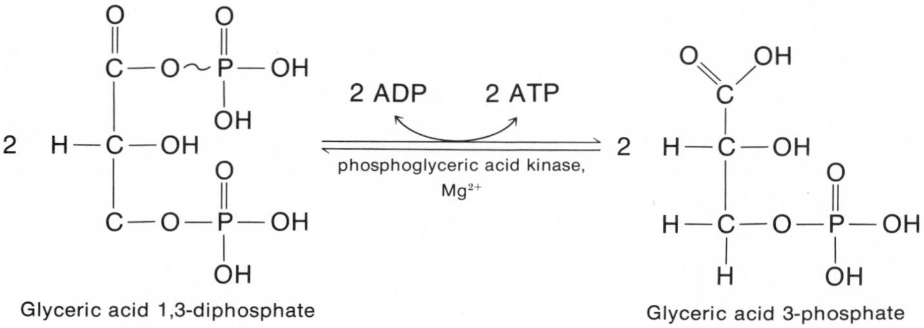

Glyceric acid 1,3-diphosphate

Glyceric acid 3-phosphate

However, the energy required for that transfer comes from the energy released as glyceric acid 3-phosphate is formed. Since one molecule of glucose is degraded to two of glyceric acid 1,3-diphosphate, step 6 produces *two* ATPs for each glucose entering glycolysis.

Step 7: Glyceric acid 3-phosphate isomerized to glyceric acid 2-phosphate The phosphate group shifts from carbon 3 to carbon 2 in preparation for subsequent reactions:

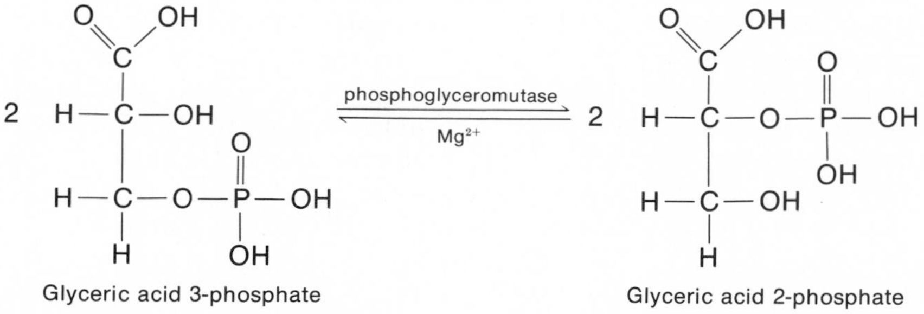

Glyceric acid 3-phosphate

Glyceric acid 2-phosphate

Like the isomerization of glucose 6-phosphate to fructose 6-phosphate of step 2, the reaction is reversible.

Step 8: Glyceric acid 2-phosphate dehydrated to phosphoenolpyruvic acid The dehydration reaction results in the formation of a double bond between carbon atoms 2 and 3, which together with the phosphate group at carbon atom 3 makes for a strained, unstable structure:

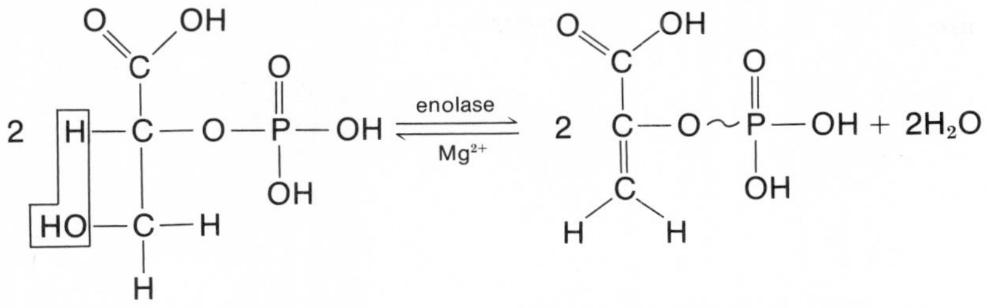

Glyceric acid 2-phosphate   Phosphoenolpyruvic acid

(enolase, Mg^{2+}) $\ \rightleftharpoons\ $ ... $+ \ 2H_2O$

Phosphoenolpyruvic acid is therefore an energy-rich structure in the sense that its hydrolysis releases a large amount of energy. The name enolpyruvic comes from the fact that pyruvic acid, which is formed in the next step, can exist in the enol form, where carbon atom 2 bears a double bond (*ene*) and an alcoh*ol* group:

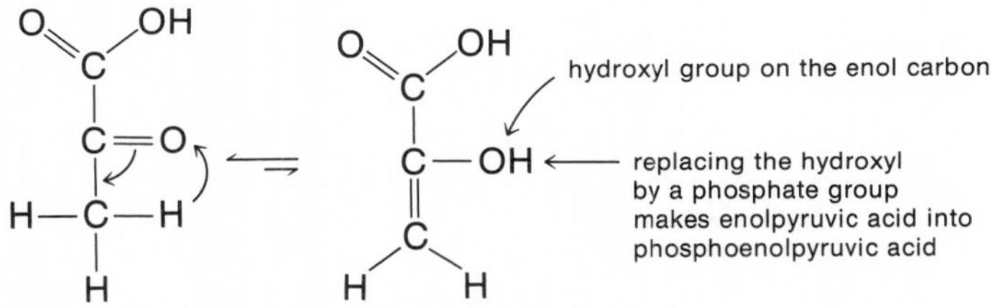

hydroxyl group on the enol carbon

replacing the hydroxyl by a phosphate group makes enolpyruvic acid into phosphoenolpyruvic acid

Pyruvic acid, a keto acid   Enolpyruvic acid, a hydroxy acid

Step 9: Phosphoenolpyruvic acid changed to pyruvic acid For the second time (see step 6) the breakdown of an energy-rich compound releases energy for the storage of chemical energy by the formation of two ATPs:

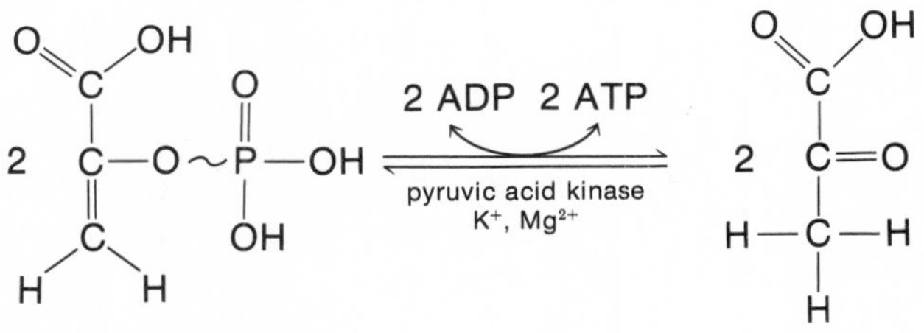

2 ADP 2 ATP

pyruvic acid kinase K^+, Mg^{2+}

Phosphoenolpyruvic acid   Pyruvic acid

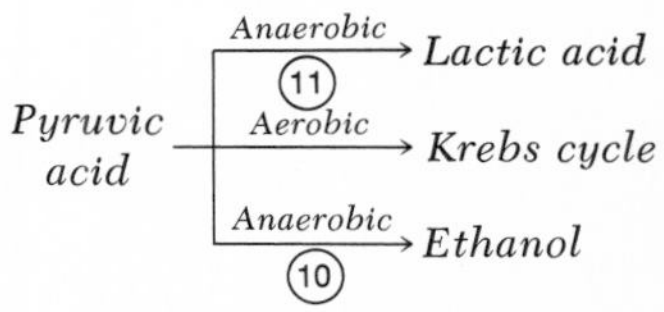

The marginal sketch recapitulates the directions pyruvic acid can follow. We first consider step 10, which takes place in yeast and is the last sequence of alcohol fermentation. It actually consists of two steps, 10*a* and 10*b*.

Pyruvic acid

Anaerobic ⑪ → *Lactic acid*

Aerobic → *Krebs cycle*

Anaerobic ⑩ → *Ethanol*

Step 10a: Pyruvic acid converted to acetaldehyde by losing CO_2
The reaction is a decarboxylation, and the splitting and rearrangement of the pyruvic acid are shown.

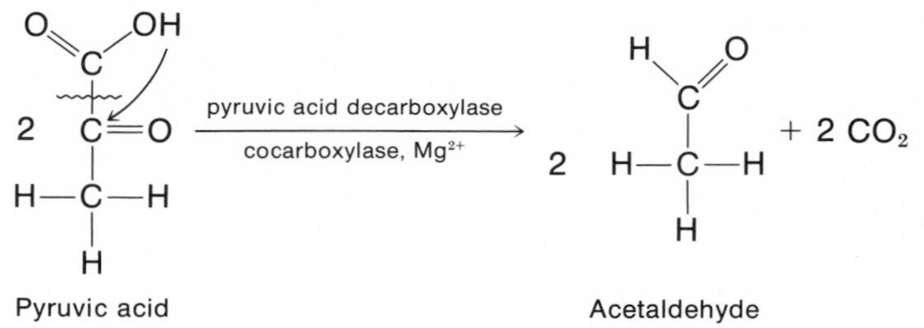

Pyruvic acid   Acetaldehyde

Step 10b: Acetaldehyde reduced to ethanol The two hydrogens coming from step 5 are delivered as $NADH + H^+$:

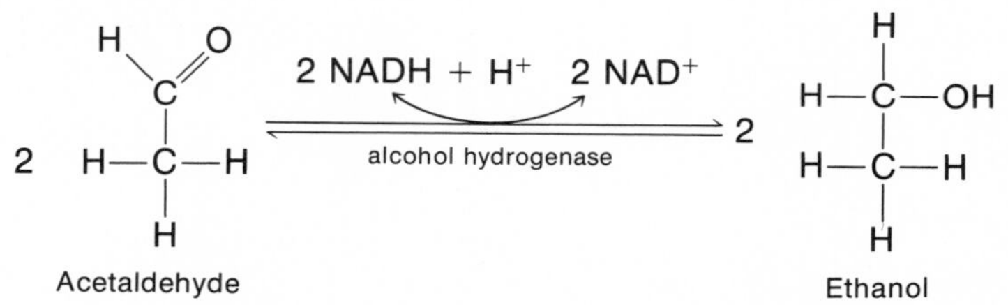

Acetaldehyde   Ethanol

After delivery is made, the resulting NAD^+ is ready to pick up another pair of hydrogens for the next acetaldehyde molecule. The overall fermentation reaction, starting with glucose and ending with two ethanol molecules, can be summarized as

Glucose + 2 phosphate + 2 ADP

$$= 2 \text{ ethanol} + 2CO_2 + 2ATP + 2H_2O$$

Although the entire sequence is anaerobic, a small but significant amount of chemical energy has been provided.

In our bodies pyruvic acid can undergo a further anaerobic reaction to lactic acid. This occurs mainly in muscle tissue, and, as mentioned before, its accumulation causes fatigue and pain by raising the level of acidity in the cells.

Step 11: Pyruvic acid is reduced to lactic acid As in the step 10*b* of fermentation in yeast cells, the pair of hydrogens required for the reduction of pyruvic to lactic acid comes from step 5, with $NADH + H^+$ again the carriers and again reverting to NAD^+ after delivery to become ready for the next round trip.

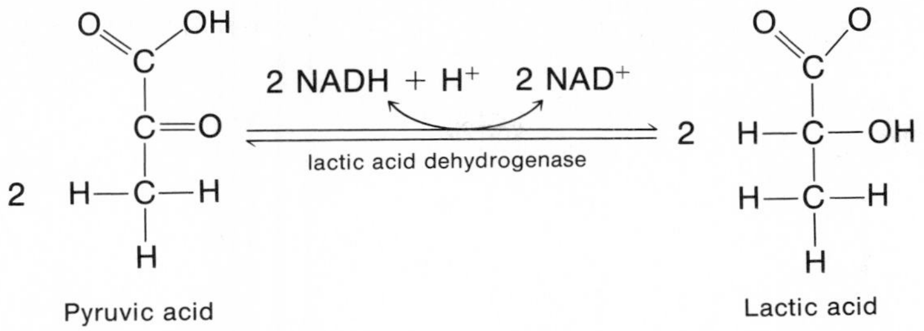

Pyruvic acid   Lactic acid

The anaerobic glycolysis of glucose to lactic acid in muscle tissue can be now summarized:

Glucose + 2 phosphate + 2 ADP = 2 lactic acid + 2 ATP + $2H_2O$

In muscle tissue lactic acid is a waste material and diffuses into the bloodstream, which carries it to the liver. There it is converted back to glucose. When there is an adequate supply of oxygen in the muscle cells, the lactic acid is oxidized back to pyruvic acid, which is then catabolized aerobically in the Krebs cycle. Before proceeding to this main pathway, we shall consider the preliminary pathway followed by the lipids before they enter the Krebs cycle.

THE CATABOLISM OF LIPIDS

Fats are present in the body essentially as triglycerides, and before any lipids can be metabolized, they are hydrolyzed by lipases to glycerol and fatty acids:

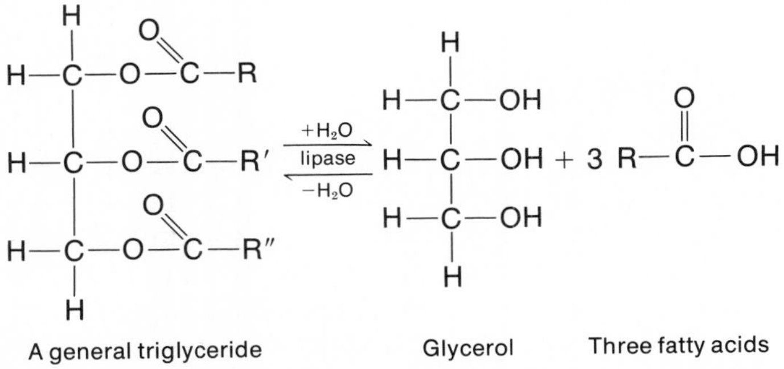

A general triglyceride   Glycerol   Three fatty acids

Glycerol undergoes the following two reactions, which prepare it for entry into the glycolysis just discussed.

GLYCEROL ACTIVATED TO α-GLYCEROPHOSPHATE

This reaction is an example of the activation of a metabolite by ATP, enabling the metabolite to enter a pathway. We met it before when glucose was activated to glucose 6-phosphate, and we shall meet it again.

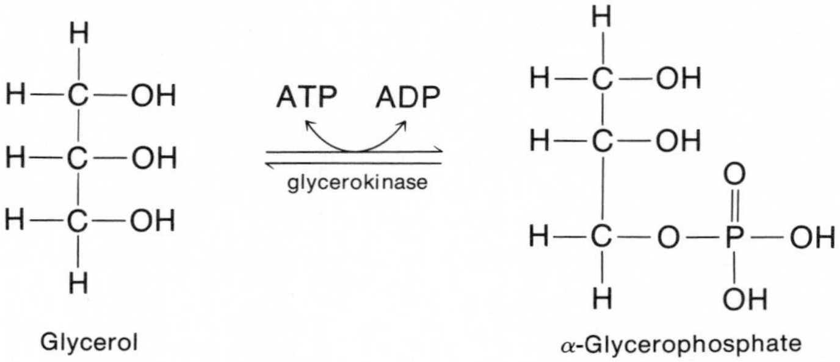

α-GLYCEROPHOSPHATE OXIDIZED
TO DIHYDROXYACETONE PHOSPHATE

Two hydrogens are removed by NAD⁺ from the hydroxyl function at carbon atom 2 to produce dihydroxyacetone phosphate, which enters the Embden-Mayerhof pathway at Step 4 of Figure 18.5.

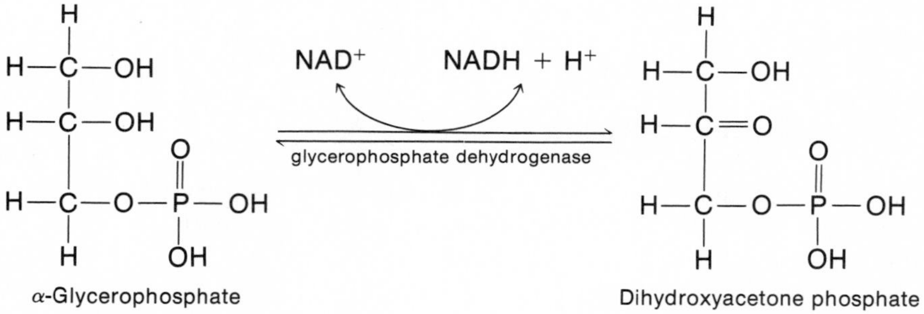

FATTY ACID METABOLISM

The breakdown of the fatty acids is accomplished by the so-called **β-oxidation process,** originally postulated by a German biochemist, F. Knoop, about the beginning of this century. Long before tagging with radioactive isotopes was available, he tagged even-numbered and odd-numbered fatty acids by adding a phenyl group at the end of the chain. When even-numbered fatty acids were fed to dogs, the metabolic end product isolated from their urine was phenylacetic acid:

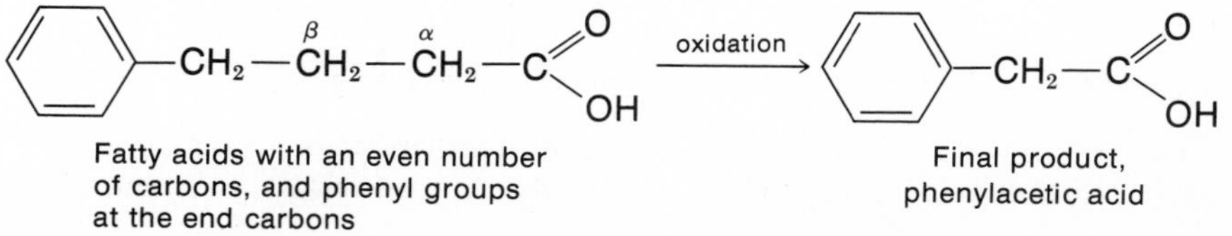

Fatty acids with an even number
of carbons, and phenyl groups
at the end carbons

Final product,
phenylacetic acid

When the fatty acids had an odd number of carbon atoms, the metabolic end product isolated was a derivative of benzoic acid:

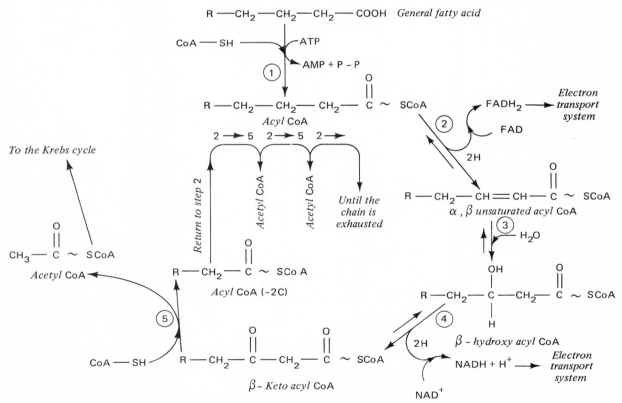

Odd-numbered fatty acids with phenyl groups at the end carbons

Final product, benzoic acid

Both results supported the idea that the breakdown of fatty acids occurs by the loss of a two-carbon structure at each stage, the bond between the α and the β carbons being cleaved. This has since been confirmed, and because the essential step is the oxidation of the β carbon atom, the process is known as the β oxidation of fatty acids. The entire sequence is shown in Figure 18.7.

FIGURE 18.7 The β oxidation of fatty acids.

STEP 1: ACTIVATION OF FATTY ACIDS

The fatty acid first reacts with coenzyme A to form a thiol ester; the energy needed is provided by ATP, so that this is really an activation step:

$$R-CH_2-\overset{\beta}{CH_2}-\overset{\alpha}{CH_2}-C\overset{\displaystyle O}{\underset{\displaystyle OH}{\big\|}} + H-S-CoA \underset{\substack{\text{fatty acid}\\ \text{thiokinase, Mg}^{2+}}}{\overset{\substack{\text{ATP} \qquad \text{AMP} + P-P}}{\rightleftharpoons}} R-CH_2-CH_2-CH_2-C\overset{\displaystyle O}{\big\|}-S-CoA$$

A fatty acid   Coenzyme A   An acyl—S—CoA ester, a thiol ester

The thiol ester formed is a high-energy structure and lends itself to the oxidation step that follows. It should be noted that *only one* ATP unit is needed for the activation of a fatty acid, regardless of the length of the chain.

STEP 2: DEHYDROGENATION OF Acyl—S—CoA

FAD = *flavin adenine dinucleotide*

The dehydrogenation is accomplished by the coenzyme FAD, which is reduced to $FADH_2$. The system FAD-$FADH_2$ plays a role analogous to that of NAD^+-$NADH + H^+$, and in Figure 18.7 it is indicated that the two hydrogens go on to the electron-transport system. This is therefore an energy-producing step.

$$R-CH_2-\overset{\boxed{H}}{\underset{H}{\overset{|}{\underset{|}{C}}}}_{\beta}-\overset{H}{\underset{\boxed{H}}{\overset{|}{\underset{|}{C}}}}_{\alpha}-C\overset{\displaystyle O}{\big\|}-S-CoA \underset{\text{acyldehydrogenase}}{\overset{\text{FAD} \qquad \text{FADH}_2}{\rightleftharpoons}} R-CH_2-C=\overset{H}{\underset{H}{\overset{|}{\underset{|}{C}}}}-C\overset{\displaystyle O}{\big\|}-S-CoA$$

Acyl—S—CoA   *trans*-α,β-Unsaturated acyl—S—CoA

The loss of a hydrogen by the α and β carbons results in a double bond between them in a trans position, and the reaction product is trans-α,β-unsaturated acyl—S—CoA, in which form it is ready for the next step.

STEP 3: HYDRATION OF α,β-Unsaturated Acyl—S—CoA

The addition of water across the double bond results in an —OH being located at the β carbon:

$$R-CH_2-\overset{H}{\underset{H}{\overset{|}{\underset{|}{C}}}}_{\beta}=\overset{}{\underset{\alpha}{C}}-C\overset{\displaystyle O}{\big\|}-S-CoA + H_2O \underset{\substack{\text{L-3-hydroxyacyl}\\ \text{S—CoA hydrolase}}}{\rightleftharpoons} R-CH_2-\overset{OH}{\underset{H}{\overset{|}{\underset{|}{C}}}}-\overset{}{\underset{H}{\overset{|}{\underset{|}{C}}}}-C\overset{\displaystyle O}{\big\|}-S-CoA$$

α,β-Unsaturated acyl—S—CoA   β-Hydroxyacyl—S—CoA, or L-3-hydroxyacyl—S—CoA

STEP 4: DEHYDROGENATION OF β-Hydroxyacyl—S—CoA

Whereas the first oxidation step 2 was FAD-dependent, this one is NAD$^+$-dependent, and again the hydrogens go to the electron-transport system. Since both hydrogens were lost at the β carbon, a β-keto group results:

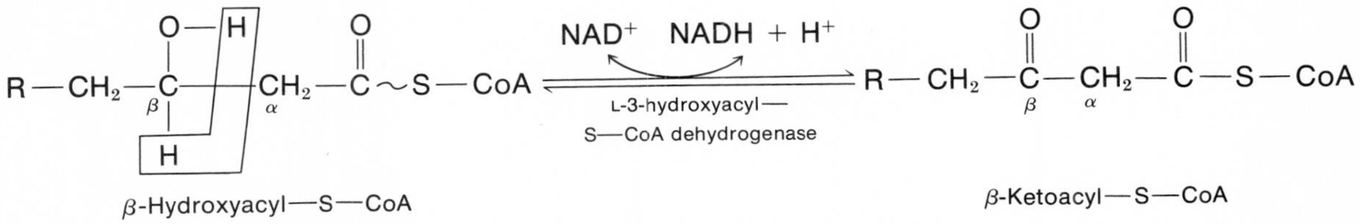

β-Hydroxyacyl—S—CoA

β-Ketoacyl—S—CoA

The enzyme L-3-hydroxyacyl dehydrogenase which catalyzes this reversible reaction has an *absolute* specificity for the substrate, so that it must be concluded that the structure produced by previous reactions has an L configuration.

STEP 5: THE THIOLYTIC CLEAVAGE OF β-Ketoacyl—S—CoA

In the last reaction of this series the bond between the α and β carbons of the ketoacyl—S—CoA is cleaved, splitting off a two-carbon fragment as Knoop postulated. The reaction requires the introduction of another molecule of coenzyme A. As shown below, the two-carbon fragment picks up the hydrogen atom of H—S—CoA to become acetyl—S—CoA, and the —S—CoA attaches itself to the β carbon of the rest of the acyl structure to reform an acyl—S—CoA:

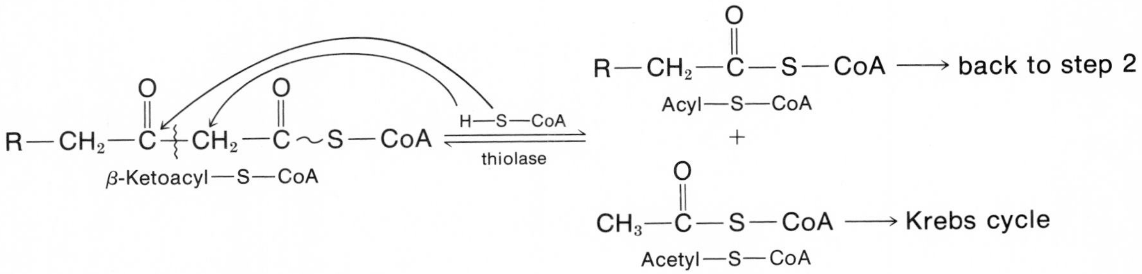

As indicated, the acetyl—S—CoA obtained goes to the Krebs cycle and further metabolism. The structure of the acyl—S—CoA formed differs from the acyl—S—CoA that originally entered into step 2 only in having *two fewer carbons*. This means that the acyl—S—CoA resulting from step 5 can go back to step 2 and repeat the step 2 to step 5 sequence with the same results: the formation of another acetyl—S—CoA and an acyl—S—CoA further diminished by two carbons. This sequence can be repeated until the fatty acid chain is completely converted into acetyl—S—CoA units. If we imagine the fatty acid as having a 10-carbon chain, its conversion to five 2-carbon acetyl—S—CoA units can be schematically represented as shown; the number of carbons is indicated by a subscript:

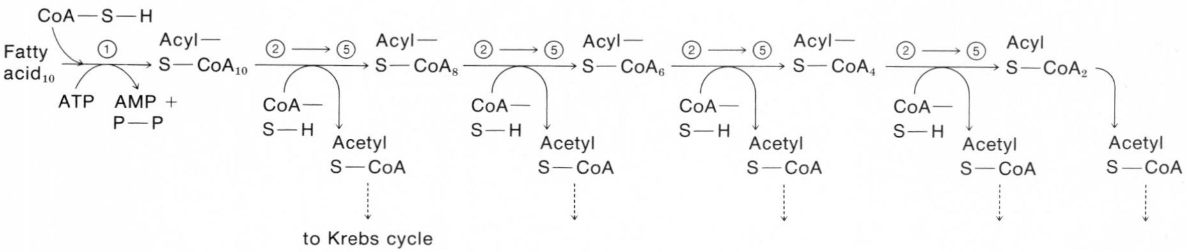

to Krebs cycle

The total score for the conversion of this 10-carbon fatty acid can now be summarized:

1 One ATP is used up in step 1.
2 Five molecules of CoA—S—H are required.
3 Five acetyl—S—CoA units are produced in four repeated sequences of steps 2 to 5; the acetyl—S—CoA's go to the Krebs cycle.
4 Since steps 2 and 4 in each sequence are oxidations and each provides a pair of hydrogen atoms, a total of four pairs of hydrogens are passed on to the electron-transport system by FAD and a total of four pairs of hydrogens are passed on by NAD^+.

Having followed lipid metabolism to the point where the products enter the Krebs cycle, we now turn to the metabolic pathways followed by the amino acids.

THE METABOLISM OF AMINO ACIDS

When the amino acids obtained by the digestion of proteins are absorbed into the body, they enter the so-called amino acid pool. The organism draws from this pool for several purposes:

1 The synthesis of the body's own proteins
2 The synthesis of other vital biological molecules
3 Metabolic degradation when not required for 1 and 2 or when required for energy

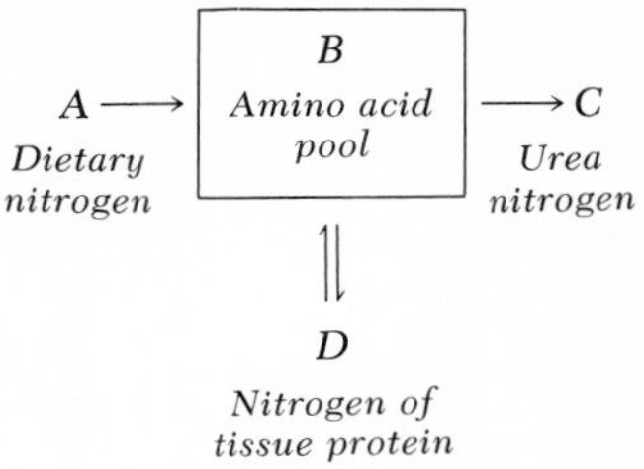

By monitoring the amino acid pool and measuring the intake of dietary nitrogen against the excretion of urea nitrogen, we can get a good picture of the condition of the body. When the nitrogen intake equals nitrogen outgo, so that $A = C$, and when the nitrogen taken by the tissues from the pool equals the amount it returns, so that $B \rightarrow D = D \rightarrow B$, there is nitrogen equilibrium, an indication of good health. A positive nitrogen balance occurs when there is a net increase in the tissue nitrogen, as during growth or during recovery from an illness. In this case $A > C$, and $B \rightarrow D > D \rightarrow B$. The reverse occurs during starvation, dieting, or illness, during which there is a net loss of tissue nitrogen, in which case $A < C$, and $B \rightarrow D < D \rightarrow B$.

Since we are primarily concerned with the catabolism of amino acids, we must examine the two basic processes that are involved: (1) the removal of the amino group, —NH$_2$, either by transamination or by deamination, after which it is converted to urea, and (2) the degradation of the carbon skeleton left by the loss of the —NH$_2$ group. There is a *pathway* for each of the amino acids, but all their end products are fed into the Krebs cycle. We shall pick up these end products later and see where they enter the Krebs cycle.

REMOVAL OF THE —NH$_2$ GROUP AND FORMATION OF UREA

In the human body the nitrogen from the α-amino acids is eliminated as the degradation product urea. This molecule contains two nitrogen atoms, so that two —NH$_2$ groups must be available for its formation. The overall and simplest reaction for the formation of urea can be written

$$2NH_3 + CO_2 = \underset{\underset{NH_2}{|}}{\overset{\overset{NH_2}{|}}{C}} = O + H_2O$$
Urea

However, since ammonia is not found as such in the body, we must determine how the nitrogen is transported. There are two pathways, **deamination** of amino acids and by **transamination.**

Deamination of amino acids It is believed today that glutamic acid is deaminated as shown below and that it provides the first nitrogen, in the form of —NH$_2$, to enter the cycle whereby urea is formed.

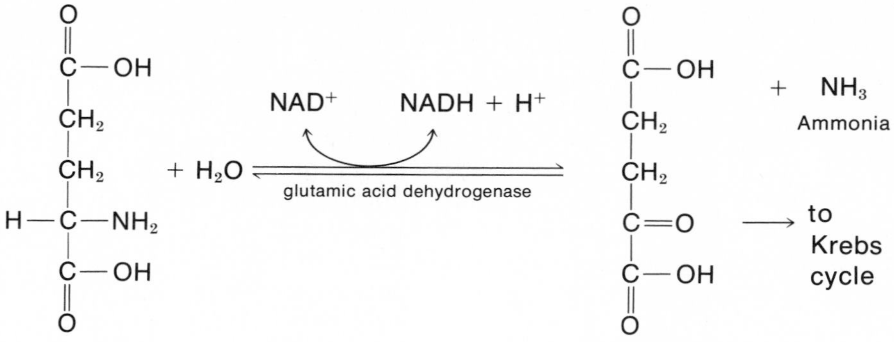

The structures show glutamic acid being converted to α-ketoglutaric acid:

Glutamic acid:
- C=O, OH (as $C-OH$ with O double bond on top)
- CH_2
- CH_2
- $H-C-NH_2$
- $C-OH$
- O

$+ H_2O$

(glutamic acid dehydrogenase, with $NAD^+ \rightarrow NADH + H^+$)

α-Ketoglutaric acid:
- $C-OH$ (O double bond)
- CH_2
- CH_2
- $C=O$
- $C-OH$
- O

$+ NH_3$ Ammonia

$\longrightarrow$ to Krebs cycle

Glutamic acid — α-Ketoglutaric acid

The two hydrogens are taken by NAD^+ for entry into the electron-transport system, and the overall reaction represents an oxidative deamination catalyzed by the enzyme glutamic acid dehydrogenase. The enzyme is specific for the substrate, and the reaction is reversible. The glutamic acid is deaminated to α-ketoglutaric acid, the whole reaction representing an interchange of water with —NH_2 to form the keto acid and ammonia. But α-ketoglutaric acid is also a component of the Krebs cycle, so that we have here a junction between the Krebs and urea cycles. The ammonia produced is converted at once into a high-energy compound, carbamylphosphate:

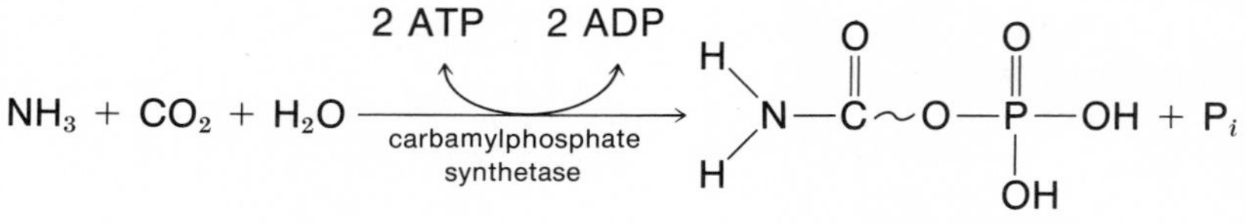

$$NH_3 + CO_2 + H_2O \xrightarrow[\text{synthetase}]{\text{carbamylphosphate}} \begin{array}{c} H \\ N-C \sim O-P-OH + P_i \\ H \quad \quad OH \end{array}$$

(with $2\ ATP \rightarrow 2\ ADP$)

Carbamylphosphate

This reaction, which is essentially irreversible and requires an energy release from two ATP units, represents a *fixation* of NH_3 with CO_2. The carbamyl phosphate enters the urea cycle by donating the carbamyl group, H_2NCO—, to ornithine.

Transamination of amino acids The second process of removing the —NH_2 group from amino acids, transamination, requires the enzyme transaminase and the coenzyme pyridoxine, which is vitamin B_6. The pyridoxine is the carrier of the —NH_2 group, which is taken by α-ketoglutaric acid to form glutamic acid:

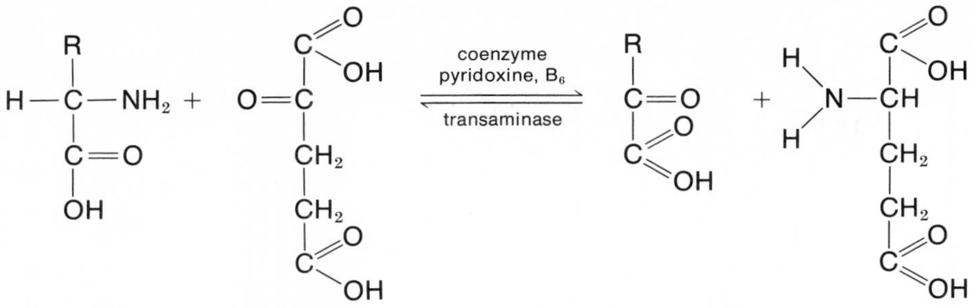

An α-amino acid   α-Ketoglutaric acid   An α-keto acid   Glutamic acid

The reaction interchanges the —NH$_2$ group and an accompanying H with the keto oxygen. Intermediate products and reactants are not shown. The keto acids produced this way are further metabolized via the Krebs cycle. The glutamic acid can now undergo the deamination reaction previously discussed. Half of the glutamic acid molecules undergo the deamination reaction; the other half undergo a second transamination by reacting with oxaloacetic acid. It is essentially the same reaction as the first transamination, with a similar —NH$_2$ and keto oxygen interchange:

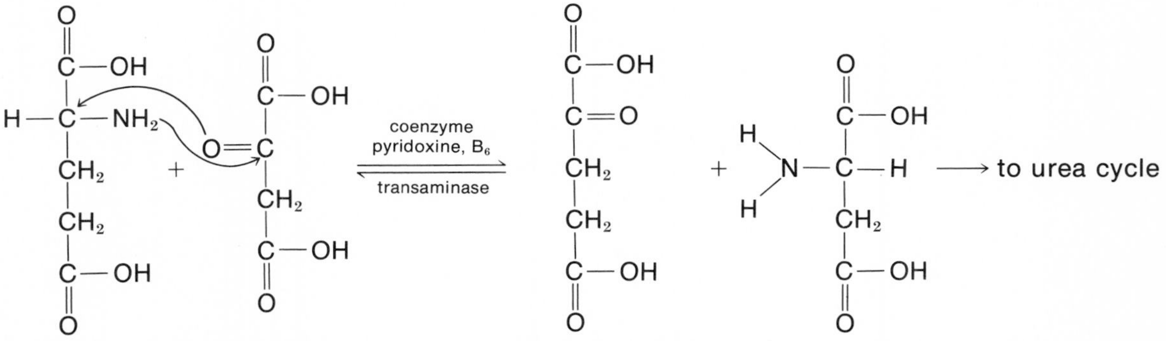

Glutamic acid   Oxaloacetic acid   α-Ketoglutaric acid   Aspartic acid

The amino group obtained in the first transamination reaction is transferred from glutamic acid to oxaloacetic acid, which is also a component of the Krebs cycle, from which it is borrowed for this purpose. The resulting α-ketoglutaric acid is now free to pick up another —NH$_2$ group from other amino acids in the amino acid pool. The second —NH$_2$ group is carried by aspartic acid to the **urea cycle,** where it reacts with citrulline, as we shall see. This sequence represents another junction between the Krebs cycle and the urea cycle. Now we know where the nitrogen and the

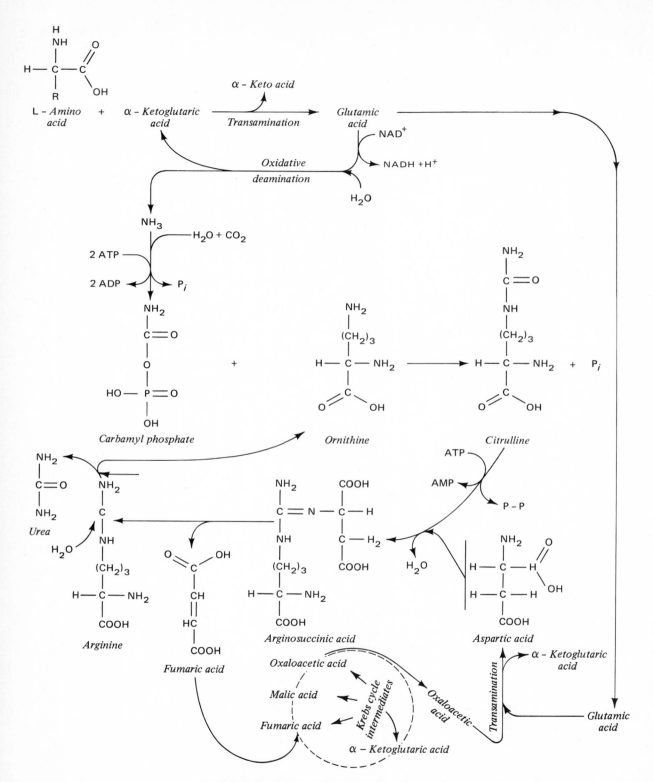

FIGURE 18.8 From an amino acid to urea.

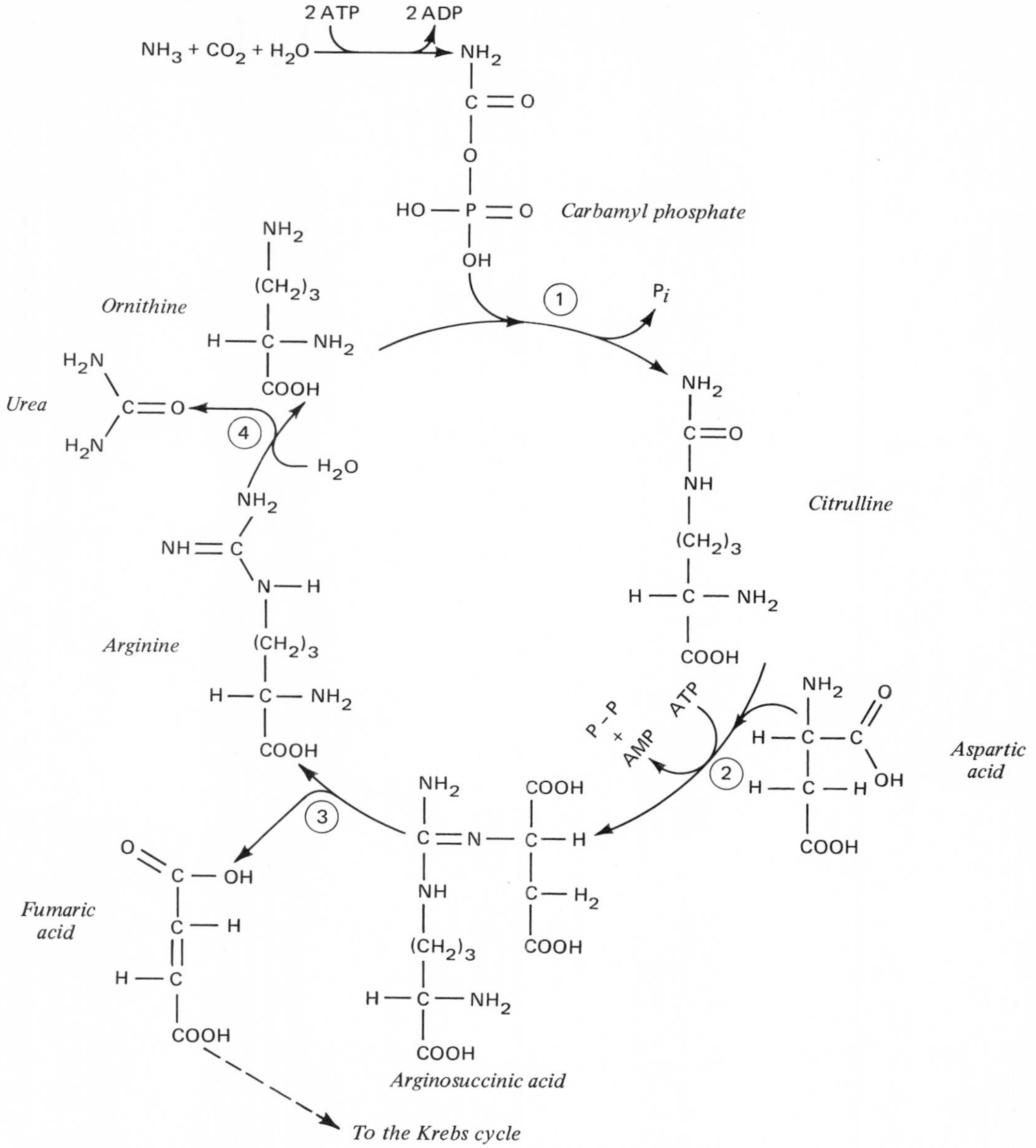

FIGURE 18.9 The urea cycle, which is part of the general scheme shown in Figure 18.8, is initiated when carbamylphosphate reacts with ornithine (it is also called the ornithine cycle). There are four steps; in the last step urea splits off and ornithine reforms, thereby regenerating the cycle.

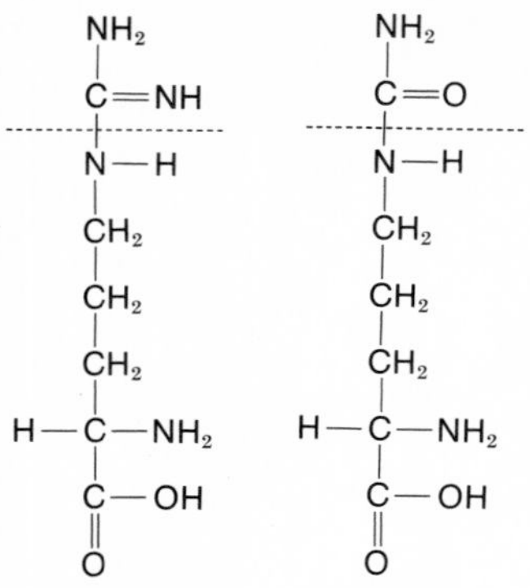

Arginine   *Citrulline*

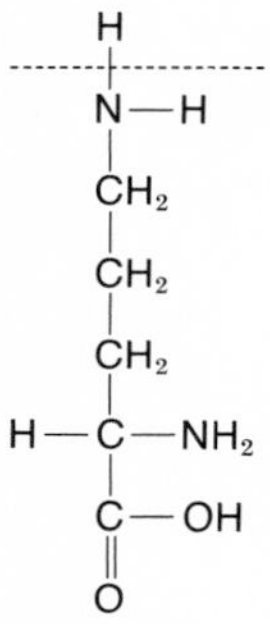

Ornithine

other atoms (C and O) come from for making urea. The urea cycle and the Krebs cycle participate in the production of urea from amino acids as shown in Figure 18.8.

THE UREA CYCLE

The urea cycle, also known as the **ornithine cycle,** contains ornithine, citrulline, and arginine, which, with the help of carbamylphosphate and aspartic acid, produce urea. If you look closely at the structures, you will note that ornithine, citrulline, and arginine are alike except for the top groups. The complete cycle (Figure 18.9) comprises four reactions, which repeat over and over to form urea.

Ornithine to citrulline   In this reaction carbamylphosphate gives the carbamyl group to ornithine to form citrulline and a phosphate:

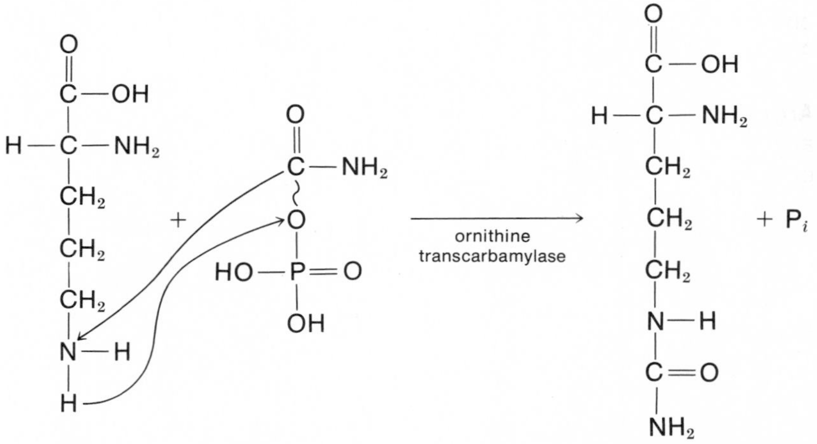

Ornithine   Carbamylphosphate   Citrulline

The irreversible reaction is catalyzed by the enzyme ornithine transcarbamylase, and the first —NH$_2$ group and carbon have entered the urea cycle.

Citrulline to argininosuccinic acid   The reaction represents a condensation of citrulline with aspartic acid. The aspartic acid is the carrier of the second —NH$_2$ group for the formation of urea:

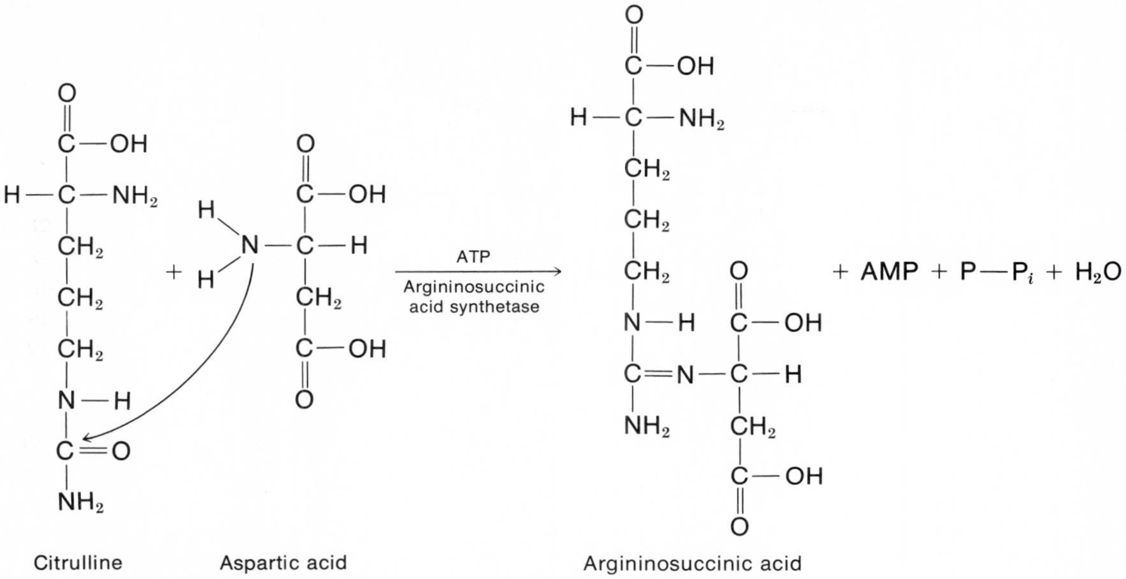

Citrulline   Aspartic acid   Argininosuccinic acid

The reaction is irreversible and requires the enzyme argininosuccinic acid synthetase. The energy for the reaction is provided by ATP.

Argininosuccinic acid to arginine This reaction is a cleavage of argininosuccinic acid to produce arginine and fumaric acid. The enzyme argininosuccinase catalyzes the reversible reaction.

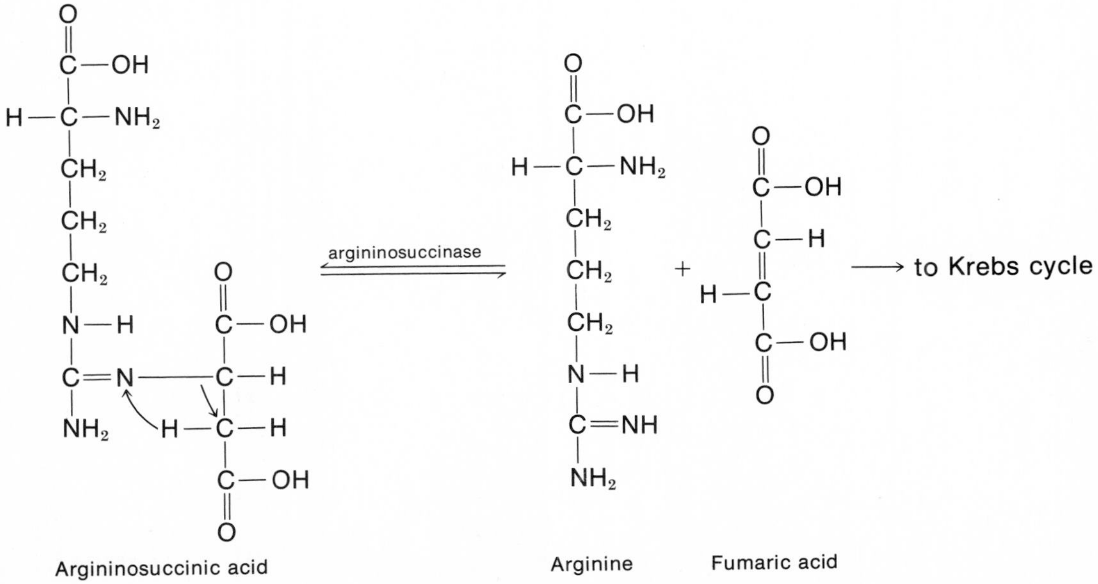

Argininosuccinic acid   Arginine   Fumaric acid

The fumaric acid enters the Krebs cycle, where it is changed to oxaloacetic acid (Fig. 18.8) and ultimately can participate in transamination again and carry the second $-NH_2$ group to the urea cycle. We see here an intimate linkage between the Krebs cycle and the urea cycle. The other product, arginine, continues in the urea cycle. Up to this point, the reaction sequence also represents a synthesis of the amino acid arginine.

Arginine to ornithine and urea This is the final reaction in the urea cycle, resulting in the formation of urea and the regeneration of ornithine, so that the cycle can begin again. The reaction is a hydrolysis of arginine by which urea is split off and ornithine regenerated:

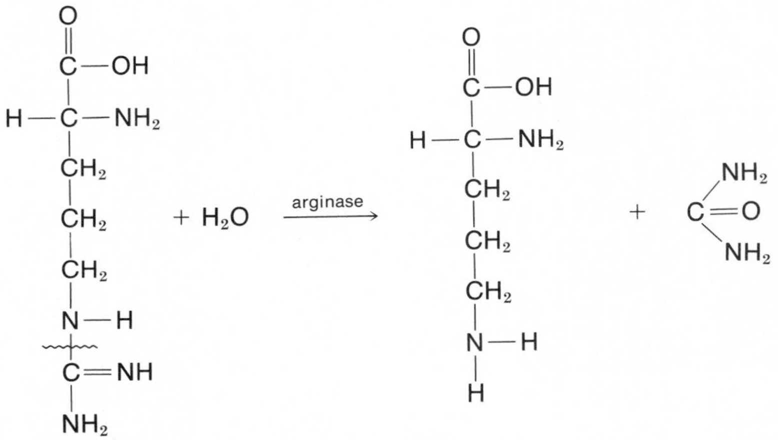

| Arginine | Ornithine, continues in the urea cycle | Urea, eliminated |

The most important site for the formation of urea is in the liver, after which it is excreted by way of the kidneys and urine. The overall process of urea formation is an *energy-consuming* process, and for each mole of urea formed 3 moles of ATP are used up, 2 moles in the fixation of CO_2 and NH_3 to make the carbamylphosphate, and a third mole in condensing citrulline with aspartic acid.

METABOLISM OF THE CARBON SKELETON OF AMINO ACIDS

The carbon skeleton of the different amino acids can be metabolized by further degradation. Each amino acid goes its own way, but they all end up in the Krebs cycle and are further metabolized. We can separate the amino acid skeletons according to the end products they produce before they enter the Krebs cycle. Amino acids are either **glucogenic** or **ketogenic.** The end products of glucogenic amino acids are either pyruvic acid or four-carbon dicarboxylic acids, which either enter the Krebs cycle or are used for the synthesis of glucose. (The name is derived from this possibility; glucogenic means can make glucose.) The ketogenic amino acids have as one of their end products keto compounds such as acetone, acetoacetic acid, or β-hydroxybutyric acid (potentially ketonic), small amounts of which can be metabolized by the body. If larger amounts are produced, they are eliminated in the urine, where they can be detected. If you are on a high-protein diet with no carbohydrates allowed, this will be the case. Sometimes, the elimination process of the keto bodies does not function well, and they accumulate in the blood-stream and cause ketosis and acidosis, increasing the acidity of the blood. This is a condition which can occur in uncontrolled diabetes. The ketogenic amino acids are leucine, isoleucine, phenylalanine, and tyrosine. All other amino acids are glucogenic. In association with the components of the Krebs cycle, we shall point out what the degradation products of the amino acids are. Where they enter the Krebs cycle will be summarized later (Figure 18.12).

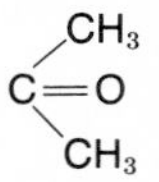

Of special interest are the amino acids phenylalanine and tyrosine, which have a common catabolic pathway and cause problems in man when genetic disorders interfere with their metabolism. The most common pathway for the metabolism of phenylalanine is its conversion to tyrosine:

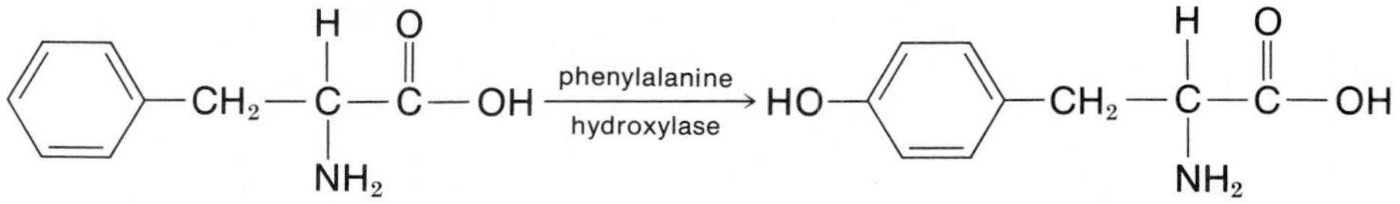

Acetone

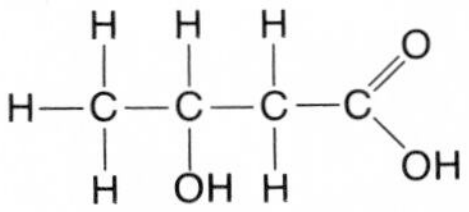

β-Hydroxybutyric acid

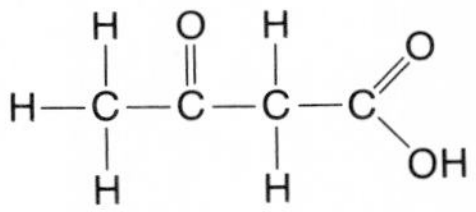

Acetoacetic acid

Phenylalanine   Tyrosine

In 1 out of 10,000 people the enzyme phenylalanine hydroxylase is missing, so that this pathway is blocked. The body then resorts to an alternate, infrequently used action on phenylalanine, transamination:

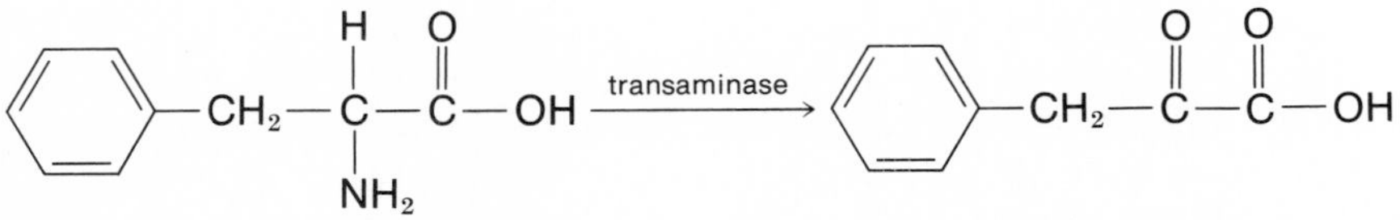

Phenylalanine   Phenylpyruvic acid

The phenylpyruvic acid is excreted in the urine, but it also accumulates in the blood if the less efficient transamination is the

TABLE 18.2 Genetic defects in amino acid metabolism

Amino Acid	Enzyme Missing	Name of Disorder
Tyrosine	Homogenistic acid oxidase	Alkaptonuria
Tyrosine	Tyrosinase	Albinism
Phenylalanine	Phenylalanine hydroxylase	Phenylketonuria (PKU)
Histidine	Histidase	Histidinemia
Isovaleric acid	Isovaleryl—S—CoA dehydrogenase	Isovaleryl acidemia
Valine	Valine transaminase	Hypervalinemia
Proline	Proline oxidase	Hyperprolinemia
Hydroxyproline	Hydroxyproline oxidase	Hydroxyprolinemia

only route of metabolism. Its presence hinders the proper development of the brain in infants, causing mental retardation. All infants born in hospitals in the United States are routinely checked for this inborn genetic deficiency, known as **phenylketonuria** (PKU). If a test confirms the presence of the defect, the infant is put on a low-phenylalanine diet to prevent mental retardation.

In a series of reactions tyrosine is degraded to a compound known as homogentisic acid, which is then converted by the enzyme homogentisic acid oxidase to fumarylacetoacetic acid.

Tyrosine is also needed by the body for the synthesis of melanins, which provide the pigmentation in the skin. One of the key enzymes in that reaction is tyrosinase. When it is missing because of a genetic defect, the synthesis of melanins is inhibited and the result is **albinism.** A number of genetic defects affecting amino acid metabolism are summarized in Table 18.2.

THE KREBS CYCLE AND THE ELECTRON-TRANSPORT SYSTEM

So far we have followed the catabolism of the carbohydrates via the Embden-Mayerhof pathway, of the lipids via β oxidation, and of the amino acids by way of deamination, transamination, decarboxylation, and the urea cycle. The products so obtained are now funneled into the Krebs cycle to be metabolized further to CO_2 and H_2O. As the Krebs cycle grinds on, the metabolites give up a pair of hydrogen atoms at several exit points for transfer to the electron-transfer system, where a series of coenzymes successively releases some of their energy as in turn they hand down the hydrogens and their electrons toward oxygen as the most effective and final acceptor. At three points in this sequence this energy is tapped for producing an ATP molecule from ADP and phosphate. The general sequence of events and the metabolites involved in the Krebs cycle are shown in Figures 18.10 to 18.12.

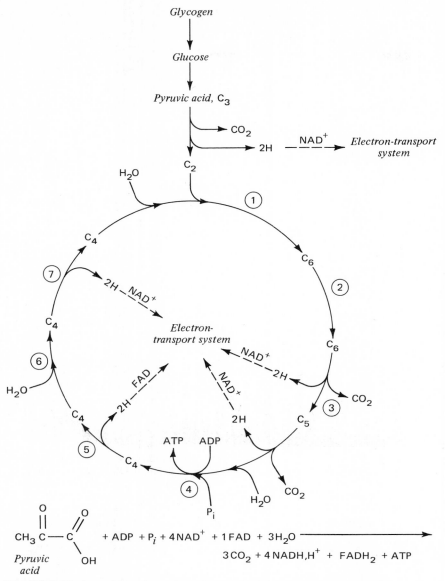

FIGURE 18.10 An outline of the Krebs cycle, which is initiated when pyruvic acid, symbolized C_3 for its three carbons, undergoes a reaction whereby it loses two hydrogens to NAD^+ and CO_2 splits off. The resulting C_2 structure enters step 1 of the cycle by condensing with the C_4 structure. This gives a C_6 structure, which in turn leads to the other intermediaries in the cycle and the re-forming of the original C_4 metabolite. The equation summarizes the net reaction. It will be shown later that some of the steps are actually multiple reactions.

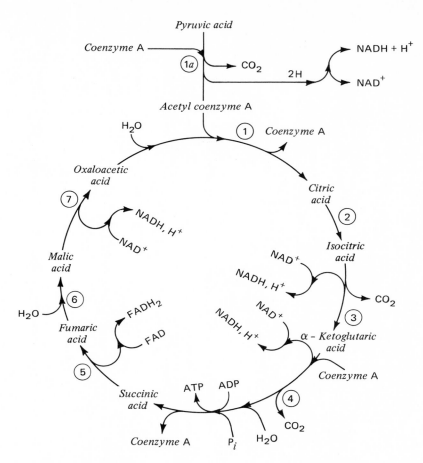

FIGURE 18.11 Reaction sequence and compounds of the Krebs cycle, showing major intermediary compounds formed. The reaction of pyruvic acid with coenzyme A, indicated as step $1a$ in the diagram, is the link between glycolysis and the Krebs cycle. The acetyl CoA formed enters the cycle by a condensation reaction with oxaloacetic acid, resulting in the six-carbon structure citric acid. In the course of one full turn of the cycle, oxaloacetic acid is re-formed, and an ADP is promoted to ATP; including step $1a$, five pairs of hydrogen atoms are delivered to the electron-transport system, four by NAD^+ and one by FAD.

The step that links the Embden-Mayerhof pathway to the Krebs cycle is the conversion of the two pyruvic acid molecules obtained from the glucose molecule to acetyl coenzyme A. The formation of this product, which is a *major* junction of metabolic pathways, takes place in several steps and with a large supporting cast of enzymes, coenzymes, and metal activators. The net overall reaction, which is essentially irreversible, adds up to an oxidation, a decarboxylation, and a union with coenzyme A:

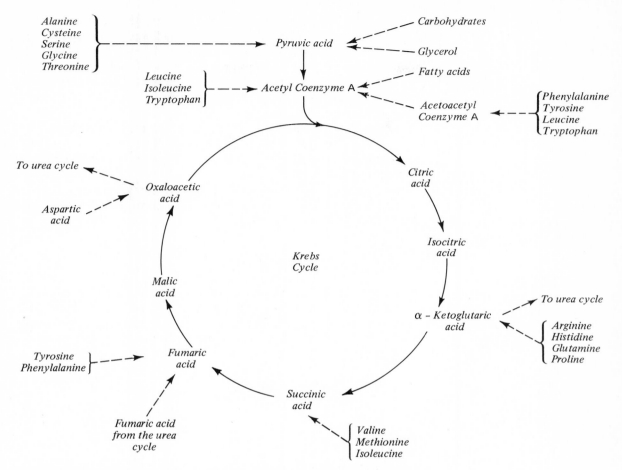

FIGURE 18.12 The Krebs cycle as the common pathway of the metabolism of the nutrients. The end products of the digestion of carbohydrates, lipids, and proteins first undergo preliminary catabolic reactions that convert them into some intermediate product of the Krebs cycle. The diagram shows the form in which they gain entry into this central pathway of metabolism.

Besides CoA, other coenzymes involved are thiamine pyrophosphate, lipoic acid, and NAD^+ as the hydrogen acceptor. The pyruvic acid dehydrogenase is a multienzyme system, and the metal activator is Mg^{2+}. In the course of the reaction the H atom of H—S—CoA is taken by NAD^+, and the rest of the structure unites with pyruvic acid at its keto carbon, as shown. The carboxyl group, —COOH, of pyruvic acid then splits off and becomes CO_2 when its H is also taken by NAD^+ for transfer to the respiratory chain. Acetyl—S—CoA is therefore the form in which both carbohydrates and lipids enter the Krebs cycle, the latter via the β oxidation of fatty acids.

The individual reactions of the Krebs cycle will now be considered. Although some of the steps are energy-consuming, the total sequence releases energy.

STEP 1: Acetyl—S—CoA AND OXALOACETIC ACID
CONDENSE TO FORM CITRIC ACID

Because the first step results in the formation of citric acid, which is tricarboxylic, the Krebs cycle is also known as the **citric acid cycle** or the **tricarboxylic acid cycle.** It was presented in its basic outline in 1937 by the biochemist Hans Krebs, who was awarded a Nobel prize in 1953.

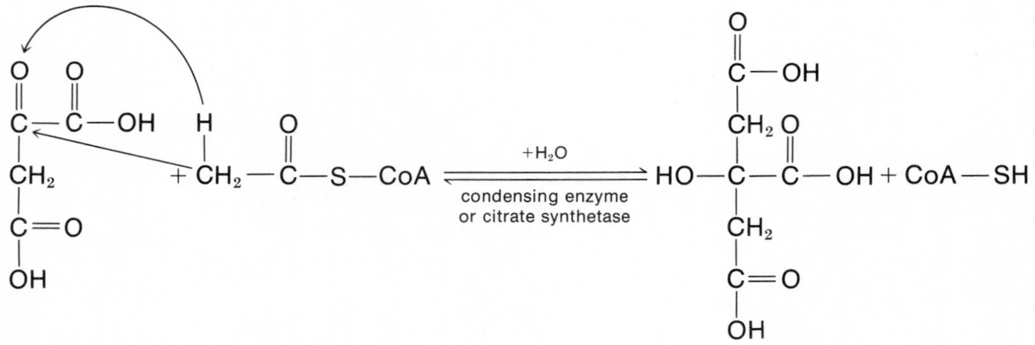

Oxaloacetic acid   Acetyl—S—CoA   Citric acid

The condensation is catalyzed by citrate synthetase as shown. One hydrogen of the acetyl—S—CoA goes to the keto oxygen, and the rest of the structure is attached to the keto carbon directly, after which the whole thing is promptly hydrolyzed, regenerating coenzyme CoA—SH and forming citric acid, a six-carbon structure. The CoA—SH is now free to react with the next pyruvic acid molecule coming down from glycolysis.

STEP 2: FORMATION OF ISOCITRIC ACID

This reaction is an isomerization whereby the —OH group moves from carbon 3 to carbon 2. The citric acid undergoes a dehydration, forming *cis*-aconitic acid, which has a double bond between the α and β carbons. Water then adds at the double bond in a stereospecific reaction to form isocitric acid; with its —OH group now on carbon 2, it is ready for the next step.

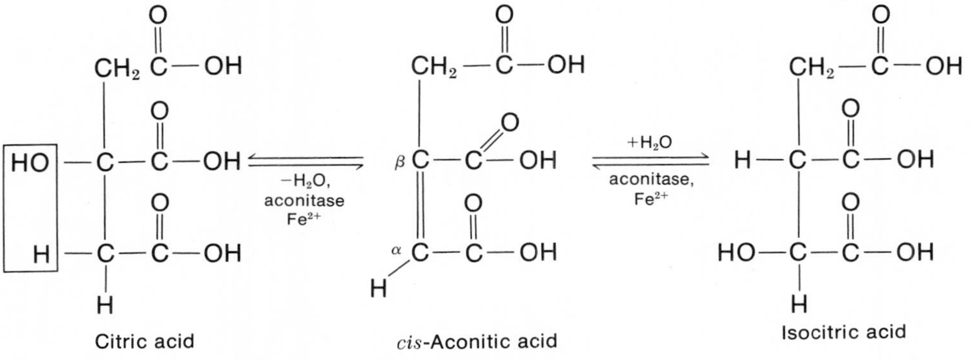

Citric acid   *cis*-Aconitic acid   Isocitric acid

The reaction is reversible and requires the enzyme aconitase and the metal activator Fe^{2+}.

STEP 3: FORMATION OF α-KETOGLUTARIC ACID

This is the first degradation reaction of the cycle; in it two hydrogens are lost and then CO_2 splits off, so that it is called an oxidative decarboxylation. The enzyme isocitric acid dehydrogenase catalyzes both steps with the aid of the metal activator Mg^{2+}.

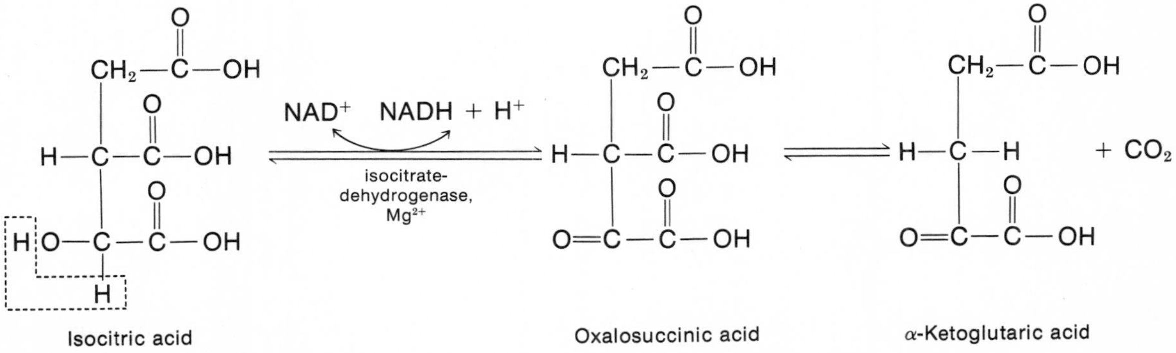

Isocitric acid   Oxalosuccinic acid α-Ketoglutaric acid

The two hydrogens removed, one from the α carbon and one from its —OH group, are carried by NAD^+ as $NADH + H^+$ to the electron-transport system. The intermediate product, oxalosuc-cinic acid, which is a keto acid, is decarboxylated to produce the α-ketoglutaric acid. Do you remember this acid? We last met it as a component of amino acid metabolism.

STEP 4: FORMATION OF SUCCINIC ACID

This reaction is similar to the conversion of pyruvic acid to ace-tyl—S—CoA in that both are oxidative decarboxylations involving the participation of several coenzymes, including CoA and NAD^+, as well as a dehydrogenase enzyme system and Mg^{2+} as a metal activator. In this case α-ketoglutaric acid is converted into suc-cinic acid by two reactions:

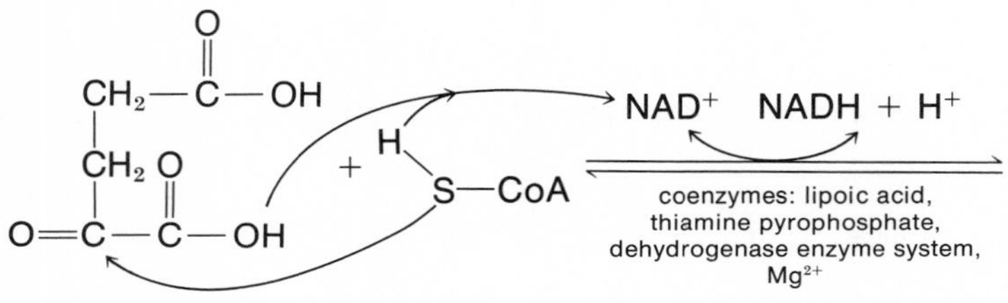

α-Ketoglutaric acid   Coenzyme A

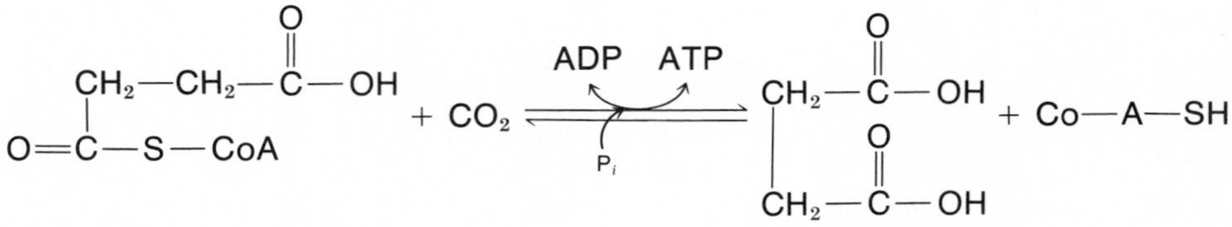

Succinyl S—CoA   Succinic acid

The first step is the formation of succinyl—S—CoA as an inter-mediary product, in the course of which NAD^+ again serves as the carrier of hydrogens to the respiratory chain and CO_2 splits off. The conversion of succinyl—S—CoA to succinic acid in the next step results in the regeneration of CoA—S—H as well as the release of sufficient free energy to make possible the only in-stance in the citric acid cycle of the formation of ATP.

STEP 5: DEHYDROGENATION OF SUCCINIC ACID
TO FUMARIC ACID

This reaction prepares for the next one, which will add another oxygen atom to the four-carbon structure. In this step dehy-drogenation removes one hydrogen each from the α and β carbons, producing a trans double bond between them:

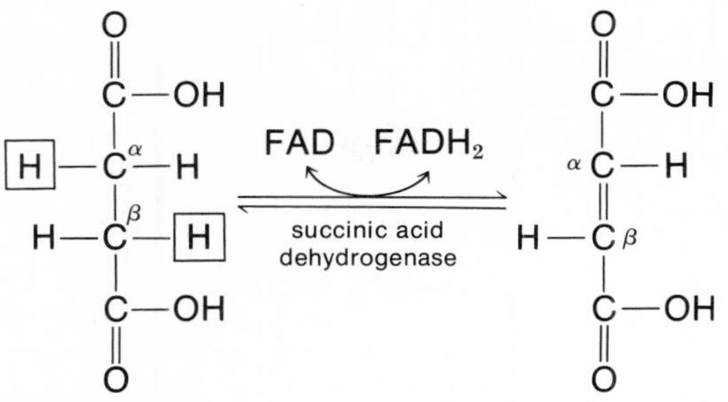

Succinic acid   Fumaric acid

The reaction is stereospecific and produces fumaric acid, which is *trans*-butenedioic acid. The pair of hydrogens released is taken here by a different electron acceptor, the coenzyme FAD, which is reduced to $FADH_2$. We shall see later that whether NAD^+ or FAD is the hydrogen acceptor makes a difference in the number of ATPs produced by the electron transport system.

STEP 6: HYDRATION OF FUMARIC ACID TO MALIC ACID

In this reaction a water molecule is added across the double bond, thereby introducing another oxygen atom into the molecule and producing malic acid, which is a hydroxy acid. This is also a stereospecific reaction, with only L-malic acid formed. We saw similar reactions in the β oxidation of fatty acids and in the conversion of citric to isocitric acid:

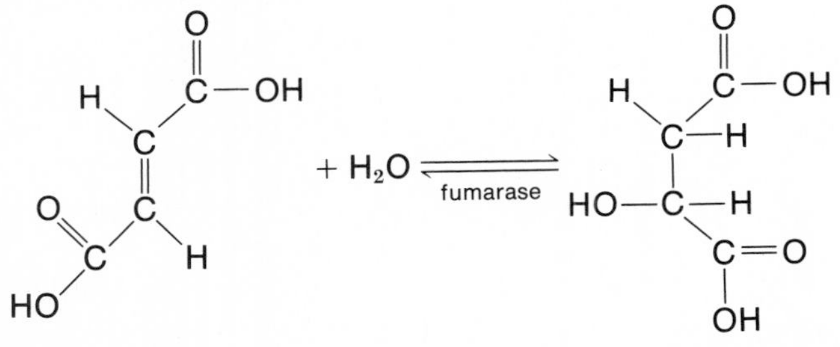

Fumaric acid   L-Malic acid

The enzyme catalyzing the reversible reaction is fumarase.

STEP 7: DEHYDROGENATION OF MALIC ACID
TO OXALOACETIC ACID

This last reaction of the Krebs cycle produces oxaloacetic acid, which is where we started. The step is again a dehydrogenation, with removal of hydrogens by NAD^+ and delivery to the electron-transport system as $NADH + H^+$:

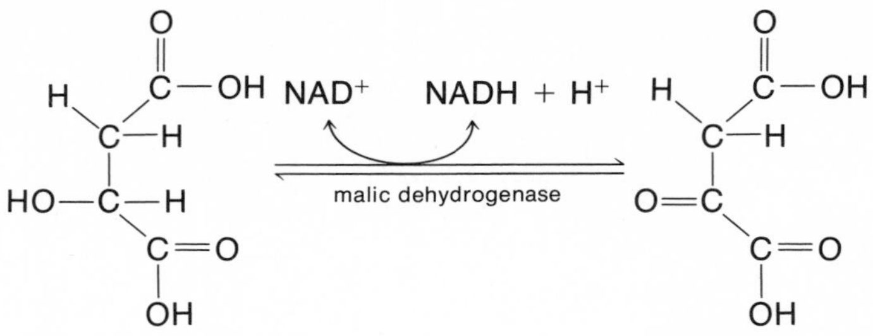

L-Malic acid   Oxaloacetic acid

The enzyme malic dehydrogenase catalyzes the reversible reaction, and the oxaloacetic acid produced can again participate in step 1; one turn of the cycle is finished, and another begins.

THE ELECTRON-TRANSPORT SYSTEM, OR RESPIRATORY CHAIN

The catch basin Krebs cycle works essentially like a meat grinder to break up received molecules into carbon dioxide and hydrogen. The hydrogens are carried off in pairs to the respiratory chain, at the end of which they combine with oxygen to form water. This is called **water of metabolism** and is the end product of a series of stepwise reactions that slowly release energy, part of which is chemically stored as ATP. The electron-transport system can be imagined as the principal power-generating plant of the cell, whose components are a series of coupled enzymes arranged in order of their increasing appetite for electrons. With each pair of hydrogens received, a pair of electrons is received, and their transfer from one coenzyme to the next constitutes an oxidation-reduction reaction, the donor coenzyme being oxidized and the acceptor coenzyme being reduced. As discussed in Chapter 6, the spontaneous movement of an electron from a donor to an acceptor is equivalent to the electron flow of an electric current, whose driving force is its electrical pressure, or voltage. The driving force behind each successive transfer of an electron pair from one coenzyme to the next is shown in Figure 18.13, which lists the oxidation-reduction potentials of the several coenzymes of the electron-transport system.

    Suppose that two hydrogens and their electrons are picked up by NAD^+ and taken to the respiratory chain as $NADH + H^+$. Waiting in line is FAD, whose affinity for electrons is 0.10 V greater $[-0.22 - (-0.32) = +0.10]$ than NAD^+, with this result:

$$NADH + H^+ + FAD \longrightarrow NAD^+ + FADH_2$$

In turn, $FADH_2$ will surrender the hydrogen pair and its electrons to the more demanding coenzyme Q, and so on until the final ac-

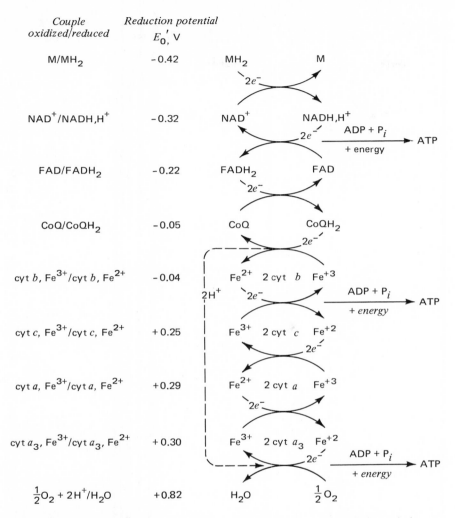

Couple oxidized/reduced	Reduction potential E_0', V
M/MH_2	-0.42
$NAD^+/NADH,H^+$	-0.32
$FAD/FADH_2$	-0.22
$CoQ/CoQH_2$	-0.05
cyt b, $Fe^{3+}/$cyt b, Fe^{2+}	-0.04
cyt c, $Fe^{3+}/$cyt c, Fe^{2+}	$+0.25$
cyt a, $Fe^{3+}/$cyt a, Fe^{2+}	$+0.29$
cyt a_3, $Fe^{3+}/$cyt a_3, Fe^{2+}	$+0.30$
$\frac{1}{2}O_2 + 2H^+/H_2O$	$+0.82$

FIGURE 18.13 The electron-transport system, also called the respiratory chain, begins when a metabolite, MH_2, is oxidized to a metabolite, M, as when malic acid is dehydrogenated to oxaloacetic acid. The two hydrogens and their electrons are taken by NAD^+, whose reduction potential E_0' is -0.32 V, as shown. This is a measure of the affinity of the coenzyme for electrons. Of the sequence of electron acceptors listed above, NAD^+ is the weakest and O_2 the strongest, so that electrons flow spontaneously toward oxygen. However, this flow takes place in steps, each step being the transfer of a pair of electrons to a coenzyme with a higher electron affinity, that is, a higher E_0'. In effect, this spontaneous flow of electrons is an electric current capable of doing work. Some of this energy goes to convert ADP into ATP, and the balance passes out as heat.

ceptor, oxygen, is reached. With each such spontaneous transfer of electrons an amount of free energy is released proportional to the positive change of the oxidation-reduction potentials.

Following NAD^+ and FAD in the respiratory chain shown in Figure 18.13 is coenzyme Q, which contains the central structure ubiquinone. The reduced form $CoQH_2$ gives up two electrons to cytochrome b, the first of the four cytochromes, all of which have iron in their structure, with the result that two H^+ ions are now available. The pair of electrons is then passed from cytochrome b to cytochromes c, a_1, and a_3 in turn by the Fe^{3+} —Fe^{2+} oxidation-reduction that occurs in each. Finally, the electron pair is taken by oxygen, so that the last reaction can be expressed as

$$\tfrac{1}{2}O_2 + 2H^+ + 2e^- = H_2O$$

The energy so generated is in part dissipated as heat, but much is chemically stored by the phosphorylation of ADP

$$ADP + P_i \xrightarrow{\text{energy}} ATP$$

The sum of the coupled reactions constituting the complete electron-transport system is an oxidative phosphorylation that fixes energy for the cell in the form of ATP. Although some mysteries remain regarding the electron-transport system, biochemists have a pretty good idea how the entire process works, and Figure 18.13 provides a general scheme of its operation.

The overall mechanism provides for the slow release of energy as one reaction follows the other. At points where enough energy is available, it is used to make ATP. The coenzyme pair $NAD^+ - NADH + H^+$ enters the sequence at its beginning, and as the electron pair introduced shuttles back and forth between the oxidized and reduced forms of the enzymes, three ATP molecules are produced before they reach oxygen and H_2O is formed. The hydrogens carried by $FADH_2$ enter at a lower energy level and thus provide only two ATP molecules as the electrons shuttle their way down to O_2 and water.

ENERGY PRODUCTION AND EFFICIENCY

By considering all the steps of the anaerobic and the aerobic pathways by which glucose is metabolized to $CO_2 + H_2O$, we can determine how many ATP molecules are produced from one molecule of glucose. From this, we then can calculate the energy fixed as ATP for use by the body. The heat produced by the combustion of a mole of glucose has been experimentally determined as 673 kcal. The ratio of the energy stored as ATP

TABLE 18.3

	ATP	
Embden-Mayerhöf pathway yields a net of 2 ATPs	2	
Reaction 5 of glycolysis produces 2 pairs of hydrogen, which are carried by NAD$^+$ to the respiratory chain, resulting in an output of 2 × 3 = 6 ATP	6	
Total ATPs, 1 glucose to 2 pyruvic acid molecules		8
For each pyruvic acid that enters and passes through the Krebs cycle 1 ATP and 5 pairs of hydrogen atoms are obtained	1	
Of the 5 pairs of hydrogen, 4 pairs are carried to the respiratory chain by NAD$^+$ as they pass down the chain; 4 × 3 = 12 ATPs	12	
FAD takes the fifth H pair to the respiratory chain, yielding 2 ATPs	2	
For each pyruvic acid molecule that enters the Krebs cycle the yield is		15
Since 2 pyruvic acid molecules are obtained from glucose, the total number of ATPs is 2 × 15 = 30		30
The metabolism of 1 mole of glucose therefore results in formation of a total of		38 moles
If the energy stored in 1 mole of ATP is taken as 7.3 kcal, the metabolic efficiency of the body is		

$$\frac{38 \times 7.3 \text{ kcal}}{686 \text{ kcal}} \times 100 \approx 41\% \text{ efficient}$$

to the total energy available from glucose indicates the body's efficiency in metabolizing carbohydrates. We add up the score in Table 18.3.

In a similar way (Table 18.4) let us follow through the production of ATPs by the metabolism of 1 mole of tristearin, a fat which hydrolyzes to 1 mole of glycerol and 3 moles of stearic acid (18-carbon chain). The efficiency of glucose metabolism is about the same as that of fat metabolism, but fewer ATP units are produced per carbon for glucose than for a fat.

$$38 \text{ ATP from 6 glucose carbons} = \frac{38}{6} \text{ or 6.3 ATP per carbon}$$

$$463 \text{ ATP from 57 tristearin carbons} = \frac{463}{57} \text{ or 8.1 ATP per carbon}$$

This simple calculation shows that although the efficiency of carbohydrate and fat metabolism is about the same, more chemical energy is available to the body per unit (weight) of fat metabolized. This might have been deduced by comparing the molecular formulas of carbohydrates and fats, which show that carbohydrates are already half-oxidized.

TABLE 18.4

	ATP

	ATP		
In converting glycerol to dihydroxyacetone phosphate, 1 ATP is used	−1		
and 2 H are released and taken by NAD⁺ to the electron-transport system, 3ATPs	3		
The passage of dihydroxyacetone phosphate through the rest of the glycolytic pathway produces 2 ATPs	2		
and a pair of hydrogens that are taken to the respiratory chain by NAD⁺ of 3ATPs	3		
From the score kept for the metabolism of glucose, we know that the breakdown of pyruvic acid to water yields 15 ATPs	15		
The metabolism of the *glycerol* portion of the fat therefore yields		22	
The β-oxidation sequence of a fatty acid begins with an activation step that uses 1 ATP	−1		
Each β-oxidation sequence results in the release of 2 pairs of H atoms; 1 pair goes to the respiratory chain by way of FAD, yielding 2 ATPs, the other via NAD⁺ and the formation of 3 ATPs, so that each sequence produces a total of 5 ATPs. The 18-carbon chain will break down to 9 acetyl—S—CoA units through 8 individual sequences (count them), producing 8 × 5 = 40 ATPs	40		
Net ATP from β oxidation of 1 stearic acid chain		39	
The passage of an acetyl—S—CoA through the Krebs cycle yields 1 ATP when succinyl—S—CoA is converted into succinic acid: 4 pairs of hydrogen are released, 3 taken by NAD⁺ to the electron-transport system, producing 3 × 3 = 9 ATPs, and 1 taken by FAD, producing 1 ATP. The acetyl—S—CoA's produced, therefore, yield a total of 9 × 12 = 108 ATPs for each 18 C chain		108	
The total number of ATP produced per chain is		147	
The hydrolysis of a tristearin results in three 18-carbon chains, whose total yield of ATPs is 3 × 147 = 441			441
Adding the subtotals of 22 obtained from glycerol metabolism to the subtotal of 441 from fatty acid metabolism yields a total of 463 moles of ATP resulting from the metabolism of a mole of tristearin			463
The combustion of a mole of tristearin has been experimentally found to be 8044 kcal; taking the energy stored in a mole of ATP as 7.3 kcal, gives the efficiency as			

$$\frac{463 \times 7.3}{8044} \times 100 = 42\%$$

As a conclusion, it should be said that all the reactions used in catabolism (condensing, splitting, hydrolysis, hydration, and dehydrogenation) are also used in the synthesis of the body's own compounds. Furthermore, the intermediates passing through the metabolic pathways are taken for synthesis wherever they fit in. We have seen some of the interactions and interrelation of catabolic pathways; they are even more pronounced in synthesis.

Finally it should be understood that biological systems usually have alternative pathways to break down food components. The **hexose monophosphate shunt** (or **pentose phosphate pathway**) is an example. Here, the glucose is metabolized so that the

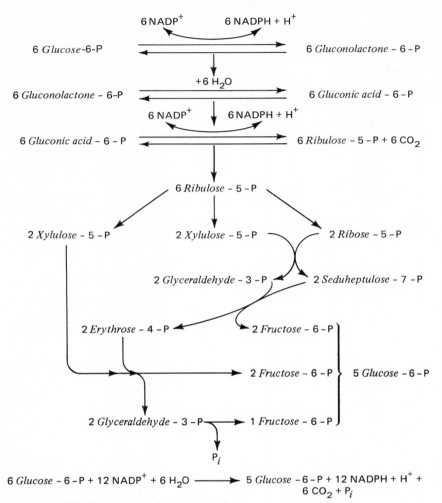

$$6\,NADP^+ \qquad 6\,NADPH + H^+$$

6 *Glucose*-6-P $\qquad\qquad$ 6 *Gluconolactone* – 6 - P

$$+6\,H_2O$$

6 *Gluconolactone* – 6-P $\qquad\qquad$ 6 *Gluconic acid* – 6 - P

$$6\,NADP^+ \qquad 6\,NADPH + H^+$$

6 *Gluconic acid* – 6 - P $\qquad\qquad$ 6 *Ribulose* – 5 - P + 6 CO_2

6 *Ribulose* – 5 - P

2 *Xylulose* – 5 - P $\qquad$ 2 *Xylulose* – 5 - P $\qquad$ 2 *Ribose* – 5 - P

2 *Glyceraldehyde* – 3 - P $\qquad$ 2 *Seduheptulose* – 7 - P

2 *Erythrose* – 4 - P $\qquad$ 2 *Fructose* - 6 - P

2 *Fructose* - 6 - P $\qquad$ 5 *Glucose* - 6 - P

2 *Glyceraldehyde* – 3 - P $\qquad$ 1 *Fructose* - 6 - P

$$P_i$$

6 *Glucose* – 6 - P + 12 $NADP^+$ + 6 H_2O $\longrightarrow$ 5 *Glucose* – 6 - P + 12 $NADPH + H^+$ +
$\qquad\qquad\qquad\qquad\qquad\qquad\qquad\qquad\qquad$ 6 CO_2 + P_i

FIGURE 18.14 The pentose phosphate pathway.

Embden-Mayerhof pathway and the Krebs cycle are by-passed. Different intermediates and reactions are involved, but the result is the same, namely, the production of energy for the body. The overall reaction is glucose 6-phosphate + 12 NADP → 6CO_2 + 12 NADPH + H$^+$ + P_i. Figure 18.14 outlines the general scheme of this pathway and the compounds involved.

NADP = *nicotinamide adenine dinucleotide phosphate, another coenzyme*

REVIEW QUESTIONS

1 What is metabolism?
2 Define anabolism and catabolism.
3 What is the purpose of metabolism?

4   What are the basic end products of metabolism, and from which compounds are the end products derived?

5   What is the respiratory chain?

6   What is the role of ATP in metabolism?

7   What are the individual products when ATP is hydrolyzed stepwise, and how much energy is released in each step?

8   Show the ATP-ADP cycle and explain its importance.

9   What is meant by a coupled reaction?

10   Make a list of all the coenzymes involved in hydrogen transport.

11   What is so important about the stepwise release of energy in metabolism?

12   What is anaerobic glycolysis?

13   Take a close look at the reactions of the Embden-Mayerhof pathway. List all the coenzymes and metal ions appearing in the reactions.

14   Which are the pacemaker reactions in glycolysis?

15   How many ATP units are produced in glycolysis, and at which reactions are they produced?

16   From memory write the structure of enolpyruvic acid.

17   Differentiate between glycolysis in animal tissue and yeast.

18   What is meant by oxygen debt?

19   What is the role of NAD^+ in glycolysis?

20   What is the function of the enzyme aldolase?

21   Make a list of all the different phospho compounds found in the Embden-Mayerhof pathway.

22   What is the function of the enzyme enolase, and how is its name derived?

23   In what different directions can pyruvic acid be further degraded?

24   Which one of the reactions in glycolysis is the first real degradation reaction, and what are the products?

25   Write the general reaction of glucose going to carbon dioxide, water, ethanol, and energy (ATP).

26   Into which metabolic pathway does glycerol enter to be metabolized?   Show reactions.

27   What was significant about Knoop's experiments.

28   How are the fatty acids activated in the β-oxidation scheme?

29   How is the oxygen introduced on the β-carbon in the β-oxidation scheme?

30   What is the thiolytic cleavage in the β-oxidation scheme, and what is its significance?

31   How many two-carbon fragments are produced when a 16-carbon fatty acid goes through the β-oxidation scheme? How many times must the β-oxidation scheme be used to break down the 16-carbon fatty acid completely? How many ATP units are needed for the indicated process of the β oxidation of the 16-carbon fatty acid?

32   Discuss the uses of amino acids in the body with examples.

33   What is the condition of a body when more dietary nitrogen goes in than urea nitrogen goes out?

34   Explain deamination and transamination.

35   What are the components of the urea cycle?

36   What compounds participate in the Krebs cycle and the urea cycle?

37 How many ATP molecules are needed to make one molecule of urea?

38 What are glucogenic and ketogenic amino acids?

39 Discuss genetic disorders which interfere with the metabolism of different amino acids.

40 Make a list of the components of the Krebs cycle, starting with pyruvic acid.

41 What is a decarboxylation reaction? Give examples.

42 How is citric acid formed in the Krebs cycle?

43 Use structures to show the difference between citric acid and isocitric acid.

44 In which of the reactions of Krebs cycle starting from pyruvic acid are hydrogens removed?

45 How is oxaloacetic acid produced from succinic acid in the Krebs cycle?

46 What type of reactions take place in the Krebs cycle?

47 What is meant by water of metabolism?

48 How many ATP units are produced when $NADH + H^+$ brings two hydrogens to the electron-transport system? How many are produced when $FADH_2$ brings two hydrogens to the electron-transport system?

49 What are the iron-containing compounds in the electron-transport system?

50 What is the significance of the electron-transport system?

51 Calculate the energy produced when 1 mole of tripalmitin becomes metabolized. Tripalmitin is made from one molecule of glycerol and three molecules of palmitic acid.

19

THE CHEMICAL FEATURES OF HEREDITY
Inner Becomes Outer

In paging through an old family album of photographs it is gratifying to recognize resemblances between ourselves and our parents, grandparents, and even great-grandparents. It gives us a sense of continuity and identity to see ourselves in those who preceded us. But if we could look back 2 or 3 billion years to the primitive organisms that first emerged on what was then a barren planet, there would be no recognition, only a sense of difference. And yet, from all that science has learned about living matter and its evolution, there *is* a continuity between that protoplasm of long ago and ourselves. This continuity is expressed in the continuous passing on of living matter by self-replication, supported by metabolism. Protoplasm experienced interactions with its surroundings. Such factors as radiant energy and chemical and temperature changes in the environment acted to bring about changes in living matter. Of course, not all types of changes could be tolerated by living matter, so that some organisms survived and others did not. However, once fixed in an organism, these changes were further transmitted by the hereditary mechanism of the organism, so that in time these changes accumulated.

Between ourselves and the first forms of life stand about 3 billion years of continuity and cumulative change, both traveling down the same path together.

The path leading from the simple organisms of the remote past to our present forms of life is not fully known, but clearly it was long and winding, with many branches, many dead ends, and some sharp turns. One of these was the emergence of **sexual reproduction,** which provided the biological advantage of increased diversification. Over a short length of this path, reckoned in generations, it is the similarities that are recognized, the continuity between parents and offspring that we call **heredity.** Over a long portion of the path, reckoned in millions of years, it is the differences that are perceived, the changes we call **evolution.** In recent years the inner mechanism in living matter that underlies both its continuity and change has been brought to light. The same chemical structures are involved in both. These substances, which are found in all living matter, will be our concern in this chapter.

Father & Son

Heredity

Modern man

Chimp

Evolution

THE NUCLEIC ACIDS

Contained in each cell of a living organism are giant molecules of **deoxyribonucleic acid** (DNA) and **ribonucleic acid** (RNA). These nucleic acids, which were mentioned earlier when the nucleoproteins were discussed, are composed of a core of nucleic acid and an outer coat of protein. The nucleic acids are the unique structures whereby each organism passes on its biological features to the next generation. The role of DNA is to *store* the specific instructions that govern the buildup of the structural and functional proteins of the organism. In turn, RNA *carries* these instructions to the **ribosomes** of the cell, where the actual synthesis of proteins takes place. Remember that there are no enzymes without proteins and no metabolic reactions (whether energy-yielding catabolism or energy-using anabolism) without the necessary enzymes. Therefore, the key to the biological characteristics of an organism lies in its ability to produce the proteins, and hence the enzymes, needed for its *specific structure and function.* In order to pass on these same biological characteristics to the next generation the organism must transmit the same instructions that guided the synthesis of its own proteins. In other words, the hereditary similarities between the generations is maintained by transmitting identical DNA from one generation to the next.

We can think of DNA as a genetic memory bank in the form of a long coded tape. Any portion of the tape that carries a specific set of instructions for a given assembly job is called a **gene.** It is

Nuclear membrane

Chromosomes composed of DNA *molecules*

Nucleoprotein

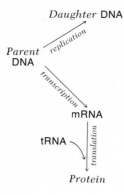

believed today that for each enzyme present in an organism, there is one specific gene responsible for its structure. This is known as the **one gene–one enzyme theory.**

DNA must replicate itself to maintain its continuity as the cells divide, but it also transcribes its instructions onto **messenger RNA** (mRNA). In turn, the transcribed message of mRNA must be read by the **transfer RNA** (tRNA) carrying the amino acids, one by one, to the site of protein synthesis; this is called **translation,** and the overall relation is shown in the margin.

COMPOSITION AND STRUCTURE OF DNA AND RNA

When nucleoproteins are hydrolyzed under suitable conditions, they are first separated into a protein fraction and a nucleic acid fraction. Further hydrolysis releases the building blocks that make up the nucleic acid chains. These are called **nucleotides,**

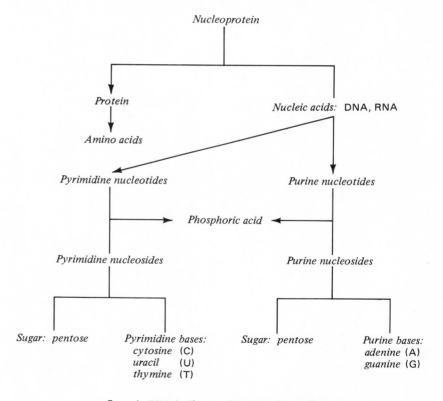

FIGURE 19.1 Hydrolysis products of nucleoproteins.

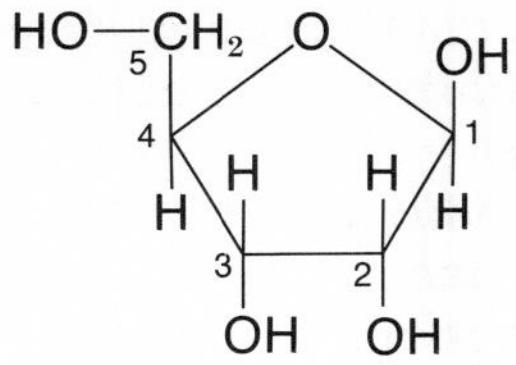

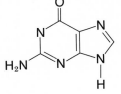

β-D-Ribose
furanose ring structure
found in RNA
(*Ribose Nucleic Acid*)

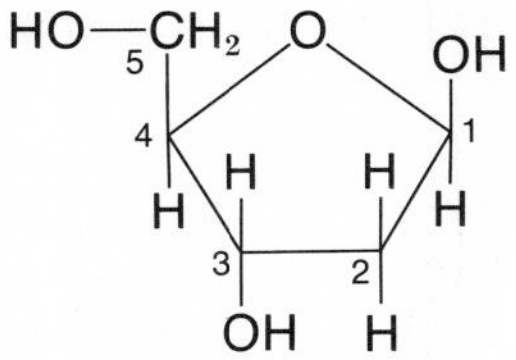

β-D-Deoxyribose
furanose ring structure
found in DNA
(*DeoxyriboNucleic Acid*)
Note the absence of oxygen
at position 2

FIGURE 19.2   The pentoses in DNA and RNA.

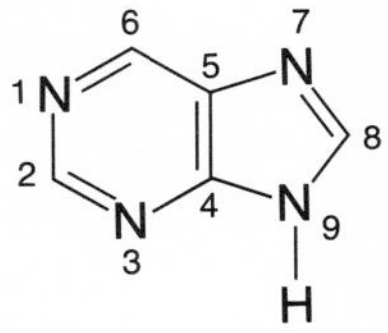

Purine

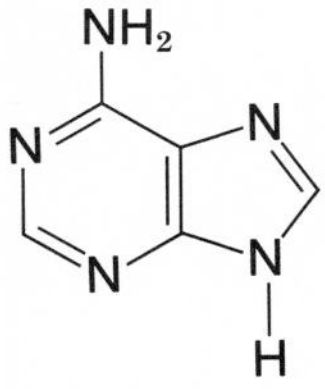

Adenine (A)
6-aminopurine
found in DNA
and RNA

Guanine (G)
2-amino-6-oxopurine
found in DNA *and* RNA

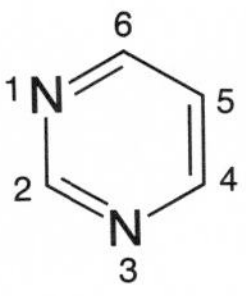

Pyrimidine

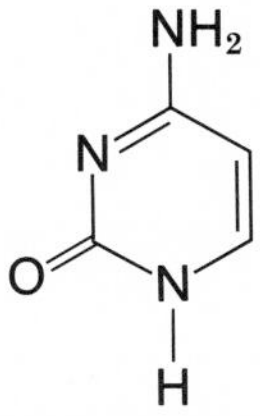

Cytosine (C), *2-oxo-6-aminopyrimidine*
found in DNA *and* RNA

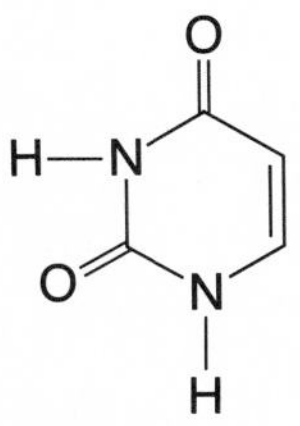

Uracil (U)
2,6-dioxopyrimidine
found in RNA

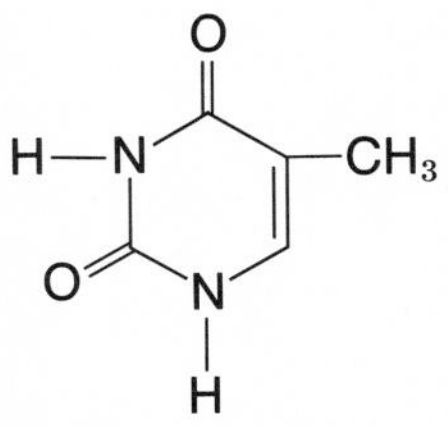

Thymine (T)
2,6-dioxo-5-methylpyrimidine
found in DNA

FIGURE 19.3   Purine and pyrimidine bases in DNA and RNA.

and all consist of a pentose sugar, a phosphate group, and an organic nitrogenous base. There are two kinds of nucleotides, pyrimidine and purine nucleotides (Figure 19.1) Further hydrolysis splits off phosphoric acid from each, leaving behind pyrimidine and purine **nucleosides,** composed of a pentose sugar and a nitrogenous base. The final hydrolysis liberates the individual sugars and several nitrogenous bases. If the nucleic acid is DNA, the sugar is 2-deoxyribose; if the nucleic acid is RNA, the sugar is ribose (Figure 19.2).

In the main, five different nitrogenous bases are found in nucleic acids (Figure 19.3). Those with the double-ring structure of purine are the purine bases, **adenine** (A) and **guanine** (G). Those with the single-ring structure of pyrimidine are the pyrimidine bases, **cytosine** (C), **thymine** (T), and **uracil** (U). Of these, adenine, guanine, cytosine, and thymine are found in DNA; the same bases are found in RNA except that uracil replaces thymine. The components of structure of DNA and RNA can now be summarized:

DNA: Phosphate—deoxyribose—A,G,C,T
RNA: Phosphate—ribose—A,G,C,U

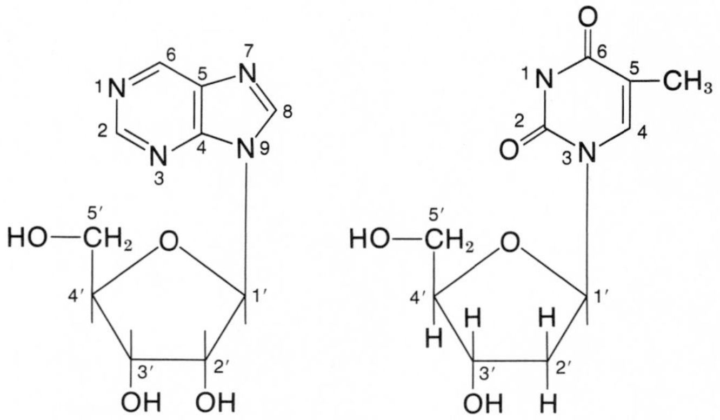

Adenosine
(ribose + adenine)

Thymidine
(deoxyribose + thymine)

Other nucleosides are: uridine (ribose + uracil), cytidine (ribose + cytosine), guanosine (ribose + guanine).
*Purine bases combined with a sugar have the ending -osine (aden*osine*); pyrimidine-combined bases with a sugar have the ending -idine (thym*idine*).*
DNA *contains: deoxyadenosine, deoxyguanosine, deoxycytidine, deoxythymidine*
RNA *contains: adenosine, guanosine, cytidine, uridine*

FIGURE 19.4 Structure and naming of nucleosides.

Adenosine 5'-monophosphate, AMP

Thymidine 5'-monophosphate, deoxy-TMP
DNA *contains:* deoxy-AMP, *deoxy*-GMP, *deoxy*-TMP, *deoxy*-CMP.
RNA *contains:* AMP, GMP, CMP, UMP.

FIGURE 19.5 Nucleotide structure and linkages.

Although the five bases shown are the principal ones found in nucleic acids, several different but related bases also occur, especially in RNA. Some of these rarer bases are 2-methyladenine, 1-methylguanine, 2-methylcytosine, and 5-hydroxymethylcytosine.

Joining any one of the nitrogenous bases to ribose or deoxyribose results in a nucleoside (Figure 19.4). If a phosphate group is now added to a nucleoside, the result is a nucleotide, a unit that consists of a sugar (S) (either ribose or deoxyribose) to which phosphate (P) is attached at one side and a nitrogenous base (B) at the other side. The details of the linkages for several nucleotide structures are indicated in Figure 19.5.

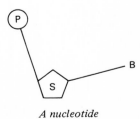

A nucleotide

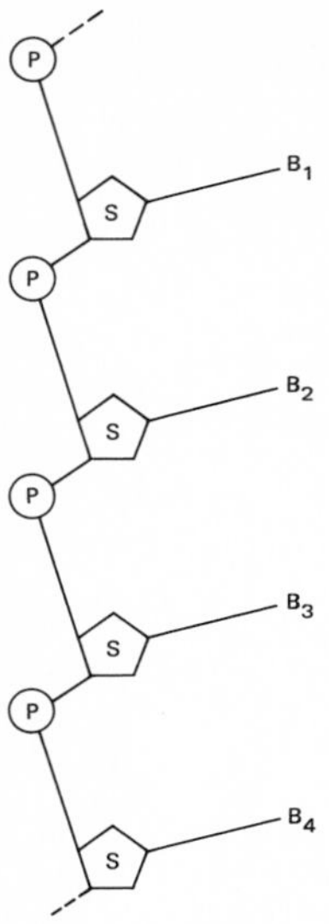

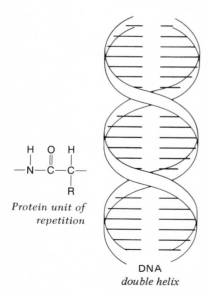

Schematic section of a nucleic acid

Protein unit of repetition

DNA
double helix

DNA AND RNA AS NUCLEOTIDE POLYMERS

The successive joining of nucleotide to nucleotide is the distinguishing structural feature of the nucleic acids. Both DNA and RNA are therefore polymers in that they are long sequences of nucleotides linked together by a common unit of repetition to form a chain, or strand. The repeating monomer is the phosphate-sugar combination, as shown in the sketch. These are the individual beads of the string, but attached to the sugar portion of each such bead is one of the five nitrogenous bases. These differing bases superimpose variety on the monotony of the repeating sugar-phosphate units of the nucleic acids. This is reminiscent of the structure of proteins, where the unit of repetition is nitrogen-carbonyl-carbon, strung together in long chains, variety being provided by the different side groups of the different amino acids. It will also be recalled that in many instances these polypeptide chains, with their dual features of uniformity and difference, arrange themselves *spontaneously* into a helical pattern in which hydrogen bonds contribute to the stability of the structure. Helical configuration is also characteristic of nucleic acids, particularly DNA. However, the **helix** formed by DNA contains two strands, one wrapped about the other, so that it is called a

double helix. In the DNA helix, the bases attached to each strand project toward the centerline of the helix so that they face each other, providing the proper position and spacing for hydrogen bonding between them.

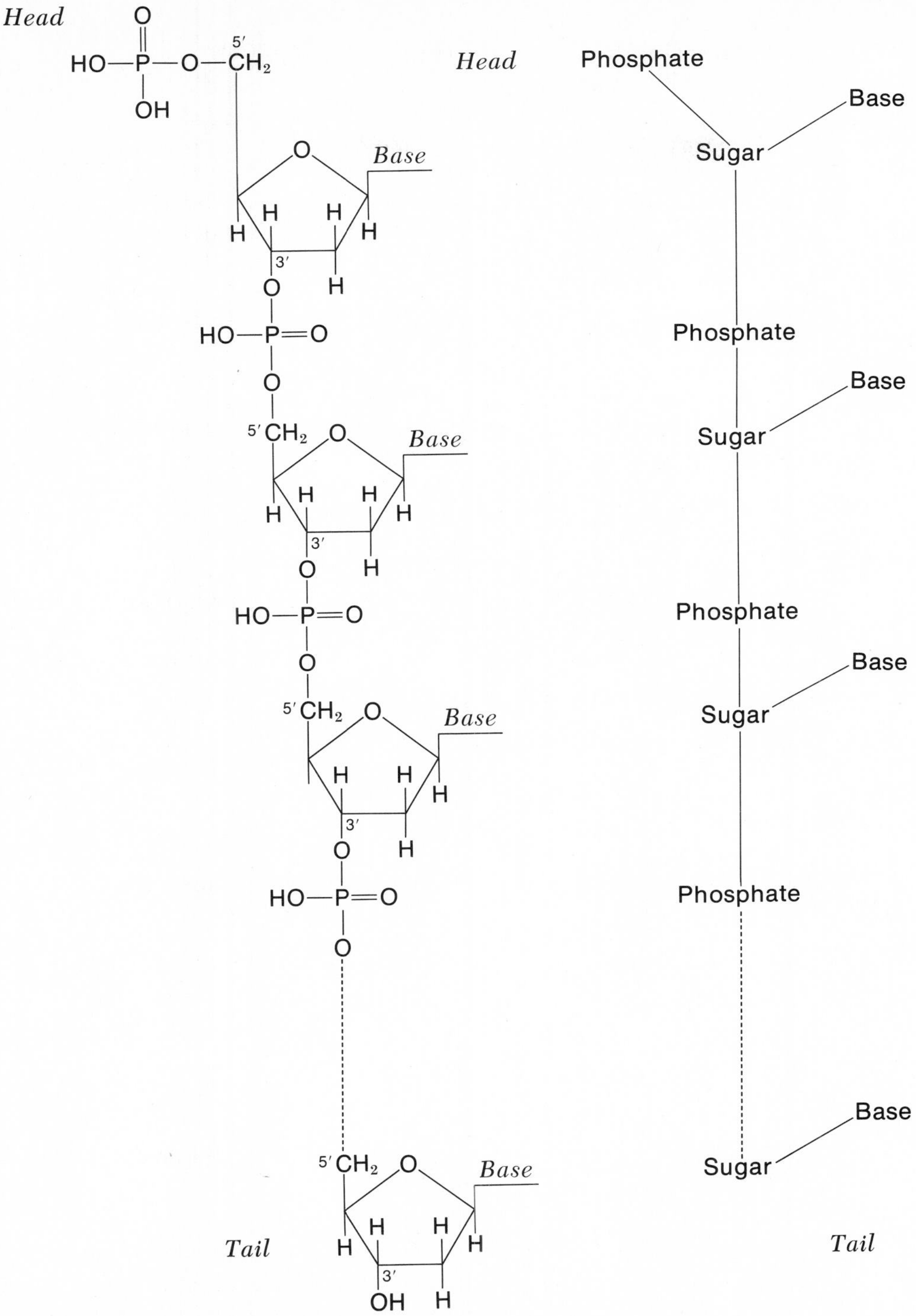

FIGURE 19.6 A DNA single strand.

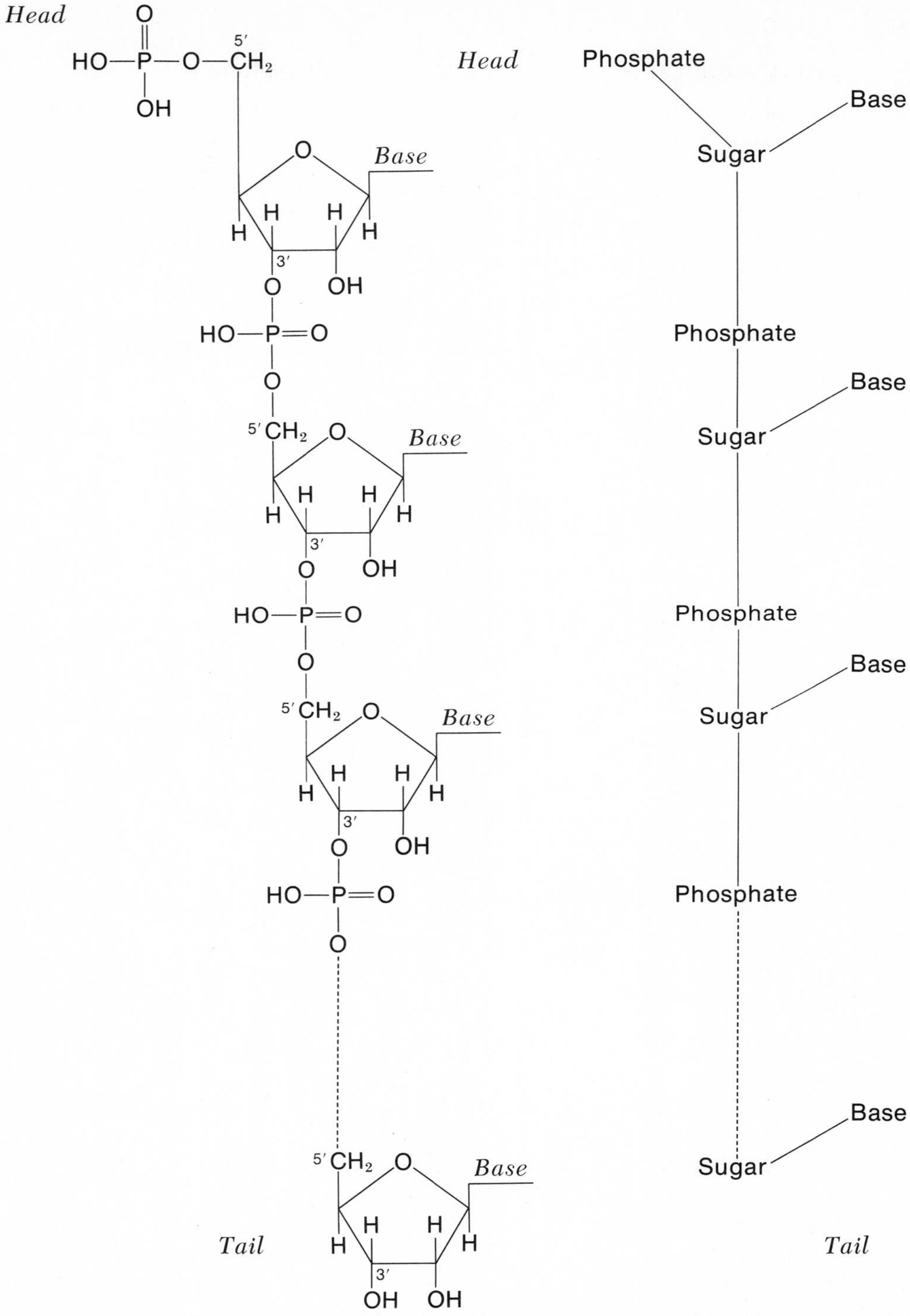

FIGURE 19.7 An RNA single strand.

DNA AND RNA STRUCTURE

Closer and more detailed representations of the DNA and RNA strand are given in Figures 19.6 and 19.7, respectively. Note that each strand has a phosphate end, called the **head** of the chain, and a sugar end, called the **tail** of the chain. By chromatography and x-ray methods, some interesting facts concerning the base composition and the structure of the DNA molecule have been obtained. By base composition is meant the relative numbers of A, G, C, T, and U present in the nucleic acid.

1 The base composition of DNA is characteristic of the organism and differs from species to species.

2 The base composition of DNA from all the enormously different tissue cells of the same organism is the same and is unaffected by age, development stage, nutritional conditions, or other physiological and environmental factors.

3 DNA exhibits certain chemical regularities in its composition called **Chargaff's rules:**

Number of adenine residues
$\qquad$ = number of thymine residues $\qquad$ or $\qquad$ A = T

Number of guanine residues
$\qquad$ = number of cytosine residues $\qquad$ or $\qquad$ G = C

so that

Sum of purines
$\qquad$ = sum of pyrimidines $\qquad$ or $\qquad$ A + G = T + C

4 Closely related organisms have similar base compositions; this is now used as a basis for taxonomy, as well as for evaluating species from the point of view of evolution.

The conclusions listed above all support the picture of DNA as the carrier of genetic information. It would be expected that the composition of DNA would be characteristic of a *specific* organism and would therefore differ from organism to organism. It would also be expected that differences in health, age, or development of a given organism would *not* bear upon the composition of its DNA; in other words, DNA composition is a *constant* in the organism.

Of particular importance to the structure of DNA is rule 3: The number of adenines equals the number of thymines, and the number of guanines equals that of cytosines. Added together, the number of larger purines in DNA equals the number of smaller

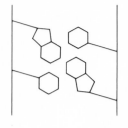

pyrimidines. That is, a cat and a rat may have a different number of purines and pyrimidines, but in each case the number of purines (A + G) equals the number of pyrimidines (T + C); further, in each case, A = T, and G = C. This suggested that some fitting and matching are involved, in which not one but two strands participate, so that a purine from one strand is always facing a pyrimidine from the other, and vice versa. Using x-ray studies, which indicated that the structure was periodic, and the overall chemical and physical data at hand, Watson and Crick proposed in 1953 what has come to be known and accepted as the **double-helix structure** of DNA.

The structure consists of two helical polynucleotide chains which coil around a common axis to form the double helix, like a circular staircase. The two chains run in opposite directions in an antiparallel arrangement. One chain goes up from tail to head and the other one comes down tail to head around the common axis (Figure 19.8).

The purine and pyrimidine bases point to the inside of the double helix. The bases from one strand are paired with bases from the other strand to permit optimum hydrogen bonding. This· would represent the steps in our circular staircase. To achieve optimum bonding, adenine from one strand is always paired with thymine from the other strand with two hydrogen bonds between

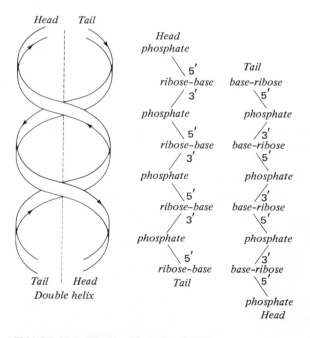

FIGURE 19.8 The double helix of DNA.

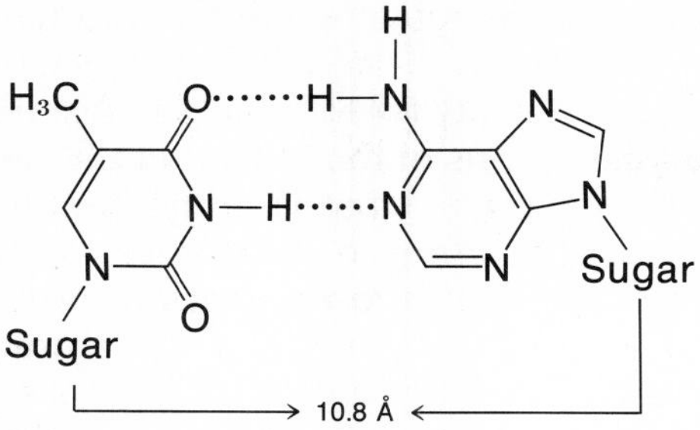

2 *hydrogen bonds*

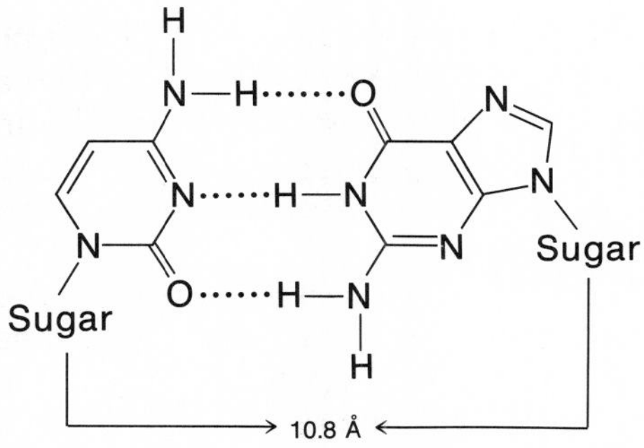

3 *hydrogen bonds*

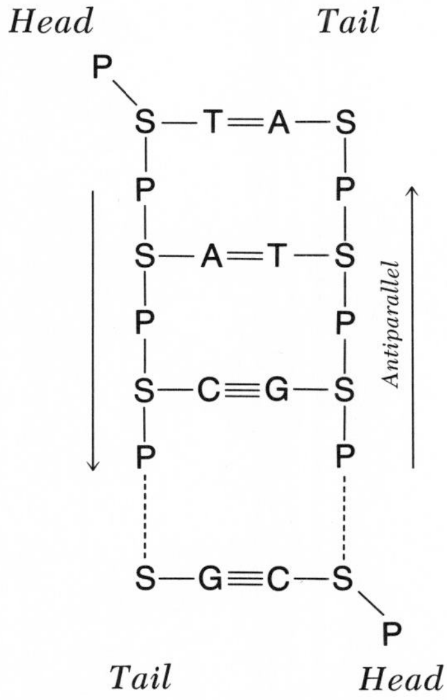

FIGURE 19.9 Base pairing and hydrogen bonding in DNA.

them. Cytosine is similarly paired with guanine with three hydrogen bonds (Figure 19.9).

The hydrogen bonds provide the stability for the double helix in conjunction with hydrophobic interactions of the bases within the structure of the helix. The two antiparallel strands of the double helix are not identical but are *complementary* to each other. The importance of this can be seen in the replication process of DNA.

REPLICATION OF DNA

The replication of DNA molecules provides for the transmission of genetic information from generation to generation by producing two identical daughter DNA molecules from one parent DNA mol-

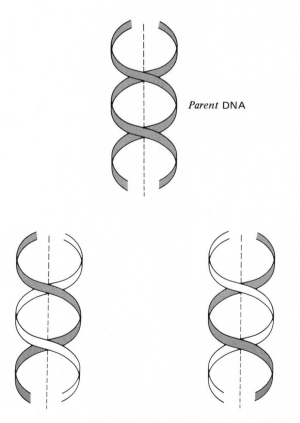

Parent DNA

Two daughter DNA *molecules
containing one strand each
from the parent.*

FIGURE 19.10 Replication of DNA.

ecule (Figure 19.10). Involved in this replication process are the parent DNA molecule, specific enzymes, and a pool of triphosphate nucleosides. The strands of the **parent DNA** act as both primer and template. The **primer DNA** is considered to be a preformed DNA needed to start the replication process and the template needed for the synthesis of the new complementary strand. The enzymes in the replication process are the poly-

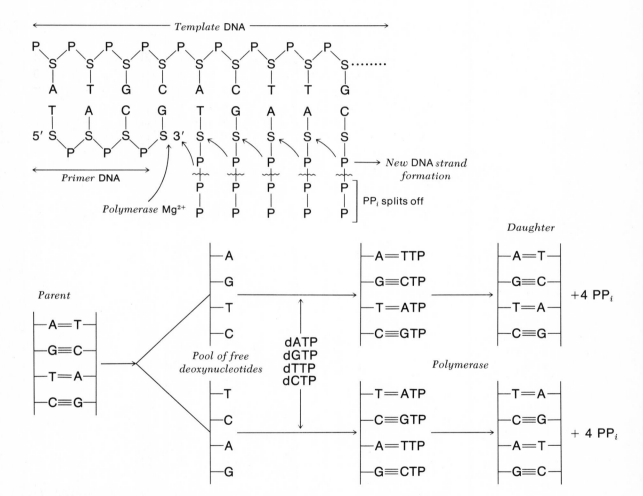

FIGURE 19.11 A closeup of DNA replication. The upper part of the figure shows the primer DNA parallel to a long section of template DNA. As each triphosphate nucleotide from the pool successively joins with the primer DNA by forming an ester with the hydroxyl group of the sugar, it drops off a pyrophosphate, PP. As the triphosphate nucleotides keep joining, a growing polynucleotide is produced that is complementary to the template DNA. The template DNA and the growing chain together replicate the original DNA. This is seen again in the lower diagram, with the two daughter DNAs formed identical to the parent DNA.

merases which join the complementary nucleoside triphosphates to the free hydroxyl groups of the primer in succession as they are required by the template to make the new strand (Figure 19.11).

Evidence indicates that the overall process begins with a partial uncoiling of the native DNA molecule and separation of the polynucleotide strands. It can be visualized as sketched in Figure 19.12. When the uncoiling and separation become sufficiently large, the formation of the **daughter strands** is initiated; they grow at about the same rate as the parent DNA uncoils and separates (Figure 19.12).

The triphosphate nucleotides line up in their complementary sequence according to the bases on the template, and the enzyme polymerase initiates the joining reaction. The new strand continues to grow as long as template DNA is available. The result is a *true* replication of the genetic information from one

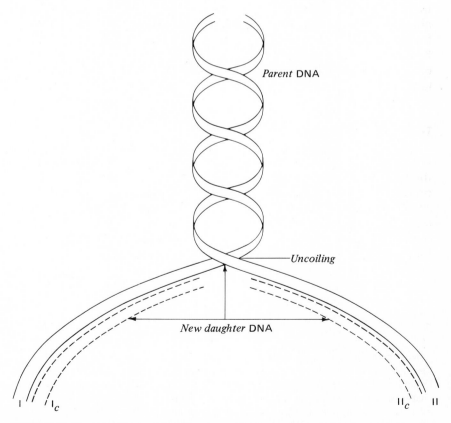

FIGURE 19.12 Uncoiling and replication of DNA. I and II are parent strands; I$_c$ and II$_c$ are new complementary DNA strands.

parent DNA molecule to the two new daughter molecules of DNA. However, there is always the chance that something can go wrong in the process. It is possible that a polynucleotide strand may break. If that happens, suitable enzymes are available to repair the break. Sometimes part of the chain is deleted, which gives a mutation. Other mutations result from the addition of a base pair or the interchange of base pairs. Such errors are reversible and can be repaired by the cell, but they are not always corrected and then show up as mutations.

Deletion of a sequence of base pairs will cause a true mutation, which in turn may result in the loss of a vital enzyme by the organism because the instructions for its production are lost or garbled. Most of these mutations do not survive. On the other hand, there is evidence that cancer cells are virulent mutants that transmit their virulence to other cells in a tissue or organ, thus multiplying increasingly. The likelihood of cell mutation may not seem so far-fetched if we realize that the body replicates cells by the millions hour after hour. Furthermore, certain physical and chemical agents can cause breaking of polynucleotide chains, fusing of bases in the chain, or deletion of bases from the chain. X-rays and gamma rays are such physical agents, and mustard gas, colchicine, coal tar dyes, and others are chemicals with mutagenic properties.

THE DNA MOLECULE, GENES, AND THE GENETIC CODE

A unicellular organism needs from several hundred to about a thousand different enzymes for its vital functions. Cells of higher organism need many more, estimated as several hundred thousand to a million different enzymes. Any single species of unicellular organisms has one DNA molecule, but any species of higher organism has many. The DNA molecules are found primarily in the nucleus of the cell, where they are identified as **chromosomes.** In a human cell there are 46 chromosomes, or 46 DNA molecules. The subunits of each chromosome are called genes, and each chromosome is made up of several thousand genes. Each gene contains the code, expressed in its sequence of nucleotide bases, for the synthesis of one enzyme. But this raises a question: DNA has four bases, A, G, C, and T, but must be able to signal for at least the 20 different amino acids that can be present in the enzyme's protein. If the 4 bases were used singly in what is called singlet coding, only 4 signals would be available and only 4 amino acids could respond. If the 4 bases are used 2 at a time, in a doublet code, then $4^2 = 16$ signals would be available but still not enough. However, if the 4 bases were used 3 at

Singlet code:
A G C T = 4 *signals*

Doublet code:
AA AG AC AT
GA GG GC GT
CA CG CC CT
TA TG TC TT = 16 *signals*

Triplet code:
A, G, C, and T *preceding each of the* sixteen *doublet codes =* 4 × 16 = 64 *signals*

TABLE 19.1 RNA triplet code for some amino acids

Amino Acid	Triplet Codon
Valine	GUA, GUG, GUC, GUU
Phenylalanine	UUU, UUC
Leucine	UUA, UUG, CUA, CUC, CUG, CUU
Glycine	GGA, GGC, GGG, GGU
Aspartic acid	GAC, GAU, GAA

a time, a **triplet code,** $4^3 = 64$ individual signals are possible. We can think of each triplet as a code word, calling for valine, glycine, and so on. The evidence at present indicates that this triplet code is universal and used by all organisms. Since there are more triplets, or **codons,** than amino acids, it is possible that more than one codon will call for a given amino acid. This has been found to be the case, and the genetic code is called **degenerate,** which refers to this duplication.

THE TRANSCRIPTION OF DNA; MESSENGER RNA

Before the genetic code can be used in the synthesis of proteins, it must first be transcribed. This transcription is done by synthesizing an RNA molecule on a DNA template strand. The process of transcription requires the DNA template, the enzyme RNA polymerase, and the nucleosides ATP, UTP, GTP, and CTP and involves a number of distinct steps. The first step is the uncoiling of the DNA double helix to expose the gene whose genetic code is to be transcribed. At the beginning of the gene is a **promoter site,** which is recognized by the enzyme RNA polymerase. The template and the enzyme become bound at this promoter site.

In the second step **transcription** is initiated by the enzyme; complementary nucleotides line up along the DNA template, and

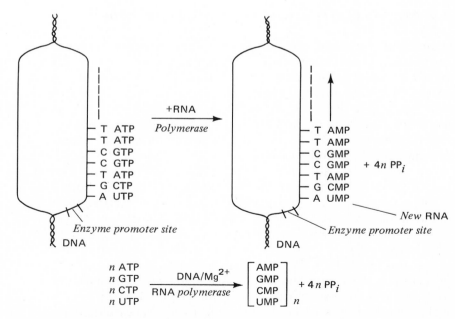

FIGURE 19.13 Transcription of RNA from DNA.

the enzyme joins one nucleotide to the next to begin the RNA molecule. The lineup of the nucleotides is such that ATP lines up with thymine on the DNA template strand. CTP lines up with guanine, UTP lines up with adenine, and GTP lines up with cytosine (Figure 19.13).

The third step is the **elongation** of the RNA chain as the complementary nucleotides continue to line up along the DNA template and continue to join up accordingly.

The final step is the **termination** of the RNA chain. The enzyme RNA polymerase recognizes a termination site on the DNA template and attaches itself there so that the RNA strand must stop at that point. When the synthesis ends at this termination point, the completed RNA molecule is released. This RNA molecule is mRNA, and it now goes to the ribosomes in the cell. It becomes attached to the ribosomes, where it acts as the template for the synthesis of one or more proteins. Note that the DNA stays secure in the nucleus; it is the mRNA which carries the transcribed message to the ribosomes for synthesis.

mRNA = *messenger* RNA

TRANSLATION OF THE GENETIC CODE; PROTEIN SYNTHESIS

The next step is translation, wherein the coded instructions of mRNA are translated into the amino acid sequence required for the synthesis of the specific protein. Participating in this synthesis are mRNA, the ribosomes, a second kind of RNA called **transfer RNA** (tRNA), and several enzymes. In addition, there must be a pool of amino acids. The first step in the protein synthesis is the *activation* of the amino acids by ATP with the aid of an enzyme called aminoacylsynthetase. There is a specific enzyme for the activation of each of the 20 amino acids used. The reaction results in the formation of activated amino acid, also called amino acid adenylate, and pyrophosphate. This is analogous to the activation of fatty acids when they enter into β oxida-

Step 1:

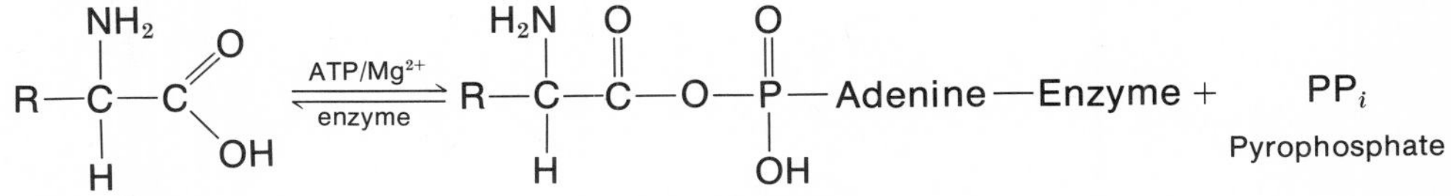

Amino acid

Activated amino acid
(aminoacyl AMP, aminoacyl adenylate)

tion. The activated amino acid remains attached to the synthetase enzyme but is subsequently passed on to a specific RNA. The enzyme aminoacylsynthetase is highly specific for the amino tRNA = *transfer* RNA

acid it activates and also for the tRNA to which it passes on the amino acid residue. The function of a tRNA molecule is to *transfer* an amino acid residue from the amino acid adenylate shown above to the site of synthesis. The tRNA molecules are of relatively low molecular weight and contain from 70 to 90 nucleotide residues per molecule.

All tRNAs contain a binding site for the enzyme aminoacylsynthetase; at the end of that site there is an adenine residue to which the amino acid can hook on later. Each tRNA will carry a specific amino acid and a specific base triplet that identifies that amino acid. This tRNA triplet, called the **anticodon,** is the complement of the mRNA codon calling for that amino acid. A given tRNA, carrying a given amino acid, will therefore situate itself along the mRNA strand only at that codon which is complementary to its anticodon and is calling for its amino acid. The polypeptide will therefore form as prescribed by the mRNA sequence of codons and as originally ordered by the DNA. It is believed that tRNAs achieve their three-dimensional structure by base pairing of the single nucleotide chain to form a double strand (Figure 19.14).

In step 2 the activated amino acid is transferred to the specific tRNA, forming an **aminoacyl-tRNA complex.** In the process the

Step 2:

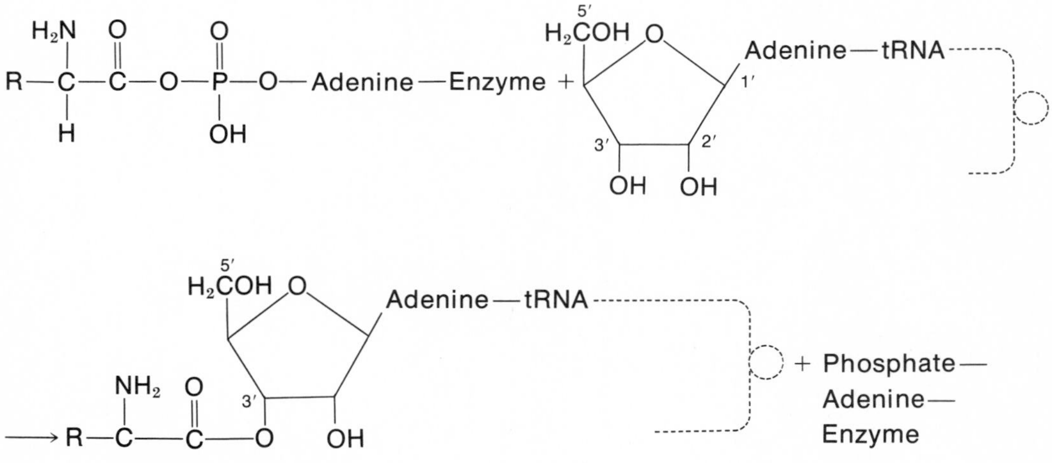

$$H_2N \quad O \quad O$$
$$R—\overset{|}{\underset{|}{C}}—\overset{\|}{C}—O—\overset{\|}{\underset{|}{P}}—O—Adenine—Enzyme +$$
$$H \qquad OH$$

(ribose structure with 5' H₂COH, O, 1', 2', 3', OH OH) Adenine—tRNA

Aminoacyl-tRNA complex

(ribose structure with 5' H₂COH, O, 3', OH) Adenine—tRNA

$$NH_2 \quad O$$
$$\longrightarrow R—\overset{|}{\underset{|}{C}}—\overset{\|}{C}—O \quad OH$$

+ Phosphate—Adenine—Enzyme

enzyme attaches itself to the binding site on the tRNA, and the amino acid residue is transferred from the phosphate of adenine to the —OH group of carbon atom 2 or 3 of ribose of the adenosine. The phosphate-adenine-enzyme complex is left behind for further duty. The aminoacyl-tRNA complex is now ready for the actual polypeptide synthesis, which involves the mRNA as the

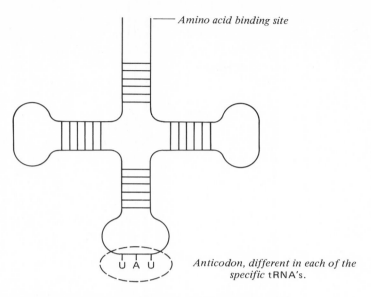

Amino acid binding site

U A U

Anticodon, different in each of the specific tRNA's.

FIGURE 19.14 The three-dimensional structure of tRNA.

template with its codons and the aminoacyl-tRNA complexes with their corresponding anticodons. The ribosomes act to stabilize the mRNA for proper alignment. The energy needed for the synthesis is provided by GTP and some cofactors. The entire synthesis can be divided into three distinct parts:

1 Initiation of the polypeptide synthesis
2 Elongation of the polypeptide chain
3 Termination of the synthesis

INITIATION OF THE POLYPEPTIDE SYNTHESIS

The process begins with the attachment of the mRNA to the ribosomes. Following this, the codon AUG on mRNA and the complementary anticodon from an aminoacyl-tRNA complex start up the process. This specific starting system contains the N-formylmethionine-tRNA system at which all protein synthesis begins but which is removed from the polypeptide chain on completion. Its *only* function is as a starter (Figure 19.15).

ELONGATION OF THE POLYPEPTIDE CHAIN

After the codon and anticodon of the initiator system are joined, elongation can take place. According to the next codon on the mRNA, the preformed aminoacyl-tRNA lines up with its anticodon facing the codon and becomes bound by use of energy derived

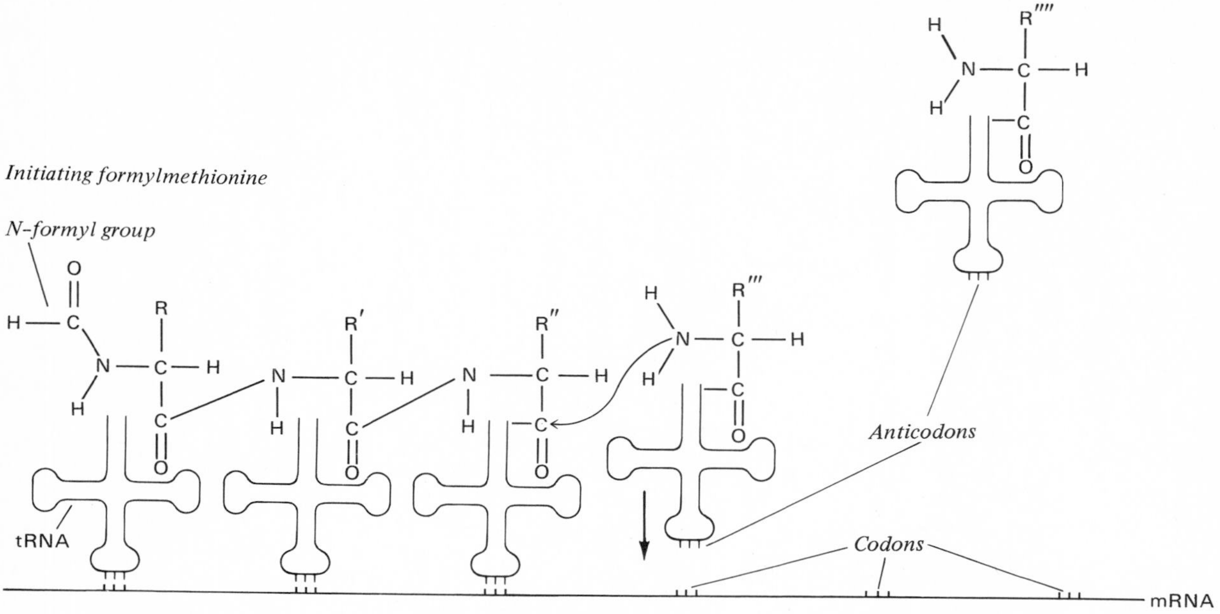

FIGURE 19.15 Polypeptide synthesis.

from GTP. The lineup will be such that the amino end faces the acid group of the N-formylmethionine. A **peptide bond** forms when the —NH$_2$ group of the next incoming amino acid links up to the carbonyl group that is attached to the first tRNA. When this bond is formed, the carbonyl carbon becomes detached from the first tRNA and the dipeptide so formed is attached to the second tRNA. The third amino acid is carried in by a tRNA responding to the next mRNA codon, and the process is repeated. A new peptide bond is formed, and the attachment of the tripeptide is now at the third tRNA. The first tRNA has left its codon on the mRNA during the second elongation step. This elongation process is repeated over and over until the end of the protein chain is reached. The formylmethionine residue which started the polypeptide chain is then hydrolyzed from the chain, leaving a free —NH$_2$ group at the amino terminal of the polypeptide chain. The function of the formylmethionine residue was to provide a proper start and ensure that an incoming —NH$_2$ group joined the carboxyl group, taking the place of the tRNA to which the carboxyl was attached.

TERMINATION OF THE POLYPEPTIDE CHAIN

The termination of synthesis is effected by the **terminator codons,** UAA, UAG, and UGA. Since none of the tRNAs contain fitting anticodons, the protein synthesis comes to a stop here. The remaining step is the release of the completed protein from the last tRNA attachment. This is achieved by hydrolysis, which produces the free carboxyl end of the protein, —COOH. Once the protein is released, it spontaneously assumes its tertiary configuration.

AGENTS AFFECTING PROTEIN SYNTHESIS

Two groups of agents affect protein synthesis in the cell.

ANTIBIOTICS

One group is the antibiotics, which are especially effective against bacteria and microorganisms. They stop protein synthesis or greatly interfere with it, so that the growth of the organism slows down or stops. Several antibiotics, such as puromycin and the tetracyclines, interfere with the normal binding of aminoacyl-tRNA, causing inhibition of the binding or improper chain growth, thereby distorting the protein being formed. Chloramphenicol and cycloheximide interfere with the peptide-bond formation. Streptomycin and other similar antibiotics interfere with the mRNA ribosome interaction, thereby causing misreading of the genetic code and disruption of protein synthesis. Actinomycin D interferes with the formation of mRNA, without which no protein synthesis is possible. Figure 19.16 gives the structures of several antibiotics.

ANTIMETABOLITES

A second group of agents is the antimetabolites, which are among the most useful chemotherapeutic agents against cancer. Antimetabolites interfere primarily with the synthesis and the utilization of the nucleic acids.

Folic acid antimetabolites Folic acid is needed by the cell to make new purine bases. Antagonists such as amethopterin competitively inhibit enzymes needed for the folic acid purine metabolism.

Purine antimetabolites 6-Mercaptopurine and 8-azaguanine resemble the purines normally used to make nucleic acids. When they are incorporated into nucleic acids instead of adenine or the other purines, faulty genetic material is produced by the cell and growth is inhibited.

Pyrimidine antimetabolites 5-Fluorouracil and azauracil are similar to the natural pyrimidines and also become incorporated into

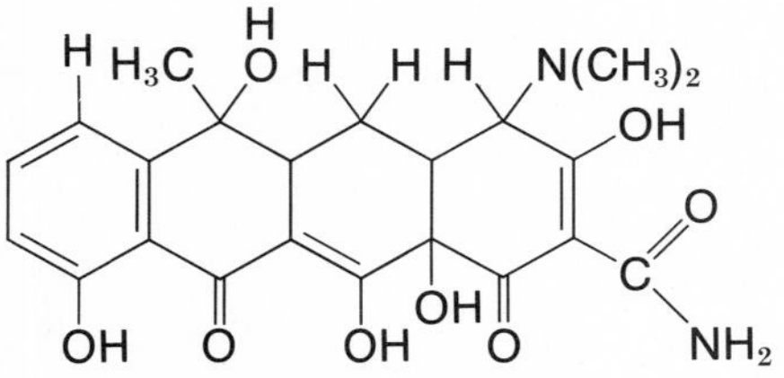

Tetracycline

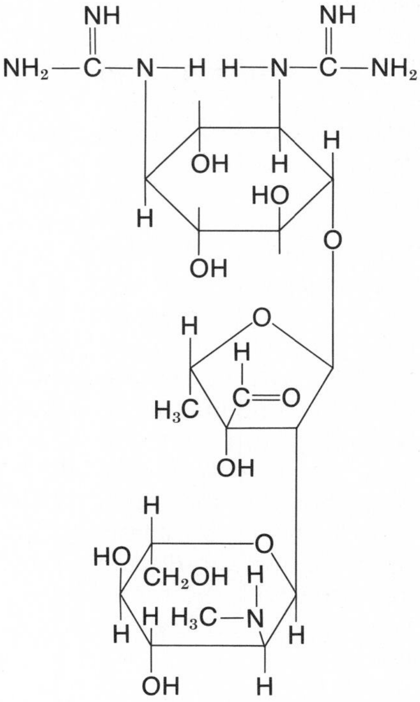

Streptomycin

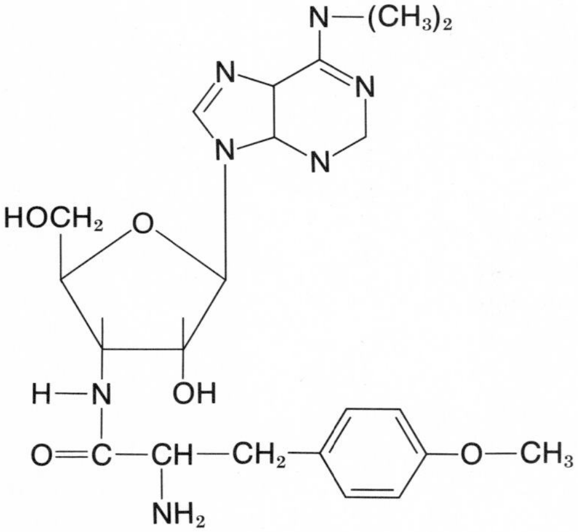

Puromycin

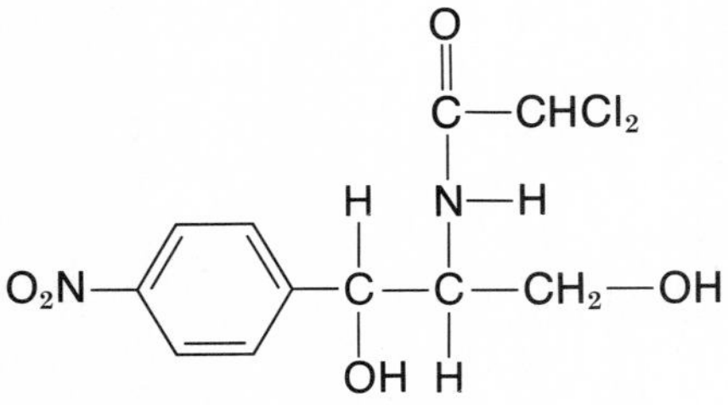

Chloramphenicol

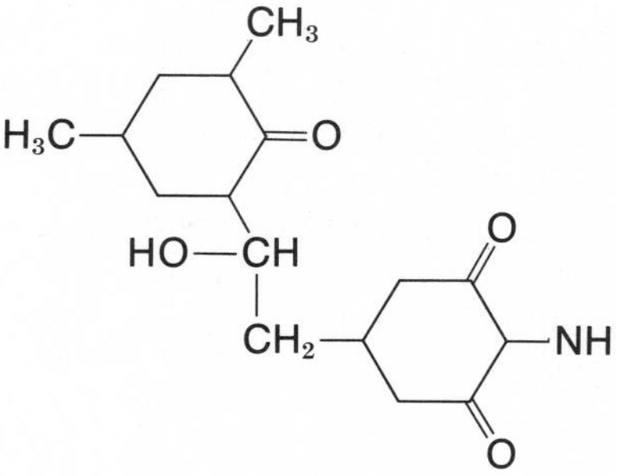

Cycloheximide

FIGURE 19.16 Structure of some antibiotics that inhibit protein synthesis.

nucleic acids. When they are used by the cell, an abnormal DNA is produced and again the growth of the cell is checked.

Most of these drugs are toxic to the whole organism if present in high concentration and must be used with great care. They usually cannot be given in large enough dosages to knock out the cancerous growth and often must be withdrawn because of undesirable side effects. This is one of the reasons they are not equally effective with all kinds of cancer.

Other special chemotherapeutic agents are compounds such as L-phenylalanine mustard (PAM), which is especially useful in

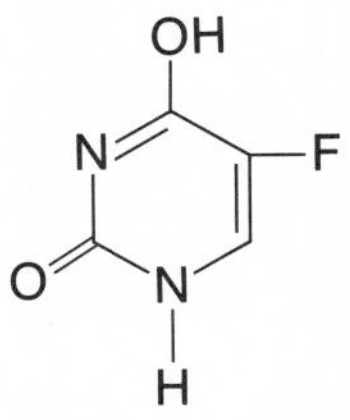

5-*Fluorouracil* 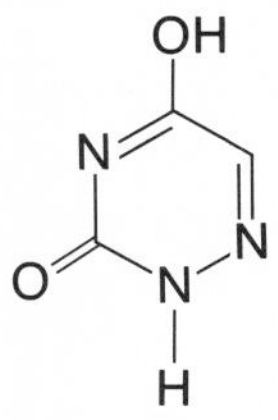   *Azauracil* 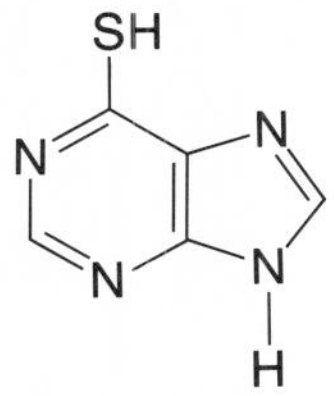   6-*Mercaptopurine*

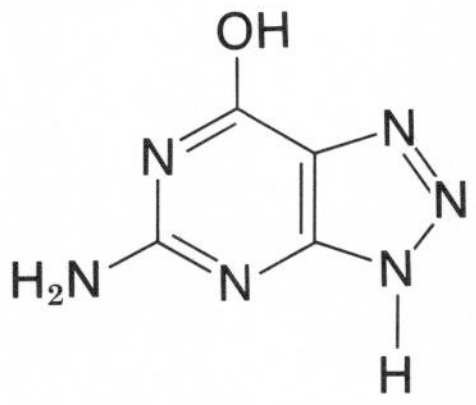

8-*Azaguanine* 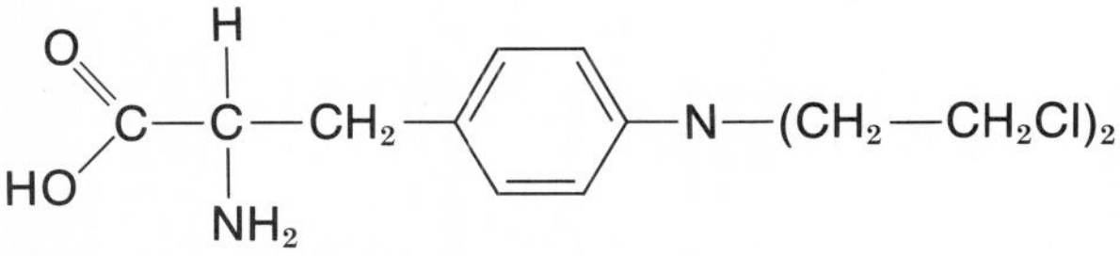   L-*Phenylalanine mustard* (PAM)

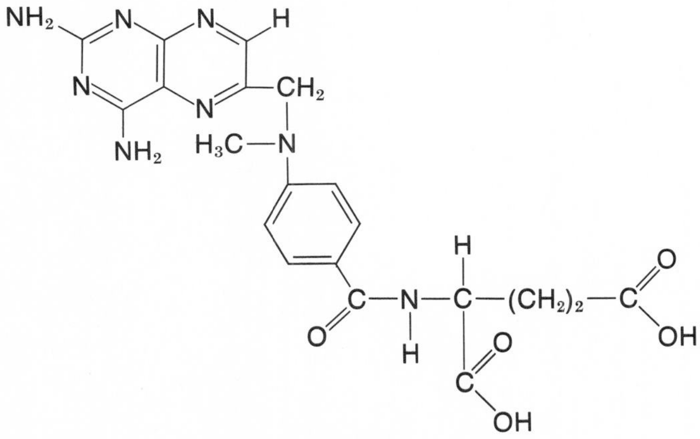

Amethopterin

FIGURE 19.17 Structure of several antimetabolites used in the treatment of cancer.

preventing recurrence of breast cancer. Estrogens are in use in the treatment of cancer of the prostate gland with excellent results. The structures of several antimetabolites are given in Figure 19.17.

REVIEW QUESTIONS

1 What are the differences between the DNA and RNA molecules in structure, composition, and function?
2 What is meant by the single gene—single enzyme theory? Can you give examples?
3 Explain the difference between nucleotides and nucleosides.
4 Describe in your own word the double-helix structure of the DNA molecule.
5 What is meant by the head and tail of the DNA chain?
6 What can you say about the base composition of DNA molecules with regard to environmental factors, species, and uses?
7 What is the importance of hydrogen bonding in the double-helix structure of DNA?
8 Explain replication, transcription, and translation.
9 Why is it important that the nucleic acid strands in DNA complementary to each other?
10 What are mutagenic agents?
11 Discuss the genetic code.
12 Explain the function of **a** mRNA and **b** tRNA.
13 What is the importance of N-formylmethionine-tRNA in protein synthesis?
14 What is the effect of antibiotics on protein synthesis?
15 How do antimetabolites work, and what are their uses? Give examples.

GLOSSARY

absolute temperature   Temperature on a scale based on a theoretical absolute zero that is 273 degrees below 0°C (also called the Kelvin scale); celsius degrees and kelvins are equal in size.

acid   A chemical entity that is a proton (H^+) donor; in water solution an acid yields H^+ ions.

acid chyme   The semifluid, strongly acid content of the stomach that is passed on to the duodenum.

activation energy   The amount of energy needed to get a reaction going. Once initiated, a spontaneous reaction will continue on its own, as when a wood fire continues to burn after it has been kindled.

active site   The portion of the enzyme molecule where close contact with the substrate occurs and where the reaction is catalyzed.

acyl halides   Organic compounds containing the acyl halide functional group (margin), where X is usually a halide such as Cl or Br.

addition reaction   The addition of another atom or group of atoms to carbons with double or triple bonds. The added substance (H_2, H_2O, etc.) divides in two (—H, —H; —H, —OH), one part attaching to one of the unsaturated carbons and the other part to the other carbon. The rest of the molecular structure is unchanged.

adenosine diphosphate (ADP)   Similar to ATP but with one less phosphate group and less energy. The ATP–ADP relation is important in metabolism.

adenosine monophosphate (AMP)   Precursor of ATP and ADP; important in metabolism and enzymatic reactions.

adenosine triphosphate (ATP)   The major chemical form in which energy is stored in living matter, to be called upon as energy is required by the organism.

adsorption   The adhesion of gaseous or dissolved materials to the surface of a solid material.   Only a very thin layer of molecules adheres to the surface, so that a good adsorbent is characterized by a large surface area.

albinism   An inherited disorder that prevents the proper metabolism of tyrosine to produce skin pigments.

alchemists   Medieval chemical practitioners who sought to change base metals into gold and to prolong life indefinitely.   Although they did not succeed, they were the forerunners of modern chemistry in that they used experimental methods, developed chemical equipment, and introduced chemical symbols.

alcohols   Hydrocarbons characterized by the presence of a hydroxyl group, —OH, in place of a hydrogen atom.

aldehydes   Organic compounds containing the aldehyde group (margin).

aldol condensation   The condensation of two molecules of an aldehyde or two molecules of a ketone.

aldose   A sugar containing an aldehyde group.

aliphatic compound   A fatty substance with an open carbon structure; also a nonaromatic carbon-ring structure.

allosteric feedback mechanism   A mechanism that stops enzymatic activity by changing the spatial arrangement of the active site; usually done by the end product of a sequence of reactions.

allylic carbon   A carbon adjacent to a carbon-carbon double bond.

alkanes   Saturated, single-bonded aliphatic hydrocarbons.

alkaptonuria   An inherited disorder that prevents the proper metabolism of phenylalanine and tyrosine.

alkenes   Unsaturated hydrocarbons containing one or more double bonds; also known as olefins.

alkyl benzenes   Organic compounds containing an aromatic structure with one or more hydrogens replaced by an aliphatic side chain (margin).

alkyl group   A hydrocarbon chain having one less hydrogen atom than its parent alkane; methyl, ethyl, propyl, etc., groups are alkyl groups corresponding to methane, ethane, propane, etc.   Alkyl groups are represented by —R.

alkyl halide   A hydrocarbon derivative with one or more hydrogens replaced by a halogen (F, Cl, Br, or I).

alkynes   Unsaturated hydrocarbons containing a triple carbon-carbon bond.

alpha (α) particle   One of the particles emitted by radioisotopes.   It is the heaviest of the particles, being identical to the nucleus of the helium atom; is the most effective in causing damage in living tissue by ionization.

amides   Organic compounds containing the amide group (margin).

amine   Organic compounds derived from ammonia by the replacement of one or more hydrogens by carbon chains.

H
|
H—N—R
Primary amine

R
|
H—N—R′
Secondary amine

amino acids   Organic structures containing the carboxyl group, —COOH, and the amino group, —NH$_2$. Amino acids are the building blocks of proteins (margin).

amphoteric   Describing a substance that can act as either an acid or a base, depending on whether it is exposed to basic or acidic conditions.

amylopectin   The branched polysaccharide found in starch; contains both α-1,4 and α-1,6 linkages, the latter providing the branching sites.

amylose   The linear polysaccharide found in starch; contains α-1,4 linkages.

anabolism   The part of metabolism involved in building up biological structures and synthesizing tissues; generally energy-consuming.

antibiotic   An organic substance that can destroy or inhibit the growth of microorganisms.

anticodon   A sequence of three nucleotide bases on transfer RNA which is complementary to the codon of a messenger RNA; the transfer RNA carries its amino acid to the site of the codon of the messenger RNA, where the amino acid is released for attachment to the growing protein chain.

antimetabolite   An organic substance that competes with the reactions of regular metabolites and curtails their use.

antioxidant   A natural or synthetic compound in fats that prevents oxidative rancidity because it is more subject to oxidation than the fat.

apoenzyme   The protein component of an enzyme.

aromatic compound   An organic compound containing one or more benzene or benzenelike ring structures; generally has an odor.

asymmetric carbon atom   A carbon atom attached to four different atoms or groups of atoms.

atom   The smallest physical form in which an element exists and which has the properties of that element.   Chemical reactions between elements are reactions between their atoms.

atomic mass unit (amu)   An arbitrary unit of mass defined as exactly one-twelfth the mass of the carbon 12 isotope, which is assigned 12 amu. On this basis the mass of a proton is 1.0073 amu, of a neutron is 1.0087 amu, and of an electron is 0.00055 amu.

atomic number   The number of positive protons in the nucleus of an atom and hence its positive charge.   This characteristic number Z identifies the atoms of each of the elements and determines its position in the periodic table.

atomic weight   The weight of an atom relative to the weight of carbon 12.

Since the mass of an atom is almost entirely the mass of its nucleus, the atomic weight, or mass, of an atom is approximately the sum of its protons and neutrons.

balanced equation An equation is balanced when the number of atoms of each kind on the left side of the equation is the same as that on the right side, although in different combinations. The principle of the conservation of mass, namely, that matter can neither be created nor destroyed, requires that an equation be balanced.

basal metabolism The rate at which an organism uses energy when at rest.

base A chemical entity that is a proton (H^+) acceptor; in water solution a base yields hydroxide, OH^-, ions. The terms base and alkali are equivalent.

beta (β) linkage A linkage between two monosaccharide portions of a disaccharide in which the condensation is between the β glycosidic hydroxyl of one monosaccharide and a hydroxyl group of the other monosaccharide.

beta (β) oxidation The process by which fatty acids are oxidized to two-carbon fragments.

bile fluid The fluid secreted by the liver and delivered to the duodenum.

bile pigments Breakdown products of hemoglobin; they appear as colored pigments in the bile fluid.

bound water A layer of water, variable in amount, that adheres to the surface of hydrophilic solids by some interaction between the water and the surface.

buffer A solution that resists changes in pH by neutralizing relatively small additions of either acid or base.

calorie (cal) The amount of heat required to raise the temperature of 1 g of water 1°C, *see also* kilocalorie.

carbanion A carbon ion carrying a negative charge.

carbohydrates Compounds composed of carbon, hydrogen, and oxygen, with the ratio of hydrogen to oxygen as in water, and containing an aldehyde or keto group. The molecular formula of carbohydrates can usually be expressed as $C_x(H_2O)_y$.

carbonium ion A carbon atom with only six outer electrons and only three bonds joining it to other atoms. The carbonium ion is positively charged and unstable.

Carboxylate ion

carboxylate ion The negative ion (margin) obtained when a carboxyl group loses a hydrogen ion, H^+.

catabolism The part of metabolism involved in the breakdown of large molecules into smaller ones, generally with a release of energy.

catalyst Any substance that increases the rate at which a reaction proceeds. Catalysts lower the energy of activation. Organic catalysts are called enzymes.

cellulose The structural polysaccharide in plants; a polymer of β-glucose connected by β-linkages.

codon The essential unit of the genetic code, made up of a given sequence of three nucleotide bases on the mRNA molecule; each such triplet is the code for a specific amino acid.

coefficient   A multiplier; in powers of 10 notation, such as 8×10^4, the coefficient is the number 8; in chemical equations, as in $2KClO_3 \longrightarrow 2KCl + 3O_2$, the numbers 2 and 3 are the coefficients. Where no number is shown, the coefficient is 1.

coenzyme   Organic molecules that combine with apoenzymes to form the complete enzyme; many coenzymes are vitamins.

colloidal state   The colloidal condition, that is, having a particle size between fine particles in true solution and coarse particles in suspension. This size range is approximately between 10 and 1000 Å.

combustion   The exothermic reaction of compounds with oxygen with the evolution of light and heat.

competitive inhibition   The condition when the active site of an enzyme can be occupied either by the normal substrate or by some other inhibiting compound similar to the substrate; substrate and inhibitor are in competition.

complex ion   A group of electrons that has gained or lost one or more electrons, thereby acquiring a positive charge (by electron loss) or a negative charge (by electron gain). Complex ions, also called radicals, participate in chemical reactions as single chemical entities, with the charge on the ion applying to the entire group of ions as a single unit, for example, CO_3^{2-} or NH_4^+.

compound   The stable union of two or more different atoms, joined together in specific and fixed ratios, so that the proportion by weight of each element in the compound is always the same. A compound has a net zero charge.

compound lipid   A triglyceride that contains other organic substances.

condensation reaction   A reaction that joins two molecules with the removal of a molecule of water or ammonia.

conformers   Isomers that differ only in the arrangement of their atoms in space.

conjugated lipid   A lipid attached to another organic molecule (a protein in the case of lipoproteins).

conjugated protein   A protein that yields amino acids and other organic molecules when hydrolyzed.

coupled reactions   Reactions (common in metabolism) in which one reaction sets another reaction into motion which sets another into motion, and so on. Generally, they are cyclic, and as long as reactants are fed into the initial reaction, final products will leave the last reaction, for example, the electron-transport system.

covalent bond   The bond formed when two atoms share a pair of electrons equally between them. True covalent bonding occurs only between atoms of the same kind, where the electronegativity of the atoms is the same and the sharing of electrons is equal.

cracking   A decomposition reaction whereby long-chain alkanes are broken down into shorter hydrocarbon chains.

crystalline solid   A solid in which the unit of structure is arranged in an orderly and repeating pattern specific to the solid. The unit of structure may be the plus and minus ions of ionic solids, the atoms of carbon in a diamond, or the molecules in ice, sugar, and other molecular crystalline solids.

cycloalkane   A saturated, cyclic, aliphatic hydrocarbon; only single bonds are present.

deamination   The removal of an amino group from an amino acid.

density   The mass in grams of 1 ml of a substance; units, g/ml.

deoxyribonucleic acid (DNA)   The nucleic acid containing deoxyribose, a pentose, in its structure; one of the genetic compounds, found primarily in the cell nucleus.

derived protein   A derivative of protein which results from the action of chemicals, heat, enzymes, and others.

detergent   An organic compound with hydrophilic and lipophilic characteristics that can act as a cleaning agent in hard water; not a soap.

detoxification   Any process by which toxic substances are rendered harmless.

dextrin   A polysaccharide resulting from the incomplete hydrolysis of starch.

dextrose   Another name for glucose, derived from its property of rotating plane-polarized light to the right.

dialysis   The process of separating colloidal particles from small molecules and ions in true solution by passing the latter through a dialyzing membrane.

dialyzing membrane   A membrane that will permit the passage of water molecules, and ions in true solution but not colloidal particles; most membranes in body tissue are dialyzing membranes.

diene   An unsaturated hydrocarbon containing two carbon-carbon double bonds.

diffusion   The spontaneous tendency of particles present in a given medium, such as air or water, to spread outward in a random manner; an expression of the tendency to uniformity.

digestion   The breakdown of ingested foods by hydrolysis into smaller molecules that can pass through the intestinal wall.

diol   An aliphatic hydrocarbon chain with two hydrogens replaced by —OH functional groups (margin).

dipolar ion   *See* zwitterion.

dipole   Generally, an electric couple with a positive charge at one end and a negative charge at the other.   A molecule has a dipole if the centers of its negative and positive charges do not coincide, as they do in polar molecules.

dipole-dipole interaction   The attractive force between polar molecules that tends to bring them together and to align them.   The negative end of the dipole of one molecule tends to align with the positive end of another molecule.   One special type of dipole-dipole interaction is hydrogen bonding.

dipole moment   An experimental value that measures the degree of polarity of a polar molecule; the higher the dipole moment, the more polar the molecule.   Ionic bonding gives the highest dipole moments since ionic bonding represents the extreme condition of polarity.

disaccharide A carbohydrate formed by condensing two molecules of monosaccharides, usually hexoses.

dissociation constant In general, an experimental value that indicates the extent to which some decomposition or dissociation takes place. Applied to acids and bases, it is a measure of the strength of the acid or base.

disulfide linkage A linkage formed by oxidation of the thiol groups from two cysteine molecules to give —S—S—, the disulfide.

double helix The arrangement of the two strands of DNA coiling around a common axis to form the double helix.

electrolyte Any substance which when dissolved in water or heated to the liquid state will allow the passage of an electric current. In general, ionic materials are electrolytes.

electron A negatively charged subatomic particle found in the region of space around the nucleus of the atom. Its negative charge is equal to the positive charge of the proton but opposite in sign; its mass is 1/1837 amu, compared with 1 amu for the proton and the neutron.

electron configuration The arrangement of an atom's electrons in their energy levels and orbitals. Electron configuration is the major guide to chemical behavior.

electronegative atom An atom that tends to gain one or more electrons and so become a negative ion. The electronegative elements are the nonmetals in the periodic table, found toward the right and up. Of all the elements, fluorine has the greatest electronegativity and most readily becomes a negative ion.

electron-transport system The principal energy-producing system in the catabolic process. In a stepwise process, the hydrogens from the Krebs cycle combine with oxygen to produce water and energy.

electronvolt (eV) A unit of energy expressed in electrical terms; specifically, the energy *one* electron will acquire in being accelerated through an electrical potential of 1 V; 1 eV is a very small unit, equal to about 4×10^{-20} cal.

electrophilic aromatic substitution The principal reaction of aromatic compounds, in which one or more atoms or groups of atoms are substituted into aromatic rings.

electrophoresis The movement of colloidal or suspended particles under the influence of a direct electric current.

electropositive atom An atom that tends to lose one or more electrons and so become a positive ion. The electropositive elements are the metals in the periodic table, toward the left and down.

electrostatic forces Forces acting between charged particles. If the particles involved are opposite in charge, like protons and electrons or positive and negative ions, an *attractive* force will act to bring them closer together. If the particles carry the same charge, like electrons and electrons or ions of the same charge, a *repulsive* force will act to separate them further.

elements The 105 known elements are the simplest pure substances, each with its own physical and chemical characteristics. Combination of these fundamental units gives the endless variety of chemical substances.

elimination reaction Removal of atoms or a group of atoms to produce a new compound.

emulsifying agent A material that serves to stabilize an emulsion by coating the finely dispersed particles and preventing them from aggregating and settling out.

emulsion A colloidal dispersion of one liquid in another liquid, the two being insoluble in each other.

emulsoid A hydrophilic sol, such as starch or gelatin, whose stability is due largely to a protective layer, or sheath, of bound water on the surface of the colloidal particles.

endothermic reaction A chemical reaction in the course of which the system absorbs heat, thereby increasing its internal energy: $E_r < E_p$.

energy The capacity for doing work, for effecting some change in a material system.

energy level Electron shell. The electrons around the atom nucleus are regarded as occupying specific energy levels, determined mainly by the quantum number n associated with that level, where n can equal 1, 2, 3, ... The larger the value of n and the higher the energy level, the greater the energy of the electrons occupying that level.

entropy A state property that is an index of the disorder, or randomness, in a substance. In the course of chemical change a net increase occurs in the entropy of the universe; symbol S.

enzyme A proteinaceous material acting as a catalyst in metabolic reactions.

enzyme specificity The property of an enzyme whereby it can act on only one or a limited number of substrates.

equation The statement of a chemical reaction, with the reacting materials on the left side and the products on the right; usually in symbols.

equilibrium A condition in which a system undergoes no net change and is therefore stable. Since it cannot change, it cannot do work, so that an equilibrium system is one that has reached a condition of minimum energy.

equivalence point That point in an acid-base neutralization where $N_a V_a = N_b V_b$, so that the number of H^+ ions from the acid and the number of OH^- ions from the base are equal.

essential amino acid An amino acid that must be supplied to the body through the diet since the body cannot make it.

essential fatty acids Fatty acids that must be supplied to the body through the diet since the body cannot make them.

essential oils Lipid-related substances that are the principal flavor and odor components in biological systems. Chemically, they are mainly terpenes and terpenoids.

ester Organic compounds containing the ester group (margin). The ester function is obtained when an alcohol reacts with an acid, with the elimination of a water molecule.

ether Organic compounds containing the —C—O—C— functional group.

excited state *See* ground state.

exothermic reaction   A chemical reaction in the course of which the energy of the system is reduced, so that heat is released to the surroundings: $E_r > E_p$.

fat-soluble vitamin   A vitamin associated with fats and soluble in them.

fermentation   The anaerobic breakdown of carbohydrates accomplished by such microorganisms as yeast.

fibrous proteins   Proteins that are generally linear; the principal structural proteins of the biological system.

formula weight   The sum of the weights of all the atoms of any chemical entity as give by its formula, for example, H^+, H_2, KCl, H_2S, OH^-, $C_6H_{12}O_6$, $Ca(OH)_2$, CH_3^-.

free energy   That portion of the internal energy of a chemical system which is available to do work. The free energy G is related to the internal energy E and the entropy S at some temperature T by $G = H - TS$, so that when a chemical system undergoes a change, the free-energy change of the reaction is $\Delta G = \Delta E - T\,\Delta S$, where ΔG is the maximum work that could be done by the reaction. A *negative* free-energy change indicates that a reaction is spontaneous.

free radical   An atom or group of atoms containing a single odd electron.

free-radical reaction   A fast chemical reaction involving free radicals.

functional group   A specific atom or arrangement of atoms attached to a carbon chain or ring. A functional group contributes specific properties, and organic compounds are often classified according to their functional groups, for example, the presence of a hydroxyl group, —OH, identifies the organic compound as an alcohol.

furanose   The ring structure of a carbohydrate in which the ring is formed from four carbons and an oxygen bridge.

gamma rays   High-energy radiation emitted by many radioactive substances. In common with the other radioactive emissions, gamma rays fog photographic plates, cause ionization, and penetrate matter. Their very high energy gives gamma rays a high penetrating power.

gastric juice   The fluid secreted by the stomach cells; it contains hydrochloric acid and proteolytic enzymes.

gene   The unit in the nucleus that carries one specific piece of hereditary or genetic information.

genetic code   The sequence of the nucleotide bases in a DNA molecule which governs the synthesis of protein molecules.

globular proteins   Proteins that are roughly spherical and generally soluble in water; they are the most biologically active proteins.

glycogen   The storage polysaccharide in animals; because of its similarity to starch it is often called animal starch, but it has a more branched structure.

glycolysis   The anaerobic breakdown of carbohydrates; also known as the Emden-Mayerhof pathway.

grignard reagent   The reaction product of an alkyl halide with magnesium metal; versatile and very reactive reagent (margin).   RMgX

ground state   The lowest possible energy state of an atom, when it is most stable. The atom's electrons all occupy the lowest energy levels they can. If an input of energy promotes one or more electrons to higher energy levels, the atom is in an excited state.

group A vertical sequence in the periodic table, starting at the top of the table and continuing down. All the atoms in a group have the same number of outer valence electrons and therefore tend to have similar chemical properties. The group number, I, II, III, etc., is simply the number of electrons in the valence shell of all atoms in that group.

gums Natural carbohydrate polymers of high molecular weight that form gel-like mixtures with water and impart viscosity to aqueous solutions.

half-life The time it takes for one-half the atoms of radioactive material to disintegrate. It is an indication of how quickly or slowly a radio-isotope loses its potency.

halogenation The introduction of one or more halogens into the structure of an organic compound.

hard water Water containing ions that combine with soap molecules to form an unpleasant, insoluble precipitate. Ca^{2+}, Mg^{2+}, and Fe^{3+} are the most serious offenders, and as long as they are available for using up the soap, no lather will form.

heat of vaporization The heat required to vaporize some mass of liquid at its boiling point to a gas; 1 g of water at 100°C requires 540 cal to be converted into 1 g of steam at 100°C.

helix The coiled arrangement found in many protein molecules.

hemicelluloses Polysaccharides that are less polymerized than cellulose; usually found in conjunction with cellulose.

heredity The biological process whereby the characteristics of a given species are transmitted from one generation to the next.

heterocyclic Characterizing an organic ring structure that contains one or more different atoms in the ring in addition to carbon (margin).

heterocyclic cleavage The breaking of a covalent bond in which one of the fragments takes both electrons that made up the covalent bond.

hexose monophosphate shunt An alternate pathway for the breakdown of hexoses.

holoenzyme A complete and biologically active enzyme.

homolytic cleavage The breaking of a covalent bond in which each fragment is left with one odd electron.

hormones Organic substances secreted by endocrine glands into the body for regulating the activities of body tissues and organs; often called chemical messengers.

hybridization A process whereby an atom rearranges orbitals of generally similar but not identical energy into new, hybrid orbitals that are equivalent in energy and shape.

hydration The addition of water to an organic molecule, for example, the addition of water across a carbon-carbon double bond to form an alcohol.

hydrocarbons Organic compounds composed solely of carbon and hydrogen.

hydrogenation The addition of hydrogen to unsaturated organic compounds; an addition reaction in which H_2 is the substance added.

hydrogen bond   A bond formed because the hydrogen atom attached to a more electronegative atom, such as oxygen, is relatively positive and therefore attracted to relatively electronegative atoms of other molecules (forming the H⋯⋯O hydrogen bond between water molecules) or in the same molecule (forming hydrogen bonds between H and both N and O atoms in proteins).

hydrogen bonding   A dipole-dipole interaction in which a bond is formed between the relatively positive H atom of one molecule and the relatively negative O (or N) atom of a neighboring molecule. In solid and liquid water the bond is between the H of one H_2O molecule and the O of a neighboring H_2O molecule. Hydrogen bonding is of great importance in the structure of proteins and nucleic acids.

hydrolysis   Splitting an organic molecule into smaller parts by reaction with water; a catalyst is generally required.

hydrolytic rancidity   Fat spoilage caused by the hydrolytic action of water on the lipid molecules.

hydrophilic   Water-loving, like ionic and polar materials and the ionic and polar regions of a generally nonpolar organic molecule.

hydrophobic   Water-hating, like nonpolar organic structures or parts of structures.

hypertonic solution   Characterizing a solution (A) that is more concentrated than another (B). The imbalance of their osmotic pressures will cause a net flow of water from B to A across a semipermeable membrane.

hypotonic solution   Characterizing a solution (A) that is less concentrated than another (B). The imbalance of their osmotic pressures will cause a net flow of water from A to B across a semipermeable membrane.

intestinal juice   The digestive juice secreted by the mucosa of the duodenum into the small intestine.

ion   An atom that has gained or lost one or more electrons, becoming a positive ion (by electron loss) or a negative ion (by electron gain).

ionic bond   The bond formed between positive and negative ions due to the electrostatic attraction between their opposite charges; also called an electrovalent bond.

ionic compound   A compound with a specific arrangement of negative and positive ions held together by electrostatic attraction. The ions may be simple, like the Na^+ and Cl^- ions of NaCl or the Ca^{2+} and the two Cl^- ions of $CaCl_2$, or they may be complex, like the CO_3^{2-} ion, which unites with the Ca^{2+} ion to make $CaCO_3$. The formula of an ionic compound is the unit of repetition from which the compound is built up (NaCl, $CaCl_2$, $CaCO_3$, etc.).

isoelectric point   The pH at which an amino acid or protein molecule has net zero charge, symbolized as pI.

isomerism   The property shared by two or more compounds composed of the same kind and number of atoms but differently arranged. For instance, butane and 2-methylpropane are isomeric; their molecular formula is the same, but their structural formula is different.

isoprene The monomer used to make natural polymers such as rubber, steroids, essential oils, and others.

isotonic Describing two solutions that have the same concentration and therefore the same osmotic pressure; there will be no net flow of water between them across a semipermeable membrane.

joule (J) The SI unit of heat; 4.184 J = 1.000 cal.

$$R-C-R'$$
$$\parallel$$
$$O$$

kernel The atom *less* its outer shell of valence electrons.

ketone Organic compounds containing a carbonyl functional group attached to an alkyl group on both sides (margin).

ketose A sugar containing a keto group.

kilocalorie (kcal) Equal to 1000 cal; also called a large calorie; used in studying foods and nutrition.

kinase An enzyme that catalyzes phosphorylation reactions.

Krebs cycle The central cyclic reaction sequence in catabolism; oxidizes a two-carbon fragment to carbon dioxide and hydrogen atoms.

levulose Another name for fructose, derived from its property of rotating plane-polarized light to the left.

alpha (α) linkage A linkage between the two monosaccharide portions of a disaccharide in which the condensation is between the α glycosidic hydroxyl of one monosaccharide and a hydroxyl group of the other monosaccharide.

lipids The general class of fats and fatlike substances insoluble in water but soluble in organic solvents.

Markovnikov's rule The rule states that when an ionized acid is added to a carbon-carbon double bond, the positive ion, H^+, attaches to the carbon having more hydrogens.

messenger RNA A high-molecular-weight RNA that carries genetic messages from DNA to the ribosomes.

metabolism The total chemical and physical processes involved in the breakdown and buildup of biological molecules.

metabolite A substance that participates in the metabolic process.

metal-ion activators Metal ions that combine with certain apoenzymes to make them fully active enzymes.

metalloid An element that has both metallic and nonmetallic properties, for example, carbon and silicon.

metastable Describing a system which is essentially unstable but which may persist for a long time if left undisturbed; if it is subjected to some input of energy or material, a metastable system will generally move to a stable, equilibrium state.

miscible Characterizing two liquids that will dissolve in each other in any possible ratio, from 0A:100B to 100A:0B, for example, ethyl alcohol and water.

mole The number 6.02×10^{23} of any chemical species. The weight of 6.02×10^{23} of any chemical species always adds up to the formula weight in grams of that species.

molecular orbital The result of the overlap and merging of two atomic orbitals, as when the $1s$ orbitals of two H atoms merge to form the bonding orbital of the H_2 molecule. Like all orbitals, molecular orbitals can hold no more than two electrons.

molecule The chemical union of two or more atoms, where the atoms can be like or unlike and the bonding between them is covalent or polar covalent, for example, H_2, O_2, H_2O, NH_3, C_6H_{14}. Whether as a gas, a liquid, or a solid, water is a collection of H_2O molecules.

mono- and diglycerides Mono- and diesters of glycerol and fatty acids.

monosaccharides The simplest carbohydrate structures; contain from three to seven carbons.

mucilages Natural carbohydrate polymers obtained from seeds and other parts of plants which form gelatinous mixtures; chemically similar to gums.

negative beta particle One of the particles emitted by radioisotopes; a negative electron.

neutral Generally, in a balanced condition; in relation to atoms and compounds, having an equal number of positive and negative charges and hence a net charge of zero; in relation to solutions, having an equal number of H^+ and OH^- ions and hence neither acidic nor basic.

neutron A subatomic particle found in the atom nucleus; it is neutral, carrying no charge, and very close to the proton in mass (*see* atomic mass unit).

nicotinamide adenine dinucleotide (NAD) An important coenzyme that participates in biological oxidation-reduction reactions by acting as either an acceptor or a donor of hydrogens.

noncompetitive inhibition The irreversible attachment of a foreign molecule to the active site of an enzyme before the normal substrate can attach to it.

nonpolar Characterizing a chemical structure or a portion of a chemical structure that has zero (or practically zero) polarity. This occurs when the structure is composed of atoms of the same kind, such as the simple diatomic molecules H_2, O_2, Cl_2, etc., or when the polarity of the covalent bonding between dissimilar atoms is internally balanced to a net value of zero due to the symmetry of the structure. The long carbon chains found in the hydrocarbons and in fats and oils, soaps, etc., are examples of nonpolar structures.

nuclear reaction Interaction between the nuclei of atoms and high-speed subatomic radiation, resulting in the formation of different atomic nuclei and different subatomic particles. Nuclear reactions yield large amounts of energy in two ways: nuclear fission (splitting the nucleus) and nuclear fusion (union of nuclei).

nucleophilic addition The typical reaction of aldehydes and ketones.

nucleophilic substitution The typical substitution reaction of the acyl compounds containing the carbonyl group (margin).

nucleus The very small, dense center of the atom; contains both protons and neutrons and carries a positive charge equal to the number of positively charged protons present.

orbital At a given energy level, the region of space around the nucleus where electrons are most likely to be found. The number of possible orbitals at a given energy level equals n^2, where n is the quantum number of that energy level. Orbitals differ in size and shape but are alike in that each can hold *no more* than *two* electrons.

organic acids Organic compounds that contain one or more carboxyl groups (margin).

osmosis The phenomenon of two solutions of unequal concentration separated by a semipermeable membrane showing a net flow of water molecules from the *less* concentrated solution to the *more* concentrated solution.

osmotic pressure The magnitude of the counterpush just needed to stop the movement of molecules across a semipermeable membrane separating pure water from a water solution; varies with the concentration of the solution.

oxidation The loss of one or more electrons, as occurs when a substance reacts with oxygen. In organic reactions oxidation is often effected by the loss of a pair of H atoms from a compound. Electropositive atoms, such as Na, K, etc., are easily oxidized to their corresponding ions, Na^+, K^+, etc.

oxidative rancidity Fat spoilage by the action of oxygen on the double bonds in the fatty acids of a lipid.

oxonium ion A group of atoms containing an oxygen atom with a positive charge.

pacemaker reaction A reaction in a sequence of coupled reactions that is essentially irreversible.

pancreatic juice The digestive fluid secreted by the pancreas.

peptide bond The linkage of the amino acids in a protein (margin), formed from the acid group of one amino acid and the amino group of the other amino acid with the elimination of water, H_2O.

period A horizontal sequence in the periodic table, beginning at the left with an alkali metal, for example, Na, and ending at the right with an inert gas, for example, Ar.

periodic table A tabulation of all the elements, in order of increasing atomic number, arranged in sequences of horizontal periods and vertical groups. Chemical differences between atoms become cumulative from left to right along a period as the valence electrons of the atoms increase from one to eight; similar properties recur down a group, all of whose atoms have the same number of outer valence electrons.

phenols Organic compounds containing an aromatic structure one or more of whose hydrogens are replaced by —OH groups. The simplest of these compounds is phenol (margin).

Phenol

phenylketonuria An inherited disorder that prevents the proper metabolism of phenylalanine.

phospholipid A triglyceride that contains a phosphate group at one of its esters.

photosynthesis The process whereby CO_2 and H_2O are fixed in green plants to form carbohydrates and O_2. This represents a storage of energy since the energy of sunlight is required; the energy is incorporated in the high energy of the carbohydrates produced. Also essential are plant chlorophyll and enzymes.

plant pigments   Substances that give color to plants and facilitate the photosynthetic process.

polar   Having an asymmetrical distribution of electrons, resulting in a higher density of negative charge in one direction and a higher density of positive charge in the other direction.

polar covalent bond   The bond formed between two unlike atoms by sharing a pair of electrons between them. The shared pair of electrons is displaced toward the atom of higher electronegativity, so that the electron sharing is unequal and the electron cloud in the molecular orbital is denser in one direction than in the other. As a result of this polarity, the molecule formed is relatively negative at one end and relatively positive at the other.

polar molecule   A molecule whose atoms are joined by polar covalent bonds, so that one end of the molecule is relatively negative and the other end relatively positive, for example, H_2O, NH_3, HCl. A molecule can have polar covalent bonds and still be nonpolar with zero polarity if its symmetrical structure offsets the electrical asymmetry of its bonds, for example, CH_4, CO_2, BCl_3.

polymerization   A reaction that links many chemical units of the same kind together into a high-molecular-weight structure called a polymer. The unit of repetition can be a single molecule, called a monomer, or a combination of two or more different molecules.

polysaccharides   Carbohydrates formed by condensing many monosaccharide molecules into long chains, both linear and branched.

positron   One of the particles emitted by radioisotopes; it has the same mass as an electron but carries a positive charge; also called a positive beta particle.

potential energy   The energy obtainable from a body that has stored-up energy due to its position, elastic strain, or composition, such as a raised weight, a wound-up spring, or a chemical substance whose arrangement of atoms adds up to high energy.

primary carbon   A carbon atom attached to only one other carbon.

primary structure of a protein   The specific sequence of amino acids in the protein.

product   The substance that results from a chemical reaction, or change.

proenzyme   *See* zymogen.

properties   Physical and chemical characteristics of a substance, such as density, melting and boiling points, solubility, specific heat, electrical conductivity, and chemical behavior.

prosthetic group   Another name for coenzyme.

proteins   Polymers formed by joining specific sequences of amino acids with a peptide bond linking each amino acid to the next. Proteins are found in all living organisms.

proton   A positively charged subatomic particle found in the atom nucleus. Its mass is close to 1 amu, and its positive charge is equal in magnitude but opposite in sign to the negative charge of the electron.

putrefaction Decay of organic matter; specifically, the breakdown of amino acids in the large intestine by microorganisms, producing compounds with unpleasant odors.

pyranose A ring structure of a carbohydrate in which the ring is formed from five carbons and an oxygen bridge (margin).

pyrolysis The decomposition of a compound by heating.

quaternary structure of a protein The arrangement of two or more proteins having tertiary structure into a larger unit called an oligomer.

quinones Aromatic organic compounds containing two keto groups.

radioactivity The spontaneous decomposition, or decay, of an unstable nucleus, resulting in emitted radiation whose loss leaves a new and different nucleus behind. Any material that undergoes such spontaneous disintegration is radioactive.

reactant A substance that enters into a chemical reaction, or change.

reduction In general, a gain of electrons, which is equivalent to a loss of positive charge. In many instances a compound is reduced by the loss of oxygen, a reaction in which one or more electrons are gained. In organic reactions reduction is often effected by the gain of a pair of H atoms. Electronegative atoms, such as F, Cl, O, are easily reduced to their corresponding ions, F^-, Cl^-, O^{2-}.

replication Passing an exact copy of the genetic information carried by parent DNA on to daughter DNA; occurs during mitosis.

ribonucleic acid (RNA) The nucleic acid that contains ribose, a pentose, in its structure. RNA is needed for the synthesis of organic molecules in the cell.

ribosomes Small structures in the cell where protein synthesis takes place.

saliva The secretion of the salivary glands into the mouth; contains the digestive enzyme ptyalin, an amylase.

salt bridge The result of interaction between oppositely charged groups in a protein molecule, as between a negatively charged carboxyl and a positively charged amino group; stabilizes the protein structure.

saturated compound an organic compound with all carbon atoms connected to each other by single covalent bonds and therefore unable to take on any additional atoms or groups of atoms; the compound is "full."

saturated solution A solution that is in equilibrium with the maximum amount of solute that can be dissolved in that solvent at the given temperature and pressure.

secondary carbon A carbon atom attached to two other carbons.

secondary structure of a protein The structure of a protein that is stabilized by hydrogen bonding, generally helical or pleated in shape.

simple proteins Proteins that yield only amino acids upon hydrolysis.

S_N1 mechanism Substitution nucleophilic unimolecular, in which the reaction rate is determined by the concentration of a single reactant.

$S_N 2$ reaction Substitution nucleophilic bimolecular, in which the reaction rate is determined by the concentration of both reactants.

soap The sodium or potassium salt of long-chain fatty acids.

sol A colloidal dispersion of a solid material in a liquid, for example, starch in water.

solution The homogeneous mixture of two or more chemical species; generally, the dissolved material is called the solute, and the material in which the solute is dissolved is called the solvent.

specific gravity The ratio of the density of a substance to the density of water; the density of water is taken as 1.

specific heat The number of calories required to raise the temperature of 1 g of a substance 1°C; symbol c; units, cal/g · °C.

starch The principal storage polysaccharide found in seeds, tubers, and roots of plants. Starch is a polymer of α-glucose, connected by α linkages.

steroid A complex hydrocarbon containing four fused rings. The steroid structure is the basis for such important biological compounds as hormones, bile acids, vitamins, and others.

substitution reaction The displacement of one or more atoms of a chemical structure by different atoms or groups of atoms.

substrate The substance on which a specific enzyme acts.

supersaturated solution A solution that is holding more solute in solution than a saturated solution of the same solute-solvent system at the given temperature and pressure. The system is metastable, and the smallest addition of solute or even mechanical disturbance will cause excess solute to precipitate out until the solution reaches saturation.

surface-active agent Any material which when added to a liquid will reduce its surface tension and hence increase its ability to spread over and wet a surface; especially important for water; also called surfactant.

surface tension The tendency of any liquid to reduce its surface area to a minimum and to resist any increase in that surface area. In effect, a liquid behaves as though it had a tight surface skin. If the molecules of a liquid adhere to each other strongly, like the molecules of water, the surface tension of the liquid is high and the skin is strong.

surroundings Everything outside a given system; for ordinary chemical systems the immediate vicinity of the system.

suspensoid Hydrophobic sols, generally metals or metal compounds, that depend for their stability on the presence of an electric charge on the surface of the colloidal particles.

system Any finite, bounded portion of the universe that is under consideration. A system can vary from very large, like a galaxy, to very small, like an atom.

temperature A measure of the intensity of heat. As the temperature of a substance increases, the increase in heat is expressed as an increase in the velocity with which the molecules of that substance are moving.

terpenes Hydrocarbons made from isoprene units; found in the essential oils.

terpenoids Oxygenated terpenes made from isoprene units; the principal odor components in the essential oils.

tertiary carbon A carbon atom attached to three other carbons.

tertiary structure of a protein The three-dimensional arrangement of the protein.

thermodynamics The study of energy transfer and the conversion of energy from one form to another.

transamination The transfer of an amino group to another organic molecule, usually a keto compound, resulting in the formation of another amino acid.

transcription The process whereby the genetic information carried by DNA is passed on to RNA.

transfer RNA Low-molecular-weight RNA molecules that recognize and carry specific amino acids to the synthesis site during protein synthesis.

transition state A reaction intermediate through which a reaction has to pass for the product to form.

translation Process whereby the genetic information carried by RNA is used to direct the synthesis of proteins from amino acids.

trienes Unsaturated hydrocarbons containing three carbon-carbon double bonds.

triglycerides The naturally occurring triesters of glycerol and three fatty acids; simple fats and oils.

triol An aliphatic hydrocarbon chain with three hydrogens replaced by —OH groups.

unsaturated compound An organic compound containing some double or triple carbon-carbon bonds, where additional atoms can be added to the structure.

unsaturated solution A solution that has not reached saturation, and can therefore dissolve additional solute.

van der Waals' forces The very weak, close-range forces of attraction acting between all atoms and molecules; serve especially to bring nonpolar substances together to form liquids and solids; effective at low temperatures.

vinylic carbons Carbons joined together by a double bond.

waxes Simple esters of long-chain fatty acids and long-chain alcohols; a group of lipids.

zwitterion Ion formed by an amino acid with both positive and negative charges on the molecule; also called dipolar ion.

zymogen or proenzyme The inactive form of an enzyme.

ANSWERS TO SELECTED
REVIEW QUESTIONS

CHAPTER 1

1 In a general sense, the reactants are the ingredients needed to make the cake, such as flour, milk, eggs, sugar, shortening (a fat), baking powder, salt, and so on. However, just putting all these things into a bowl does not make a cake. The cake results from the mingling and interaction of the many chemical and physical changes that take place in the course of preparation and baking. These changes require energy, not only the heat energy used in baking, but also the mechanical energy required for mixing, stirring, kneading, folding, cutting, grating, and so on, all of which contribute to the final result.

3 Water turning to ice on a cold day is a physical change, often called a change of "state." In the course of freezing there is no change in the composition of the water nor in its chemical nature. If the cold day turns warmer, the ice melts back to the original water.

7 Boric acid: a mild antiseptic, commonly used as an eye wash

Vinegar: a tart flavoring material, used in salad dressings, mayonnaise, sauces, etc.

Baking soda: the essential ingredient in baking powder that causes doughs and batters to rise, used as an anti-acid and as a cleaning aid

Alcohol: the alcohol used for drinking is called ethanol, which is also useful as a germicide (it is what is swabbed on the skin before getting an injection); other kinds of alcohols are used as "rubbing alcohol," as solvents, and for many other purposes

Kerosene: the fuel for kerosene lamps, used as a solvent and a paint thinner

Ammonia water: used for household cleaning, has a characteristically pungent odor

Sodium bicarbonate: this is the chemical name for baking soda, and is the "bicarb" used against "that acid feeling" (an antacid)

Table salt: we all know what salt is because we all use it; its chemical name is sodium chloride, and in addition to its use in foods, it is used as a preservative, in pickling, as a quenching bath for steels, and for the making of many other chemicals; in the body it is found in blood, sweat, and tears

Aspirin: the most widely used pain reliever, or analgesic

Sugar: the common food sweetener we all use; actually there are many "sugars," and table sugar is called sucrose

Milk of magnesia: a laxative

Copper metal: copper is an excellent conductor of electricity and is, therefore, used for electrical cables, wiring, and contacts; copper is also a good conductor of heat making it useful for pots and pans; in the past, copper was used as a roofing material, which on exposure to the weather turned green; today less expensive materials are used for that purpose

Oxygen gas: more than one-fifth of the air around us is oxygen gas (we can't do without it); as long as we live oxygen is carried to all our cells by the bloodstream

CHAPTER 2

1   **a** they share the prefix milli ($= \frac{1}{1000}$)
    b millimeter: a measure of length; milliliter: a measure of volume; milligram: a measure of mass, or weight; millisecond: a measure of time

3 $8882 \text{ m} \times \dfrac{39.37 \text{ in}}{\text{m}} \times \dfrac{1 \text{ ft}}{12 \text{ in}} = 29{,}140 \text{ ft}$

5   **a** $320 \text{ m} \times \dfrac{39.37 \text{ in}}{\text{m}} \times \dfrac{1 \text{ yd}}{36 \text{ in}} = 350 \text{ yards}$

    b $330 \text{ ft} \times \dfrac{12 \text{ in}}{\text{ft}} \times \dfrac{1 \text{ m}}{39.37 \text{ in}} = 100.6 \text{ m}$

    the dimensions of the square are therefore $100 \times 100 \text{ m}$

    c $1 \text{ in} = 2.54 \text{ cm}$; therefore $22 \text{ cm} \times \dfrac{1 \text{ in}}{2.54 \text{ cm}} = 8.7 \text{ in}$

7   **a** $1 \text{ l} = 10^3 \text{ ml} = 1000 \text{ ml}$, 1 l of wine $= 1000 \text{ ml}$ of wine
    b From Table 2.1, $1 \text{ l} = 1.057 \text{ qt}$, so that a liter of wine $= 1.057 \text{ qt}$ of wine

c since 4 cups = 1 qt and 1 l = 1.057 qt,

$$1\,l = 1.057\,\cancel{qt} \times \frac{4\ \text{cups}}{\cancel{qt}} = 4.2\ \text{cups}$$

9 a $850\ ml = 850\,\cancel{ml} \times \dfrac{1\ l}{1000\,\cancel{ml}} = 0.850\,\cancel{ml}$

b $1600\ ml = 1600\,\cancel{ml} \times \dfrac{1\ l}{1000\,\cancel{ml}} = 1.600\,\cancel{ml}$

c $1\ \mu m = 10^{-6}\ m;\ 1\ ml = 10^{-3}\ m$

$$55\ \mu m = 55\,\cancel{\mu m} \times \frac{10^{-6}\,\cancel{m}}{1\,\cancel{\mu m}} \times \frac{1\ ml}{10^{-3}\,\cancel{m}} = 55 \times 10^{-3}\ ml$$

$$= 5.5 \times 10^{-2}\ ml$$

11 $10\ lb = 10\,\cancel{lb} \times \dfrac{1\ kg}{2.2\,\cancel{lb}} = \dfrac{10\ kg}{2.2} = 4.5\ kg$

13 a $0.85\ kg = 0.85\,\cancel{kg} \times \dfrac{1000\ g}{\cancel{kg}} = 850\ g$

b From a above, 0.85 kg = 850 g, but 1 mg = 10^{-3} g; therefore,

$$0.85\ kg = 850\,\cancel{g} \times \frac{1\ mg}{10^{-3}\,\cancel{g}} = 850 \times 10^{-3}\ mg = 8.5 \times 10^{-5}\ mg$$

15 If 1 tablet contains 5 grains of aspirin, 2 tablets contain 10 grains. Since 15.4 grains (apothecary weight) = 1 g, we can write

$$\frac{15.4\ \text{grains}}{1\ g} = 1 = \frac{1\ g}{15.4\ \text{grains}}$$

Using the latter fraction as the suitable multiplier

$$10\,\cancel{\text{grains}} \times \frac{1\ g}{15.4\,\cancel{\text{grains}}} = \frac{10}{15.4}\ g = 0.65\ g$$

17 41°C = ? F. Again, °F = 32 × 9/5°C, so

$$°F = 32 + (\tfrac{9}{5} \times 41) = 32 \times 73.8 = 105.8°F.$$

This is 7.2°F more than our own normal 98.6°F.

19 From Table 2.5, the specific heat of water = 1, and of soil = 0.20, which means that 1 cal of heat is required to increase the temperature of 1 g of water 1°C, but only 0.20 cal will increase the temperature of 1 grain of soil 1°C. This means that the soil will warm up more quickly than the water, which is why in northern latitudes swimming in a lake in late spring and early summer can be a cold experience. But what happens when summer turns to autumn?

21 a Heat will flow from the hot metal bar to the cooler water.
b Two bodies are in thermal equilibrium when there is no net flow of heat between them; this means that they are both at the same temperature. In this case, the bar and the water are both at 86°F.
c The temperature change of the water is (86 − 68)°F = +18°F. From Table 2.6, a change of 18°F = a change of 10°C, since a °C is 9/5 as large as a °F. Therefore, the calories absorbed by the water = ml water × △Tc = 11,000 ml × 10°C = 110,000 cal or 110 kcal. Assuming that the water-bar system neither gained nor lost heat (i.e., assuming perfect insulation) the heat absorbed by the water could

only come from the bar as it cooled. That is, the 110 kcal gained by the water was lost by the bar.

23

$$\text{Density} = \frac{\text{mass (in g)}}{\text{volume (in ml)}}$$

$$\text{Density of A} = \frac{17.6 \text{ g}}{4.2 \text{ ml}} = 4.2 \text{ g/ml}$$

$$\text{Density of B} = \frac{9.3 \text{ g}}{3.8 \text{ ml}} = 2.4 \text{ g/ml}$$

$$\text{Density of C} = \frac{128.7 \text{ g}}{10.5 \text{ ml}} = 12.3 \text{ g/ml}$$

$$\text{Density of D} = \frac{65.0 \text{ g}}{8.8 \text{ ml}} = 9.4 \text{ g/ml}$$

CHAPTER 3

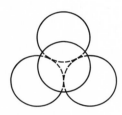

1 Put three of the balls together on a flat surface so that they form a triangle as shown. Each ball touches the other two. The fourth ball is then set on top of the three balls in the hollow at the center of the triangle. Each ball will touch all the other three.

3 **a** The third subatomic particle, the electron, is found around the nucleus.

b Proton: 1 amu, +1 charge
Neutron: 1 amu, 0 charge
Electron: 1/1837 amu, −1 charge

When clothes are tumbled in a clothes dryer, the sliding and rubbing cause charged behavior. When taken from the dryer, some clothes will cling together (opposite charges) and some will fly apart (same charge).

5 The proton is in the center of the atom and is 1837 times as heavy as the electron. It is quite unlikely, therefore, that a proton will be dislodged from the atom simply by rubbing action. The much lighter electron, in the outer portion of the atom, will leave home much more readily.

7

Element	Atomic Number	Number of Neutrons	Atomic Weight	Number of Electrons
Beryllium	4	5	9	4
Helium	2	2	4	2
Carbon	6	6	12	6
Sodium	11	12	23	11
Phosphorus	15	16	31	15

9 **a** H atom: $1s^1$ there is 1 electron in the s orbital

quantum number
$n = 1$

the orbital is the spherical s orbital

b H atom: $2\,s^1$

quantum number
$n = 2$

The atom has received an input of energy, and its single electron has been excited to the next higher energy level, $n = 2$. The orbital is larger than for $n = 1$, but is similarly spherical and is occupied by the 1 electron.

11 Two electrons as far apart as possible, forming a straight line; three electrons as far apart as possible, forming an equilateral triangle.

CHAPTER 4

1 $_3$Li, $_{11}$Na, and $_{19}$K all have 1 electron in their outermost shell.

3 With 4 electrons in its outer shell, carbon will generally share its electrons with other atoms rather than lose electrons (as does Na, a metal) or gain electrons (as does Cl, a nonmetal). Being neither metal nor nonmetal, carbon is considered a metalloid. In period 2, Li with 3 protons is the most electropositive and the most metallic atom; F with 9 protons is the most electronegative and the most nonmetallic atom. Although Ne has the most protons (10) in their period, it is also the most inert atom because its outer electron shell has a full complement of 8 electrons so that Ne tends to neither gain, lose, nor share electrons.

5 Fluorine has 7 outer electrons and 2 F atoms can be represented as joining as shown. The covalent bond between the 2 atoms of a diatomic molecule is the shared electrons between them. Since both atoms are alike, the electrons are shared equally, and there is no polarity.

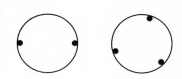

7 H$_2$S

11 Oxidation is electron loss so that the Na atom which is losing an electron is being oxidized. Reduction is electron gain so that the Cl atom which is gaining an electron is being reduced.

15 The ions of MgSO$_4$ are: Mg^{2+} and SO$_4^{2-}$

NaOH:	Na$^+$ and OH$^-$
H$_2$SO$_4$:	2H$^+$ and SO$_4^{2-}$
Na$_2$SO$_4$:	2Na$^+$ and SO$_4^{2-}$
Ca(NO$_3$)$_2$:	Ca^{2+} and 2NO$_3^-$
NH$_4$Cl:	NH$_4^+$ and Cl$^-$
Na$_3$PO$_4$:	3Na$^+$ and PO$_4^{3-}$
Na$_2$CO$_3$:	2Na$^+$ and CO$_3^{2-}$
CH$_3$COONa:	Na$^+$ and CH$_3$COO$^-$

CHAPTER 5

1 The atmosphere contained little, or no oxygen, and it permitted the passage of high energy, ultraviolet radiation. As oxygen-using organisms who are sensitive to ultraviolet light we could not exist under these conditions.

3 Carbon-containing compounds (including those which today we call amino acids), inorganic minerals, and water.

5 Since the bond is between two atoms of the same kind the shared electrons are attracted equally to each atom, and are therefore as likely to be in the vicinity of one atom as the other. This is equivalent to saying the shared electrons, which is the bond, are equidistant from the two atoms.

8 From Table 5.8 in the margin O_2 (zero polarity) is covalently bonded, H_2S (moderate polarity) is polar covalently bonded, and NaF (very high polarity) is ionically bonded.

9 Each F atom has a half-filled $2p_z$ orbital. The 2 orbitals merged into a nonpolar bond (see top sketch of Figure 5.5, page 89).

11 **a** By promoting an e^- in $2s^2$ to $2p_z$

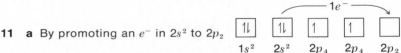

b There are 4 half-filled orbitals, and carbon can form 4 bonds.
c The changeover is called hybridization, and the 4 equivalent orbitals formed are symbolized sp^3.

13 There are 2 nonbonding sp^3 orbitals at the oxygen atom in the H_2O molecule (see page 94), each containing 2 electrons. These act to modify the tetrahedral geometry of sp^3 orbitals so that the angle between the bonding orbitals is reduced to 104.5°. Since O is more electronegative than H, the bonding is polar covalent. The unpaired negative electrons in the 2 nonbonding orbitals also contribute to the polarity of the H_2O molecule by adding to the relative negativity of the oxygen end of the molecule.

15 From Figure 5.7, $BeCl_2$ is the linear molecule, and BCl_3 is triangular. In each instance the electron pairs being shared are as far apart as possible, thus minimizing electron-electron repulsion. Any change from the linear structure of $BeCl_2$ or the triangular structure of BCl_3 would bring the electron pairs closer, and increase repulsion forces.

Ethylene

17 The double bond is made up of one sigma and one pi bond. The hydrogenation yields ethane, which has sigma bonds only.

19 The strength of a bond is the work required to rupture the bond.

21 Although there are some important and interesting exceptions, solid covalent and polar covalent compounds are, in general, mechanically weaker and softer than ionic materials. Covalent compounds are much less resistant to heat and will melt or decompose at temperatures that will not affect ionic solids. Although small covalent molecules, such as sugar, are arranged in the solid in an orderly and specific pattern, and even large covalent molecules may show some tendency to form a crystalline pattern, most covalent compounds are not crystalline in nature, unlike ionic compounds.

23 The NaCl we ingest is present in the body as Na^+ and Cl^- ions. The salt dissolves either in the food to which it is added or in our body fluids, beginning with the saliva.

25 The weak van der Waals forces become significant (1) at low temperatures, and (2) when the molecules are large and composed of many atoms.

27 **a** Webster's 7th New Collegiate Dictionary defines diffusion as "the process whereby particles of gases, liquids or solids intermingle as the result of their spontaneous movement and in dissolved substances move from a region of higher to one of lower concentration."

b A gas diffuses rapidly because the attractive forces tending to hold them together are negligible compared to their kinetic energy of motion which tends to scatter them in all directions.

c A drop of ink added gently to a beaker will diffuse, but rather slowly, (1) because the attraction between the ink particles is not negligible, and (2) because in diffusing they must make their way through the water. Stirring will, of course, speed up the ink particles and speed up their diffusion. Diffusion ends when the ink is uniformly distributed throughout the water.

d The sugar must first dissolve, after which the molecules diffuse through the liquid; here, too, stirring speeds things up. Also, the smaller the solid particles of sugar, the faster the solution and diffusion.

29 **a** Of the 4 imaginary materials in Figure 5.2, the ionic compound at the bottom has the maximum polarity. However, its structure is derived from $+$ ion to $-$ ion attraction rather than from dipole-dipole attraction between polar covalent molecules as shown in the margin. The compound above the ionic example is the most strongly polar covalent and these molecules would exhibit the strongest dipole-dipole attraction between them.

b Of the 4 hydrogen halides shown, HF has the highest dipole moment and the highest boiling point. Those molecules with the highest polarity (as measured by their dipole moment) will have the highest dipole-dipole attraction and will hold on to each other most strongly. In order to make the HF molecules let go of each other, as occurs at the boiling point, a high temperature is needed.

c Ammonia NH_3 is a strongly polar molecule (Table 4.4) and would be expected to dissolve in a polar solvent such as water. In fact, ammonia is very soluble in water.

d CCl_4 has zero polarity (see Table 4.4) and would not be expected to dissolve in polar water. In fact, CCl_4 does not dissolve in water. On the other hand, being a nonpolar liquid, CCl_4 will dissolve nonpolar material such as fats, oils, grease, etc.

CHAPTER 6

1 Oxygen is more electronegative than hydrogen so that electrons shared between them are displaced toward the oxygen. Also, oxygen has two nonbonding sp^3 orbitals each of which contains an unshared pair of electrons; this further adds to the relative electronegativity of the oxygen end of the molecule.

3   A surfactant would be added to water in order to reduce its surface tension so that the water would spread more readily over the surface and enter cracks and crevices.

5   Carbon tetrachloride, CCl_4, is not soluble in water because the CCl_4 molecule is nonpolar whereas water is polar.

7   If the winter temperature falls below 0°C for any length of time the water in the radiator would freeze, the ice would expand, and the radiator would crack. The purpose of the antifreeze is to reduce the freezing point of the liquid in the radiator.

9   Since the skin is dry the "cooling by evaporation" system is not working. Evaporation of water takes a great deal of heat from the body and so helps keep it cool.

11   Pure water becomes acidic by adding to it a proton donor, or acid, such as HCl, etc. It becomes basic by adding a proton acceptor, or base, such as NH_3, CO_3^{2-} ion, OH^- ion, etc.

13   **a** The coarse particles remaining on the filter paper have a diameter of over 1000 Å units.
   b The coarse particles could be allowed to settle to the bottom and the liquid above it poured off, or decanted.
   c If a clear filtrate shows a Tyndall effect it can be concluded that it contains particles in the colloidal size range, about 10 to 1000 Å units in diameter.
   d If the filtrate shows good electrical conductance it indicates the presence of ions from the dissolved minerals in the soil. The ions are about 0.5 to 10 Å units in diameter.
   e The boiling point of the filtrate would be above 100°C, since the presence of a dissolved nonvolatile solute raises the boiling point of water.

15   Seawater contains dissolved salts and is denser than fresh water.

17   Red blood cells are isotonic with respect to a 0.9% solution of NaCl; that is, there will be no gain or loss of water between the cells and the 0.9% solution. The NaCl concentration of a hypotonic solution will be less than 0.9%, so that osmosis will cause water to flow from the hypotonic solution into the cells causing them to swell and burst.

19   To the extent that water contains dissolved minerals it is "hard." Hard water wastes soap because ions such as Ca^{2+} and Mg^{2+} combine with the soap molecules to form an unpleasant precipitate.

21   A colloidal system is one that contains particles within the colloidal size range, about 10 to 1000 Å units. The colloidal particles are the dispersed phase and the medium within which they are dispersed is called the continuous phase.

23   Adsorption is a surface phenomenon, i.e., it occurs on the surface, and the more surface available the more effective is the adsorbing agent. Colloidal materials have a high ratio of surface area to mass and thus provide a large surface area on which foreign particles can be adsorbed. Adsorbents are therefore often colloidal in their particle size.

25   For a given colloidal material the electric charge on the surface of each colloidal particle is the same. Since like charges repel, the colloidal particles stay away from each other, so that the system is kept stable. If these surface charges are neutralized by adding ions of opposite charge, the colloidal particles would no longer repel each other and would aggregate into larger particles. That is, the colloidal state would be lost and would give way to a suspension.

27 **a** The protective sheath of water referred to in the above question is also called "bound water"; the "binding" is generally due to the interaction between the polar water molecules and polar sites of the colloidal material. A material in the form of a long string or fiber has more surface area than the same material rolled into a ball. Since the binding of water to a colloid occurs at the surface, a fibrous colloid such as gelatin will generally hold more bound water than a globular colloid.

b A hydrophobic colloid in a water medium will not be stable if ions are present in the water since these ions may neutralize the surface charges on the hydrophobic particle. However, if a strongly hydrophilic colloid, such as gelatin, is added it can protect the hydrophobic colloid by surrounding it and so keep away the ions that would otherwise neutralize its surface charge and cause precipitation.

29 An example of an emulsion is oil (nonpolar) in water (polar). Vigorous shaking or stirring will break up the oil into small globules, but when the mechanical action stops, the oil will collect and form a layer over the heavier water. The polar water and the nonpolar oil reject each other, and the two layers form spontaneously.

31 Emulsifiers generally consist of a long nonpolar portion and a smaller polar portion, thereby being attractive to both polar and nonpolar liquids. Soap qualifies as an emulsifier because of its long nonpolar carbon chain that ends in an ionic, and therefore strongly polar, structure.

33 In addition to water being held as bound water, gelatin also mechanically entraps water in its tangled brush heap of fibers. When cooled the liquid sol becomes a semirigid gel.

35 A bubble is a gas (usually air) surrounded by a liquid (usually water) whose surface tension has been reduced by a surfactant (often soap). A foam is a mass of bubbles; foams in foods, however, use surfactants other than soap.

37 Dialyzing membranes will allow ions and small molecules to pass out, but will keep in the larger protein particles. If a mixture of proteins and smaller molecules and ions are put in a dialyzing bag around which water is circulating, the smaller particles will pass out and be washed away, leaving behind the larger proteins.

CHAPTER 7

1 Chemical change proceeds by breaking existing bonds between atoms and/or forming new bonds. Since the making and breaking of atom-to-atom bonds involves energy input or output, a specific energy charge accompanies a given chemical reaction.

3 The hydrogen ion H^+ is the smallest of all the ions. Since the loss of the hydrogen atom's single electron leaves a proton, H^+ is no more than a proton.

5 Acids are sour, turn litmus paper red, and yield H^+ ions in water solution. Bases are bitter in taste, slippery to the touch, turn litmus paper blue, and yield OH^- ions in water solution. Both acids and bases are electrolytes because when added to water they dissociate, to a lesser or greater extent, contributing ions that can conduct an electric current through the liquid.

7 In terms of proton transfer an acid is a proton donor and a base is a proton acceptor.

9 A strong acid is one that dissociates almost completely into its ions when dissolved in water, e.g., HCl, HNO_3, H_2SO_4, etc. A weak acid dissociates to its ions only to a limited extent, e.g., acetic acid (CH_3COOH), boric acid (H_3BO_3), carbonic acid (H_2CO_3).

11 $CH_3COOH \rightleftharpoons CH_3COO^- + H^+$. Acetic acid is a weak acid because its dissociation into acetate ion, CH_3COO^-, and hydrogen ion proceeds only to a small extent.

13 These nonmetallic oxides are gases and when passed through water react with the water to form compounds that are proton donors (acids). See page 149, Eqs. 7.3a, 7.3b, and 7.3c.

15 **a** Dissolved $CO_2 > H_2CO_3 > HCO_3^- > CO_3^{2-}$. In addition, there are H^+ ions from the dissociation of H_2CO_3 and HCO_3^-. The contribution from HCO_3^- is so small as to be negligible; therefore the H^+ ion concentration is about equal to the HCO_3^- ion concentration.

b As the CO_2 concentration is increased by being pushed into the water by pressure, Eq. 7.4 is "tipped" as shown and the reaction flows to the right; this increases H_2CO_3 formation which in turn increases H_2CO_3 dissociation into H^+ and HCO_3^- ions. The increase in H^+ ions means an increase in acidity.

c Boil the water; the CO_2 will be driven off. After which pour the water into a narrow-necked flask and stopper it.

17 When dissolved in water, these hydroxides increase the OH^- ion concentration by dissociating, e.g., $NaOH \rightarrow Na^+ + OH^-$. Since a base is a material that will serve to increase the OH^- concentration when added to water, these hydroxides are considered bases.

19 Acid-base neutralization takes place by adding an acid, or proton donor, to a basic solution having an excess of OH^- ions over H^+ ions. Whether this excess of OH^- is derived from the dissociation of an hydroxide, such as $NaOH$, or from the interaction between a basic material, such as Na_2CO_3, with water, the acid-base neutralization is the same. It is the affinity of OH^- for H^+ that is the basis for acid-base neutralization.

21 **a** $H_2SO_4 + 2NaOH \rightarrow 2H_2O + Na_2SO_4$
b $2HNO_3 + Mg(OH)_2 \rightarrow 2H_2O + Mg(NO_3)_2$
c $H_2SO_4 + Ca(OH)_2 \rightarrow 2H_2O + CaSO_4$
d $H_3PO_4 + 3KOH \rightarrow 3H_2O + K_3PO_4$

23 Excess HCl in the stomach can be neutralized by using $NaHCO_3$, a mild alkali,

$$HCl + NaHCO_3 \rightarrow H_2CO_3 + NaCl$$
i.e.,
$$HCl + NaHCO_3 \rightarrow H_2O + CO_2\uparrow + NaCl$$

Referring to Figure 7.6, since the reaction is really between H^+ and HCO_3^-, the "tipping" is due to H^+ ions entering one step to the left rather than as shown.

25 **a** A buffer system serves to protect a solution from sharp changes in its pH due to the addition of an acid or a base.
b H_2CO_3–HCO_3^- is the carbonic acid buffer system in the blood
$HCO_3^- + H^+ \rightleftharpoons H_2CO_3$: protection against an acid
$H_2CO_3 + OH^- \rightleftharpoons H_2O + HCO_3^-$: protection against a base

27 If a clear solution of limewater becomes increasingly turbid as a gas is bubbled through it, this is a confirmation that the gas is CO_2. The turbidity is due to the formation of insoluble $CaCO_3$ as follows: $CO_2 + H_2O \rightarrow H_2CO_3$, and limewater is $Ca(OH)_2$. Therefore the reaction is $H_2CO_3 + Ca(OH)_2 \rightarrow 2H_2O + CaCO_3\downarrow$.

29 Epsom salt, $MgSO_4 \cdot 7H_2O$, is a hydrate. Salts that are normally found as hydrates (e.g., $CuSO_4 \cdot 5 H_2O$) can be converted to the anhydride ($CuSO_4$) by heating, but they tend to revert to the hydrate by taking H_2O molecules from the moisture in the atmosphere. $CaCl_2$ is used as a desiccant because it will remove water from the air about it and become $CaCl_2 \cdot 2H_2O$.

31 The electrons made available in the oxidation reaction $Zn - 2e = Zn^{2+}$ will be accepted by the oxidized form of all the couples listed below Zn in Table 7.8. In so doing, these materials will be reduced. For example: $Zn + Cu^{2+} \rightarrow Zn^{2+} + Cu^0$. The Zn is oxidized to Zn^{2+}, the Cu^{2+} is reduced to Cu^0.

33 $C + O_2 \rightarrow CO_2\uparrow$ $2H_2 + O_2 \rightarrow 2H_2O$
The rusting of iron, $4Fe + 3O_2 \rightarrow 2Fe_2O_3$, is a slow oxidation, as is the oxidation of the food nutrients we ingest.

35 I: addition; II: dehydration; III: hydration; IV: substitution; V: decomposition; VI: condensation; VII: hydrolysis

CHAPTER 8

1 The arithmetical ratio in which any two atoms will combine depends essentially on the number of outer electrons that each atom can gain, lose, or share.

3

	H_2S	H_2SO_4	CO_3^{2-}	$Mg(OH)_2$	C_2H_5OH
Formula wt.	34	98	60	58	46
Wt. of 1 mole, g	34	98	60	58	46
Wt. of 0.1 mole, g	3.4	9.8	6.0	5.8	4.6

	H_2S	H_2SO_4	CO_3^{2-}	$Mg(OH)_2$	C_2H_5OH
Wt. of 4 moles, g	136	392	240	232	184
Number of that chemical species in 4 moles	24.08×10^{23} of H_2S	24.08×10^{23} of H_2SO_4	24.08×10^{23} of CO_3^{2-}	24.08×10^{23} of $Mg(OH)_2$	24.08×10^{23} of C_2H_5OH

5 Population density gives the number of people within a given area; the concentration of a solution indicates the weight of solute particles in 100 ml of the solution or the number of solute particles in 1 l (1000 ml) of the solution.

7 a Formula wt. of $Ca(OH)_2$ is $40 \times 2 (16 \times 1) = 74$

$$1.48 \text{ g of } Ca(OH)_2 = \frac{1.48 \text{ g}}{74 \text{ g/mole}} = 0.02 \text{ mole}$$

$$\text{Solution molarity} = \frac{0.02 \text{ mole}}{10 \text{ l}} = 0.002 \, M$$

b Formula wt of KCl is 74.5; 1 mole of KCl weighs 74.5 g

$$\text{Molarity} = \frac{\text{moles of solute}}{\text{liters of solution}}$$

substituting,

$$0.1 \, M = \frac{\text{moles of KCl}}{2 \text{ l}}$$

Therefore,

moles of KCl $= 0.2$ mole $= 0.2$ mole $\times \dfrac{74.5 \text{ g}}{\text{mole}} = 14.9$ g of KCl

c $C_6H_{12}O_6$: $\dfrac{180 \text{ g}}{\text{mole}}$

$$\frac{90 \text{ g}}{180 \text{ g/mole}} = 0.5 \text{ mole of } C_6H_{12}O_6$$

$$M = \frac{\text{number of moles}}{\text{volume in liters}}$$

that is,

$$2 \, M = \frac{0.5 \text{ mole}}{\text{volume in l}}$$

$$\text{volume} = \frac{0.5 \text{ l}}{2} = 0.25 \text{ l} = 250 \text{ ml}$$

d CH_3COOH: $\dfrac{60 \text{ g}}{\text{mole}}$ $\quad \dfrac{10 \text{ g}}{60 \text{ g/mole}} = 1/6 \text{ mole of } CH_3COOH$

$$M = \frac{\text{moles}}{\text{volume in l}}$$

Therefore, $M = \dfrac{1/6 \text{ mole}}{0.1 \text{ l}} = 10/6 = 1.67 \, M$

9 a Solution 3 is most acidic; **b** Solution 4 is most basic; **c** Solution 2 is most nearly neutral

11 a 0.1 N NaOH =

$$\frac{0.1 \text{ mole NaOH}}{1} = \frac{0.2 \text{ mole NaOH}}{2 \text{ l}} = \frac{0.2(40 \text{ g})}{2 \text{ l}} = \frac{8.0 \text{ g}}{2 \text{ l}}$$

Weigh out 8.0 g NaOH and dilute to 2 l.

b 0.8 N $CH_3COOH = \dfrac{0.8 \text{ mole acid}}{1} = \dfrac{0.08 \text{ mole acid}}{100 \text{ ml}}$

$$= \frac{0.08 (60 \text{ g})}{100 \text{ ml}} = \frac{4.8 \text{ g}}{100 \text{ ml}}$$

100 ml of a 0.8 N CH_3COOH solution contains 4.8 g of the acid.

c 0.001 N $Ca(OH)_2$ = 0.0005 M $Ca(OH)_2$ which contains 0.0005 mole $Ca(OH)_2$ per liter or 0.00025 mole $Ca(OH)_2$ per 500 ml. Since 1 mole $Ca(OH)_2$ = 74 g, weigh out $0.00025 \times 74 = 0.0185$ g $Ca(OH)_2$ and dilute to 500 ml.

13 a 0.05 N H_3BO_3 has 0.05 mole H^+/liter of soln = 0.01 mole H^+/200 ml soln.

Since 1 mole H^+ ions = 6.02×10^{23} ions, there will be

$$\frac{(0.01) \times (6.02 \times 10^{23}) \text{ } H^+}{200 \text{ ml}} = \frac{6.02 \times 10^{21} \text{ } H^+}{200 \text{ ml}}$$

b Yes, they are the total number of *available* H^+ ions.
c The same number of OH^- ions, i.e., 6.02×10^{21} OH^- ions.

15 $N_a \times V_a = N_b \times V_b$
$N_a = 0.1$ N HNO_3 $N_b = 0.001$ N $Ca(OH)_2$
$V_a = ?$ $V_b = 500$ ml $Ca(OH)_2$
0.1 $N \times V_a = 0.001$ $N \times 500$ ml

$$V_a = \frac{0.001 (500 \text{ ml})}{0.1} = 5 \text{ ml}$$

17 A 6% w/w solution contains 6 g NaCl in 100 g of prepared solution, or 12 g of NaCl in 200 g of the final solution. To prepare this weigh out 12 g of NaCl, then add enough distilled water to make a total weight of 200 g. Since 1 g of water can be considered as having a volume of 1 ml this means that the volume of distilled water added to the 12 g of NaCl would be $200 - 12 = 188$ ml. The volume of the 200 g of 6% w/w NaCl solution will, therefore, be well below 200 ml, since the 12 g of NaCl, when dissolved in water, will add only a small volume to the 188 ml of water.

19 A 4% v/v solution contains 4 ml of solute (acetone in this case) per 100 ml of final solution.

$$4\% = \frac{4 \text{ ml solute}}{100 \text{ ml solution}} = \frac{40 \text{ ml solute}}{1 \text{ l solution}}$$

That is, measure out 40 ml of acetone and add enough distilled water to make up a final volume of 1 l.

21   1 mg = 10^{-3} g = 0.001 g
   1 kg = 10^3 g = 1000 g
1 mg/kg = 0.001/1000 = 0.0001/100 = 0.0001%
1 mg/kg = 0.001/1000 = 1/1,000,000 = 1 ppm

CHAPTER 9

1 The energy of moving water and the energy of moving wind. Both are still in use today, especially the energy of moving water which is utilized in hydroelectric generating stations throughout the world.

3 In striking a match the act of rubbing the match across an abrasive surface converts mechanical energy into heat energy by friction. This heat energy raises the temperature of the match so that it ignites and burns. In turn, the burning of the match is the conversion of chemical energy into heat energy, i.e., the flame.

5 Since the addition of heat *to* the $CaCO_3$ is necessary for its decomposition the reaction is endothermic; i.e., the system absorbs heat as the reaction proceeds. Since thermal energy is being absorbed the products have a higher internal energy than the reactants.

9 In general the heat energy that is not converted to work in a heat engine passes into the surroundings and eventually into the atmosphere.

11 Entropy, S, can be considered as a measure of the disorder in a given material or system. The more random the arrangement of the particles of a system, the greater its entropy.

13 **a** Yes
b The work done varies directly with the vertical distance h between A and B.
c It will be converted into heat energy resulting from friction, vibration, etc.
d No
e As between A and B, B is the equilibrium condition.

15 ΔG is the free energy change of the reaction
ΔE is the heat of reaction of the reaction
T is the temperature of the reaction
ΔS is the entropy change of the system in the course of the reaction
a ΔE is negative and ΔS is positive because the system increases in disorder coming to equilibrium. Then the negative value of ΔG is increased beyond the negative value of ΔE by the numerical value of $-T\Delta S$. The reaction will be spontaneous.
b If the reaction is spontaneous ΔG must be negative even if ΔE is positive. From $\Delta G = \Delta G - T\Delta S$ this is possible only if $-T\Delta S$ is a greater negative value than ΔE is a positive. A high negative value of $-T\Delta S$ will be favored by a large increase in disorder of the system and hence a large value of ΔS, and by a high temperature T; under these circumstances $-T\Delta S$ will be very negative.

17 The heat of activation is C. The net energy released by the reaction is B.

19 System I would consist mostly of reactants; system III would consist mostly of products; system II would contain about equal amounts of reactants and products.

21 Protium has no neutron in the nucleus; atomic weight = 1. Deuterium has 1 neutron in the nucleus; atomic weight = 2. Tritium has two neutrons in the nucleus; atomic weight = 3.

23 When an atom disintegrates, or decays, it transforms into some other element by emitting from its nucleus one of three different particles, often also accompanied by high-energy radiation called gamma rays.

25 A positron, or a positive beta particle, has the same mass as an electron but has a positive rather than a negative charge. A negative beta particle has the mass and charge of a negative electron, i.e., 1/1837 amu and a charge of −1.

27 A high-velocity stream of alpha or beta particles is also referred to as alpha rays or beta rays.

29 "Hard" radiation has a high penetration power because it possesses high energy. The energy of radiant energy is related to its frequency, the speed with which the waves oscillate up and down. The more rapid the oscillation the harder the radiation. Gamma radiation has more energy and more penetrating power than x-rays.

31 The effects of radioactive materials on living matter are related to their half-life, to the degree of penetration of their emitted particles and radiation, and to their ionizing effect.

33 Alpha particles are by far the most effective in causing ionization in living matter. Ionization damage results in a swelling of the cell and its nucleus, causing the cell fluid to become more viscous and the cell wall to function improperly. Also, the chromosomes carrying the hereditary genes may be altered.

35 Radioisotopes are in common use in radiation therapy, in medical diagnosis and research, in treatment of food, insect control, and in varied other ways.

CHAPTER 10

1 A belief that organic compounds can only be made by living organisms

2 NH$_2$
|
C=O
|
NH$_2$

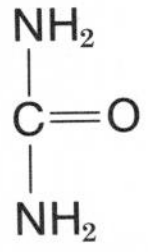

3 Biochemistry, Medicinal Chemistry, Polymer Chemistry

5 Single, double, triple bonds; straight chains; branched chains; cyclic structures

6 Tetrahedral, 109°   Trigonal, 120°   Planar, 180°

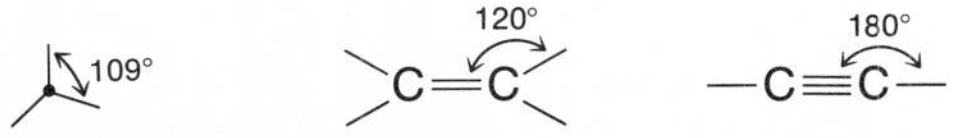

7 Single bond: free rotation
Double bond: restricted rotation
Triple bond: restricted rotation

9 **a** A heterocyclic compound is one that has a cyclic structure containing carbon and one or more different elements in the cyclic structure.

b A carbon chain with one or more carbon branches.
c Carbon ring structures based on benzene with alternating single and double bonds.
d Cyclic structures containing only carbons in the cyclic structure.
e Electronegativity is the measure of electron pull an atom has. It will result in polarity within a molecule due to the uneven sharing of electrons in a covalent compound.

13 a $CH_3CH_2C{\overset{\displaystyle O}{\underset{\displaystyle H}{}}}$ Propanal, an aldehyde

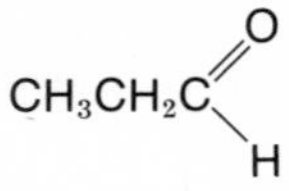

b $CH_3\overset{\displaystyle O}{\overset{\|}{C}}CH_3$ 2-Propanone, a ketone

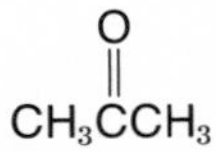

c $CH_2{=}CHCH_2OH$ 2-Propen-1-ol (allyl alcohol), an unsaturated alcohol

d $CH_3CH \quad CH_2$ 1,2-Epoxypropane, a cyclic ether
$ \overset{\displaystyle}{\underset{O}{\diagdown\diagup}}$

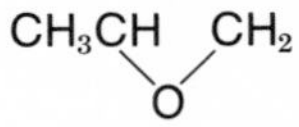

CHAPTER 11

1 a 2,2-Dimethylpropane
b 2,2-Dimethylbutane
c 3-Methylpentane
d 3,4,5-Trimethylheptane

3 a $CH_3CH_2CH_2CH_2CH_3$ n-Pentane

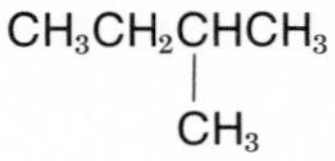

$CH_3CH_2\underset{\displaystyle |}{\overset{\displaystyle}{C}}HCH_3$ 2-Methylbutane
CH_3

$CH_3\underset{\displaystyle CH_3}{\overset{\displaystyle CH_3}{\overset{|}{\underset{|}{C}}}}CH_3$ 2,2-Dimethylpropane

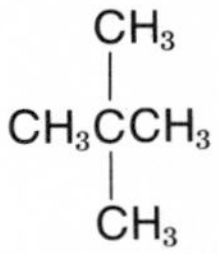

b $CH_3CH_2CH_2CH_2CH_2CH_3$ n-Hexane

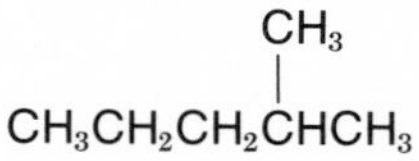

$CH_3CH_2CH_2\underset{\displaystyle}{\overset{\displaystyle CH_3}{\overset{|}{C}}}HCH_3$ 2-Methylpentane

$CH_3CH_2\overset{\displaystyle CH_3}{\overset{|}{C}}HCH_2CH_3$ 3-Methylpentane

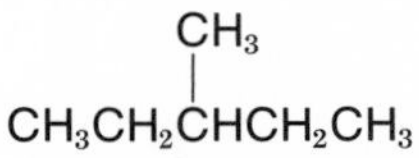

$CH_3CH_2\underset{\displaystyle CH_3}{\overset{\displaystyle CH_3}{\overset{|}{\underset{|}{C}}}}CH_3$ 2,2-Dimethylbutane

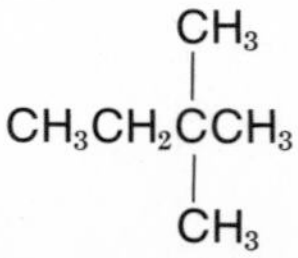

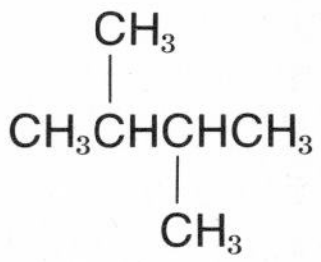 2,3-Dimethylbutane

c $CH_3CH_2CH_2CH_2CH_2CH_2CH_3$ *n*-Heptane

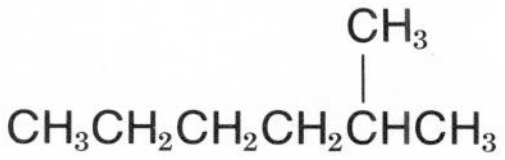 2-Methylhexane

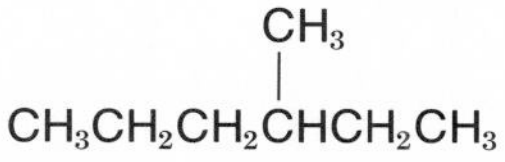 3-Methylhexane

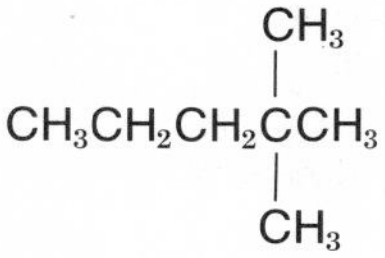 2,2-Dimethylpentane

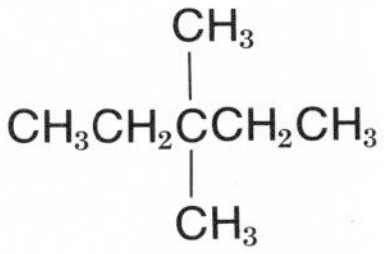 3,3-Dimethylpentane

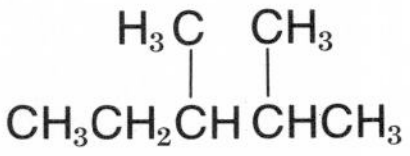 2,3-Dimethylpentane

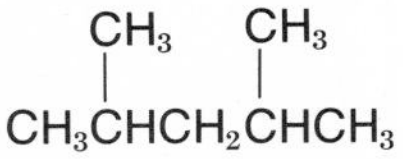 2,4-Dimethylpentane

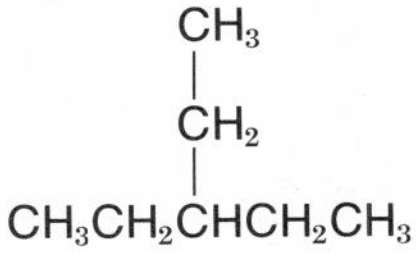 3-Ethylpentane

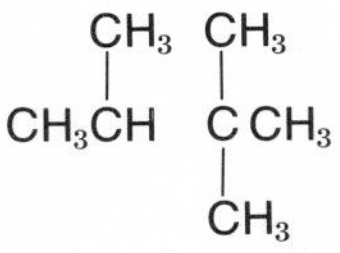 2,2,3-Trimethylbutane

d $CH_3CH_2CH_2CH_2CH_2CH_2CH_2CH_3$ *n*-Octane

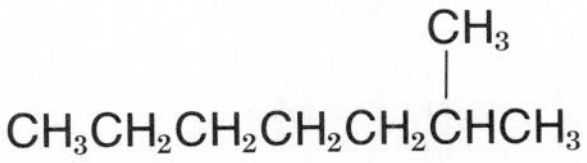 2-Methylheptane

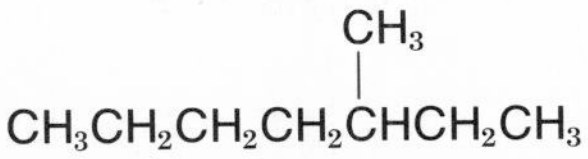 3-Methylheptane

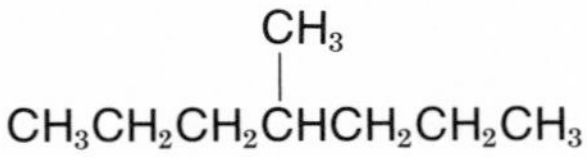

$$CH_3$$
$$|$$
$$CH_3CH_2CH_2CHCH_2CH_2CH_3$$ 4-Methylheptane

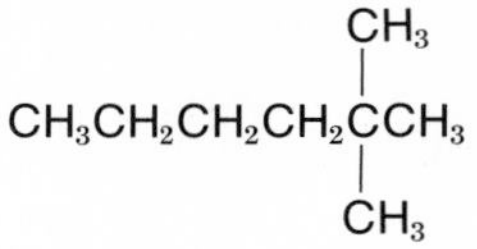

$$CH_3$$
$$|$$
$$CH_3CH_2CH_2CH_2CCH_3$$ 2,2-Dimethylhexane 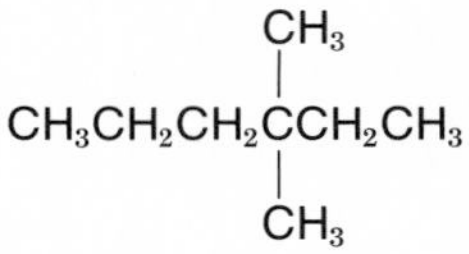
$$|$$
$$CH_3$$

$$CH_3$$
$$|$$
$$CH_3CH_2CH_2CCH_2CH_3$$ 3,3-Dimethylhexane 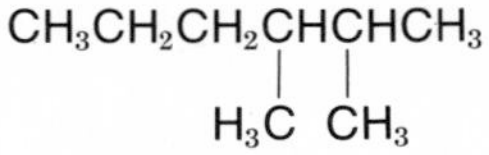
$$|$$
$$CH_3$$

$$CH_3CH_2CH_2CHCHCH_3$$ 2,3-Dimethylhexane 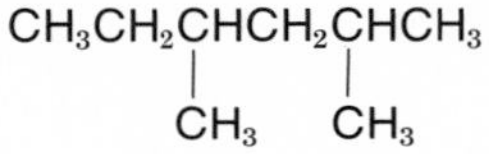
$$| \quad |$$
$$H_3C \quad CH_3$$

$$CH_3CH_2CHCH_2CHCH_3$$ 2,4-Dimethylhexane 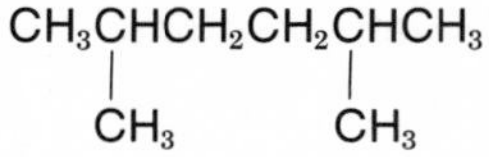
$$| \qquad |$$
$$CH_3 \quad CH_3$$

$$CH_3CHCH_2CH_2CHCH_3$$ 2,5-Dimethylhexane 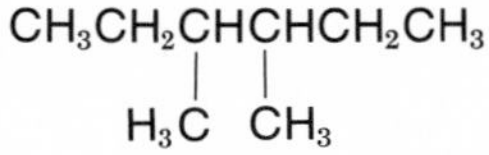
$$| \qquad |$$
$$CH_3 \qquad CH_3$$

$$CH_3CH_2CHCHCH_2CH_3$$ 3,4-Dimethylhexane
$$| \quad |$$
$$H_3C \quad CH_3$$

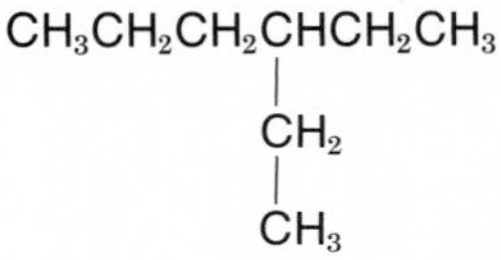

$$CH_3CH_2CH_2CHCH_2CH_3$$ 3-Ethylhexane
$$|$$
$$CH_2$$
$$|$$
$$CH_3$$

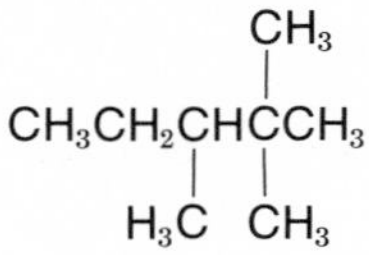

$$CH_3$$
$$|$$
$$CH_3CH_2CHCCH_3$$ 2,2,3-Trimethylpentane 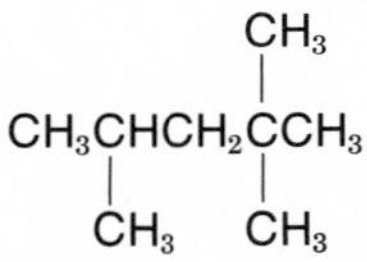
$$| \quad |$$
$$H_3C \quad CH_3$$

$$CH_3$$
$$|$$
$$CH_3CHCH_2CCH_3$$ 2,2,4-Trimethylpentane 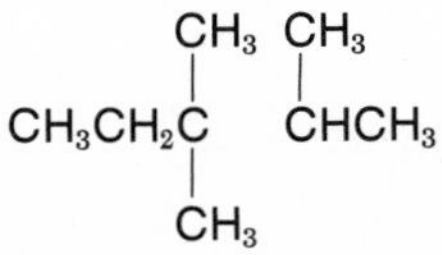
$$| \qquad |$$
$$CH_3 \quad CH_3$$

$$CH_3 \quad CH_3$$
$$| \qquad |$$
$$CH_3CH_2C \qquad CHCH_3$$ 2,3,3-Trimethylpentane
$$|$$
$$CH_3$$

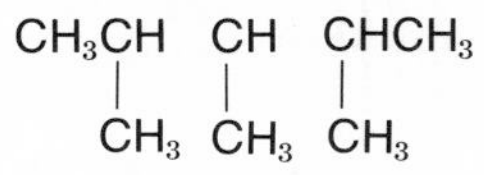

CH₃CH CH CHCH₃   2,3,4-Trimethylpentane 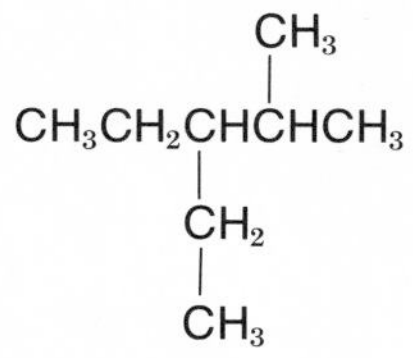
  | | |
  CH₃ CH₃ CH₃

       CH₃
       |
CH₃CH₂CHCHCH₃   2-Methyl-3-ethylpentane 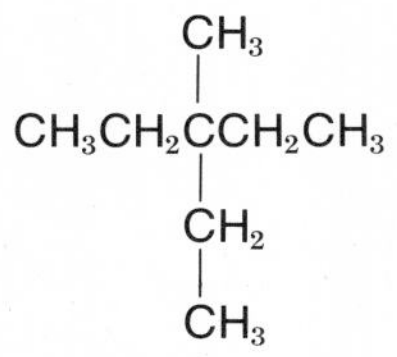
      CH₂
      |
      CH₃

      CH₃
      |
CH₃CH₂CCH₂CH₃   3-Methyl-3-ethylpentane 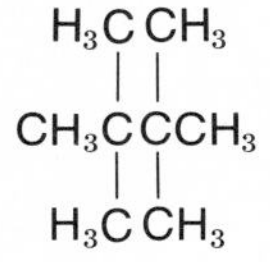
      CH₂
      |
      CH₃

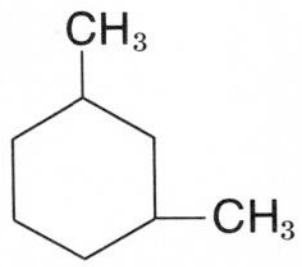

H₃C CH₃
 | |
CH₃CCCH₃   2,2,3,3-Tetramethylbutane
 | |
H₃C CH₃

5 a (methylcyclohexane with CH₃ groups) **b** (1,1-dimethylcyclobutane with CH₃ groups)

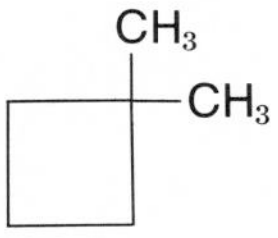

7 a Propane; **b** 2-methyl-3-hexene; **c** 2-pentyne; **d** 5-methyl-3-heptyne

9 a Bromochloromethane; **b** 2,2-dichlorobutane; **c** chlorocyclohexane; **d** 1,1,2,2-tetrachloroethane; **e** 2-methyl-2-chlorobutane

11 a Ethoxypropane (ethyl-*n*-propyl ether); **b** 3-pentanol; **c** 1,2-butanediol; **d** 1,2-pentanediol; **e** 3-butene-1-ol

13 a Propanal; **b** 4-hydroxybutanal; **c** 2-butanone; **d** 3-methyl-2-pentanone; **e** cyclobutanone

15 a 2-Butenoic acid; **b** hexanedioic acid (adipic acid); **c** 4-hydroxyhexanoic acid; **d** 2-methylpropanedioic acid (methylmalonic acid); **e** *trans*-2-butenedioic acid (fumaric acid)

17 a 2-Propanamine (isopropylamine); **b** *N*-methylethanamine; **c** *N,N*-dimethyl-2-propanamine

19 a Propanoyl bromide; **b** ethylbutanoate; **c** propanoic anhydride; **d** butylmethanoate; **e** *N,N*-dimethylethanamide

21 a Naphthalene; **b** ethylbenzene; **c** 2-propylnaphthalene; **d** 1,2-dimethylbenzene (*o*-xylene); **e** 1-ethyl-4-methylbenzene (*p*-ethyltoluene); **f** bromobenzene

23 a Phenol; **b** 4-methylphenol (*p*-cresol); **c** 1-naphthol (*α*-naphthol); **d** 2-chlorobenzenecarbaldehyde (*o*-chlorobenzaldehyde); **e** propionophenone (ethylphenyl ketone); **f** benzophenone (diphenyl ketone); **g** 4-methyl-*N,N*-dimethylbenzamine (*N,N*-dimethyl-*p*-tolu-idine)

25 a

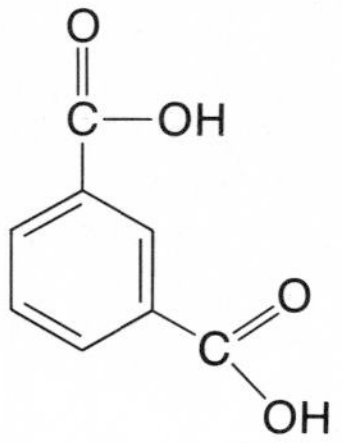

b

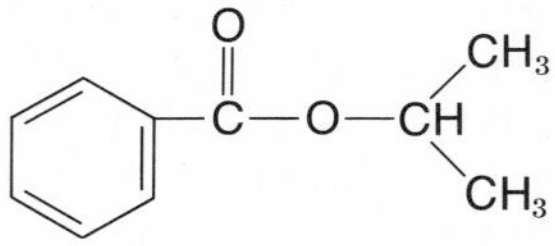

c

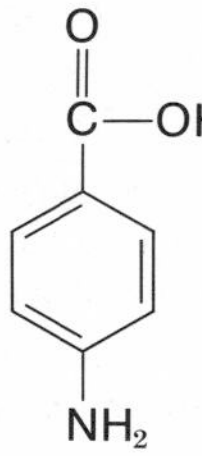

d

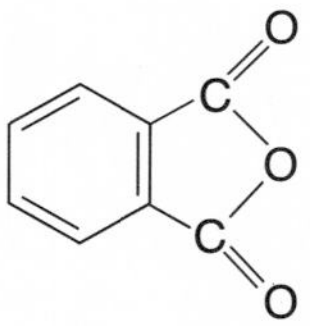

e

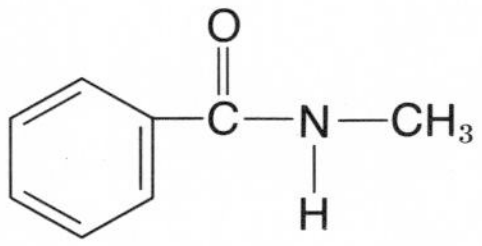

f

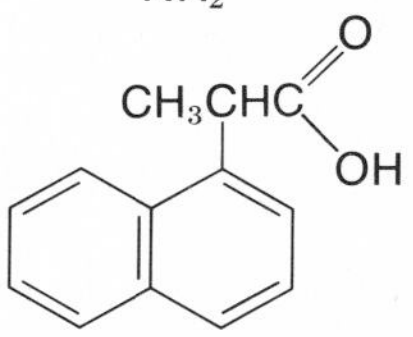

g

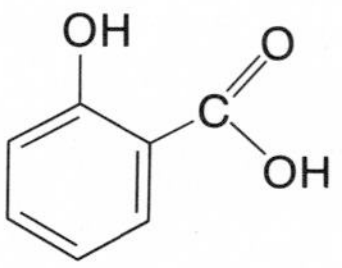

h

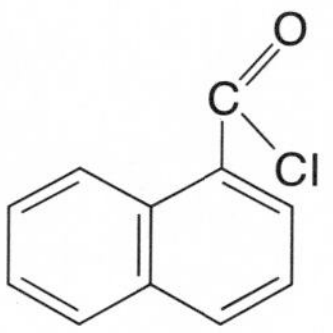

27 a

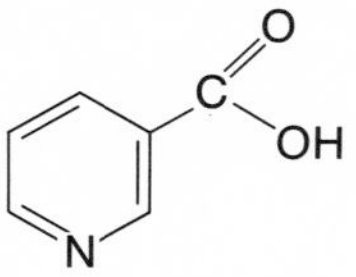

b

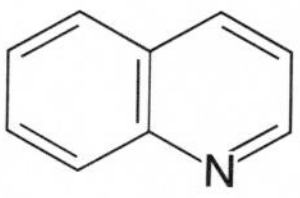

c

d $CH_3CH_2CH_2CCH_2C$

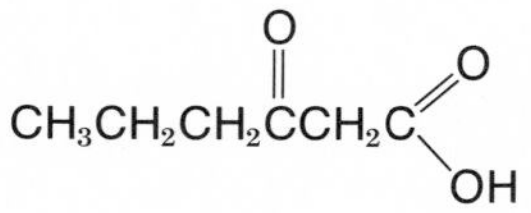

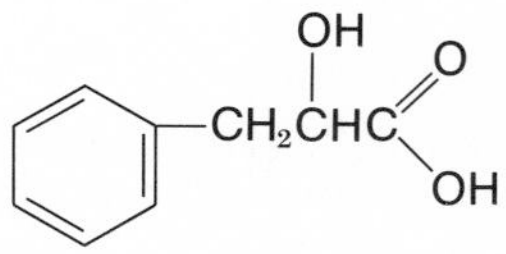

e

f HO—

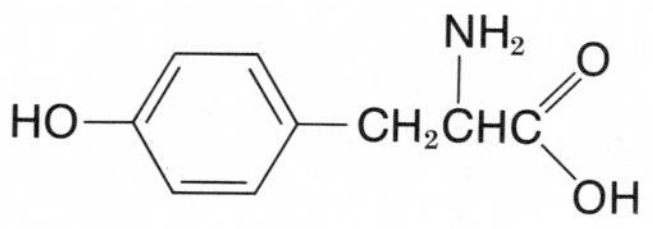

CHAPTER 12

1 Organic reactions take place at or near functional group(s).

3 $C_7H_{16} + 11O_2 \rightarrow 7CO_2 + 8H_2O$

5 Petrochemicals are the organic chemicals made very often from petroleum and natural gas. They are alcohols, aldehydes, and others.

7 Cracking is the thermal decomposition of alkanes at elevated temperatures and the exclusion of oxygen. This will produce alkanes and alkenes of shorter chain length and also carbon and hydrogen gas.

9 Homolytic cleavage is the breaking of a covalent bond in which each fragment carries one electron from the covalent bond. Heterolytic cleavage is the breaking of a covalent bond in which both electrons from the covalent bond are carried by one fragment.

11 Reactions of alkenes. See Figure 12.2 (page 293), reactions of alkanes and alkenes.

13 A two-step mechanism:

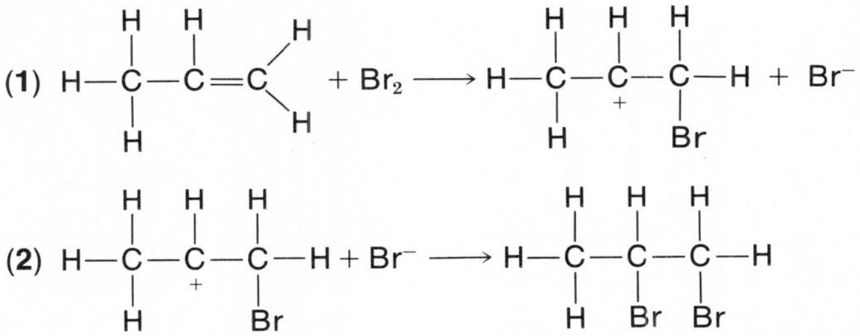

In (1) the double bond polarizes the bromine molecule bond and the positive end attaches to a carbon of the double bond, yielding a carbonium ion and a bromide ion. In (2) the carbonium ion and bromide ion combine to form the product, a dihalide.

15 A two-step reaction mechanism:

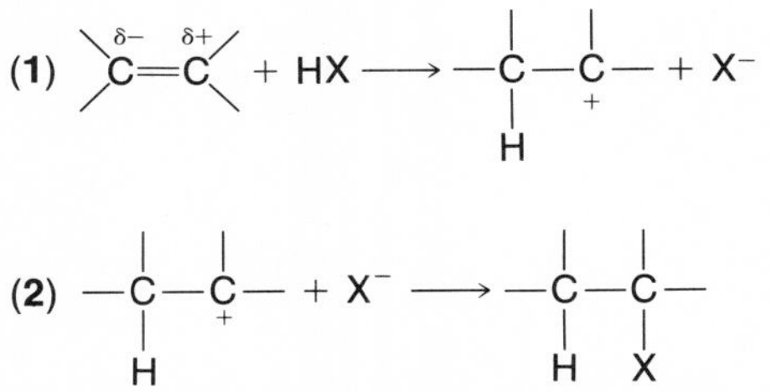

In (1) the double bond becomes polarized and hydrogen ion adds to the partial negative carbon producing a carbonium ion. In (2) the carbonium ion combines with the halide ion to form the product, an alkyl halide.

17 A carbonium ion is a carbon atom in a group of atoms that carries only 6 electrons and a positive charge. It is a very reactive particle and exists only for a very short time in the course of a reaction.

19 Good sources are the *Merck Index,* which is an encyclopedia of chemicals and drugs, and such dictionaries as the *Condensed Chemical Dictionary* (published by Van Nostrand Reinhold Company).

21 $CH_2{=}CHCHCH_2CH_3$
        |
        Cl

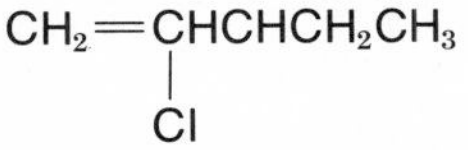

23 From Table 12.1 we can see that there are different groups which, when attached to a benzene ring, will influence further substitution. Examples of ortho- and para-directing groups:

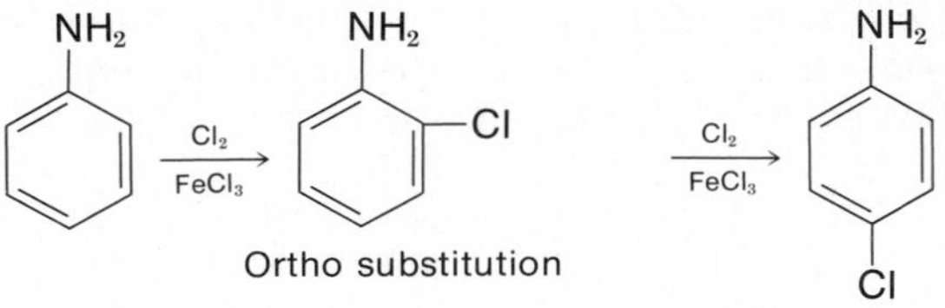

Aniline $\xrightarrow[\text{FeCl}_3]{\text{Cl}_2}$ 2-chloroaniline (Ortho substitution)

Aniline $\xrightarrow[\text{FeCl}_3]{\text{Cl}_2}$ 4-chloroaniline (Para substitution)

Benzoic acid $\xrightarrow[\text{FeCl}_3]{\text{Cl}_2}$ 3-chlorobenzoic acid (Meta substitution)

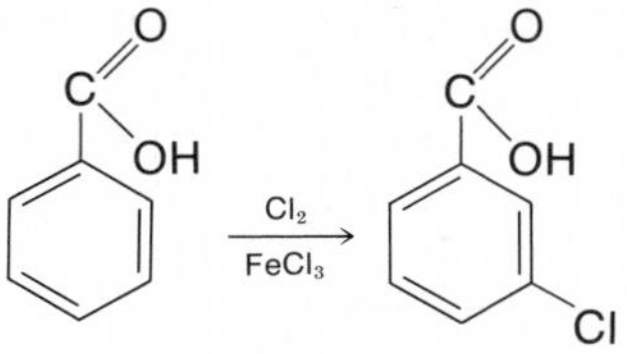

27 S_N1 is a unimolecular mechanism, S_N2 is a bimolecular mechanism. S_N1 is distinctly a two-step mechanism with the formation of a carbonium ion as the intermediate. Mechanism works best in the case of tertiary alkyl halides. S_N2 mechanism can be thought of as a one-step mechanism with an intermediate called the transition state. Mechanism works best for primary alkyl halides.

28 E_1 mechanism is unimolecular and again has as an intermediate a carbonium ion, but now there is the elimination of a hydrogen adjacent to the carbonium ion and an alkene is formed. It is the preferred path for tertiary alkyl halides. E_2 mechanism is bimolecular dependent but can be considered a one-step reaction. It competes with the S_n2 mechanism.

29 The transition state can be considered a reaction intermediate found in S_n2 and E_2 reaction mechanisms. It has the highest energy of any chemical entity encountered in the reaction and exists only for a very short time.

31 A tertiary alcohol has the alcohol group attached to a tertiary carbon.

33 **a** $CH_3CH{=}CH_2$ **b** $CH_3CH_2CH_2ONa$

c $CH_3\overset{O}{\overset{\|}{C}}{-}O{-}\underset{CH_3}{\overset{CH_3}{\underset{|}{\overset{|}{C}}}}{-}CH_3$ **d** $CH_3CH_2CH_2{-}O{-}CH_2CH_2CH_3$

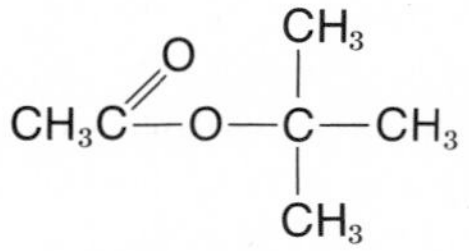

e no reaction

35 **a** $CH_3\underset{OH}{\underset{|}{CH}}CH_3 \xrightarrow[\text{H}_2\text{SO}_4]{160°C} CH_3CH{=}CH_2$

$CH_3CH{=}CH_2 + HBr \longrightarrow CH_3\underset{Br}{\underset{|}{CH}}CH_3$

b $CH_3\underset{OH}{\underset{|}{CH}}CH_3 \xrightarrow[160°C]{\text{H}_2\text{SO}_4} CH_3CH{=}CH_2$

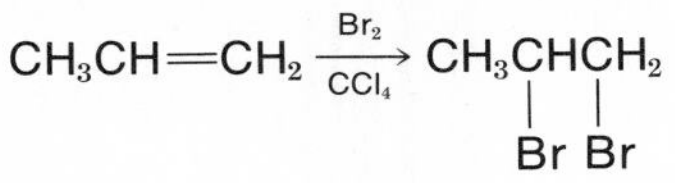

$$CH_3CH{=}CH_2 \xrightarrow[CCl_4]{Br_2} CH_3\underset{\underset{Br}{|}}{C}H\underset{\underset{Br}{|}}{C}H_2$$

c $CH_3\underset{\underset{OH}{|}}{C}HCH_3 \xrightarrow[160°C]{H_2SO_4} CH_3CH{=}CH_2$

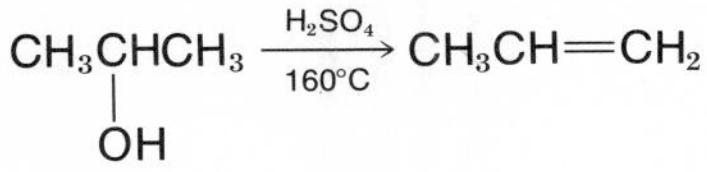

$$CH_3CH{=}CH_2 \xrightarrow[heat/pressure]{H_2/Pt} CH_3CH_2CH_3$$

d Product from **a** $\xrightarrow[ether]{Mg} CH_3\underset{\underset{MgBr}{|}}{C}HCH_3$

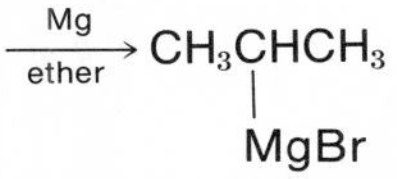

39 Formaldehyde is used as a disinfectant, a preservative for biological specimen, an embalming fluid, and in the manufacture of plastics.

41 Nucleophilic addition is the typical reaction of aldehydes and ketones. It is the attack of an electron-rich nucleophilic reagent on the electron-deficient carbonyl carbon.

43 a $CH_3CH_2CH_2\underset{\underset{CH_3}{|}}{\overset{\overset{OH}{|}}{C}}{-}C{\equiv}N$

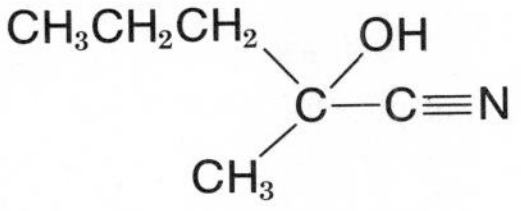

b $CH_3\underset{\underset{CH_3}{|}}{\overset{\overset{OH}{|}}{C}}{-}CH_3$

c $\underset{\underset{CH_3}{|}}{\overset{\overset{CH_3}{|}}{C}}{=}N{-}OH$

d $CH_3CH_2\underset{\underset{OH}{|}}{C}HCH_3$

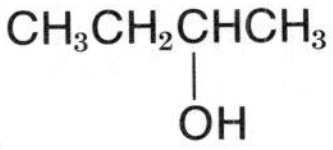

47 The mechanism of ester formation is a five-step mechanism.

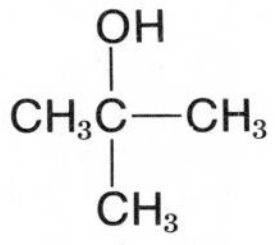

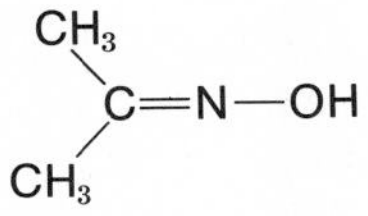

(1) 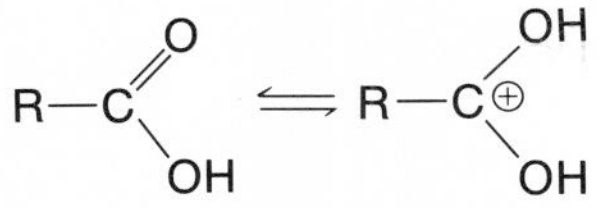 $R{-}\overset{\overset{O}{\|}}{C}{-}OH \;\rightleftharpoons\; R{-}\overset{\overset{OH}{|}}{C}{}^{\oplus}{-}OH$

(Carbonium ion)

(2) 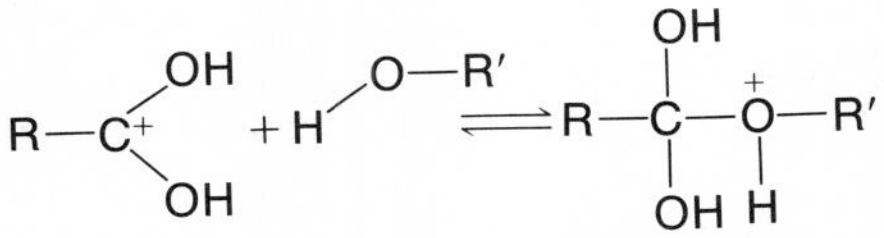 $R{-}\overset{+}{\underset{\underset{OH}{|}}{C}}{-}OH +H{-}O{-}R' \;\rightleftharpoons\; R{-}\underset{\underset{OH}{|}}{\overset{\overset{OH}{|}}{C}}{-}\overset{+}{O}{-}R' \;\; \overset{|}{H}$

Oxonium ion

(3) 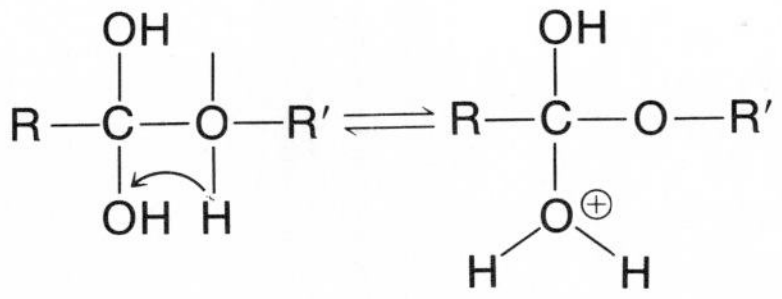 $R{-}\underset{\underset{OH}{|}}{\overset{\overset{OH}{|}}{C}}{-}O{-}R' \;\; \overset{|}{H} \;\rightleftharpoons\; R{-}\underset{\underset{H\;\;H}{\overset{\oplus}{O}}}{\overset{\overset{OH}{|}}{C}}{-}O{-}R'$

Oxonium ion

(4) 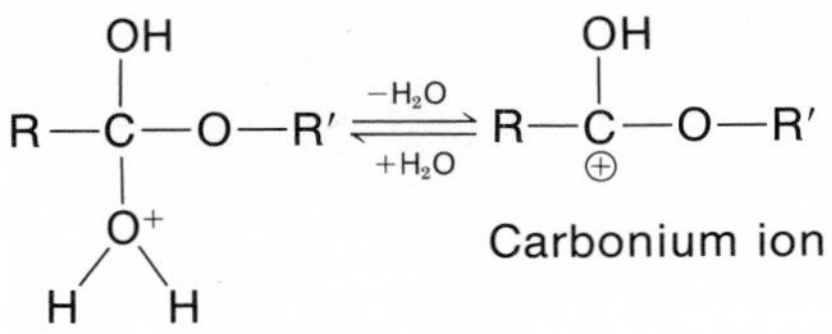
Carbonium ion

(5) 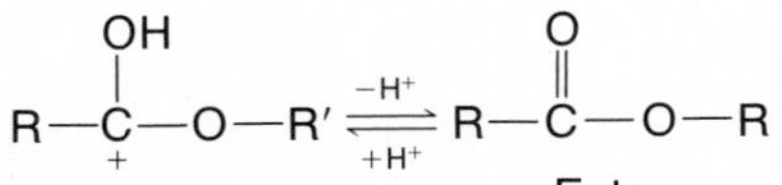
Ester

CHAPTER 13

3 Photosynthesis is the fixation of CO_2 and H_2O plus the energy of sunlight into biological compounds. This fixed energy is then used by other biological systems, e.g., humans.

5 $2^7 = 2 \times 2 \times 2 \times 2 \times 2 \times 2 \times 2 = 128$ possible isomers

7 **a** Aldose; **b** pyranose; **c** beta; **d** 5; **e** carbon 1 and 6; **f** carbon 2, 3, and 4; **g** carbon 1

9 **a** 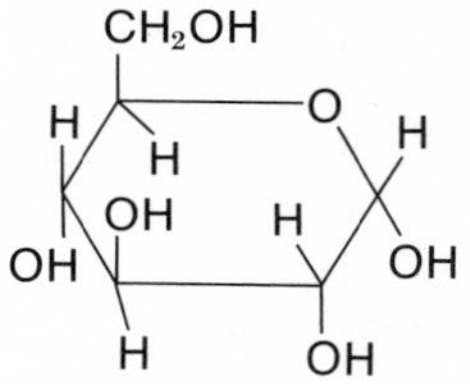
α-Glucopyranose

b 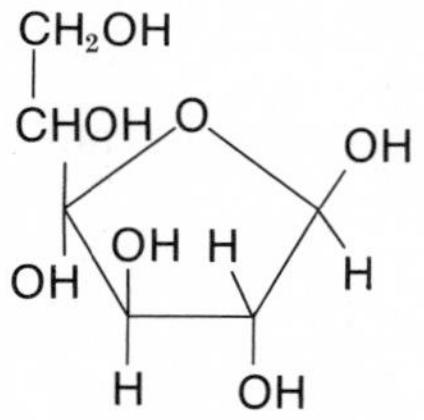
β-Glucofuranose

11 Sucrose exists only in one form and therefore crystallizes readily.
13 See Table 13.2. **a** Seliwanoff test; **b** Mohlish test positive, iodine test negative, Benedict test negative; **c** iodine test blue-black color; **d** iodine test reddish color

15
or

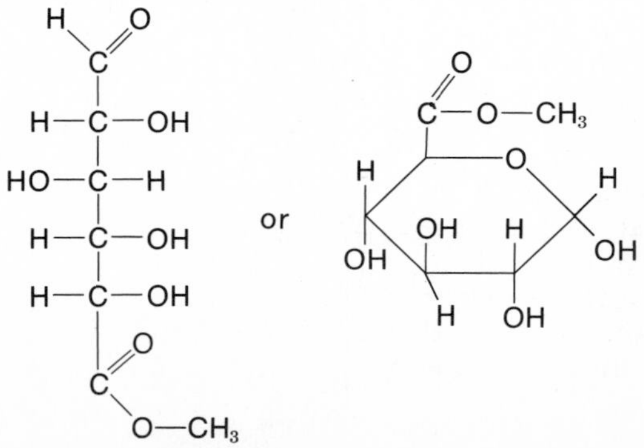

17 Starch, dextrin, cellulose are homopolysaccharides and yield only one building block, namely, glucose, upon hydrolysis. Mucilages, gums, pectins, etc. are heteropolysaccharides because they yield more then one building block upon hydrolysis.

19 When the starch I_2 complex is subjected to higher temperatures the complex disintegrates due to expansion of the starch molecule and the I_2 molecule no longer fits. Upon cooling the complex is restored.

21 Retrograde starch is obtained because of the tendency of amylose to come out of solution upon standing. Amylose forms a gel which slowly precipitates.

23 Glycogen

25 Provide bulk and roughage necessary for the normal functioning of the digestive tract

29 We use hemicellulose primarily to isolate the components from which they are made, such as xylose, mannose, etc.

CHAPTER 14

2 A wax is a simple ester made by the interaction of a long-chain fatty acid and a long-chain fatty alcohol. Waxes protect plants against water loss and the invasion by microorganisms.

5 Saponification is the hydrolysis of a fat molecule by use of an alkali such as NaOH. The end products of the saponification are a glycerin molecule and 3 molecules of the sodium salts of the fatty acids, which are 3 soap molecules.

7 Oils and fats are called neutral fats since the nature of the fatty acids is lost when the triglyceride molecule of the fat is made.

9 Fats and oils mimic the physical characteristics of the fatty acids from which they are made. Fats contain mostly long-chain saturated fatty acid residues. The parent acids are solids at room temperature and so are the fats. Oils contain many long-chain unsaturated fatty acid residues. The parent acids are liquid at room temperature and so are the oils. Similar is the behavior of short-chain fatty acids residues where the parent acids are liquids at room temperature.

11 Margarines contain fats and oils from different sources. They are usually rich in unsaturated fatty acids and also contain vitamins and flavoring agents.

13 Acrolein is an unsaturated aldehyde and is formed by the dehydration of glycerin. It appears in fats or oils as the result of hydrolytic rancidity which produces glycerin. Upon heating the glycerin becomes dehydrated and acrolein is formed. The presence of acrolein is used to identify hydrolytic rancidity.

15 Soap acts as an emulsifying agent. When oily substances are removed from hands or clothing by mechanical action, these substances are emulsified by the soap. The suspended particles can now be carried away by water.

19 Mono- and diglycerides are glycerin molecules connected by ester linkages with either one or two fatty acid residues. We use them as emulsifying and dispersing agents in foods, oils, cosmetics, etc.

21   The difference between lecithin and lysilecithin is that in lysolecithin the β-fatty acid residue is missing. Instead there is the free —OH group from glycerin.

25   The genins are glycosides.

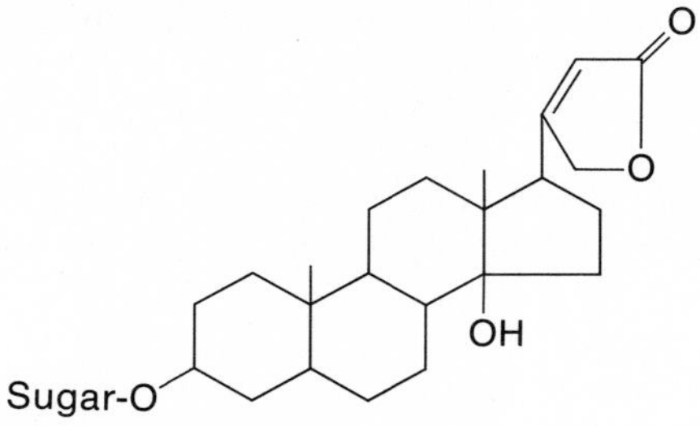

Digitoxigenin

27   A mineralocorticoid is a steroid hormone produced by the adrenal cortex. It regulates the mineral metabolism of the body and affects also the water balance in the body.

29

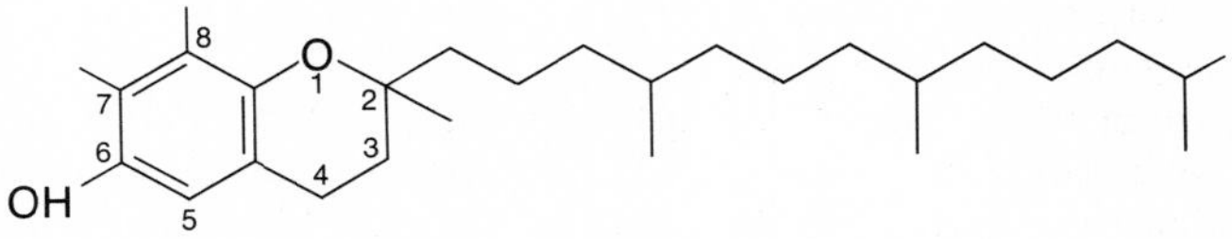

α-Tocopherol

31

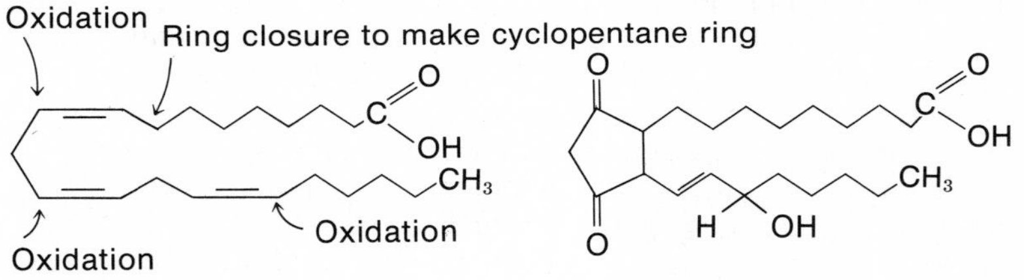

Homo-γ-linolenic acid   PGE

CHAPTER 15

1   Carbon, hydrogen, nitrogen, oxygen, sulfur

5   A β-amino acid. The amino group is one carbon removed from the carboxyl group.

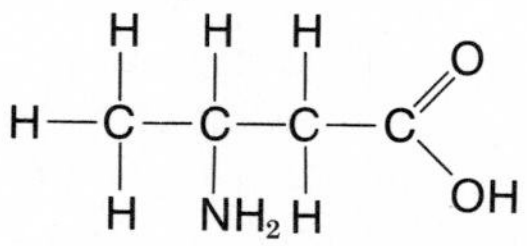

7   Thyroxine is a hormone found in the thyroid gland. Citrulline and ornithine are two amino acids found as intermediates in the urea cycle.

9   Amphoteric applies to compounds that can act both as acids or bases. Amino acids can do this. The carboxyl group functions as an acid and the amino group as a base.

11

Alanylglycylphenylalanine

15 The type of bond found in the secondary structure is the hydrogen bond. It involves the carbonyl oxygen from one peptide bond and the hydrogen from the amino group of another peptide bond

21 Denaturing of a protein means the disturbance of its native biologically active conformation. This can be done reversibly and irreversibly. Proteins are denatured by heat, sound, vigorous shaking and stirring, the addition of chemical reagents such as acids, bases, and salts, and by organic solvents. Each agent will have an effect on certain bonds such as hydrogen bonds, salt bridges, etc.

23 The general test for proteins is the biuret reaction. In a strong alkaline medium proteins react with dilute copper sulfate to form a copper peptide color complex. The intensity of the color produced is indicative of the concentration of the protein.

CHAPTER 16

1 A catalyst is an agent that will speed up chemical reactions by lowering the energy of activation. In the chemistry of living matter the catalysts are the enzymes.

3 Zymogens are complete enzymes, composed of protein, but in an inactive form. Apoenzymes are the protein portions of enzymes that require coenzymes or metal activators to function as complete and active enzymes.

5 Metal activators are ions such as Ca^{2+}, Fe^{2+}, Cu^{2+}, Mg^{2+}, Mn^{2+}, Cl^-, which must be present with certain apoenzymes to make them complete and active. It is believed that the metal activators aid in the proper formation and the stabilization of the enzyme-substrate complex.

11 The activated complex formed from enzyme and substrate is the intimate joining of enzyme and substrate. It is the transition state in which the reaction is initiated and carried out to form the products and regenerate the enzyme.

13 The three-point attachment is needed to provide a secure and stable fit between the enzyme and substrate.

15 Low temperatures will retard and slow down enzymatic reactions. Increasing the temperature to some optimum temperature will speed up enzymatic reactions. The reaction rate will however decline with further increase in temperature because the enzyme will be inactivated.

17 Competitive inhibition is the reversible combination of the active site of the enzyme with a substrate, other than the one it normally acts upon. Examples are the biological antagonists and antimetabolites which compete for the active site of a bacterial or microbial enzyme in the body. By doing so, they will slow down the vigor of the bacteria or microbe and the body defenses can act decisively against them. Drugs such as the sulfa drugs, antibiotics, and others act in such a manner.

19 Allosteric feedback mechanism is a reversible safety mechanism in the body. In the case of overproduction of a final product of a sequence of enzyme reactions, this final product will attach itself to the enzyme that starts the reaction sequence. The attachment will not be at the active site of the enzyme, but it will distort the geometry of the active site and no further reaction can be catalyzed by the enzyme. When the end product excess is sufficiently removed, the attached end product leaves and the reaction sequence is resumed.

21 By dialysis

23 Enzymes are used by humans in many ways: in fermentations, the manufacture of penicillin and other antibiotics, in medicine, the ripening of fruit.

25 A meat tenderizer hydrolyzes the tough connective tissue in meat and so makes it more tender.

CHAPTER 17

1 **a** Amino acids; **b** monosaccharides; **c** glycerol and fatty acids; **d** glucose; **e** phosphoric acid, ribose, deoxyribose, purines, and pyrimidines

3 In many foods predigestion has taken place due to cooking, baking, fermentation, marinating, and aging. The changes are primarily in carbohydrates and proteins. Starch can be hydrolyzed to dextrins and disaccharides to monosaccharides. Proteins are tenderized by softening of the tough connective tissue. Some proteins are coagulated and can thus be digested easier.

5 **a** Aroma of food, appearance of food, the addition of spices; **b** the flow of saliva and the flow of gastric juices; **c** the use of beverages such as wine, beer, cocktails, etc.

7 All enzymes used in Question 6 can be classified as hydrolases according to the reaction they catalyze. They can also be classified according to substrate, so that we have proteoses, lipases, and carbohydrases. Specifically we have amyloses and disaccharidases such as maltase, sucrase, and lactase; dipeptidases, exoproteases and endoproteases, lipases and nucleases.

9 Stomach: Pepsin hydrolyzes proteins to form proteoses and peptones. Rennin coagulates the protein casein to paracasein. Gastric lipase hydrolyzes lipids to glycerol and fatty acids. The stomach mucus protects the stomach wall against the effects of hydrochloric acid and against digestion by pepsin.

Small intestine: Pancreatic juice enzymes are proteases which hydrolyze protein and protein fragments into proteoses, peptones and dipeptides, amylase that hydrolyzes starch into maltose, lipase that hydrolyzes lipids into glycerol and fatty acids, and nucleases that hydrolyze RNA and DNA into nucleotides. The enzymes from the intestinal juice are disaccharidases which produce monosaccharides from disaccharides, dipeptidases which hydrolyze dipeptides into amino acids, and phosphatase which hydrolyzes nucleotides into nucleosides and phosphate.

11 Pepsinogen by HCl to pepsin; trypsinogen by enterokinase to trypsin; trypsin activates chymotrypsin.

13 Exopeptidases hydrolyze the peptide-bond starting from the end of a protein or a protein fragment. The result is a shorter polypeptide chain and a free amino acid.

15 Small food components are absorbed in the stomach. They are monosaccharides, free amino acids, alcohol, and water-soluble medications. The main entry for digested food components into the body is in the small intestine. The absorption takes place through the villi into the bloodstream. Free amino acids are absorbed directly. Monosaccharides are either transported by direct diffusion or by an as yet not fully understood process called active transport. This process involves some chemical agent. Lipids and phospholipids are transported after combining with the bile acids into the body.

17 **a** Reaction of benzoic acid with glycine to form hippuric acid. An amide acid is formed that is less toxic and more soluble so that it can more easily be removed. **b** Phenol reacts with the sugar acid glucuronic acid to form a glycoside which again is less toxic and more soluble.

CHAPTER 18

1 The total chemical events taking place in the body are called metabolism.

3 The purpose of metabolism is to provide energy to the biological system and the means for the maintenance and repair of that organism.

5 The respiratory chain or electron transport system is a series of oxidation-reduction processes in which H atoms are carried by coenzymes. It ends with the combination of the H atoms with oxygen to form H_2O.

7 $ATP + H_2O = ADP + phosphate + 7.3$ kcal/mole
$ADP + H_2O = AMP + phosphate + 7.3$ kcal/mole
$AMP + H_2O = adenosine + phosphate + 3.4$ kcal/mole

9 Coupled reactions are the combinations of two or more reactions in a sequence by which the final outcome is the production of energy. Many reactions in biological systems require coupling to make them go.

11 It prevents overheating of the body. It does not waste energy which

would be the case by rapid release of energy. It provides energy transfer by ATP in this way so energy can be used when and where needed.

13 Metal ions: Mg^{2+}, K^+
Coenzymes: NAD^+

15 A total of 4 ATP molecules are produced; 2 ATP molecules are produced in step 6 and 2 ATP molecules are produced in step 9.

17 In animal tissue the pyruvic acid can be converted by addition of 2 H atoms into lactic acid. In yeast cells the pyruvic acid is decarboxylated yielding CO_2 and acetaldehyde. The acetaldehyde is further converted by the addition of 2 H atoms to ethanol.

19 The role of NAD^+ is to accept 2 H atoms in step 5 and make them available in animal tissue to form lactic acid, and in yeast cells to form ethanol.

21 ATP, ADP, glucose 6-phosphate, fructose 1,6-diphosphate, glyceraldehyde 3-phosphate, dihydroxyacetone phosphate, glyceric acid 1,3-diphosphate, glyceric acid 3-phosphate, glyceric acid 2-phosphate, phosphoenol pyruvic acid.

23 Pyruvic acid can be made into **a** lactic acid, **b** CO_2 and ethanol, **c** can go to the Krebs cycle for further degradation.

25 $C_6H_{12}O_6 + 2$ phosphate $+ 2$ ADP $= 2$ ethanol
$+ 2 CO_2 + 2$ ATP $+ 2 H_2O$

27 Knoops' experiment elucidated the metabolic breakdown of fatty acids. By tagging them he could isolate and identify the final product of their metabolism. From the results obtained he concluded that in each case 2 carbons are removed in succession from the fatty acids. Since the break is at the β carbon the process was called β oxidation.

29 The oxygen is introduced into the fatty acid molecule by the addition of water across the double bond in the α,β position.

31 Eight 2-carbon fragments are produced. The β-oxidation scheme is used seven times. One ATP molecule is needed for activation.

33 The body is growing or is recovering from an illness.

35 Ornithine, citrulline, arginine

37 3 ATP molecules

39 Several of the amino acids have been identified for which some metabolic disorders occur. These disorders are genetically controlled which means that in the genetic makeup of that person a gene or a number of genes produce faulty enzymes and the specific amino acid cannot be properly metabolized. For examples see Table 18.2.

41 Decarboxylation is the removal of the carboxyl group from an organic acid. It results in the shortening of the carbon chain by one carbon. Examples are the decarboxylation of pyruvic acid in yeast cells to produce acetaldehyde, the decarboxylation of pyruvic acid to produce subsequently acetyl-S-CoA, the decarboxylation of isocitric acid to give α-ketoglutaric acid, and the decarboxylation of α-ketoglutaric acid to give succinic acid.

43

```
        CH₂COOH                    CH₂COOH
          |                          |
HO—C—COOH                    H—C—COOH
          |                          |
   H—C—COOH                 HO—C—COOH
          |                          |
          H                          H
     Citric acid              Isocitric acid
```

45 Succinic acid is dehydrogenated to form a double bond and the new acid fumaric acid. Fumaric acid is hydrated by the addition of water to the double bond to introduce an oxygen atom into the structure by way of OH group. The new acid formed, malic acid, is dehydrogenated to form the keto acid oxaloacetic acid.

47 When 2 H atoms are carried from metabolic reaction to the electron-transport system they combine with oxygen at the final step to form water, called water of metabolism.

49 Cytochrome a, a_3, b, and c

51

From glycerol:	22 ATP units
From each palmitic acid by way of β-oxidation:	35 ATP units
From Krebs cycle for each palmitic acid:	96 ATP units
Total for each palmitic acid:	131 ATP units
For the three palmitic acid molecules:	393 ATP units
Total from glycerol + 3 palmitic acids:	415 ATP units

CHAPTER 19

1 DNA and RNA differ in structure by virtue of DNA being a double-stranded helix whereas the RNA is a single-stranded molecule. They differ since DNA contains the pentose deoxyribose and RNA contains the pentose ribose. They differ in base composition since DNA contains the base thymine and RNA contains the base uracil. They differ in function since the DNA stores the genetic information and replicates itself. The RNA is made by transcription from DNA and its function is the transmittance and translation of the genetic information from the DNA to be used in protein synthesis.

3 Nucleotides contain phosphate, a pentose, and a purine or pyrimidine base. Nucleosides contain a pentose and a purine or pyrimidine base.

5 The head of the DNA chain is where the chain begins with a phosphate residue. The tail of the chain is the pentose-base component.

7 Hydrogen bonding connects the two DNA strands that make up a complete DNA molecule in its helical configuration. The hydrogen bonds connect the bases from one strand with the bases from the other. Optimum hydrogen bonding is achieved when adenine is hydrogen bonded to thymine with two hydrogen bonds, and cytosine is bonded to guanine with three hydrogen bonds.

9 The two DNA strands are complementary to each other, ensuring the formation of the two daughter DNA molecules that are identical to the parent DNA.

11 The genetic code is a triplet code. A combination of three bases serves as the code instruction for a specific amino acid to be incorporated into a protein at its synthesis. Since there are more possibilities in the triplet code (64 combinations of bases) and there are only some 20 amino acids, some amino acids are coded more than once; because of this the code is called degenerative. The triplets are called codons. The triplet code is universal and used by all organisms as is indicated by evidence obtained so far.

13 The *N*-formylmethionine-tRNA complex is the starter for any protein synthesis. It provides for the initiation of the synthesis and the proper direction in the elongation of the protein chain. Upon completion of the protein molecule the complex is removed.

15 Antimetabolites are chemotherapeutic agents which interfere with the synthesis and utilization of nucleic acids and other biological compounds. They are used in the treatment of cancer. They can cause the formation of faulty folic acid or different purines and pyrimidines. By doing so, cancer cells, which grow at a faster rate than normal cells, are slowed up and the natural defense mechanism in the body can act more effectively against them. Amethopterin is a folic acid antimetabolite. 6-Mercaptopurine and 8-azaguanine are purine antimetabolites, and 5-fluorouracil and azuracil are pyrimidine antimetabolites.

INDEX

INDEX